The Geotraveller

The multi-coloured rocks in the walls of the Yellowstone Canyon illustrate the derivation of the name of the Yellowstone National Park. The brightest patches of colour are due to hot springs and fumaroles that have hydrothermally altered the thick sequences of rhyolitic volcanic ashes and pyroclastics erupted from the Yellowstone Volcano during the most recent of the caldera events. Thin basaltic flows (a subordinate component of the volcanism) that display prominent columnar jointing are intercalated with the rhyolitic ashes and pyroclastics. The hydrothermal features, including the famous geysers, are driven by heat associated with a shallow magma chamber located beneath the Yellowstone Caldera

Roger N. Scoon

The Geotraveller

Geology of Famous Geosites and Areas of Historical Interest

 Springer

Roger N. Scoon
Department of Geology
Rhodes University
Grahamstown, South Africa

ISBN 978-3-030-54692-2   ISBN 978-3-030-54693-9   (eBook)
https://doi.org/10.1007/978-3-030-54693-9

Cover photograph: Delicate Arch is the most photogenic of the sandstone monoliths in the Arches National Park, northern Utah. Differential erosion of the gently-dipping Entrada sandstone (Middle Jurassic) is a key component in formation of arches

This Springer imprint is published by the registered company Springer Nature Switzerland AG
The registered company address is: Gewerbestrasse 11, 6330 Cham, Switzerland

Preface

The Geotraveller describes the geology of famous geosites and areas of archaeological and historical interest from the USA, Africa, and Europe. Geological descriptions are supported by simplified geological maps and colour photographs. Many of the geosites occur in national parks and reserves, some of which have been upgraded to world heritage sites, while others are located in newly formed geoparks. A Geopark is a unified area that advances the protection and use of geological heritage in a sustainable way and promotes the economic well-being of the people who live there (definition from Wikipedia). There are Global Geoparks and National Geoparks. Many of the geosites not located in parks or reserves could be protected as areas of special interest. The creation of geoparks in areas of outstanding natural landforms is indicative of a growing interest in geological heritage and geotourism.

The Geotraveller is largely based on published geological articles that are not readily accessible to the non-specialist. A short commentary on historical and archaeological features is provided where appropriate. The book is directed at both the professional geologist and the educated layperson. A short glossary is included to assist with the increasingly complex geological nomenclature. The geosites are located on a map (Map A) with the rock sequences indicated on a stratigraphic chart (Map B).

The first section of the book deals with geosites located in North America. Amongst these, the state of Utah in the southwestern USA contains some of the most spectacular landforms on Earth. Giant canyons, free-standing monoliths, and natural sandstone arches are associated with thick sequences of mostly flat-lying Mesozoic strata. Differential erosion of resistant sandstone formations that are interbedded with relatively soft shales and mudstones is a key process in development of landforms at the Canyonlands National Park, as well as at Monument Valley. The natural rock arches at the Arches National Park, possibly one of the most recognizable geological phenomena on Earth, have formed in thick sequences of jointed sandstones associated with the Jurassic-age Entrada Formation (Cover). The Dinosaur National Park contains a museum built over a quarry face where an unusual concentration of dinosaur fossils can be observed in situ within mudstones and shales of the Jurassic-age Morrison Formation. The landscapes of the Yosemite National Park in California have been carved out of Mesozoic granitic batholiths by the action of glaciers and ice sheets. The Late Pleistocene Ice Ages were a global phenomenon, and the deep valleys, waterfalls, and Alpine lakes of Yosemite offer an idealized version of a glaciated landscape. The Yellowstone National Park is not only the world's most well known national park but is also the oldest. The creation of Yellowstone to protect the world's greatest concentration of geothermal features provided the impetus to conserve wilderness areas throughout the world. The geysers, hot springs, and mud pots are driven by heat from an active magma chamber situated beneath the Yellowstone Caldera (Frontispiece). The Yellowstone National Park also offers the opportunity to examine the relationship between geology and large mammals that include bears and bison. The landforms and mountain scenery of the Canadian Rocky Mountains attract millions of tourists annually. The 232 km-long Icefields Parkway in the Banff and Jasper National Parks reveals captivating views of giant U-shaped valleys, snow-clad peaks, icefields and glaciers, and lakes and waterfalls. The Yoho National Park includes exposures of marine fossils associated with the "Cambrian Explosion of Life".

Descriptions of the spectacular geology of East Africa make up the second part of the book. Commentary is provided on the way the geology has influenced the evolution of life, including early hominids. Many of the national parks and reserves in southern Uganda are associated with the Albertine Rift, the western branch of the East African Rift System. The rifting severely impacted drainage patterns within the heart of the African continent, including the upper reaches of the Victoria Nile. The rifting and associated volcanism, a relatively recent phenomena, also impacted the speciation and distribution of fauna and flora, examples of which can be observed in the Murchison Falls and Queen Elizabeth National Parks. The endangered Mountain gorilla is postulated to have evolved in the relatively ancient Bwindi Forest (Uganda), spreading to the younger forests of the Virunga Mountains (Democratic Republic of the Congo, Rwanda, Uganda) in more recent times. The Nyiragongo Volcano in the southern part of the Virunga National Park, Democratic Republic of the Congo, includes a summit crater that contains probably the world's largest and most active lava lake. The Ngorongoro Conservation Area in northern Tanzania reveals a diverse range of landforms, including regional plateaus, volcanic uplands with giant calderas, alkaline lakes, and active and quiescent volcanoes. The biannual migration of grazers on the Serengeti Plains is in part related to nutrient-rich grasses which grow on ashes associated with the Oldoinyo Lengai Volcano. The coexistence of multiple hominid species is an intriguing feature of the palaeoanthropological sites of Oldupai Gorge and Laetoli, localities which have greatly influenced our understanding of human evolution. Initiation of new species by Darwinian evolution during the Pliocene and Pleistocene epochs (5.3 Ma-11,500 BP), including hominins, occurred in remarkably short time intervals. In East Africa, the two epochs were characterized by intense volcanism and extreme climatic cycles (e.g., the Ice Ages).

The third part of the book deals with the complex geology of the Mediterranean. Large parts of the central and eastern Mediterranean remain tectonically active and the volcanoes of southern Italy and the Italian Islands, including Etna and Stromboli, have helped shape the science of volcanology. Parts of southern Italy and eastern Sicily are subjected to severe, even catastrophic earthquakes related to crustal extension. The volcanoes, however, are driven by proximity to an active subduction zone related to collision of the African and Eurasian Plates. The geological features of this region may in part be associated with historical and classical sites, some of which are described in Greek mythology, including Homer's *Odyssey*. The region proximal to the Bay of Naples contains three active or dormant volcanoes, Campi Flegrei, Ischia, and Vesuvius. The historical eruption of Vesuvius in 79 AD affected the Roman cities of Herculaneum and Pompeii, which are of special interest to volcanologists and archaeologists. Many of the famous historical sites and antiquities of southeast Greece, including the Acropolis of Athens, Delphi, and Thermopylae, occur in regions subjected to active tectonism. Tectonism has in part reshaped landscapes, and recent geomorphological reconstructions generally substantiate the classical descriptions. Northwest Greece includes mountainous terrains related to the Alpine Orogeny, in which antiquities and monasteries, including the Meteora, occur in seemingly inaccessible locations. The ancient settlements of the Peloponnese, including Mycenae, typically occur in valleys created by grabens associated with the crustal extension which has followed the Alpine Orogeny. The juxtaposed Alpine tectonic zones are dominated by limestone mountains which supply groundwater to settlements in the valleys. Active and dormant volcanoes of the Hellenic Volcanic Arc include the Methana Volcanic Complex in the Peloponnese and the islands of Milos and Santorini in the Aegean Sea. Milos offers an unparalleled opportunity for geotourism and mining heritage, with geotrails accessing sites including deposits of pumice and agglomerate in spectacular sea cliffs. The Santorini archipelago reveals a large, partially submerged caldera, the formation of which is ascribed to one of the most explosive volcanic eruptions in the historical record. The eruption buried the Minoan city of Akrotiri and may have initiated Plato's legend of the lost continent of Atlantis.

The ancient city of Troy in western Turkey, which reveals nine levels, including the Homeric city of the *Iliad*, is similar to the Greco-Roman antiquities of Ephesus and Pergamum in being located in an active graben. The grabens are subjected to relatively rapid rates of sedimentation which adversely affected harbours in ancient times. The Hierapolis-Pamukkale archaeological and geosite in southwest Turkey includes one of the world's largest deposits of travertine. The travertine, which is a hard, compact variety of limestone, was deposited from hot springs or geothermal waters associated with active graben faults. The travertine has covered large sections of the Hierapolis antiquity. The Cappadocia region of central Turkey contains innumerable natural rock monoliths and pinnacles, together with the largest concentration of underground rock dwellings reported in the world. Many of the most remarkable landforms in this region, including the "fairy chimneys" in the Göreme Historical National Park, are related to differential erosion of recent volcanic deposits.

The Lake District National Park in northwest England is dealt with in the fourth section of the book. This is the most widely visited rural area in the British Isles, with tourists attracted by outdoor activities including a network of mountain footpaths. The national park includes rugged peaks associated with extinct volcanoes and ice-sculptured landforms made famous by painters and writers. The Lake District has a rich mining heritage with the opportunity to visit historical slate and copper mines.

Two examples of layered mafic igneous intrusions are described in the fifth and final part of the book. Layered mafic igneous intrusions are large plutons characterized by sub-horizontal rock layers. The first example is the Skaergaard Intrusion in Greenland, which, despite its relatively small size, has had a substantial influence on the development of conceptual thinking regarding layered intrusions. The other example of this category of geological phenomena is the giant Bushveld Igneous Complex in South Africa. There is no consensus on the formation of igneous layering, despite intensive studies. The Bushveld Igneous Complex includes important layered orebodies, or "reefs", including the platinum-rich Merensky Reef, together with layers of chromitite and vanadium-rich Ti-magnetite. The history of the discovery of the platinum in the Eastern Limb makes a fascinating read.

Grahamstown, South Africa   Roger N. Scoon

The original version of the book was revised: Belated corrections have been incorporated. The corrections to the book are available at https://doi.org/10.1007/978-3-030-54693-9_18

Map A Location of geosites

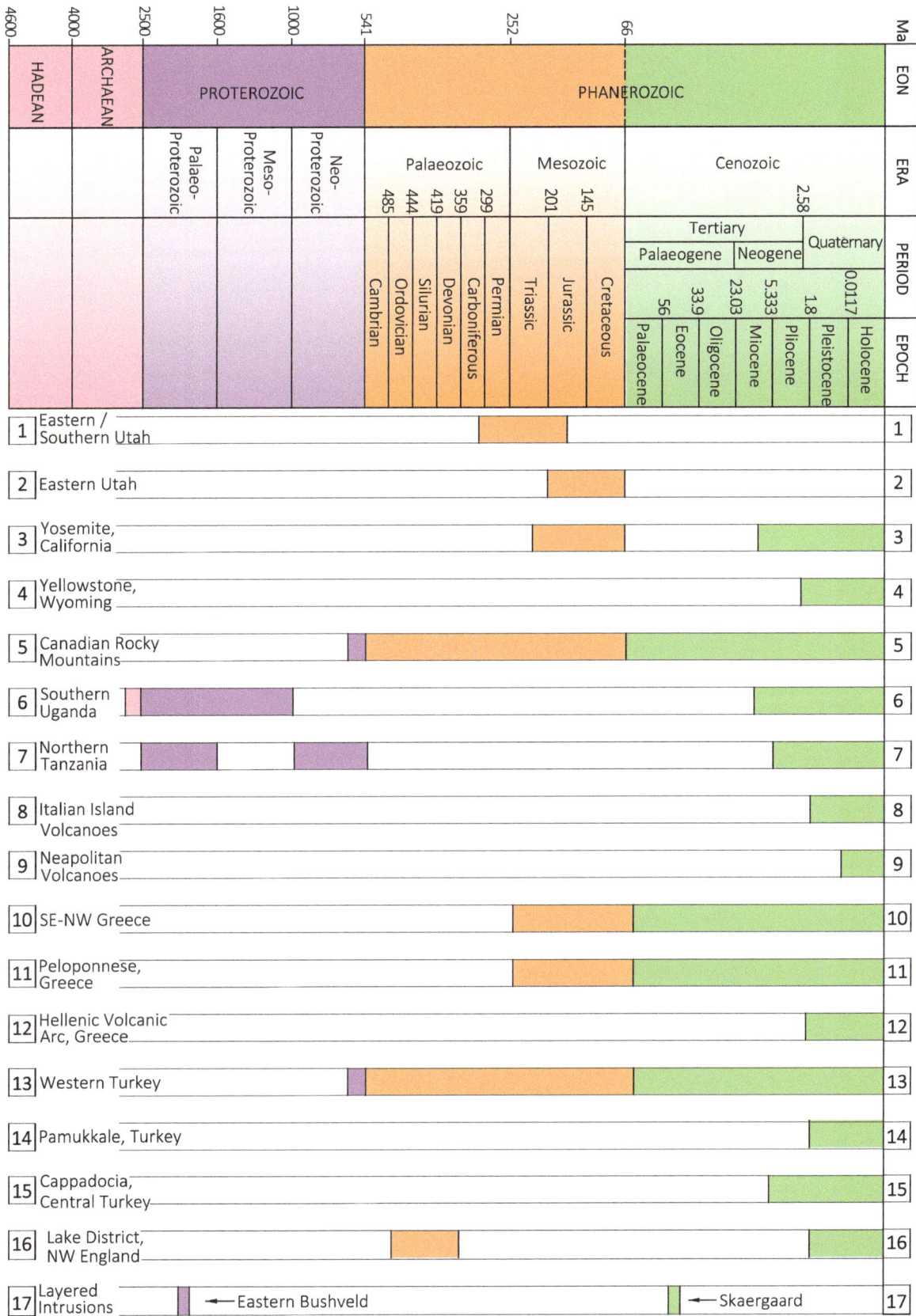

Map B Stratigraphic chart. The Lower Carboniferous in North America is referred to as the Mississippian and the Upper Carboniferous is the Pennsylvanian period. The base of the Pleistocene is located at 1.8 Ma (older usage), rather than 2.58 Ma as is currently recommended to indicate the start of the Quaternary

Acknowledgements

This contribution would not have been possible without the detailed research of geologists, archaeologists, and historians. Maps produced by Geological Surveys and other government institutions are similarly essential. The simplified maps and diagrams were draughted by Lyn Whitfield and her patience in attending to the seemingly endless corrections is greatly appreciated. The satellite images were provided and processed by Philip Eales (Planetary Visions/DLR). Belinda Boyes-Varley is thanked for her assistance with labelling the photographs and also for the layouts of the original Geotraveller articles. Reviews by Prof. Carl Anhaeusser and Dr. Andrew Mitchell greatly improved the manuscript and scientific content. The support and encouragement of Springer Books was also invaluable, particularly the assistance of Margaret Deignan (Netherlands) and Ritu Chandwani (India), as was that of the Geological Society of South Africa. Steve Prevec as editor of the Geobulletin when the first of the series of articles was published is thanked for his encouragement. My wife, Amelia, accompanied me on almost all of the field excursions and geological hikes and has supported me whole-heartedly throughout the preparation of the manuscript. We both gratefully acknowledge the many guides and rangers who assisted us in our travels.

The Geotraveller started as a regular contributor of short articles on areas of geological interest to the Geobulletin, a quarterly publication of the Geological Society of South Africa. The first article was published in March 2010 and the most recent contribution was published in December 2020.

Contents

1   Canyonlands National Park and Monument Valley, Eastern and Southern Utah 1
   1.1   Introduction 1
   1.2   Colorado Plateau 3
   1.3   Regional Geology 5
         1.3.1   Upper Carboniferous 6
         1.3.2   Lower Permian 6
         1.3.3   Triassic 7
         1.3.4   Jurassic 8
         1.3.5   Mineral Resources 8
   1.4   Canyonlands National Park 9
         1.4.1   Stratigraphy 9
         1.4.2   Salt Domes 10
         1.4.3   Additional Features 11
         1.4.4   Trails 11
   1.5   Monument Valley 11
         1.5.1   Navajo Tribal Park 11
         1.5.2   Stratigraphy 12
         1.5.3   Monuments 15
   References 16

2   Arches National Park and Dinosaur National Monument, Eastern Utah 19
   2.1   Introduction 19
   2.2   Regional Geology 21
   2.3   Moab Fault 22
   2.4   Arches National Park 22
         2.4.1   Geological Setting 23
         2.4.2   Middle Jurassic Sandstone 26
         2.4.3   Formation of Arches 27
         2.4.4   Delicate Arch 29
         2.4.5   Landscape Arch 29
         2.4.6   Balanced Rocks 30
         2.4.7   Trails 30
         2.4.8   Stone Tools 31
   2.5   Dinosaur National Monument 31
         2.5.1   Geological Setting 32
         2.5.2   Morrison Formation 33
         2.5.3   Quarry Sandstone 34
   References 36

3   Yosemite National Park, California   37
      3.1   Introduction   37
      3.2   History of the Park   38
      3.3   Ecosystems   39
      3.4   Park Entrances   40
      3.5   Geological Setting   40
      3.6   Glaciations   42
      3.7   Geosites   43
               3.7.1   Formation of Granitic Domes   46
               3.7.2   Alpine Lakes   46
               3.7.3   Waterfalls   46
      3.8   Giant Sequoia   46
      References   50

4   Yellowstone National Park, Wyoming   51
      4.1   Introduction   51
      4.2   History and Physiography   52
      4.3   Fauna   53
      4.4   Rocky Mountains   53
               4.4.1   Grand Teton National Park   53
      4.5   Older Volcanism   55
      4.6   Yellowstone Volcano   56
      4.7   Late Pleistocene Ice Ages   56
      4.8   Yellowstone Caldera   57
               4.8.1   Yellowstone Tuff   57
               4.8.2   Plateau Rhyolite   57
               4.8.3   Obsidian   57
               4.8.4   Basalt Flows   60
      4.9   Geothermal Activity   60
               4.9.1   Circulation of Groundwater   61
               4.9.2   Hot Springs   61
               4.9.3   Mudpots and Fumaroles   62
               4.9.4   Geysers   62
               4.9.5   Travertine   64
      4.10   Yellowstone Lake and Canyon   64
      References   66

5   National Parks of the Canadian Rocky Mountains   69
      5.1   Introduction   70
      5.2   Rocky Mountain Fold-and-Thrust Belt   71
      5.3   Erosion and Ice Ages   72
      5.4   Geological Setting   73
      5.5   Jasper National Park   73
               5.5.1   Lake Maligne   73
               5.5.2   Maligne Gorge   73
               5.5.3   Mounts Edith Cavell and Geraldine   75
               5.5.4   Waterfalls on the Athabasca River   77
               5.5.5   Columbia Icefield   77
               5.5.6   Athabasca Glacier   77
      5.6   Banff National Park   80
               5.6.1   Bow Valley   82
               5.6.2   Lake Louise   82
               5.6.3   Moraine Lake   82

5.7 Yoho National Park 85
    5.7.1 Natural Bridge and Emerald Lake 85
    5.7.2 Cambrian Explosion of Life 87
    5.7.3 Burgess Shale 87
References 88

6 National Parks, Rivers and Lakes of Southern Uganda 89
  6.1 Introduction 90
  6.2 National Parks and Speciation Patterns 91
  6.3 Regional Geology 92
    6.3.1 Central African Craton 95
    6.3.2 Proterozoic Belts 95
    6.3.3 Regional Plateau 95
    6.3.4 East Africa Rift System 95
  6.4 Lakes and Rivers 97
    6.4.1 Lake Albert and Lake Edward 99
    6.4.2 Lake Victoria 99
    6.4.3 Victoria Nile 99
  6.5 Quaternary Climatic Regimes 99
  6.6 Lake Victoria and the Victoria Nile 100
  6.7 Murchison Falls National Park 100
    6.7.1 Lake Albert 103
  6.8 Queen Elizabeth National Park 104
    6.8.1 Lake Edward and Lake George 104
  6.9 Kibale National Park 104
  6.10 Crater Fields 105
  6.11 Rwenzori Mountain and Semliki National Parks 105
  6.12 Lake Mburo National Park 105
  6.13 Bwindi Impenetrable National Park 107
  6.14 Mgahinga National Park and the Virunga Mountains 108
    6.14.1 Virunga Volcanic Province 108
    6.14.2 Nyiragongo 109
  6.15 Evolution of Mountain Gorilla 111
References 115

7 Lake Natron and the Ngorongoro Conservation Area, Northern Tanzania 117
  7.1 Introduction 117
  7.2 Regional Geology 119
  7.3 Lake Natron 120
    7.3.1 Geological Setting of Lake Natron 120
  7.4 Oldoinyo Lengai 121
  7.5 Ngorongoro Conservation Area (NCA) 122
    7.5.1 Geological Setting 122
    7.5.2 Ngorongoro Caldera 123
    7.5.3 Other Volcanoes of the Ngorongoro Volcanic Complex 126
    7.5.4 Shifting Sands 127
    7.5.5 Serengeti Migration 128
  7.6 Oldupai Gorge 128
    7.6.1 Geological Setting 128
    7.6.2 Oldupai Group 130
    7.6.3 Discovery of the Hominin Fossils 130

7.7   Laetoli   133
7.8   Evolution of Hominins   134
7.9   Treks   135
References   137

8   **Mediterranean Basins and Italian Island Volcanoes**   139
8.1   Introduction   140
8.2   Ancient Greeks and Romans   140
8.3   Regional Geology   141
        8.3.1   Alpine Orogeny   142
        8.3.2   Accretionary Convergent Margin   143
        8.3.3   Apennine-Mahgreb Thrust   143
8.4   Mediterranean Basins   145
8.5   Earthquakes   145
8.6   Volcanism   146
        8.6.1   Explosiveness of Eruptions   146
        8.6.2   Magma Compositions   147
8.7   Aeolian Islands   149
        8.7.1   Volcanic Island Arc   149
        8.7.2   Stromboli   150
        8.7.3   Vulcano   150
        8.7.4   Fossa Cone   150
8.8   Etna   152
        8.8.1   Early Volcanism   153
        8.8.2   Historical Activity   153
        8.8.3   Recent Activity   157
8.9   Pantelleria   158
8.10   Archaeological Sites of Magna Graecia   158
8.11   The Odyssey   160
References   166

9   **Neapolitan Volcanoes, Southern Italy**   169
9.1   Introduction   170
9.2   Volcanology   172
9.3   Regional Geology   172
9.4   Neapolitan Volcanoes   172
9.5   Campi Flegrei Volcano   173
        9.5.1   Historical Tectonism   175
        9.5.2   Magma Chamber   176
9.6   Ischia Volcano   176
        9.6.1   Cliff Sections   177
        9.6.2   Recent Activity   177
9.7   Somma-Vesuvius Massif   177
        9.7.1   Prehistoric Eruptions of Somma-Vesuvius   180
        9.7.2   Periodicity   181
        9.7.3   79 AD Event   181
        9.7.4   Roman Cities   182
        9.7.5   Roman Beach Resorts   185
        9.7.6   Details of the 79 AD Volcanic Deposits   185
9.8   Vesuvius National Park   188
References   191

10 Antiquities and Archaeological Sites of Southeast and Northwest Greece 193
 10.1 Introduction 193
 10.2 Regional Geology 194
  10.2.1 Alpine Zones 195
  10.2.2 Basins and Grabens 198
 10.3 The Acropolis of Athens 198
  10.3.1 Tourkovounia Limestone 199
  10.3.2 Antiquities 199
 10.4 Plains of Marathon 201
 10.5 Delphi 202
  10.5.1 Regional Geology 203
  10.5.2 Delphic Landscape 204
  10.5.3 Antiquities 204
  10.5.4 Delphic Oracles 204
 10.6 The Pass of Thermopylae 205
  10.6.1 Modern Landscape 207
  10.6.2 Geomorphological Reconstruction 207
 10.7 Pindus Mountains 208
  10.7.1 Regional Geology 208
  10.7.2 South Pindus Mountains 210
  10.7.3 North Pindus Mountains 211
 10.8 Meteora 214
  10.8.1 Regional Geology 214
  10.8.2 Meteora Group 214
 References 217

11 Antiquities and Archaeological Sites of the Peloponnese and Zakynthos, Greece 219
 11.1 Introduction 220
 11.2 Early Civilizations 221
 11.3 Regional Geology 223
  11.3.1 Alpine Zones 223
  11.3.2 Basins and Grabens 224
 11.4 The Argolid 224
  11.4.1 Ancient Humans 225
  11.4.2 Antiquities 225
 11.5 Corinthia and Achaea 226
  11.5.1 Corinth Basin 226
  11.5.2 Antiquities 227
  11.5.3 Corinth Canal 229
  11.5.4 Earthquakes and Tsunamis 230
 11.6 Elis 230
  11.6.1 Olympia 231
 11.7 Arcadia 233
 11.8 Laconia and Messenia 235
  11.8.1 Antiquities in Laconia 235
  11.8.2 Antiquities in Messenia 237
 11.9 Zakynthos 240
  11.9.1 Geological Framework 240
  11.9.2 Earthquake of 1953 241
  11.9.3 Springs 242
 References 243

12 Volcanoes of the Hellenic Volcanic Arc, Greece 245
   12.1   Introduction 246
   12.2   Regional Geology 247
          12.2.1   Aegean Basin 248
          12.2.2   Subduction Zone 248
          12.2.3   Island Arc Volcanism 249
   12.3   Methana Peninsula 249
          12.3.1   Geological Framework 249
          12.3.2   Methana Volcanic Complex 250
          12.3.3   Mavri Petra Volcano 251
          12.3.4   Pausanias Submarine Volcanism 253
   12.4   Milos 253
          12.4.1   Geological Framework 254
          12.4.2   Volcanism 254
          12.4.3   Geotrails 255
          12.4.4   Historical Mining 255
          12.4.5   Archaeological Sites 257
          12.4.6   Current Mining 258
          12.4.7   Geosites 259
   12.5   Santorini 262
          12.5.1   Geological Framework 264
          12.5.2   Explosive Cycles 265
          12.5.3   Minoan Event 267
          12.5.4   Historical and Archaeological Sites 268
          12.5.5   Kameni Islands 269
          12.5.6   Columbo Seamount 270
          12.5.7   Recent Activity 271
   References 271

13 Antiquities and Archaeological Sites of Western Turkey 273
   13.1   Introduction 274
   13.2   First Century Christianity in Asia Minor 275
   13.3   Regional Geology 276
          13.3.1   Regional Sutures 276
          13.3.2   Southwest Migration of the Anatolian Microplate ... 277
          13.3.3   Basins and Grabens 278
          13.3.4   Geological Terrains 278
          13.3.5   Sea Level Changes 278
          13.3.6   Geomorphological Changes 280
   13.4   Troy 280
          13.4.1   Geological Setting 281
          13.4.2   Ancient Troy 281
          13.4.3   Historical Reconstruction 283
   13.5   Gallipoli Peninsula National Park 284
   13.6   Assos and Pergamum 284
          13.6.1   Geological Setting 285
          13.6.2   Assos 285
          13.6.3   Pergamum 285
   13.7   Ephesus 288
          13.7.1   Geological Setting 288
          13.7.2   Antiquities 292
          13.7.3   Geomorphological Reconstruction 295

13.8   Antiquities in the Denizli Basin 295
       13.8.1   Geological Setting 296
       13.8.2   Aphrodisias 296
       13.8.3   Laodicea 298
       13.8.4   Colossae and Kibyra 298
       13.8.5   Marbles 299
   References 299

**14   The Hierapolis-Pamukkale Archaeological and Geosite,
Southwest Turkey** 301
   14.1   Introduction 302
   14.2   Travertine 302
   14.3   Geological Setting 303
       14.3.1   Denizli Basin 303
       14.3.2   Pamukkale Travertine Deposits 304
       14.3.3   Formation of the Travertine Deposits 304
   14.4   Pamukkale Geosite 305
   14.5   Hierapolis Archaeological Site 307
   14.6   Earthquakes 310
   References 310

15   Cappadocia, Central Turkey 313
   15.1   Introduction 313
   15.2   Historical Activity 315
   15.3   Regional Geology 315
   15.4   Cappadocia Volcanic Province 316
       15.4.1   Petrogenesis and Volcanic Centres 317
       15.4.2   Ignimbrite Sheets and Ash-Fall Deposits 317
   15.5   Göreme Historical National Park 318
   15.6   Differential Erosion 320
   15.7   Human Settlement 321
       15.7.1   Üchisar Castle 321
       15.7.2   Ihlara Gorge 322
       15.7.3   Underground Cities 322
   15.8   Current Hazards 325
   References 326

16   The Lake District, Northwest England 327
   16.1   Introduction 328
   16.2   Human Settlement and Literary Connections 329
   16.3   Topography 330
   16.4   Regional Geology 331
       16.4.1   British Caledonides 331
       16.4.2   Caledonian Orogeny 333
       16.4.3   Skiddaw Group 335
       16.4.4   Eycott Volcanic Group............................ 336
       16.4.5   Borrowdale Volcanic Group 336
       16.4.6   Windermere Supergroup........................... 338
       16.4.7   Ordovician Intrusions............................. 338
       16.4.8   Devonian Granite................................. 338
   16.5   Late Pleistocene Glaciation 340
   16.6   Finger Lakes 340

16.7   Archaeology and Mining Heritage.   343
       16.7.1   Lead Mines.   345
       16.7.2   Coppermines Valley, Coniston   345
       16.7.3   Graphite   345
       16.7.4   Tungsten   347
       16.7.5   Slate   347
       References.   350

17   Skaergaard Intrusion, Greenland and Eastern Bushveld Complex,
     South Africa.   353
     17.1   Introduction   354
     17.2   Skaergaard Intrusion.   355
            17.2.1   Geological Research   357
            17.2.2   Geological Setting   357
            17.2.3   Igneous Layering.   357
     17.3   Eastern Bushveld Complex   358
            17.3.1   Human Habitation and Mining   361
            17.3.2   Discovery of Platinum in the Eastern Limb   364
            17.3.3   Dunite Pipes   364
            17.3.4   Zones and Igneous Layering.   364
     17.4   Origin of Igneous Layering.   369
     References.   373

Correction to: The Geotraveller C1

Glossary.   375

Index   385

Abbreviations

BP	Years before present
BVG	Borrowdale Volcanic Group
DAD	Debris Avalanche Deposit
DRC	Democratic Republic of the Congo
EARS	East African Rift System
Ga	Billions of years
IAFZ	Izmir-Ankara Fault Zone
Ma	Millions of years
NAFZ	North Anatolian Fault Zone
NCA	Ngorongoro Conservation Area
PGE	Platinum Group Element
RAMSAR	Conservation on wetlands signed in Ramsar, Iran in 1971

List of Figures

Fig. 1.1   The heavily incised Colorado Canyon viewed from Dead Horse Point. The thick succession of flat-lying Upper Palaeozoic and Mesozoic strata include the distinctive bench developed on the White Rim Sandstone (Lower Permian). The mesa is capped by the Kayenta Sandstone (Lower Jurassic) 2

Fig. 1.2   Map showing location of some of the national parks in Utah, together with Monument Valley which overlaps the boundary with Arizona 3

Fig. 1.3   The dominant physiographic and geologic feature of southern and eastern Utah is the Colorado Plateau. The Colorado Plateau is bordered to the north and east by the Rocky Mountains and to the south and west by the Basin and Range terrain. *Source* Simplified after Baars (1993) 4

Fig. 1.4   Geological map of southern and eastern Utah showing location of the Canyonlands and Arches National Parks together with the Monument Valley. *Source* Simplified from regional map of Geological Survey of Utah (2005). 5

Fig. 1.5   Simplified stratigraphic sections for Canyonlands and Monument Valley. The thickness of the strata is entirely schematic as there is considerable regional variation. The relatively resistant nature of the different formations is shown schematically. *Source* Modified after Baker (1936) and Harris et al. (2004) 6

Fig. 1.6   Steeply-dipping Lower Permian strata on the flanks of the Monument Uplift, viewed from Highway 163 southern Utah 7

Fig. 1.7   The Petrified Forest Member constitutes the colourful rock layers exposed in the Painted Desert, Arizona. *Source* Wikipedia. 8

Fig. 1.8   The canyons and mesas at Canyonlands National Park reveal multiple beds of flat-lying, mostly red sandstones. 9

Fig. 1.9   Upheaval Dome, Canyonlands. The pale grey rocks in the core are the Organ Rock Mudstone (Lower Permian) and the red-brown rocks in the rim are the Wingate Sandstone (Lower Jurassic) 10

Fig. 1.10   **a** Chaotic cross-bedding indicative of wind-blown sands, Navajo Sandstone, Canyonlands; **b** Mud cracks consistent with a fluvial environment at the same locality. 11

Fig. 1.11   The desert landscape of the Monument Valley Navajo Tribal Park includes free-standing monoliths such as West Mitten Butte (left), East Mitten Butte, and Merrick Butte (right) 12

Fig. 1.12   Image showing location of the Monument Valley with some of the well known mesa and monoliths. Rain God Mesa is located just to the south of the image. The pediment is dominated by the Organ Rock Mudstone (light brown) and the highest mesas and monoliths are capped by the resistant Shinarump Conglomerate (pale green). *Source* Google Earth 13

Fig. 1.13   Idealized cross section of the regional plateau (left) and Monument Valley (right), the latter including mesas, buttes, pinnacles, and spires. The pediment is underlain by the relatively soft Organ Rock Mudstone. The near-vertical cliffs are associated with the resistant De Chelly Sandstone. The contact between the De Chelly Sandstone and the Moenkopi Sandstone is a major unconformity, although the Permian-Triassic boundary (the greatest mass extinction event known) is not found in Utah. The plateau, mesas and buttes are capped by the extraordinarily resistant Shinarump Conglomerate 13

Fig. 1.14   View looking south through the "North Window" at Monument Valley. The characteristic stratification of the Lower Permian and Triassic sediments can be observed in Cly Butte (left) and Elephant Butte (right). Pinnacles of the Rain God Mesa occur in the background with the Thunderbird Mesa in the far distance 14

Fig. 1.15   **a** The massive nature of the De Chelly Sandstone can be observed in the near-vertical rock faces at Monument Valley; **b** Pinnacles of De Chelly Sandstone form due to closely-spaced vertical jointing 15

Fig. 1.16   The West Mitten Butte at Monument Valley with its distinctive pinnacle. 16

Fig. 1.17   The pinnacles and spires grouped to the south of Spearhead Mesa include the 137 m-high "Totem Pole" (right) 16

Fig. 2.1   Delicate Arch, Arches National Park, is comprised of gently-dipping Entrada Sandstone (Middle Jurassic). The snow-capped La Sal Mountains are visible in the distance 20

Fig. 2.2   Map showing location of the Arches National Park and Dinosaur National Monument, eastern Utah. 21

Fig. 2.3   Simplified stratigraphic section for the Arches National Park and Dinosaur National Monument. The thickness of the strata is schematic. *Source* Modified after Baker (1936), Harris et al. (2004) and others. 22

Fig. 2.4   Monoliths of the Entrada Sandstone (Middle Jurassic) project from the pediment at Arches National Park. Localized areas of badlands erosion on the flanks of the Salt Valley (centre) reveal pale buff coloured cliffs of the Wingate Sandstone (Lower Jurassic). The snow-capped La Sal Mountains in the background are associated with intrusive igneous rocks 23

Fig. 2.5   **a** View looking south along Highway 191 from the entrance road to Arches National Park with Moab and the La Sal Mountains visible in the distance. The western escarpment is capped by the Wingate Sandstone and the eastern escarpment by the Entrada Sandstone; **b** Details of the Moab Fault in an information board at the Arches National Park with the up-faulted block to the left and the down-faulted block to the right (view looking north). Note the repetition of the Wingate Sandstone. 24

Fig. 2.6   Map of the Arches National Park showing the location of the Visitor Centre and selected geosites. The southern boundary of the park abuts against the Colorado River 25

Fig. 2.7   The high-altitude desert (grey and pale mauve) of the Arches National Park contrasts with the narrow vegetated strip (green) associated with the sinuous Colorado River. Salt Valley (brown) is a prominent physiographic feature in the centre of the park. The majority of arches occur on the flanks of this structure (dark mauve). Highway 191 follows a prominent NW-SE aligned valley (red) associated with the

Moab Fault: the national park is situated east of the fault. *Source* Google
Earth   26

Fig. 2.8   Geological map and section of the Arches National Park. The area
proximal to the Salt Valley contains numerous small faults which are not
shown for reasons of scale. *Source* Simplified from maps of the Utah
Geological Survey.   27

Fig. 2.9   The three members of the Entrada Formation are exposed in cliff faces
at Park Avenue, a small canyon at Arches National Park   28

Fig. 2.10   Formation of arches is initiated by erosion of broad sandstone fins
(background) in the Slickrock Member, as seen in the Devils Garden,
Arches National Park   29

Fig. 2.11   Holes develop in slightly softer sandstone beds of the Slickrock Member
(lower and top) which may be separated by more resistant cross-bedded
units (centre), as seen on the trail to Delicate Arch.   30

Fig. 2.12   A window located in a sandstone fin comprised of the Slickrock
Member next to the trail to Delicate Arch.   31

Fig. 2.13   Landscape Arch is reported to have the longest span of any natural arch.
The arch is developed on a prominent fin or buttress comprised
of massive sandstones of the Slickrock Member   32

Fig. 2.14   Balanced Rock, Arches National Park, consists of the Slickrock Member
perched on a relict of the Dewey Bridge Member (both Entrada
Sandstone) resting on a pedestal of Navajo Sandstone   33

Fig. 2.15   The Quarry Museum, Dinosaur National Monument is constructed over
the "Quarry Sandstone", a steeply-dipping and partially contorted rock
face anomalously rich in dinosaur fossils   34

Fig. 2.16   The impure sandstone, shales and mudstones of the Morrison Formation
are steeply-dipping in an exposure near the Dinosaur Museum.   35

Fig. 2.17   Geological map of Dinosaur National Monument. *Source* simplified
from maps of the Geological Survey of Utah (2005)   35

Fig. 2.18   **a** The Quarry Museum exposes a bedding plane of the Quarry Sandstone
with more than 1,600 dinosaur fossils in situ; **b** Fossils of sauropods
in the Quarry Sandstone   36

Fig. 3.1   Half Dome is one of the most well known of the granitic monoliths
at the Yosemite National Park. The 600 m-high, planar north face,
which hangs above the upper sections of the Yosemite Canyon, is
defined by a set of near-vertical joints. Exfoliated scree was removed by
glacial activity and ice sheets scraped clean large sections of the granite
plutons which dominate the Sierra Nevada in this region   38

Fig. 3.2   Map showing location of the Yosemite National Park relative to coastal
ranges in the state of California   39

Fig. 3.3   Yosemite is well known for black bear which can be observed on granite
pavements.   40

Fig. 3.4   Image of the Yosemite National Park showing the mountainous
terrain and entrance gates and geosites referred to in the text. *Source*
Google Earth.   41

Fig. 3.5   Tenaya Lake, located between the Tuolumne and Yosemite Canyons
is an example of a high-altitude or Alpine lake located in a hanging
valley at an elevation of 2,484 m   42

Fig. 3.6   Graph showing estimated temperatures during the previous 20 centuries.
The most pronounced cycles (with estimated durations) are the Medieval
Warm Period (950–1250 AD) and the Little Ice Age (1300–1850 AD).
The relatively rapid temperature changes during the Little Ice Age

              promoted numerous advances and recessions of ice sheets at Yosemite.
              Source https://commons.wikimedia.org   43

Fig. 3.7   **a** The Yosemite Canyon is the principal feature of the national park. The
              U-shape is typical of ice-carved valleys although the canyon initially
              developed due to an older phase of erosion; **b** The view from Glacial
              Point includes pastures associated with nutrient-rich moraine deposits
              in the floor of the Yosemite Canyon.   44

Fig. 3.8   El Capitan is the highest of the granitic domes with a near-vertical
              face of 1,095 m. Just visible are sections of younger granitic intrusives
              and cross-cutting granite pegmatite dykes and veins (darker)   45

Fig. 3.9   The near-vertical-face of Half Dome includes dark vertical lines
              associated with water seepage.   47

Fig. 3.10   Many of the granite domes proximal to the Yosemite Canyon reveal
              prominent joints. The rock slabs have been swept clean by glacial
              activity   47

Fig. 3.11   The idyllic scenery of Mirror Lake occurs in a hanging valley perched
              hundreds of metres above the Yosemite Canyon.   48

Fig. 3.12   **a** The upper section of the Yosemite Falls, a tributary of the Merced
              River, plunges 436 m over granitic cliffs into the Yosemite Canyon;
              b The Vernal Falls on the Merced River is associated with a glacial
              step cut into the basement granite by the Yosemite Glacier   49

Fig. 3.13   The stands of giant sequoia in the Mariposa Grove, Yosemite
              encouraged protection of the area   50

Fig. 3.14   Historical photograph of the Wawona Tree, a giant sequoia
              in the Mariposa Grove, which fell in 1969 (internet)   50

Fig. 4.1   The multi-coloured rocks in the walls of the Yellowstone Canyon
              illustrate the derivation of the name of the Yellowstone National Park.
              The brightest patches of colour are due to hot springs and fumaroles that
              have hydrothermally altered the thick sequences of rhyolitic volcanic
              ashes and pyroclastics erupted from the Yellowstone Volcano during
              the most recent of the caldera events. Thin basaltic flows (a subordinate
              component of the volcanism) that display prominent columnar jointing
              are intercalated with the rhyolitic ashes and pyroclastics. The
              hydrothermal features, including the famous geysers, are driven by heat
              associated with a shallow magma chamber located beneath the
              Yellowstone Caldera   52

Fig. 4.2   Map showing location of the Yellowstone National Park, in the states
              of Wyoming, Idaho and Montana, northwest USA   53

Fig. 4.3   Image showing the location of the Yellowstone National Park together
              with some of the mountain ranges, including the Tetons. *Source* Google
              Earth   54

Fig. 4.4   Views of wildlife, including bison, in association with active geothermal
              features is a unique feature of the Yellowstone National Park   54

Fig. 4.5   Deeply incised U-shaped valleys in the Beartooth Mountains formed
              during the Late Pleistocene Ice Ages   54

Fig. 4.6   View from Jenny Lake of the triangular-shaped and snow-capped peaks
              in the Grand Teton National Park.   55

Fig. 4.7   Information board located close to the Teton Fault, looking west
              towards the uplifted Teton Range   55

Fig. 4.8   The volcanism of the Columbia River basalts and the Yellowstone
              Volcano is related to a single deep-seated hot spot which appears
              to have "migrated" eastward due to the westward drift of the North
              Atlantic Plate   56

Fig. 4.9   Geological map of the Yellowstone National Park showing extent of the
              youngest caldera, a scarp related to an older caldera, and the larger
              geyser basins. *Source* Simplified after Keefer (1975)   58
Fig. 4.10   View of the desolate landscapes typical of geyser basins at Yellowstone
              with the rim of the youngest caldera visible in the background   59
Fig. 4.11   Simplified cross-section of the Yellowstone Volcano illustrating a
              schematic magma chamber at depth. The position of the magma
              chamber relative to the caldera is offset to the east due to the apparent
              migration of the hot spot.   59
Fig. 4.12   Obsidian (black volcanic glass) occurs in pink-coloured banded rhyolite
              at Obsidian Cliff, Yellowstone   60
Fig. 4.13   Columnar jointing typical of basaltic lava flows at Yellowstone   60
Fig. 4.14   The Norris Geyser Basin contains numerous geothermal features
              including geysers, fumaroles and hot springs   61
Fig. 4.15   Section showing how geothermal features work The geyserite (grey) is
              underlain by rhyolite depicted schematically as getting hotter with depth
              (pink, orange, and red). *Source* based entirely on White (1967)   62
Fig. 4.16   **a** Blue pools are relatively hot; **b** Brown pools are cooler; **c** Orange
              pools are rich in thermophiles; **d** Mud pools occur in areas where
              groundwater is sparse; and **e** A colour chart (from an information board
              in the park) illustrating the relationship between colour and temperature
              and/or the presence of thermophiles   63
Fig. 4.17   **a** The column of hot water associated with Old Faithful, the most
              famous of the geysers at Yellowstone, erupts on average every 92 min;
              b Castle Geyser in the Upper Geyser Field has built up a substantial
              mound of geyserite and includes a relatively long steam phase at the end
              of each eruptive cycle   65
Fig. 4.18   Terraces of travertine at the Mammoth Hot Springs reveal constantly
              changing forms as they are shaped by subtle changes in the springs'
              plumbing system   66
Fig. 4.19   The Yellowstone River includes several large waterfalls   66
Fig. 5.1   View looking south along the Icefields Parkway in the U-shaped valley
              of the North Saskatchewan River, Banff National Park. The ice-scoured
              mountains of the Front Ranges consist of well-bedded, eastward-dipping
              Palaeozoic strata   70
Fig. 5.2   Map showing location of the three national parks in the Canadian
              Rockies, the continental divide, major rivers, and access routes   71
Fig. 5.3   Stratigraphy of the Canadian Rocky Mountains. *Source* Simplified
              after Leckie (2017)   72
Fig. 5.4   **a** Selected geosites in the Banff and Jasper National Parks accessed by
              the Icefields Parkway and sites in the Yoho National Park close to the
              Trans-Canadian Highway. Only icefields in the national parks are shown
              (slope glaciers are too small to show). The provincial boundary
              corresponds to the "Great Divide"; **b** Geological map of the area
              proximal to the Icefields Parkway. *Source* Simplified after Yorath
              and Gadd (1995); Leckie (2017), and maps of the Geological Survey
              of Alberta   74
Fig. 5.5   The northern part of the Icefields Parkway, Jasper National Park,
              follows the course of the Athabasca River with views of the Main
              Ranges   75
Fig. 5.6   The narrow gorge of the Maligne River carved into flat-lying limestone
              of the Palliser Formation (Devonian)   76
Fig. 5.7   **a** The upper section of the Maligne Canyon reveals evidence of a cave
              system in the sidewalls; **b** The lower sections of the Maligne River
              includes springs exiting the cave system.   77

Fig. 5.8   **a** Mount Edith Cavell is comprised almost entirely of resistant quartzite
             of the Gog Group (Lower Cambrian); **b** The crenulated face of Mount
             Geraldine includes cirques that contain small glaciers (foreground left
             and centre)   78

Fig. 5.9   **a** The gorge below the Athabasca Falls occurs in well-bedded quartzite
             of the Gog Group (Lower Cambrian); **b** Large potholes on the walls
             of the Sunwapta Gorge are relics of higher water stands. The gorge cuts
             into limestone of the Cathedral Formation (Middle Cambrian)   79

Fig. 5.10   Terminal moraines left by the receding Athabasca and Saskatchewan
             Glaciers, part of the irregular-shaped Columbia Icefield, are visible on
             this satellite image. *Source* Modified Copernicus Sentinel data of August
             2018 processed by Philip Eales, Planetary Visons/DLR   80

Fig. 5.11   **a** The Athabasca Glacier is located in a valley fed by the Columbia Icefield
             (vehicles on the glacier are specialized tour buses); **b** Recession of the
             Athabasca Glacier reveals extensive terminal and lateral moraines   81

Fig. 5.12   Geothermal deposits, Cave Spring, Banff. *Source* https://
             www.google.com/ gonewiththewynns.com   82

Fig. 5.13   **a** View of Lake Louise looking southwest towards the high peaks
             of Mount Victoria. A barren rock face comprised of Cog Quartzite
             separates the two glaciers on the lower flanks; **b** View of Lake Louise
             from the trail to Lake Agnes shows the renowned turquoise colour with
             gently dipping quartzites of the Cog Group exposed in the mountain
             face   83

Fig. 5.14   **a** The turquoise-coloured Moraine Lake is one of the most scenic of the
             glacial lakes in the Canadian Rockies; **b** Cone-shaped scree deposits
             occur at the base of the cliffs above Moraine Lake   84

Fig. 5.15   Terminal moraine and landside deposit with large blocks of Cog Group
             quartzite dams Moraine Lake   85

Fig. 5.16   The debris fan associated with the unstable face of Cathedral Mountain,
             on the northern side of Kicking Horse Pass, Yoho National Park, affects
             the Canadian Pacific Railway and Trans-Canada Highway.
             Source Original photograph by Lukas Arenson and annotation of Leckie
             (2017)   86

Fig. 5.17   Natural Bridge, Yoho National Park, is associated with a knickpoint on
             the Kicking Horse River. The calcareous slate of the Chancellor
             Formation (Middle Cambrian) reveals prominent cleavage   87

Fig. 5.18   A slab of Burgess Shale in the interpretative centre at Field contains
             typical Middle Cambrian marine fossils, including a large trilobite,
             Ogygopsis (width of approximately 4 cm)   88

Fig. 6.1   The Nyinambuga Crater viewed from Ndali Lodge in the Ndali-Kasenda
             Crater Field, southern Uganda   90

Fig. 6.2   Map showing location of national parks and major lakes in southern
             Uganda, together with the two components of the White Nile (Victoria
             Nile and Albert Nile) and parks in the DRC and Rwanda in the vicinity
             of the Virunga Mountains   91

Fig. 6.3   Image of southern Uganda showing the extensiveness of the wooded
             savannahs on the regional plateau (light green), the forested rift
             shoulders and areas of central Africa west of the Albertine Rift (dark
             green), and the rift valley with its finger-shaped lakes. Lake Victoria and
             Lake Kyoga are associated with shallow warps on the regional plateau.
             The Victoria Nile exits Lake Victoria at Jinja, flows northwest via Lake
             Kyoga, and exits Lake Albert as the Albert Nile. *Source* Google Earth . . .   92

Fig. 6.4   **a** African elephant are a common site in the Queen Elizabeth National
             Park and can be approached to within a few metres in the Kazinga

Channel; **b** Cape buffalo can be observed in swamps at the Murchison
Falls National Park; **c** The Uganda kob is the most abundant grazer on
grassy savannahs of the Queen Elizabeth National Park; **d** Chimpanzee
are protected in forests of the Kibale National Park; **e** Mountain gorilla
occur in Afromontane forests of the Virunga Mountains (Volcanoes NP,
Rwanda); **f** Aquatic birds such as yellow-billed stork and little egret are
a common site on the banks of the Kazinga Channel 93

Fig. 6.5   Simplified geological map of southern Uganda. Each of the
finger-shaped lakes of the Albertine Rift occurs in a separate
sedimentary basin. Five rift-related volcanic terrains are recognized
in the Albertine Rift: **a** Fort Portal-Kasekere; **b** Ndali-Kasenda;
c Katwe-Kikorongo; **d** Bunyaraguru; **e** Virunga. The Mount Elgon
volcanic field is part of the Gregory Rift (inset). *Source* Simplified from
the 1:1,500,000 scale map compiled by R. MacDonald and published by
the Department of Geological Survey and Mines, Uganda (1966). 94

Fig. 6.6   Large sections of the regional plateau in southern Uganda are covered
by thick palaeosols and recent deposits of gravels and alluvium,
as seen in this quarry near Mbarara 95

Fig. 6.7   Simplified geological map of East Africa depicting the three branches
of the EARS. Some rifts are truncated by older, Cretaceous-age rifts
(e.g., Anza Graben). Volcanism associated with the Ethiopian and
Gregory Rifts is notably extensive. The Albertine Rift is dominated by
sedimentary basins. *Source* Compiled from the 1:1,000,000 Geological
Map of Kenya, the 1:2,000,000 Geological Map of Tanzania, and the
1:1,500,000 Geological Map of Uganda 96

Fig. 6.8   The effects of the Late Cenozoic rifting on river systems and lakes in
southern Uganda. **a** Pre-rifting; **b** Formation of a large palaeo-lake in the
Albertine Rift; **c** Isolation of lake basins in the Albertine Rift;
d Formation of the Albert Nile and Lake Victoria; **e** Upwarping between
the Albertine Rift and Lake Victoria; **f** Development of the Victoria
Nile. *Source* Compiled from Beadle (1981), Pickford et al. (1993),
Talbot and Williams (2008) and others 98

Fig. 6.9   **a** The Victoria Nile exits Lake Victoria at Jinja; **b** View of the Owens
Dam on the Victoria Nile with hills (background) comprised of the
Buganda-Toro System; **c** View from Wildwaters Lodge, Kangulumira
Island, showing resistant outcrops of the Uganda Gneiss Complex
in a fast-flowing, wide section of the Victoria Nile. 100

Fig. 6.10   The Uganda Gneiss Complex is intruded by dark amphibolite dykes
of the Buganda-Toro System, Kangulumira Island 100

Fig. 6.11   **a** The Victoria Nile at Murchison Falls is constricted from a wide
channel to a narrow gorge as the river plunges into the Albertine Rift;
b The gorge is carved out of resistant rocks of the Uganda Gneiss
Complex 101

Fig. 6.12   **a** In the Murchison Falls National Park, the Victoria Nile is fringed
by riverine forest with hardwoods, including African mahogany and
borassus palms; **b** The Victoria Nile enters Lake Albert by a delta that
includes a myriad of channels and small islands. The water hyacinths
have survived the passage from Lake Victoria 102

Fig. 6.13   View looking south along the Albertine Rift. Lake Albert is constrained
between two escarpments. *Source of Image* Christian Hormann 2012
(http://earth.imagico.de). 103

Fig. 6.14 **a** The tranquil setting of Lake Edward viewed from Uganda; **b** The sluggish, meandering Kazinga Channel which connects Lakes George and Edward is located in the Queen Elizabeth National Park 104

Fig. 6.15 **a** Kitagata Crater Lake and **b** Kikeronga Salt Pan. Both localities occur in the Katwe-Kikorongo Crater Field, Queen Elizabeth National Park 106

Fig. 6.16 Outcrops of amygdaloidal alkali lavas and agglomerate cap the walls of the Kyemengo Crater, Katwe-Kikorongo Crater Field 107

Fig. 6.17 Rwenzori Mountains. *Source* en.Wikipedia.org 107

Fig. 6.18 Monoliths of the Uganda Gneiss Complex are a feature of the Lake Mburo National Park 108

Fig. 6.19 The Bwindi Impenetrable National Park is situated in an area of uplands (on the shoulders of the Albertine Rift) covered by dense Afromontane and lowland forest 109

Fig. 6.20 The volcanic peak of the Sabinyo Volcano in the Virunga Mountains, viewed from the Volcanoes National Park, Rwanda 110

Fig. 6.21 Image of the Nyiragongo Volcano showing location in the southern part of the Virunga National Park (DRC) and proximity to Goma and Lake Kivu. The ascent of the cone starts at the Kibati gate and includes sections located on the 2002 lava. *Source* Google Earth 110

Fig. 6.22 The steep-sided and rugged Nyiragongo cone (foreground) as compared with the smoother and gentler slopes of Nyamulagira (background). Both volcanoes are more-or-less continuously active 111

Fig. 6.23 View of Goma airport, partly covered by the 2002 lava flow, looking south over the city which is situated next to Lake Kivu. *Source* Guido Potters (http://www.gnu.org/copyleft/ fdl.html 112

Fig. 6.24 **a** The active summit crater of the Nyiragongo Volcano contains a lava lake (left) and an active spatter cone (right). Lava fountains associated with the spatter cone attain a height of approximately 50 m; **b** The rim of the near vertical-walled summit crater is capped by lava from the major eruption of 1977 113

Fig. 6.25 **a** The lava lake (left) and active vent (right) in the summit crater of the Nyiragongo Volcano are a spectacular sight in the twilight; **b** A night time view of the lava lake shows the occurrence of cracks and segments with magma flares 114

Fig. 7.1 The Western Escarpment of the Gregory Rift near Lake Natron is constructed of multiple, near-horizontal layers of lava and tephra 118

Fig. 7.2 Map showing location of selected national parks and wilderness areas in northern Tanzania 119

Fig. 7.3 Geological map of the northern Tanzanian divergence. The rifted terrains are dominated by volcanics (light green) and include the extinct volcanic complex associated with the Ngorongoro Highlands (dark green). Some of the larger cones and calderas are shown, but for reasons of scale the subordinate sedimentary basins are omitted. *Source* Simplified after Dawson (2008) 119

Fig. 7.4 The unusual colour of the northern and central parts of Lake Natron is due to the red photosynthesizing pigment in the cyanobacteria. The rugged terrain on either sides of the lake is associated with deeply-eroded Pleistocene volcanics. View looking north; width of lake in southern part of image is approximately 10 km. *Source* NASA Terra-ASTER image for 2003, processed by Philip Eales, Planetary Visions/DLR 121

Fig. 7.5   The symmetrical cone of the Oldoinyo Lengai Volcano rises abruptly
            from the floor of the rift valley near Lake Natron................... 122

Fig. 7.6   The eastern part of the Ngorongoro Conservation Area (NCA) is
            constrained to the Eyasi Half-graben, a discrete structural block (located
            between the Rift Valley and Eastern Serengeti Plains) defined by
            prominent escarpments. The Rift Valley contains Lake Natron and Lake
            Manyara; Lake Eyasi is situated in the Eyasi Half-graben. The contrast
            between the lush Ngorongoro Highlands (dark green) and arid terrains
            of the Salei Plains, Gol Mountains, Rift Valley and Eastern Serengeti
            Plains (beige) is pronounced. The Ngorongoro Highlands is associated
            with the extinct Ngorongoro Volcanic Complex (EC: Empakaai
            Caldera; LE: Lemagrut; LO: Loolmalasin; NC: Ngorongoro Caldera;
            OC: Olmoti Caldera; OL: Oldeani; SA: Sadiman). Other components
            of the NCA are dominated by wind-blown ash from the active Oldoinyo
            Lengai Volcano. The palaeoanthropological sites of Oldupai and Laetoli
            are located in the Eastern Serengeti Plains. The Oldupai River peters out
            in the Olbalbal Swamps (OS). Approximate width of view is 150 km.
            Source NASA Landsat 7 ETM + image mosaic for the year 2000
            sourced from the University of Maryland Global Land Cover Facility,
            processed by Philip Eales, Planetary Visions/DLR.................... 123

Fig. 7.7   The eastern approach to the NCA ascends the Western Escarpment near
            Lake Manyara (background, left). The escarpment is composed of
            multiple, near-horizontal volcanic layers of lavas and tephra associated
            with the Ngorongoro Volcanic Complex............................ 124

Fig. 7.8   Ngorongoro Caldera and Lake Magadi from the viewpoint on the
            southeast of the caldera during the wet season...................... 124

Fig. 7.9   Geological map of the Ngorongoro Conservation Area. Outcrop of the
            Oldupai and Laetoli Basins are not shown for reasons of scale. *Source*
            Simplified after Pickering (1958; 1964; 1965), Orridge (1965) and
            Dawson (2008)... 125

Fig. 7.10   Simplified geological section of the NCA and adjacent terrains.
            Section centred on the Ngorongoro Caldera (latitude $3°\ 10'$ South).
            Localities in brackets (e.g., Lake Ndutu, Oldupai Gorge, Kilimanjaro
            etc.) are located north of the section line. The thickness of the rift-related
            volcanics is schematic... 125

Fig. 7.11   Geological map of the Ngorongoro Caldera and part of the Olmoti
            Caldera. *Source* simplified from the Geological Survey of Tanzania
            1:125,000 quarter degree sheet 53 by Pickering (1965).............. 126

Fig. 7.12   The lush paradise of the Ngoitokitok Springs, Ngorongoro Caldera.
            The northern wall of the caldera is dwarfed by the flanks of the Olmoti
            Volcano.. 127

Fig. 7.13   A Maasai herdsman on a trail leading to the Elanairobi Ridge. The
            background shows the thickly forested inner slopes and lake of the
            Empakaai Caldera.. 127

Fig. 7.14   **a** Isolated, barchan-style dune of black volcanic ash at Shifting Sands
            with the western rampart of the Ngorongoro Highlands visible in the
            background; **b** The front of the dune (which is migrating towards
            the right of the photograph) reveals a steep slope and distinctive
            outer horns.. 129

Fig. 7.15   Migration of wildebeest and zebra in the vicinity of Lake Ndutu
            on the Eastern Serengeti Plains................................. 130

Fig. 7.16   **a** The stratigraphy at Oldupai Gorge includes distinctive, near-horizontal
                 layers of lacustrine sediments and volcanic lavas and ashes (view close
                 to the Visitors Centre). The principal hominin discoveries were made in
                 Beds I and II, which are separated by the Marker Tuff; **b** A view of the
                 red-brown buttes associated with Bed III shows the proximity of the
                 extinct volcanoes associated with the Ngorongoro Highlands
                 (background). 131
Fig. 7.17   Stratigraphy of the Oldupai Gorge with column on right showing details
                 of Bed I and the lowermost part of Bed II. *Source* Modified after Hay
                 (1976) and Stollhofen and Stanistreet (2012) with radiometric data
                 of Deino (2012) 132
Fig. 7.18   The FLK site where the cranium OH-5 (*Zinjanthropus boisei*) was
                 discovered in Bed I at Oldupai Gorge is commemorated by a stone cairn
                 and plaque. The section in the cliff face includes resistant tuff layers
                 (pale grey) interbedded with clays (brown) in the upper part of Bed
                 I. The uppermost tuff layer in the section is the Marker Tuff which is
                 overlain by clays (dark brown) in the lower part of Bed II. 133
Fig. 7.19   **a** Reconstruction of the cast of the 1.85 Ma old cranium of
                 Zinjanthropus boisei (OH-5) discovered in Bed I at Oldupai Gorge,
                 together with the lower jaw (the Peninj mandible) found near Lake
                 Natron. Photograph by Lillyundfreya of the Westfälisches Musuem;
                 b Replica of the 1.9 Ma old cranium of *Homo habilis* (KNM-ER1813)
                 from Koobi Fora, Kenya. Photograph by Locutus Borg 134
Fig. 7.20   Laetoli is located in an area of badlands erosion on the Eastern Serengeti
                 Plains. The museum is just visible in background. 134
Fig. 7.21   Cast of part of the footprints made by *Australopithecus afarensis*,
                 Laetoli Museum 135
Fig. 7.22   **a** Hiking in the Engare Sero Gorge near Lake Natron reveals exposures
                 of volcanic debris avalanche deposits (foreground, right) as well as thick
                 sequences of lavas and tephra; **b** Nephelinitic breccia dyke cuts a
                 basaltic lava flow, Engare Sero Gorge 136
Fig. 8.1   Lava fountains and small lava flows associated with strombolian
                 activity, eruption of the southeast summit crater Etna (24th December
                 2018). *Source* Emanuela Carone, VolcanoDiscovery Italy (https://
                 images.app.goo.gl/) 140
Fig. 8.2   Image of the central and eastern Mediterranean, including bathymetry,
                 showing the seas and volcanoes described here. *Source* Satellite Image
                 of Europe based on NASA MODIS data processed by Philip Eales,
                 Planetary Visions/DLR 141
Fig. 8.3   Image of the central and eastern Mediterranean showing the tectonic
                 framework of plates, microplates and sutures, together with the
                 bathymetry. *Source* Geological boundaries and faults simplified from
                 articles referenced in the text; Satellite Image of Europe based on
                 NASA MODIS data, processed by Philip Eales, Planetary
                 Visions/DLR 143
Fig. 8.4   Cross-section showing the position of active fore-arc and back-arc
                 basins in an accretionary convergent margin. *Source* By Zyzzy2 at the
                 English language Wikipedia, CC BY-SA 3.0, https://commons.
                 wikimedia.org/w/index.php?curid=54512237. 144
Fig. 8.5   Simplified tectonic map of southern Italy and the Italian Islands showing
                 location of the Quaternary volcanoes. The Apennine-Maghreb Thrust is
                 an active subduction zone. The Tyrrhenian Sea is a back-arc basin, the
                 Aeolian Islands are associated with a volcanic island arc, and the
                 Calabrian Wedge is a fore-arc basin. The Etna Volcano is a discrete

stratovolcano located proximal to the plate boundary. The Pantelleria Volcano is situated on a transform fault on the convergent plate boundary. The fault zone located between the west coast of Calabria and eastern Sicily is associated with the Siculo-Calabrian Rift. *Source* After Panza and Suhadolc (1990). 144

Fig. 8.6   Volcanoes that generate pyroclastic fall deposits can be divided into well-defined fields on a plot of fragmentation index (percentage of deposit with grains finer than 1 mm) against the area of dispersal. Coloured areas include the majority of eruptions on Earth. Ash-fall deposits associated with the relatively quiescent Hawaiian eruptive style have restricted distribution. Most eruptions of Etna and Pantelleria plot in this field. The catastrophic Plinian eruptions (which are divided into three separate fields) have larger components of pyroclastics that disperse over great distances. Many of the eruptions associated with the Neapolitan volcanoes plot in these fields. Strombolian eruptions define an intermediate field between Hawaiian and Sub-plinian. Eruptions located above the shaded area are associated with either short-lived explosive activity (Vulcanian) or systems that have reacted with water (Surtseyan or Phreatoplinian). *Source* After Walker (1973) and Wright et al. (1980) 147

Fig. 8.7   Plot of total alkali (sodium and potassium oxides) versus silica ("TAS diagram") is a popular method of categorizing alkaline volcanic rocks. The line X–Y is used to separate the alkali and sub-alkali groups. The basalt subdivision can be divided into three groups: alkali basalt, transitional basalt (shaded area straddling the division), and subalkali basalt. The alkali-rich foidolite group includes two principal variants, nephelinite (potassium-rich) and leucitite (sodium-rich). Also shown are three fractional crystallization sequences, A1–A2 (potassium or sodium alkali basalt trend), B1–B2 (calc-alkaline basalt trend) and C1–C2 (foidolite trend). 148

Fig. 8.8   Map of the Aeolian Islands showing location of the two active volcanoes 149

Fig. 8.9   **a** The near-symmetrical cone of the island of Stromboli (September 2004). *Source* By Steven W. Dengler, CC BY-SA 3.0, https://commons.wikimedia.org/; **b** Erupting incandescent molten lava fragments, Stromboli (December 1969). *Source* By B Chouet from the website of the United States Geological Survey. 151

Fig. 8.10   The northern side of the island of Vulcano viewed from Lipari. The island is comprised of several accreted cones and calderas. Extinct features are located on the left and the active Fossa cone with the Grand Cratere occur in the centre above the town. *Source* By Brisk g—Own work, Public Domain, https://commons.wikimedia.org/. 152

Fig. 8.11   **a** Blocks and veins of sulphur crystals located on the rim of the Fossa cone, Vulcano, with Stromboli visible in the background (right); **b** The Gran Cratere includes veins and pockets of sulphur crystals. 152

Fig. 8.12   The Etna Volcano constitutes a large massif in northeast Sicily. Also visible are the Aeolian Islands. *Source* 3D Image based on NASA MODIS data, processed by Philip Eales, Planetary Visions/DLR. 153

Fig. 8.13   **a** The Etna massif towers threateningly above the city of Catania, northeast Sicily; **b** Southern flank of Mount Etna showing lateral cones and a lava flow from the 2001 eruption. *Source* Wikipedia (Wilson 44691 own work) 154

Fig. 8.14 **a** The ascent of Etna includes hiking on recent (unvegetated) lava flows with cinder and agglomerate; **b** The summit craters include active geothermal vents and steam is more or less continuously being emitted . . . 155

Fig. 8.15 **a** Subsidiary cones of cinder and ash occur near Etna's four summit craters; **b** Active geothermal vents have deposited silica and sulphur 156

Fig. 8.16 The city of Catania was rebuilt after the 1669 eruption of Etna and the 1693 earthquake with classical buildings, squares, and wide streets to enable rapid evacuation. Parts of the city were developed on foundations constructed on Roman buildings (including part of an amphitheatre built of blocks of Etna lava) destroyed by earlier eruptions and earthquakes . . . 157

Fig. 8.17 **a** The archaeological site of Paestum is situated on a broad coastal plain at the base of the Apennines. The Sele Plain is underlain by Quaternary sediments. Hills in the background consist of Mesozoic carbonates; **b** Paestum includes three well-preserved temples, including the Temple of Athena built in the Doric style (possible age of 500 BC) 159

Fig. 8.18 **a** The Velia archaeological site includes Roman walls damaged by earthquakes; **b** Excavations at Velia have exposed an alluvial sequence which includes historical remains (including terracotta Roman tiles) from the 2^{nd}C AD until present. 160

Fig. 8.19 **a** The Neapolis Archaeological Park at Syracuse is situated on a limestone pavement (Syracuse Limestone) that erodes to form a karstic landscape; **b** The *Teatro-Greco* is constructed into the limestone bedrock. Natural caves and quarries in the limestone are visible in the background 161

Fig. 8.20 **a** The flat-lying nature of the marine limestones can be observed at the *Latomie* quarries; **b** The "Ear of Dionysus" is located in the Syracuse Limestone 162

Fig. 8.21 Image of the Eastern Mediterranean showing some of the possible locations associated with Homer's *Odyssey*. *Source* Satellite Image of Europe based on NASA MODIS data, processed by Philip Eales, Planetary Visions/DLR 163

Fig. 8.22 A map located in the village of Stavrós, Ithaca, shows a highly speculative route that *Odysseus* may have endured on his voyage from Troy (location 1) to Ithaca (location 14) 163

Fig. 8.23 **a** The village and sheltered harbour of Kioni is one of the many idyllic locations on the island of Ithaca; **b** "Homer's School", located in the limestone (Mesozoic) hills near Stavrós, Ithaca, includes a pulpit-like rock. 164

Fig. 8.24 **a** The *Ciclopi* are black pillars of fine-grained dolerite that rise from the sea close to the harbour at Acitrezza. The dolerite reveals prominent columnar jointing; **b** The dolerite occurs in near-surface sills injected into the Syracuse Limestone 165

Fig. 8.25 Waterspouts are common phenomena in the Straits of Messina and other parts of the Mediterranean (example here photographed in the Adriatic Sea, between Albania and southern Italy). They typically occur as clusters with multiple centres that drop instantaneously over several square kilometres 166

Fig. 9.1 The two peaks of the Somma-Vesuvius Volcano (Vesuvius on the right) tower above the city of Naples with mountains of the Southern Apennines (left) and Sorrento Peninsula (far right) visible in the background 170

Fig. 9.2   **a** The Neapolitan volcanoes are located on the margins of the broad
               Campanian Plain between the Southern Apennines and the Gulf of
               Naples. The islands of Capri and Procida are included in the image, but
               Ischia is located to the west of the area shown. Pliny the Younger
               observed the 79 AD eruption from Misenum. Pliny the Elder died at
               Stabaie; **b** Three-dimensional image looking east showing the
               more-or-less continuous urban conurbation on the shores of the Gulf of
               Naples. The dome-shaped Somma-Vesuvius massif and the nested
               calderas of the Campi Flegrei Volcano are prominent features. *Source*
               3D Image based on NASA MODIS data, processed by Philip Eales,
               Planetary Visions/DLR 171
Fig. 9.3   Geological map of the area around the Gulf of Naples. *Source*
               Simplified after Orsi et al. (2004) 173
Fig. 9.4   **a** The scenic town of Amalfi is overshadowed by steep cliffs of
               Mesozoic limestone; **b** The western side of the island of Capri includes
               cliffs and sea stacks consisting of Mesozoic limestone 174
Fig. 9.5   Chronology of eruptions of the principal Neapolitan Volcanoes. *Source*
               Modified after Orsi et al. (2004) and Di Vito et al. (2011) 175
Fig. 9.6   View from Camaldoli Hill shows the city of Pozzuoli nestling between
               the rim of the Campi Flegrei caldera. 176
Fig. 9.7   The Verdolino Quarry at Pozzuoli exposes the two principal products
               of the Campi Flegrei Volcano, the Campanian Ignimbrite (39,000 BP)
               and the Neapolitan Yellow Tuff (15,000 BP) 177
Fig. 9.8   **a** Roman columns at the Serapeo Market, Pozzuoli, buried in the Gulf of
               Naples for many years due to ground subsidence, abruptly reappeared in
               the 11thC AD due to localized uplift; **b** The harbour at Pozzuoli includes
               an old wall used by the British Navy in 1799 (left of the car). The
               location relative to the current sea level is indicative of several metres of
               uplift during the past 200 years. 178
Fig. 9.9   **a** The Solfatara, Pozzuoli, is a crater located in the Campi Flegrei
               caldera complex. The crater contains sulphur deposits and active
               fumaroles; **b** Lake Averno occurs in a scenic crater in the Campi Flegrei
               calderas 179
Fig. 9.10   Sequence of lavas and pyroclastics at Punta Imperatore, Ischia. 1: Lavas
               (< 75,000 BP); 2: Breccia deposits (75,000–55,000 BP); 3: Pyroclastic
               rocks (75,000–55,000 BP); 4: Mount Epomeo Green Tuff (55,000 BP);
               5: Deposits of Scarrupo Di Panza Volcano; 6: Pyroclastic rocks
               (28,000–18,000 BP) 180
Fig. 9.11   Trachybasaltic scoria cone and feeder dyke at Grotta di Terra, Ischia. 1:
               Lavas (<75,000 BP); 2: Pyroclastic rocks (75,000–55,000 BP); 3:
               Trachybasaltic dyke and sill (28,000 BP); 4: Pyroclastic rocks (<10,000
               BP). The feeder dyke has spread laterally to form a sill part way up the
               cliff, exploiting the primary layering of the older volcanic deposits 181
Fig. 9.12   Deposits associated with Plinian and sub-Plinian deposits of
               Somma-Vesuvius are exposed in the Traianello Quarry. The Basal
               Pumice (22,000 BP) and the Verdoline Pumice (19,000 BP) are capped
               by the Principali Pumice, the latter associated with the Campi Flegrei
               Volcano. The Mercato Pumice (8,800 BP) and Avellino Pumice
               (3,800 BP) are discordantly overlain by ash-flows associated with the 79
               AD event 182

Fig. 9.13   **a** The Roman city of Herculaneum is located in the modern suburb
             of Ercolano, at the base of the Somma (left) and Vesuvius (right) cones;
             b Roman house at Herculaneum partially filled with volcanic ash
             (mostly of the A4 and A5 deposits) from the 79 AD eruption 183

Fig. 9.14   **a** House at Herculaneum with original wood beams which survived
             the 79 AD eruption; **b** Part of the reconstructed Oplontis Villa 184

Fig. 9.15   Part of the reconstructed city of Pompeii with views of
             Somma-Vesuvius in the background 185

Fig. 9.16   Ruins of Roman beach villas at the Marina di Equa, Sorrento Peninsula.
             The beach is comprised of volcanic sand. Deposits from the 79 AD
             eruption occur at the base of the cliff and partially infill the buildings   . . . 186

Fig. 9.17   **a** The Pompeii Pumice, the earliest deposit associated with the 79 AD
             eruption, is exposed at the Traianello Quarry (Photograph: Amelia
             Scoon); **b** Coarse-grained ash (A3 layer) associated with the 79 AD
             eruption, Herculaneum 187

Fig. 9.18   Section at the Oplontis Villa shows the air fall lithic-rich ash (A7) which
             is sharply overlain by the S5 surge. The S6 surge and F6 pyroclastic
             flow completely buried the villa 188

Fig. 9.19   Graded beds of air-fall ash (A7) at the Oplontis Villa 189

Fig. 9.20   The Vesuvius crater with part of the Somma caldera visible in the
             background. *Source* https://images.app.goo.gl/SLuoKJRayTHaKjHU9 190

Fig. 9.21   The northern wall of the Vesuvius crater is capped by deposits of the
             1944 eruption. Lava overflowed the crater during this eruption, partially
             destroying the town of San Sebastiano to the east of Naples. Scale
             provided by a group of people (just visible on the horizon) 191

Fig. 10.1   The monastery of St. Nicholas Anapausas in the Meteora, near
             Kalambaka, northwest Greece, is perched precariously on the crest
             of a near-vertical pinnacle of molasse sediments (Miocene) 194

Fig. 10.2   Map showing location of antiquities, archaeological sites and geosites
             in southeast and northwest Greece 195

Fig. 10.3   Simplified geological map of Greece showing the dominance of the
             NW-SE trending Alpine zones. *Source* Simplified from regional
             geological maps and references in the text 196

Fig. 10.4   **a** Thinly-bedded limestones and marbles with near-vertical dips are
             characteristic of the Alpine zones of northwest Greece (locality near
             Igoumenitsa); **b** The flysch of northwest Greece includes
             steeply-inclined deposits of schist (locality near Kalarrytes) 197

Fig. 10.5   Geological map of part of the city of Athens (1-Acropolis; 2-Temple
             of Rome and Augustus; 3-Theatre of Dionysus; 4-Klepsydra Spring)
             and section of the Acropolis. *Source* Map and section simplified from
             Gaïtanakis (1982), Andronopoulos and Koukis (1990) and Higgins
             and Higgins (1996) 198

Fig. 10.6   **a** The conical-shaped Lycabettus Hill (with the Church of St. George),
             Athens, is comprised of resistant Tourkovounia Limestone;
             b Monumental buildings of the Acropolis of Athens are built on a
             distinctive flat-topped hill of Tourkovounia Limestone 200

Fig. 10.7   The Areopagus is a prominent outcrop of the Tourkovounia Limestone,
             northwest of the Acropolis 201

Fig. 10.8   The Tourkovounia Limestone is a grey, blocky rock, in part crystalline,
             as observed on the Areopagus 201

Fig. 10.9   **a** The flat-topped Acropolis Hill is dominated by the
partially-reconstructed Parthenon; **b** The colonnade of Doric columns
in the Parthenon consists of white Pentelic Marble emplaced
on a three-step platform constructed of large blocks of grey marble 202

Fig. 10.10   **a** The monumental buildings on the Acropolis are emplaced into
the bedrock limestone; **b** The Temple of Rome and Augustus is situated
on a limestone pavement on the crest of the Acropolis 203

Fig. 10.11   The Theatre of Dionysus is excavated into the Tourkovounia Limestone
on the southern side of the Acropolis 204

Fig. 10.12   The modern town of Delphi is overlooked by the antiquities perched
on the lower slopes of the Parnassus Mountains 205

Fig. 10.13   **a** The Sacred Way and Temple of Apollo, two of the more important
antiquities at Delphi, are enclosed by steep, limestone cliffs of the
Parnassus Mountains; **b** The rugged nature of the Delphic Landscape
is apparent from the Theatre of Delphi which is located in the upper part
of the archaeological site. 206

Fig. 10.14   The Temple of Apollo, Delphi is constructed near several springs which
contained vapours with hallucinogenic gases during the Classical
Greece and Hellenistic periods 207

Fig. 10.15   The commemoration to the Battle at the Pass of Thermopylae (480 BC)
is located on a broad floodplain at the head of the Gulf of Malia 208

Fig. 10.16   Map of Thermopylae showing sea-level changes and location of the
army camps relative to the three gates. *Source* After Kraft et al.
(1987) and information boards at the site 209

Fig. 10.17   The Pindus Mountains, northwest Greece, reveal multiple ranges
aligned parallel to the Ionian Sea coast and the island of Corfu
(Kerkira) 209

Fig. 10.18   Geological map and section of the Pindus Mountains, northwest
Greece. *Source* Simplified from regional maps and Higgins and Higgins
(1996) 210

Fig. 10.19   The Dodona antiquity includes a large theatre carved into the hill
slopes 211

Fig. 10.20   The village of Kalarrytes is approached via a road that snakes up
the flanks of the South Pindus Mountains 212

Fig. 10.21   The Kipina Monastery, Kalarrytes, is cut into a cliff face comprised
of resistant Triassic limestone 212

Fig. 10.22   Metsovo is dwarfed by the high peaks of the North Pindus
Mountains 213

Fig. 10.23   Limestone pavement, Vikos-Aoös Geopark. 213

Fig. 10.24   **a** A pinnacle consisting of Lower Meteora Formation (Miocene)
with view westward of the Peneois Graben and South Pindus
Mountains; **b** Cliffs of resistant molasse sediments tower above
the village of Kalambaka 215

Fig. 10.25   **a** Monasteries of the Meteora are perched on the crests of near-vertical,
smooth cliff faces comprised of the Lower Meteora conglomerate;
b Pinnacles and crenulated cliff faces at Meteora are associated with
closely-spaced vertical joints in the Lower Meteora conglomerate 216

Fig. 10.26   **a** Upper Meteora conglomerate may include small hollows;
b The conglomerate includes both rounded and angular boulders
and pebbles 217

Fig. 11.1   The narrow entrance to Navarinou Bay, southwestern Peloponnese, is
             the site of two famous battles in ancient and modern Greek history. The
             bay is associated with a sub-graben of the Kalamata Graben, which is
             partially infilled by poorly-consolidated sediments (Neogene). The
             islands of Pylos (upper left) and Sfaktira (upper right) occur on the
             western flanks of the sub-graben and consist of resistant limestone
             (Upper Cretaceous) 220

Fig. 11.2   Location map showing antiquities and historical sites of the
             Peloponnese. The Corinth Canal transects the Isthmus of Corinth.
             Source Satellite Image of Europe based on NASA MODIS data,
             processed by Philip Eales, Planetary Visions/DLR 221

Fig. 11.3   Ancient districts and capitols of the Peloponnese during the Archaic
             and Classical Greece periods. 222

Fig. 11.4   Simplified geological map of the Peloponnese and Zakynthos showing
             the Alpine trend of the four tectonic zones and the younger basins and
             grabens. *Source* Geological Map of Europe (https://geoviewer.bgr.de)
             and articles referenced in the main text. 224

Fig. 11.5   Lion gate, Mycenae. *Source* Andreas Trepte (Own work, CC BY-SA
             2.5, https://commons.wikimedia.org) 225

Fig. 11.6   **a** The limestone rock that *Theseus* was required to lift prior to
             commencing his epic journey to Athens can be examined at Troezen;
             b Marble slab at Troezen showing ptygmatic veins of calcite 226

Fig. 11.7   The antiquity of Epidaurus is situated on the eastern slopes of Mount
             Arachneo (Triassic-Jurassic limestone) and includes the Asklepieion,
             one of the best preserved theatres of the Hellenistic period. 227

Fig. 11.8   **a** Marine and lacustrine terraces on the southern shores of the Gulf of
             Corinth include extensive badlands (Neogene-Quaternary sediments)
             located beneath interior mountains (Upper Cretaceous limestone);
             b Terraces observed next to the highway between Corinth and Patras
             form deeply eroded, pale buff-coloured cliffs 228

Fig. 11.9   View looking northeast of a section of the Rion-Andírrion Bridge which
             crosses the narrows between the Gulfs of Corinth and Patras 229

Fig. 11.10   The Vouraikos Gorge on the northern slopes of Mount Chelmos occurs
             in Upper Cretaceous limestone 229

Fig. 11.11   Location map showing historical sites in the vicinity of the modern town
             of Corinth. The Corinth Canal cuts through the narrowest point of the
             Isthmus 230

Fig. 11.12   **a** Ancient Corinth is situated on a marine terrace (Pleistocene) at the foot
             of a prominent limestone hill (Jurassic); **b** View of the archaeological
             site of Acrocorinth located on the hill of Jurassic limestone. View
             overlooks the marine terrace with the Gulf of Corinth visible in the
             background 231

Fig. 11.13   **a** Pale buff-coloured marls, sandstones, and conglomerates (Upper
             Pliocene) are exposed in the central part of the Corinth Canal; **b** Dark
             brown sandstone and limestone (Pleistocene) unconformably overlie
             the Upper Pliocene strata in the extremities of the canal, as seen here
             under the railway bridge. Both views looking southeast towards the
             Saronic Gulf 232

Fig. 11.14   **a** Excavations next to the Corinth Canal at Posidhonia have uncovered
             parts of the 6[th]C BC stone carriageway, the *Diolkos*; **b** Recent deposits
             of beachrock covering the stone carriageway accumulated during the
             previous 2,500 years when the carriageway was submerged. The

          carriageway has subsequently been uplifted by more than 1 m.
          The Posidhonia Floating Bridge is visible in the background 233

Fig. 11.15 **a** The antiquity of Olympia is situated in the Alpheios Graben, which is
          enclosed by hills and ridges of shelly limestone (Lower Pliocene); **b** The
          stadium is situated in an old, alluvium-filled river channel (Quaternary)
          and overlooked by the Hill of Kronos (Upper Pleistocene terrestrial
          sediments) 234

Fig. 11.16 **a** Slab of shelly limestone (Lower Pliocene) at Olympia; **b** The
          monumental western entrance to the stadium at Olympia 235

Fig. 11.17 **a** The road linking Sparti and Kalamata crosses the Taygetos Mountains,
          which consist of resistant limestone and marble (Triassic-Eocene);
          b Ancient Sparta is situated on a low knoll (right) consisting of marl and
          clay (Plio-Pleistocene). Part of modern Sparti and the western edge
          of the Eurotas Graben are visible in the background. 236

Fig. 11.18 **a** The historical site of Mystras is constructed on Kastro Hill, a resistant
          block of dolomite (Triassic-Jurassic); **b** The view from Mystras (with
          the dolomite visible in the foreground) overlooks the Eurotas Graben
          with Sparti visible in the background 237

Fig. 11.19 The historical town of Monemvasia is constructed on the steep southern
          slopes of a rocky outcrop consisting of resistant dolomite (Triassic). 238

Fig. 11.20 **a** The Venetian Castle of Methoni at the entrance to the Gulf of Messene
          stands on a prominent wave cut platform consisting of limestone (Upper
          Cretaceous); **b** Pylos Island at the entrance to Navarinou Bay includes
          sea caves and arches cut into limestone (Upper Cretaceous) 238

Fig. 11.21 The Mycenaean walls (13thC BC) of the Palace of Nestor are protected
          by a roof in an archaeological site which overlooks a broad plateau
          situated above Navarinou Bay. 239

Fig. 11.22 Shipwreck Beach is enclosed by steep cliffs of white limestone
          (Cretaceous), typical of the north coast of Zakynthos 240

Fig. 11.23 Simplified geological map of the island of Zakynthos. *Source* Avramidis
          et al. (2017a). 241

Fig. 11.24 **a** View of the town of Zakynthos showing the large harbour with
          limestone hills (Triassic) of the Vasilikos Peninsula in the background.
          The Ionian Thrust is aligned parallel with the shoreline (foreground) and
          extends southward to Lagana Bay (background, right); **b** The pale grey
          cliffs of poorly consolidated sandstone and marl (Pliocene-Pleistocene)
          located above the town of Zakynthos are severely eroded and form
          extensive areas of badlands. 242

Fig. 11.25 The seepage of bitumen into Keri Lake, Zakynthos was described by the
          Greek historian Herodotus. 243

Fig. 12.1   Pedestals on a wave-cut platform at Cape Pelekouda, Milos, consist
          of boulders of resistant andesite lava (dark grey-green) deposited
          by a volcanic debris avalanche on a base of soft pyroclastics (pale
          grey-white) 246

Fig. 12.2   Satellite image showing the extent of the Hellenic Volcanic Arc. Active
          or quiescent volcanoes are located in the Methana Peninsula, the islands
          of Milos and Santorini in the Cycladic archipelago, and the island
          of Nisryos in the Dodecanese. *Source* Satellite Image of Europe
          based on NASA MODIS data, processed by Philip Eales of Planetary
          Visions/DLR 247

Fig. 12.3   Cross-section centred on Santorini showing the subducted slab of
             oceanic crust and thinned nature of the continental crust (dominated
             by supracrustal rocks) beneath the Hellenic Volcanic Arc. Plumbing
             system of the volcanoes associated with the Hellenic Volcanic Arc (e.g.,
             Santorini) may include shallow crustal chambers. *Source* Simplified
             after various sources including Meier et al. (2007) 248
Fig. 12.4   The town of Methana nestles on the lower slopes of rugged volcanic
             hills, typical of the Methana Peninsula. Discoloration in the harbour is
             caused by geothermal water discharged from hot springs 250
Fig. 12.5   Geological map of the Methana Peninsula. *Source* Simplified from
             Gaitanakis and Dietrich (1995), Pe-Piper and Piper (2013). 251
Fig. 12.6   **a** The village of Agios Nikolaos is located at the base of thickly
             vegetated hills associated with the Mantling Ash (foreground) with lava
             domes forming the higher hills (background) in the northern part of the
             Methana Peninsula; **b** Sub-horizontal beds of red dacitic ashes and
             pyroclastics (Mantling Ash) are exposed near Palaeo Kastro, Methana
             Peninsula. 252
Fig. 12.7   **a** Boulders of andesite (dark grey-green) and dacite (red) in the seawall
             at Agios Nikolaos; **b** Dry stone walls are constructed of andesite and
             dacite lava; **c** Blocks of andesitic agglomerate include plagioclase laths
             (white) and small volcanic bombs (dark) 253
Fig. 12.8   The historical lava flow of the Mavri Petra Volcano forms a dark
             swathe on the hill slopes above the village of Kaimeni Hora, Methana
             Peninsula. 254
Fig. 12.9   The unusual shape of Milos is related to accretion of multiple volcanic
             centres. The two youngest volcanoes, Fyriplaka and Trachilas, reveal
             relatively barren landforms. The irregular patches of white dotted over
             large parts of the island are quarries associated with production of
             industrial minerals. *Source* Google Earth 255
Fig. 12.10   View looking east over the Bay of Adamas (an active graben) towards
             the town of Plaka on the island of Milos. Plaka is situated above the
             potential impact of tsunamis on a volcanic plateau. The cone-shaped
             hills on the plateau are lava domes. The harbour town of Adamas
             (centre right) and the north-east coastline of the island (background) are
             also visible 256
Fig. 12.11   Geological map of Milos including geosites and mines referred to in the
             text. *Source* Simplified from Fytikas et al. (1986b). 257
Fig. 12.12   **a** Extensive spoil heaps at Nychia, Milos illustrate the remarkable scale
             of obsidian mining during Palaeolithic times; **b** Obsidian occurs in
             banded rhyolite (Lower Pleistocene); rubble and flakes also visible 258
Fig. 12.13   **a** The sulphur mines at Paliorema Bay occur in a narrow ravine on the
             south coast of Milos. Multi-coloured cliffs reveal hydrothermal
             alteration of lavas and ashes (Lower-Middle Pleistocene); **b** Veins and
             fractures infilled by yellow crystals of sulphur 259
Fig. 12.14   **a** Active fumarole at Kalamos, Fyriplaka Volcano; **b** Fyriplaka Beach is
             rimmed by multi-coloured cliffs of hydrothermally altered rhyolitic
             pyroclastics (Upper Pleistocene) 260
Fig. 12.15   **a** Cape Pelekouda is comprised of lavas and pyroclastics capped by the
             Green Lahar (a debris avalanche deposit with large boulders)
             (Lower-Middle Pleistocene); **b** The cleft and caves at Papafrangas are
             comprised of agglomerates with clusters of welded fragments, or
             volcanic bombs (Lower Pleistocene). 261

Fig. 12.16   **a** Dark grey-green andesite lavas (Lower-Middle Pleistocene) forms the cliffs near the Fylakopi archaeological site. The andesite is overlain by a thin debris avalanche deposit; **b** Columnar jointed andesite lavas, Glaronisia Islet 261

Fig. 12.17   **a** The cliffs at Fyropotamos reveal a section of lavas, ashes, and pyroclastics (Lower-Middle Pleistocene) in which the distinctive white rocks are subaqueous pumice beds; **b** Pumice beds are intercalated with layers of altered reddish-brown diatomite, road cutting near Fyropotamos 262

Fig. 12.18   **a** At Sarakiniko, the white pumice deposits (Lower-Middle Pleistocene) form a barren, unvegetated coastline; **b** Graded beds in the pumice are indicative of multiple, subaqueous eruptions 262

Fig. 12.19   The partially submerged Santorini Caldera is located between the crescent-shaped island of Thera (east) and the subsidiary island of Therasia (west). The Kameni Islands are situated in the centre of the caldera. The steep walls of the caldera are clearly visible on the western side of Thera. *Source* Image from an ASTER instrument on NASA's Terra Satellite (21st November 2000) of the Santorini Islands (18 km in width) processed by Philip Eales of Planetary Visions/DLR and first published in "Map: satellite" by Dorling Kindersley (2007) 263

Fig. 12.20   **a** The caldera wall on Thera (view looking south) reveals multiple layers of volcanic ashes, lavas, and pumice deposits; **b** The road from the new harbour exposes some of the older caldera events. Deposits associated with the Minoan Event are represented by the distinctive pale-grey or white layer capping the plateau in both (**a**) and (**b**) 264

Fig. 12.21   Geological map of the Santorini archipelago. The majority of the deposits depicted as the Thera Pyroclastics are associated with the Second Explosive Event. The plateaus that cap the islands of Thera and Therasia are primarily underlain by tephra associated with the Minoan Event. *Source* Simplified after Druitt et al. (1999) 265

Fig. 12.22   **a** Chaotic volcanic deposits in the walls of Santorini Caldera, Thera, are associated with some of the older caldera events; **b** Agglomerate associated with one of the pre-Minoan events, Red Beach near Akrotiri 266

Fig. 12.23   The white or pale grey tephra correlated with one of the pre-Minoan events contains numerous lithic fragments and small volcanic bombs, indicative of a highly explosive, Plinian-style eruption. Exposure in road cutting in cliffs above Port Athinias 267

Fig. 12.24   **a** Part of the Minoan city of Akrotiri which was entirely buried by the Minoan Event, Thera; **b** The fractured stair case at Akrotiri is attributed to a powerful earthquake several months before the main eruption 268

Fig. 12.25   Ruins at Mesa Vouno or Ancient Thera showing the use of decorative slabs of red agglomerate in some of the limestone walls and columns . . . 269

Fig. 12.26   **a** Satellite image of the Kameni islands, Santorini. *Source* Image processed by Philip Eales of Planetary Visions/DLR; **b** Simplified geological map showing ages of eruptions on the Kameni Islands. *Source* Druitt et al. (2015) 270

Fig. 12.27   The Mikra Kameni Crater on the island of Nea Kameni. 271
Fig. 13.1   The Great Theatre at the antiquity of Ephesus is partially built into the
              side of Mount Pion, a horst block located within the Küçük Menderes
              Graben. In the background is a ridge of resistant metamorphic rocks,
              part of Mount Preon, the horst block located on the southern side of the
              graben. 274
Fig. 13.2   Map showing location of antiquities and archaeological sites in western
              Turkey described here. Thrace is a historical name for the European
              section of Turkey. The historical name for the Asian section of Turkey
              is Anatolia. The approximate eastern extent of the Roman Empire
              is depicted. 275
Fig. 13.3   The regions and towns of Asia Minor in the 1stC AD. The letters that the
              apostle Paul (a native of Tarsus in the region of Cilicia) wrote to
              Christian communities in the region are included in the New Testament.
              The seven churches of Ephesus, Smyrna, Pergamum, Thyatira, Sardis,
              Philadelphia and Laodicea (indicated in red) were addressed by the
              apostle John in the book of Revelation. John was in exile on the island
              of Patmos. The majority of the region formerly known as Asia Minor is
              incorporated into the modern country of Turkey. The city of Smyrna is
              renamed Izmir. Patmos and the other islands in the Aegean Sea and
              eastern Mediterranean are part of Greece. *Source* Map draughted by
              Andrew Mitchell. 276
Fig. 13.4   The tectonic framework of western Turkey is dominated by major
              curvilinear sutures ascribed to the differential movement of the
              Anatolian Microplate and the surrounding terrains. For details of the
              plate boundaries and trenches the reader is referred to Fig. 8.3. *Source*
              Satellite Image of Europe based on NASA MODIS data, processed by
              Philip Eales of Planetary Visions/DLR with tectonics in part after
              Bozkurt (2001) 277
Fig. 13.5   Simplified geological map of part of western Turkey showing alignment
              of major sutures and grabens relative to the antiquities and historical
              sites of Troy, Gallipoli Peninsula National Park, Assos, Kazdağı
              National Park, and Pergamum. The Dardanelles Graben demarcates the
              southern limit of the Thrace Basin. The remainder of the area confined
              by the NAFZ is correlated with the Rhodope-Serbomacedonian Massif
              of northwest Greece. The area south of the NAFZ is part of the Sakarya
              Tectonic Zone. *Source* Geological Map of Europe (*https://geoviewer.*
              bgr.de) and articles referenced in the main text. 279
Fig. 13.6   Simplified geological map of part of the Menderes Massif, western
              Turkey, showing alignment of major sutures and grabens relative to the
              antiquities and historical sites of Ephesus, Aphrodisias, Laodicea,
              Colossae, and Pamukkale-Hierapolis. *Source* Geological Map of Europe
              (*https://geoviewer.bgr.de*) and articles referenced in the main text 280
Fig. 13.7   **a** View from the antiquity of Troy looking northwest over the coastal
              plain toward the Dardanelles with hills of the Gallipoli Peninsula visible
              in the background; **b** The Schliemann Trench cut down from the Roman
              period (red bricks, upper right) to some of the oldest parts of Troy (low
              walls in the base of the excavation) 282
Fig. 13.8   Simplified geological map of the area surrounding the antiquity of Troy
              with reconstruction of the coastline in the Early Bronze Age (2500 BC),
              in the Late Bronze Age (1250 BC), and in Roman times (1 AD). *Source*
              Modified after Higgins and Higgins (1996) and Kraft et al. (1980) 283

Fig. 13.9   The antiquity of Troy contains nine cities, Level I through Level IX. The
             Homeric city is correlated with Level VI and the Greek and Roman
             cities with Level VIII and Level IX, respectively 283

Fig. 13.10   **a** One of the few walls remaining of the Homeric city (Level VI) at
             Troy; **b** Parts of the Greek Level VIII (lower left) and the Roman
             Level IX (centre and upper right) occur in the Sanctuary at Troy 283

Fig. 13.11   During the Late Bronze Age or Homeric times (1250 BC) Troy was
             perched on a hill located adjacent to Troia Bay. The city was enclosed
             on the northern and southern sides by marshes associated with the
             Scamander and Simois Rivers. Historical sites include Achilleion
             (Besika Bay), the landing place of the Greek armies, the Kesik Cut, and
             the ford over the Scamander River. The Trojans mounted their defence
             at Batieia (Thorn Hill). *Source* Kraft et al. (2003) 284

Fig. 13.12   View looking north from Anzac Cove, Gallipoli Peninsula National
             Park, showing badlands associated with rapid erosion of
             Neogene-Quaternary sediments 284

Fig. 13.13   The dark-coloured trachytic lava at Assos contains prominent
             phenocrysts of plagioclase (white). 285

Fig. 13.14   **a** The coastal resort of Assos is located at the base of a steep hill
             consisting of dark-coloured trachytic lava. The antiquity is visible on the
             crest of the hill; **b** The antiquity of Assos overlooks the Aegean Sea with
             the island of Lesbos visible in the background 286

Fig. 13.15   The Temple of Athena at Assos is built on the trachytic lava bedrock . . . 287

Fig. 13.16   The antiquity at Pergamum is located on a volcanic plateau on the
             northern side of the Bakırçay Graben. The acropolis occurs on the
             highest point of the mesa which, despite the relatively steep sides,
             includes natural terraces where many of the larger monumental
             buildings were constructed. *Source* Google Earth image 287

Fig. 13.17   View of part of the rugged topography characteristic of the volcanic
             plateau at Pergamum. 288

Fig. 13.18   **a** Reconstruction of part of the acropolis complex at Pergamum (view
             looking east) with the theatre situated on a lower terrace (foreground)
             and the Pergamum Altar on an upper terrace (centre right); **b** View
             of the theatre at Pergamum looking southwest over the modern city of
             Bergama (situated wholly within the Bakırçay Graben). Outcrops of
             grey-green andesite lava visible on the mesa 289

Fig. 13.19   **a** The foundations of the Pergamum Altar; **b** The foundations and bases
             of the columns of the Temple of Trajan at Pergamum are constructed
             from dark grey-green andesite lava; the columns are of white marble 290

Fig. 13.20   **a** The extensive portico is part of the reconstructed Asklepion at
             Pergamum; **b** The Lower Agora at Pergamum is situated at the base
             of the acropolis complex. 291

Fig. 13.21   The archaeological site of Ephesus is located situated close to the
             southern margins of the Küçük Menderes Graben. The foundations
             of the Artemision occur in swampy part of the valley but many of the
             other monumental buildings occur on the flanks of the more resistant
             horst block. *Source* Google Earth image. 292

Fig. 13.22   **a** View looking west along Harbour Street, Ephesus, towards a grassy
             area which corresponds with the historical location of the Sacred
             Harbour. The harbour was originally part of a large estuary, but the
             Aegean Sea (upper right) is now several kilometres to the west. This part
             of the archaeological site is located in the Küçük Menderes Graben, the

            southern margin of which is fringed by resistant horst block, such as
            Mount Preon (upper left); **b** The Great Theatre at Ephesus is partially cut
            into the slopes of Mount Pion, a horst block located within the graben.
            View includes Marble Street in the foreground 293

Fig. 13.23 **a** View of a section of Marble Street at Ephesus; **b** The Library of
            Celsius at Ephesus includes columns built of marble breccias 294

Fig. 13.24 **a** Marble slabs were used for decorative purposes in the housing
            complex at Ephesus; **b** Book-matched slabs of marble breccia at
            Ephesus may date from the Archaic Greek period 295

Fig. 13.25 Geological map mounted in a display board at Ephesus showing
            westward migration of the Sacred Harbour in the region to the north
            of the main archaeological site (width of view approximately 5 km). The
            Great Theatre is located near the southern end of the Sacred Way.
            Source Map after Kraft et al. (2003) 296

Fig. 13.26 **a** The archaeological site of Aphrodisias includes sections of several
            temples located in a broad valley (the Morsynus Graben) enclosed by
            rolling hills (horst blocks); **b** The Bouleuterion at Aphrodisias includes a
            colonnaded stage constructed of locally-quarried marble 297

Fig. 13.27 The unusual ovoid-shaped stadium at Aphrodisias includes seats built
            into raised earthworks 298

Fig. 13.28 The theatre at Kibyra. *Source* https://images.app.goo.
            gl/ykVDc99aXmwY8Wqx5 299

Fig. 14.1   The snow white deposits of travertine form an extensive ridge above the
            village of Pamukkale 302

Fig. 14.2   The Hierapolis-Pamukkale archaeological and geosite is situated near
            the regional city of Denizli, southwest Turkey. Most of the antiquities at
            Hierapolis are from the Roman period 303

Fig. 14.3   Generalized map showing the structural setting of travertine deposits in
            the vicinity of Pamukkale. *Source* Simplified after Altunel and Hancock
            (1993) and Geological Map of Europe (https://geoviewer.bgr.de) 305

Fig. 14.4   Image showing the travertine deposits at Pamukkale. The modern
            village of Pamukkale is located at the base of the travertine ridge near
            the small lake (lower left). The Hierapolis archaeological site occurs on
            the crest of the travertine ridge (centre) with the necropolis extending
            northward on Frontinus Street (a Roman road). The crest of the ridge
            corresponds to the alignment of hot springs where travertine is actively
            forming, including the modern pool which may correlate with the
            location of the Sacred Pool. The eastern part of the ridge is dominated
            by older sections of the travertine deposit. *Source* Google Earth 306

Fig. 14.5   **a** The ridge at Pamukkale consists almost entirely of white deposits of
            travertine; **b** The new deposits (white) drape downward from the
            plateau, covering the older deposits (grey) at the base of the ridge 307

Fig. 14.6   The fissure-ridge deposits of travertine reveal pale orange or buff colours
            at Pamukkale 308

Fig. 14.7   **a** Vermicular-textures demonstrate how the travertine layers in the
            fissure-ridge deposits at Pamukkale build up incrementally cm by cm;
            b Botryoidal form of travertine at Pamukkale 308

Fig. 14.8   The terrace-mound deposits of travertine at Pamukkale are continuously
            forming as the hot waters supersaturated in calcium carbonate drip down
            the flanks of pools 308

Fig. 14.9   Historical map of the antiquity of Hierapolis. The travertine deposit is depicted by area "B". Other sites of interest are the Agora (7), Frontinus Street (8), Theatre (11), and the Southern Gate (14). *Source* Information Board at the site 309

Fig. 14.10   The Roman theatre at Hierapolis is built from blocks of light grey marble with white travertine used for decorative purposes (notably in the gallery and stage) 309

Fig. 14.11   Part of Frontinus Street which connects the Agora and Northern Necropolis at Hierapolis. The street was covered with a 2 m-thick deposit of travertine prior to recent excavations 309

Fig. 14.12   **a** Some burial chambers adjacent to Frontinus Street at Hierapolis remain partially covered by deposits of terrace-mound travertine; **b** The Southern (Byzantine) Gate at Hierapolis includes a monolithic arch supported by walls built from large blocks of marble (light grey) with selective use of travertine (white) 310

Fig. 15.1   The "Fairy Chimneys" at Paşabaĝ in the Göreme Historical National Park, Cappadocia, are natural rock pinnacles that consist of relatively soft volcanic ash and pumice (pale grey). The distinctive caps consist of horizontal layers of resistant basaltic lava (dark). 314

Fig. 15.2   Map showing location of the Cappadocia Plateau in central Turkey 315

Fig. 15.3   The deeply dissected Cappadocia Plateau contains snow- and ice-capped volcanic peaks, including Mount Erciyes, and is cut on the northern side by the Kizilirmak River. *Source* Google Earth 316

Fig. 15.4   The Cappadocia Plateau contains extensive areas of badlands dominated by natural rock monoliths and pinnacles 317

Fig. 15.5   The snow- and ice-capped peak of Mount Erciyes with views of the deeply eroded Cappadocia Plateau in the foreground 317

Fig. 15.6   The tranquil setting of Ihlara Gorge contains a small river in a wooded setting. The gorge is carved into thick sheets of ignimbrite. The ignimbrite sheets reveal poorly developed columnar jointing 318

Fig. 15.7   **a** The pale grey ignimbrite at Ihlara Gorge includes welded fragments; **b** The volcanic ash from Göreme contains abundant lithic fragments, indicative of Plinian-style eruptions 319

Fig. 15.8   **a** Some of the flat-lying layered deposits of volcanic ash exposed in underground rock dwellings at the Ihlara Gorge reveal knife-sharp contacts; **b** Cross bedding in finely laminated deposits of volcanic ash in the Ihlara Gorge is enhanced by orange-brown discoloration due to weathering. 320

Fig. 15.9   Map of the Göreme Historical National Park and surrounding areas of the Cappadocia Plateau. 321

Fig. 15.10   Landforms at the Göreme Historical National Park are dominated by flat-lying strata which outcrop in elongate ridges. Extensive areas of badlands erosion occur in the valleys. The yellow and pinkish-brown colours reflect changes in the iron oxide content of the volcanic ash, pumice, and lava. 322

Fig. 15.11   **a** Göreme is situated on the edge of an extensive area of badlands erosion protected in a national park; **b** The town of Göreme is a mix of ancient and modern structures carved out of natural volcanic monoliths and pinnacles, with some hotel rooms excavated into the soft ash and pumice 323

Fig. 15.12   The Camel Rock is a distinctive rock monolith located in the Dervent
              Valley. Lava (grey) forms the resistant capping to the pinkish volcanic
              ash which dominates the feature 324
Fig. 15.13   Üchisar Castle is associated with an outcrop of ignimbrite. The castle
              was probably first inhabited by the Hittites but subsequently used by
              recent civilizations. 325
Fig. 15.14   Numerous structures have been carved out of the cliff faces at Ihlara
              Gorge, including dovecotes 326
Fig. 16.1   Coniston Water viewed from Brantwood House, the home of Victorian
              writer and philosopher John Ruskin, is an example of the tranquil views
              which have attracted visitors to the Lake District views for many years.
              The gentle landforms near the lake are associated with metasedimentary
              rocks. The rugged hills in the background, including the broad massif
              of the Old Man of Coniston, consist of the resistant Borrowdale
              volcanics 328
Fig. 16.2   Location map showing the Lake District National Park in Cumbria,
              northwest England. Also shown are the larger of the finger-lakes and
              some of the higher summits. Archaeological and mining sites are as
              follows: 1-Carrock Fell; 2-Castlerigg; 3-Greenside Lead mine; 4-High
              Street; 5-Tilberthwaite Slate quarries; 6-Coniston Copper mines;
              7-Honister Slate mine; 8-Seathwaite Graphite mine 329
Fig. 16.3   The Lake District has been a popular destination for hiking, or fell
              walking, for many years. The background includes views of Derwent
              Water and the Skiddaw massif 331
Fig. 16.4   The topography of the Lake District is dominated by a mountainous core
              with relatively gentle landforms on the perimeter. The dashed line shows
              the regional unconformity demarcating the contact of the SE-dipping
              Borrowdale Volcanics and the overlying Windermere Supergroup to the
              south. The Solway Firth is a regional suture which separates the
              Avalonian and Laurentian terrains of the British Caledonides. *Source*
              Landsat 7 ETM+ image for 2000 of Cumbria processed by Philip Eales
              of Planetary Visions/DLR 332
Fig. 16.5   The regional unconformity between the Borrowdale Volcanic Group
              (rugged hills in foreground) and the Windermere Supergroup (gentle
              landforms proximal to the lake) is associated with a marked change in
              the topography. View looking southeast from Coppermines Valley and
              overlooking Coniston Water 333
Fig. 16.6   **a** The steep scree slopes above Wastwater (Wasdale) were formed by
              glacial erosion during the Late Pleistocene Ice Ages; **b** The town of
              Coniston is situated at the convergence of several hanging valleys which
              contain streams with small waterfalls 334
Fig. 16.7   Geological map of the Lake District inlier showing the dominance of
              Lower Palaeozoic strata. The Borrowdale Volcanic Group (BVG) is
              subdivided into lower (plain) and upper (stippled) components, the latter
              includes four calderas, Dutton (D), Gosforth (G), Helvellyn (H), and
              Scafell (S). Intrusive bodies include the Ordovician-age Carrock Fell
              gabbro (1), the Devonian-age Skiddaw granite (1); the Devonian-age
              Shap granite (2), the Ordovician-age Eskdale granophyre (3) and the
              Ordovician-age Ennerdale microgranite (4). *Source* Simplified from
              maps of the British Geological Survey 335

Fig. 16.8   Location of the Caledonide terrains (pale brown) relative to
               palaeo-continents at the end of the Caledonian Orogeny (Early
               Devonian). *Source* By Woudloper—Own work, CC BY-SA 1.0,
               https://commons.wikimedia.org/w/index.php?curid=5038110 336

Fig. 16.9   **a** The rounded massif of Skiddaw is typical of outcrops of the Skiddaw
               Group (viewed from the south); **b** This view of Derwent Water shows the
               town of Keswick located beneath the rounded slopes of Blencathra 337

Fig. 16.10   **a** Striated or banded outcrops are typical of the terrestrial deposits of
               volcanic ash found in the lower part of the Borrowdale Volcanics
               (Seathwaite, Borrowdale); **b** A close up of the same outcrop shows the
               irregularity of the banding. 339

Fig. 16.11   **a** Columnar jointing in an ignimbrite layer, Yewdale Breccia, in the
               upper part of the Borrowdale Volcanics (Coniston Fells); **b** Columnar
               joints in an ignimbrite layer, upper part of Borrowdale Volcanics
               (Tilberthwaite); **c** Finely laminated volcanic ash characteristic
               of subaqueous deposits of the Borrowdale Volcanics (Honister
               slate mine) 340

Fig. 16.12   **a** Steep-sided glaciated valleys in the Lake District contain relatively
               small streams, such as Styhead Gill, Borrowdale, and yet are partially
               filled by fertile deposits of alluvium; **b** Drumlin fields occur in many
               of the high moorlands of the Lake District (northwest of the
               Langdale Pikes). 341

Fig. 16.13   **a** The tarn below Goat Fell, part of the massif associated with the Old
               Man of Coniston, is an example of a hanging valley with an Alpine lake
               (Coniston Water visible in the background); **b** Streams typically exit
               hanging valleys via a series of cascades and small waterfalls, as seen
               below Levers Water on the Coniston Hills 342

Fig. 16.14   Buttermere is one of the most scenic lakes in the Lake District
               National Park 343

Fig. 16.15   **a** The Neolithic stone circle at Castlerigg, near Keswick, is located on
               Skiddaw slates at the foot of Blencathra; **b** The standing stones consist
               of volcanic rocks derived from outcrops of the Borrowdale Volcanic
               Group in the rugged hills several kilometres to the south 344

Fig. 16.16   **a** High Street is a Roman road and wall (right) located on the crest of
               several intersecting cirques; **b** View from High Street looking west over
               Hayeswater, an Alpine lake located in a hanging valley at the base of
               Gray Crag (footpath on the ridge), with the massif of Helvellyn in the
               background (left). 346

Fig. 16.17   **a** The large spoil heaps in the Coppermines Valley attest to the
               extensiveness of the Victorian age copper mines near Coniston; **b** The
               copper mineralization is contained in quartz veins emplaced in the
               Borrowdale volcanics 347

Fig. 16.18   **a** The alignment of spoil heaps at Seathwaite, southwest of Keswick,
               traces the surface expression of the subvertical veins of quartz-graphite;
               b Entrance to one of the old graphite mines in the Borrowdale
               Volcanics, Seathwaite 348

Fig. 16.19   **a** Some of the adits (tiny dark features in the photograph) at the Honister
               slate mine are situated on almost inaccessible hill sides; **b** The main
               entrance to the Honister slate mine, used for both tourism and
               production, shows the steeply-dipping axial planar cleavage typical
               of the Borrowdale volcanics in this region 349

Fig. 16.20   **a** A slab of thinly bedded subaqueous volcanic ash shows the finely
              banded nature of the high quality Honister slate (left), although the
              presence of fractures and agglomerate (right) can affect the value;
              b Volcanic agglomerate from the Honister mine makes an attractive
              dimension stone 350
Fig. 16.21   **a** Quarries at Tilberthwaite, near Coniston, exploited pockets of slate
              from the upper part of the Borrowdale volcanics; **b** The slate at
              Tilberthwaite includes finely-bedded volcaniclastics with prominent
              sedimentary textures 350
Fig. 17.1   The intercalation of thin layers or stringers of chromitite (black) in a
              thick layer of anorthosite (fawn) at the Dwarsriver National Monument,
              Eastern Bushveld (Geosite G10), is a classic example of igneous
              layering. The thin layers and stringers of chromitite bifurcate from the
              base of the metre-thick layer of UG1 chromitite (top left) 354
Fig. 17.2   Wager Peak reveals the bare rock faces that characterize large parts
              of the Skaergaard Intrusion. The view exposes part of the Lower Zone,
              the entire Middle Zone, and part of the Upper Zone. The "Triple Group"
              is just visible below the snow line. The peak is named after geologist
              Lawrence Rickard Wager ("Bill"), a world-renowned mountaineer who
              ascended to a few hundered metres of the summit of Everest in 1933 355
Fig. 17.3   **a** Generalized map of the Skaergaard Intrusion; **b** W-E section of the
              Skaergaard Intrusion. *Source* Irvine et al. (2001) 356
Fig. 17.4   The contact between a remnant of Eocene basalt and Archaean granite
              gneiss is exposed on Kraemer Island. Gently dipping igneous layering
              of the Layered Series of the Skaergaard Intrusion is visible in the
              background. Ice floes and a small berg are visible in the fjord 358
Fig. 17.5   **a** Macrorhythmic modal layering of ferrogabbro in the Middle Zone,
              Kraemer Island; **b** Rhythmic alternation of thin, modally graded layers
              in uniform ferrogabbro in the Upper Zone, Home Bay. The basal parts
              of the modally graded layers are rich in dark minerals (iron-rich olivine
              and Fe-Ti oxides) and the light-coloured upper parts are dominated by
              plagioclase 359
Fig. 17.6   **a** Trough layering (Trough G) in the Upper Zone, Home Bay. The dark
              layers are rich in Fe-rich olivine and Fe-Ti oxides; **b** Neil Irvine pointing
              out trough layering in the Upper Zone on the expedition to the
              Skaergaard Intrusion in 2001 360
Fig. 17.7   **a** A small discordant body of anorthosite (pale brown) situated in a
              layered sequence of troctolite (dark brown) and anorthosite (pale brown)
              in the Lower Zone, Uttental Plateau; **b** Blocks or xenoliths of gabbroic
              troctolite (pale grey) in rhythmically layered magnetite gabbro (dark
              grey) in the Middle Zone, Kraemer Island 361
Fig. 17.8   Regional geological map of the Bushveld Igneous Complex showing the
              three main limbs of the Rustenburg Layered Suite (satellite bodies not
              shown). Each limb is an inclined sheet, or sill-like feature, which
              subcrop at depth beneath the felsites and granites in the interior of the
              Transvaal Basin. The boxed area of the Eastern Limb contains most
              of the well known geosites. *Source* Simplified from the 1:1,000,000
              scale map published by the Council for Geoscience 362
Fig. 17.9   A three dimensional satellite image (two times vertical exaggeration) of
              part of the Eastern Limb of the Bushveld Igneous Complex looking
              north showing the deeply-incised Groot and Klein Dwarsriver Valleys.
              Most of the famous geosites occur in the Steelpoort Valley and the

Olifants Valley. The Bushveld Escarpment and Leolo Mountains are localized features of the Eastern Limb. The extreme irregularity of the floor contact between the towns of Steelpoort (S) and Burgersfort (B) is related to a trough-like body of Lower Zone. The mountainous terrain to the north and east of the Olifants Valley is associated with resistant rocks in the metamorphic aureole. *Source* Landsat 7 image for the year 2000 sourced from the University of Maryland Global Land Cover Facility and processed by Philip Eales of Planetary Visions/DLR 363

Fig. 17.10   Statue of King Sekhukhune situated on a koppie comprised of gabbronorite, at the Tsate Memorial in the Eastern Limb (Geosite G29) 363

Fig. 17.11   Geological map of part of the Eastern Limb of the Bushveld Igneous Complex showing zones, reefs, and selected geosites described by Scoon and Viljoen (2019). *Source* Simplified from the 1:250,000 scale maps published by the Council for Geoscience 365

Fig. 17.12   **a** The platinum-rich core-zone at the Driekop pipe (was mined from a glory hole accessed by a vertical shaft (headgear remains intact). The majority of the hill is comprised of the normal (unmineralized) dunite in the pipe. The background shows the Bushveld Escarpment which consists of resistant gabbronorite (Geosite G9); **b** Relict of the 1926–1930 glory hole at the Onverwacht pipe includes reticulate veining of secondary magnesite in the barren dunite. A protuberance of the platinum-rich core-zone can be observed (slightly darker rock with less veining) (Geosite G24) 366

Fig. 17.13   Section of the Onverwacht platinum mine aligned approximately N-S and looking east (four times horizontal exaggeration). The high-grade ore included examples such as A1 (214 g/t Pt in core-zone rich in phlogopite); A2 (31 g/t average of central part of core-zone); A3 (1,186 g/t at contact between core-zone and lump of chromite); A4 (29 g/t average of 211 m^2 area of core-zone); A5 (17 g/t average of 39 m^2 area of core-zone); A6 (9.5 g/t average of 39 m^2 area of core-zone). The shaft was developed to a final depth of 320 m. *Source* Original mine plan of Onverwacht Platinum Limited with additional details after Wagner (1929) 367

Fig. 17.14   Generalized stratigraphic column for the Rustenburg Layered Suite showing zones, lithologies, reefs and marker layers. The irregular basal contact is emphasized (the lowermost zones are not developed in many areas) and only the Upper Zone (noting the planar nature of the roof contact) is persistent over the entire strike length of the Eastern Limb . . . 368

Fig. 17.15   Simplified cross section of the Rustenburg Layered Suite in the Eastern Limb showing the irregular basal contact associated with syn-Bushveld domes 369

Fig. 17.16   **a** The migmatite in the Marginal Zone consists of micropyroxenite (dark grey) and fine-grained norite (pale grey) (Geosite G6); **b** A rheomorphic breccia in the Marginal Zone contains blocks of micropyroxenite (dark grey) embedded in a pinkish matrix of quartz-rich granophyre (Geosite G12) 370

Fig. 17.17   Aerial view of igneous layering in the Lower Zone (view looking north over the Olifants River Trough, Geosite G14). The deeply-eroded harzburgite subzone (pale) is sandwiched between resistant ridges associated with the Lower Bronzitite and Upper Bronzitite subzones. The pale colour of the olivine-rich layers is related to secondary magnesite. Photograph by M J Viljoen 371

Fig. 17.18   The MG2 and MG3 chromitite layers straddle an important boundary in
              the centre of the Critical Zone, Annex Grootboom (Geosite G2). The
              Lower Critical Zone consists entirely of ultramafic lithologies (and is
              dominated by feldspathic orthopyroxenite), but the Upper Critical Zone
              includes well-defined layers of norite and anorthosite. Photograph by M
              J Viljoen 372

Fig. 17.19   The PGE mineralization in the UG2 Reef at Hackney (Geosite G11) is
              largely constrained to the layer of chromitite. Features of note include
              the irregular contact with the underlying pegmatoidal feldspathic
              orthopyroxenite, the small xenoliths of anorthosite (white) in the
              chromitite (black), and the relatively planar upper contact of the
              chromitite layer. Photograph by M J Viljoen 372

Fig. 17.20   The planar nature of the upper chromite stringer in the Merensky Reef at
              Winnaarshoek (Geosite G33) contrasts with the irregularity of the
              contact between the feldspathic orthopyroxenite (brown) and the
              overlying layer of leuconorite (pale grey). The spotted appearance of the
              leuconorite is characteristic of this lithology (dark spots are
              orthopyroxene; white matrix is plagioclase). Photograph by Qin Wang . . . 373

Fig. 17.21   The Main Magnetite Layer (black) at Magnetite Heights (Geosite G21)
              has a width of approximately 2 m and is underlain by a prominent layer
              of anorthosite (fawn-white) 373

List of Boxes

Box 3.1   Late Pleistocene and Holocene Ice Ages.   42
Box 4.1   Four Stages of Geysers.   64
Box 5.1   Turquoise Lakes   80
Box 6.1   East African Rift System   97
Box 8.1   Ocean Trenches and Back-arc Basins   141
Box 8.2   Plinian Eruptions.   146
Box 11.1   Historical Overview   222

Abstract

The eastern and southern parts of the state of Utah, southwest USA, contain some of the world's most spectacular desert landforms. The principal feature of the Canyonlands National Park is the occurrence of deep, near-vertical walled, sinuous canyons. The mesas and rock monoliths at Monument Valley are universally known as they have been shown in many films and television series. The dominant physiographic and geologic feature of the region is the Colorado Plateau, an extensive, high-altitude hinterland that extends into Arizona, Colorado, and New Mexico. The plateau is underlain by mostly Upper Palaeozoic-Mesozoic sedimentary rocks which encompass a broad period of the Earth's history. Regional uplift during the Late Cenozoic initiated a period of intensive erosion and the paucity of younger strata is a characteristic feature. Moreover, the Palaeozoic-Mesozoic strata reveal only modest levels of deformation and are mostly flat-lying. The arid or semi-arid environment of large parts of the southwestern USA is essential for preservation of canyons, mesas (flat-topped hills), and rock monoliths which would probably be destroyed in wetter climates. Canyons develop where the uplifted, flat-lying strata is being rapidly incised by major rivers that experience seasonal flooding. Annual flooding of the Colorado River system in eastern and southern Utah is enhanced by snow melt derived from the Rocky Mountains. The deeply-incised nature of canyons, together with the verticality of the rock monoliths, is related to differential erosion of the sedimentary strata. The steepest cliff sections in the canyon walls and monoliths, as well as the capping to mesas, typically consist of resistant sandstone formations. The softer impure sandstone, mudstone, and shale are more readily eroded. The canyons of the Colorado River and Green River at the Canyonlands National Park expose a succession of sedimentary strata which extend from the Upper Carboniferous, through the Permian-Triassic, and into the Lower Jurassic. The intercalation of resistant sandstone beds with more readily eroded impure sandstone, mudstone, and shale is particularly well exposed at Dead Horse Point. The Dead Horse State Park forms an adjunct to the national park. The Shafer Trail provides vehicle access to a prominent bench located low down in the canyon. The bench is associated with the White Rim Sandstone (Lower Permian), an anomalously resistant stratum that can be delineated over large parts of the national park. The monoliths at the Navajo Tribal Park, Monument Valley, include buttes, pinnacles, and spires with names such as Rain God Mesa, West and East Mitten Butte, and Totem Pole. Most of the steepest cliff sections consist of the resistant De Chelly Sandstone (Lower Permian). This formation is the lateral-equivalent of the White Rim Sandstone. The mesas together with some of the higher monoliths are capped by the Shinarump Conglomerate. This Upper Triassic formation forms a resistant capping over large parts of southern Utah and northern Arizona.

Keywords

Buttes • Canyons • Colorado Plateau • Desert landforms • Differential erosion • Sandstone monoliths

1.1 Introduction

The Colorado Plateau is a vast arid or semi-arid hinterland that covers large parts of the state of Utah, together with parts of Arizona, Colorado, and New Mexico, in the southwest USA. The plateau occupies an area of approximately 130,000 km^2 and contains some of the most iconic landforms on Earth. This has resulted in the highest concentration of national and state parks in the USA. Views of

Photographs not otherwise referenced are by the author

Fig. 1.1 The heavily incised Colorado Canyon viewed from Dead Horse Point. The thick succession of flat-lying Upper Palaeozoic and Mesozoic strata include the distinctive bench developed on the White Rim Sandstone (Lower Permian). The mesa is capped by the Kayenta Sandstone (Lower Jurassic)

the Colorado Canyon at the Canyonlands National Park, eastern Utah, may surpass the more famous Grand Canyon, Arizona. The flat-lying strata encompass hundreds of millions of years of the Earth's history and include stratigraphic sections that extend from the Lower Permian and Triassic into the Lower Jurassic (Fig. 1.1). The views of the mesas and monoliths at Monument Valley, southern Utah, are universally known. The steepest cliff sections at both Canyonlands and Monument Valley are associated with resistant sandstone formations. The intercalated sequences of mudstones and shales are more readily eroded.

The two parks described here, the Canyonlands National Park and the Navajo Tribal Park at Monument Valley, are located in relatively remote sections of the Colorado Plateau (Fig. 1.2). Canyonlands occurs in the eastern part of Utah, close to the regional town of Moab, and Monument Valley is located proximal to the border with Arizona. The Colorado

Plateau is transgressed by the Colorado River system, which includes several major tributaries, such as the Green River and San Juan River. The rivers, which flow southwest or south, are sourced in the Rocky Mountains and localized areas of hill relief.

The Colorado Plateau is an uplifted geologic terrain characterized by mostly flat-lying sedimentary rocks. The intercalation of near-horizontal beds with significant physical-chemical differences is fundamental to the formation of many of the desert landforms, including both canyons and monoliths. The development of canyons and monoliths requires relatively long periods of erosion. The arid or semi-arid environment of large parts of the southwestern USA is essential for preservation of features which would probably be destroyed in wetter climates. An additional key parameter in formation of canyons is the occurrence of major rivers that experience annual floods. The Colorado River

Fig. 1.2 Map showing location of some of the national parks in Utah, together with Monument Valley which overlaps the boundary with Arizona

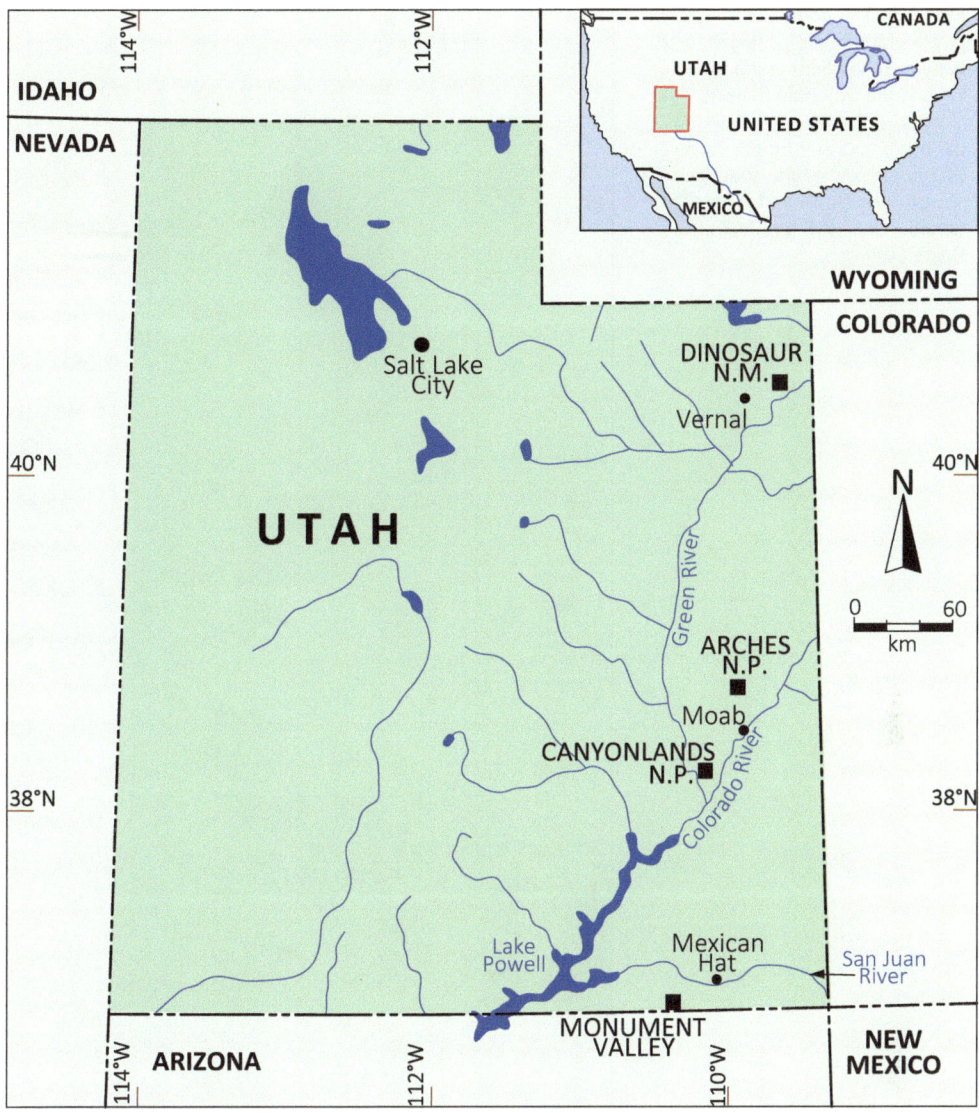

1.2 Colorado Plateau

The Colorado Plateau formed as a consequence of regional uplift, a relatively uncommon geologic process in comparison with the widespread development of cordilleras or fold mountains (e.g., the Rocky Mountains). The mechanisms driving the uplift are possibly related to a deep-seated thermal event which caused the upwelling mantle to invade and

system is swollen annually by snow melt derived from the Rocky Mountains. In recent years, however, floods have been tempered by construction of giant dams and manmade lakes, such as the Hoover Dam and Lake Powell.

under-plate the lithosphere (e.g., Flowers 2010; Karlstrom et al. 2014). Expansion of the cooling asthenosphere caused the Crust to uplift over several kilometres. Uplift probably commenced during the Early Miocene, at approximately 20 Ma, although the majority of the process may have occurred during the previous 6 Ma (Baars 1993). The Colorado Plateau is underlain by sedimentary strata that encompass more than 600 Ma of the Earth's history, i.e., from the Proterozoic through the Palaeozoic to the Mesozoic. The primary, flat-lying nature of the sedimentary strata is preserved and most of the rocks exhibit modest levels of deformation (these are characteristic features of uplift). The Colorado Plateau has a slight westward slope, which is ascribed to regional tilting.

Fig. 1.3 The dominant physiographic and geologic feature of southern and eastern Utah is the Colorado Plateau. The Colorado Plateau is bordered to the north and east by the Rocky Mountains and to the south and west by the Basin and Range terrain. *Source* Simplified after Baars (1993)

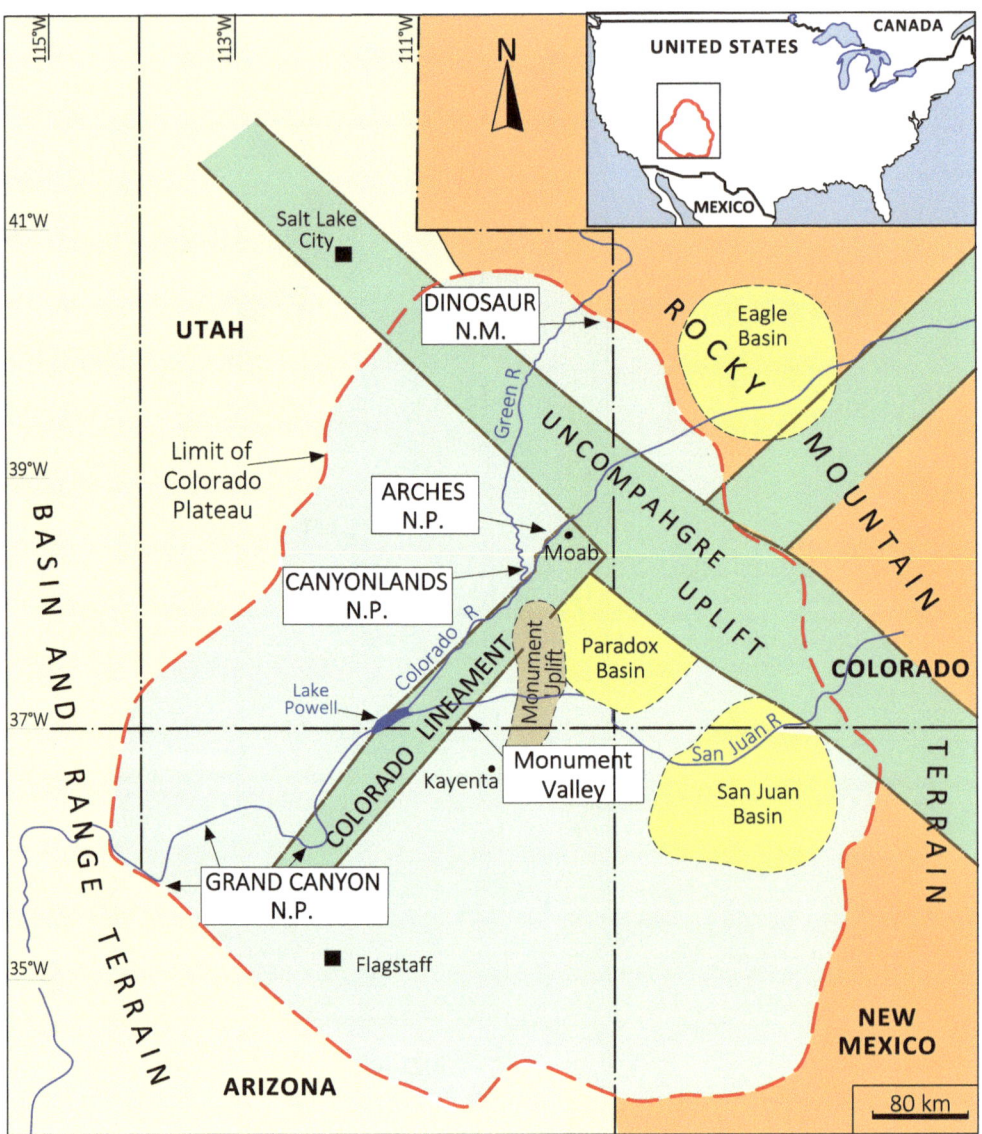

The extent of the Colorado Plateau is shown on a simplified structural map (Fig. 1.3). The approximately ellipsoidal-shaped terrain is bordered to the north and east by the Rocky Mountains and to the south and west by the Basin and Range terrain. Minor north-south compression of the Colorado Plateau has resulted in development of asymmetrical monoclines and deep basins in which remarkable amounts of uplift (and depression) can be observed. Location of the monoclines and basins is controlled by two structural lineaments (Baars 1993; Faulds et al. 2008; Baars 1993). The Uncompahgre Uplift is aligned NW-SE and the Colorado Lineament trends NE-SW. Mesozoic strata associated with the Uncompahgre Uplift reveal structural relief of 2,500 m. The juxtaposed Paradox Basin contains Proterozoic rocks which have been structurally depressed by 6,000 m. The Monument Uplift is a north-south trending monocline associated with the Colorado Lineament which reveals 2,000 m of uplift.

The flat-lying nature and relatively undeformed nature of the sedimentary strata in the Colorado Plateau may be compared with the steeply-dipping, often contorted strata, typical of the two juxtaposed terrains. The Rocky Mountain terrain is dominated by plutonic and high-grade metamorphic rocks associated with the long-lived and poorly constrained Laramide Orogeny (80–35 Ma). The Basin and Range terrain developed from lateral stretching and crustal thinning that commenced in the Miocene at approximately 17 Ma (Flowers 2010).

1.3 Regional Geology

The Colorado Plateau in eastern and southern Utah is underlain by sedimentary formations assigned to the Upper Palaeozoic and Mesozoic (Fig. 1.4). This area was originally mapped by Baker (1936), who noted that the combined thickness of sedimentary rocks exceeds 2,000 m. The dominant structural feature is the Monument Uplift. Upper

Fig. 1.4 Geological map of southern and eastern Utah showing location of the Canyonlands and Arches National Parks together with the Monument Valley. *Source* Simplified from regional map of Geological Survey of Utah (2005)

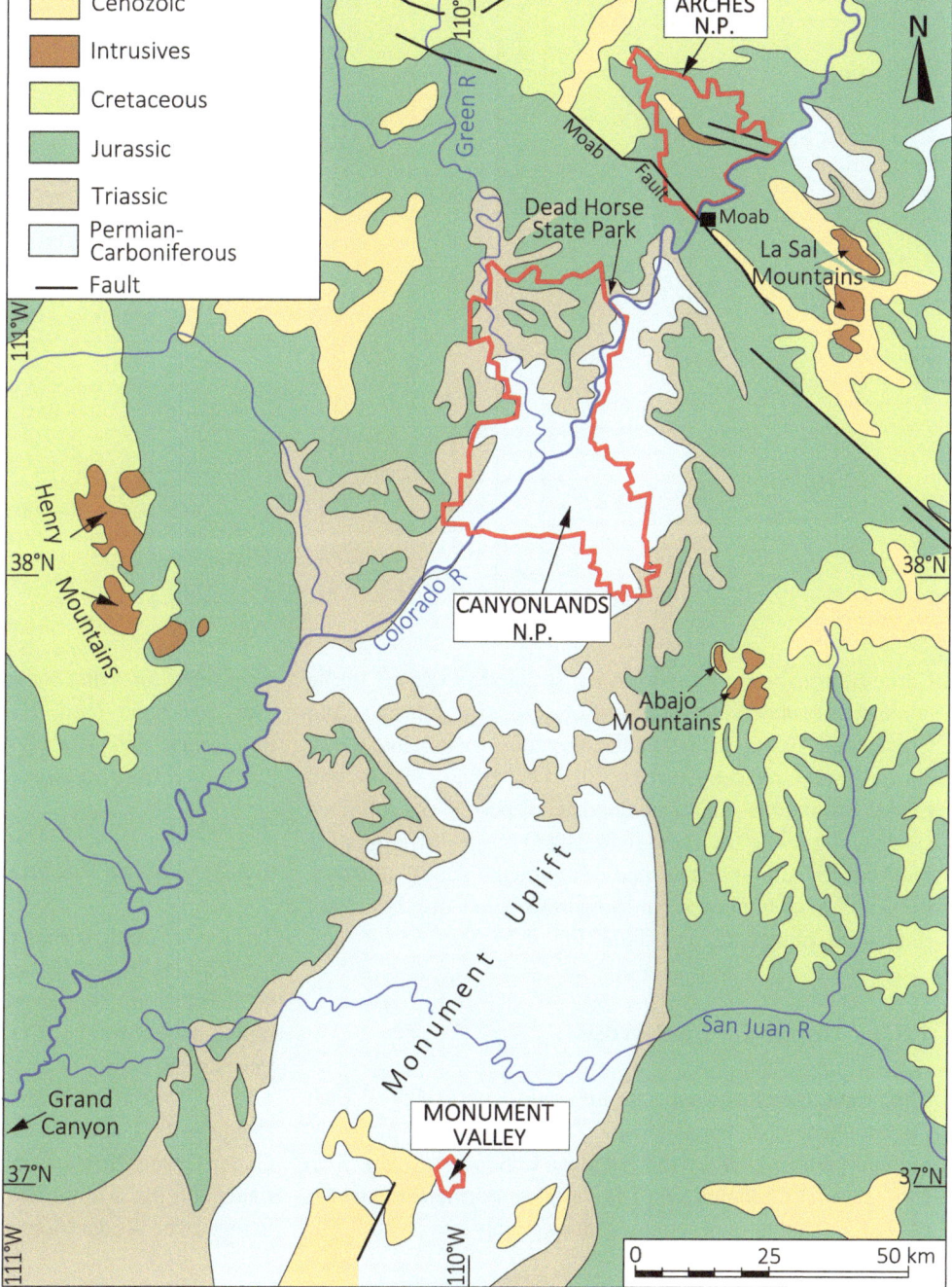

Fig. 1.5 Simplified stratigraphic sections for Canyonlands and Monument Valley. The thickness of the strata is entirely schematic as there is considerable regional variation. The relatively resistant nature of the different formations is shown schematically. *Source* Modified after Baker (1936) and Harris et al. (2004)

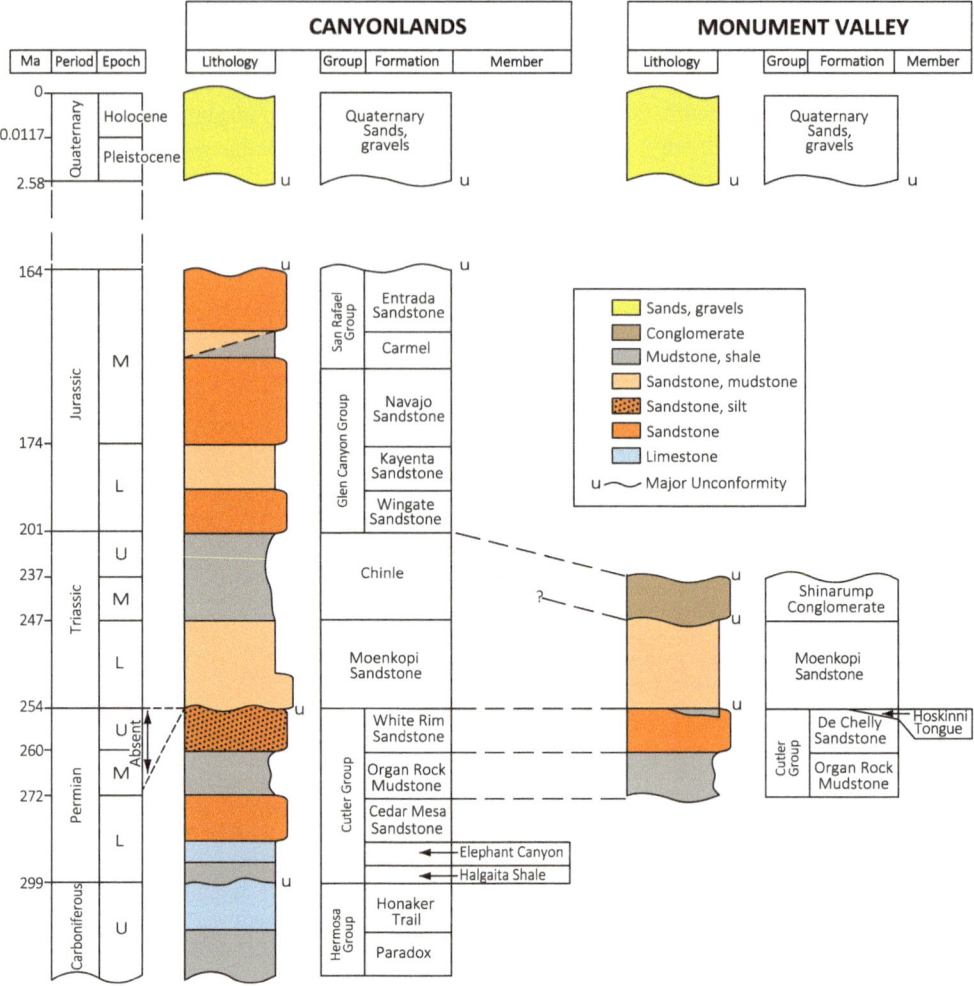

Carboniferous and Lower Permian strata are exposed in the core, with successive outcrops of Triassic, Jurassic, and Cretaceous strata on the flanks. Igneous rocks (mostly of Jurassic and Cretaceous age) occur in the Abajo, Henry, and La Sal Mountains. Cenozoic deposits of poorly-consolidated sands and gravels are found in localized basins. Stratigraphic sections for Canyonlands and Monument Valley illustrate the dominance of successive formations consisting largely of sandstone, mudstone, and shale (Fig. 1.5).

1.3.1 Upper Carboniferous

The strata associated with the uppermost part of the Upper Carboniferous (or Pennsylvanian) are represented in southern and eastern Utah by the Hermosa Group (Fig. 1.5). Two formations are recognized. The Paradox Formation consists of black shales that contain thick evaporite beds. The

evaporites, or salts, include anhydrite, gypsum, and halite. The Honaker Trail Formation consists of fossiliferous limestones and shales. Both formations were deposited during marine conditions (Baars 1993; Harris et al. 2004).

1.3.2 Lower Permian

The palaeo-environment of the Permian marks a change from marine conditions in the Upper Carboniferous to terrestrial sedimentation. This is recorded in the geologic record as a major unconformity (Fig. 1.5). The terrestrial nature of the Permian is a global characteristic, ascribed to the concentration of the Earth's land masses into the supercontinent of Pangaea. In Utah, the Lower Permian is represented by the Cutler Group. Strata associated with the Cutler Group on the flanks of the Monument Uplift reveal steep dips that are unusual for the Colorado Plateau (Fig. 1.6). The lowermost

Fig. 1.6 Steeply-dipping Lower Permian strata on the flanks of the Monument Uplift, viewed from Highway 163 southern Utah

components of the Cutler Group are the Elephant Canyon, and Halgaito Formations. The Cedar Mesa Sandstone is of particular interest as it consists of thick deposits of resistant, windblown (aeolian) sand which accumulated in a palaeo-desert. This environment was followed by deposition of the Organ Rock Mudstone, relatively soft, pale grey mudstones and shales which formed in fluvial conditions. At Canyonlands, the uppermost formation of the Cutler Group is the White Rim Sandstone. The silt-rich nature of this distinctive lithology is ascribed to proximity to the Uncompahgre Uplift. At Monument Valley, however, the uppermost formation of the Cutler Group is the De Chelly Sandstone. This marks a return to desert conditions in an area which was not influenced by localized uplift. The Lower Permian strata are terminated in Utah by a second major unconformity: the Middle and Upper Permian is entirely absent.

1.3.3 Triassic

The Triassic records a change from the terrestrial conditions of the Permian to shallow marine or estuarine environments. The Triassic strata in southern and eastern Utah are not as well defined as the Lower Permian. Two formations are recognized, both of which should possibly be awarded group status (Fig. 1.5). Thick sequences of impure, silt-rich sandstones and mudstones constitute the Moenkopi Sandstone. These sediments were deposited in a Lower Triassic palaeo-sea located on the western edge of the North American continent. By the Middle Triassic the palaeo-sea had retreated northwest, and the marine conditions were replaced by fluvial and aeolian environments. The Middle-Upper Triassic strata are represented by the Chinle Formation. This thick sequence of mudstones and shales is widely developed

Fig. 1.7 The Petrified Forest Member constitutes the colourful rock layers exposed in the Painted Desert, Arizona. *Source* Wikipedia

in large parts of southwestern USA (Blakey and Ranney 2008). In Arizona, the lowermost member of the Chinle Formation (which is not present at Canyonlands) contains the colourful rock layers preserved in the Petrified Forest National Park (Fig. 1.7). At Monument Valley, the Chinle Formation is entirely absent. There are no strata associated with the Middle Triassic at Monument Valley, but the Upper Triassic is represented by the Shinarump Conglomerate.

1.3.4 Jurassic

The Lower and Middle Jurassic in southern and eastern Utah is dominated by thick units of mostly aeolian sandstone, indicative of a return to desert conditions (Fig. 1.5). Some of the sandstone formations are extremely pure, resulting in particularly resistant units. The principal examples are the Wingate, Navajo, and Entrada Formations. The Kayenta Sandstone, however, includes mudstones and shales in the upper part, indicative of localized fluvial conditions.

1.3.5 Mineral Resources

Southern and eastern Utah contains important mineral and energy resources. The San Juan Basin hosts large oil and gas reserves and is currently exploited for shale gas methane. Areas near Canyonlands and sections of the Monument Uplift and were mined between 1945 and 1967 for uranium. Deposits of vanadium and copper occur to the south of Monument Valley. These minerals occur in the Shinarump Conglomerate and can be envisaged as part of a regional orefield associated with the Colorado Lineament. The Paradox Formation is exploited for salts, including potash, in the vicinity of Canyonlands.

1.4 Canyonlands National Park

Canyonlands is Utah's largest national park (area of 1,366 km^2) and includes deep, sinuous canyons associated with the Colorado and Green Rivers, southwest of Moab (Fig. 1.2). Downstream from the confluence of the Colorado and Green Rivers there is a 22 km stretch of white water rapids. This section of the Green River is also known as the Stillwater Canyon. Canyonlands includes panoramic views and the geology is far more accessible than the Grand Canyon. The "Island in the Sky" is a plateau or mesa at Canyonlands wedged high above the confluence of the Colorado and Green Rivers. Dead Horse Point State Park is an additional block of ground in the northeastern corner of the national park, perched above a pronounced bend in the Colorado River (Fig. 1.1). The unusual name of this mesa arose as it is rimmed on three sides by the canyon and forms a natural corral formerly used to constrain wild horses. In the 19thC, many of the corralled horses died of exposure. This locality has been used in several films, including "Thelma and Louise" (for the "Grand Canyon" scene).

The sinuous nature of the Colorado River at Canyonlands is unusual, as meanders typically occur in flood plains where rivers overlie soft alluvial deposits. The meanders at Canyonlands occur in hard, resistant rocks which have been preserved due to the rapidity of the incision (a response to the regional uplift). Many of the sandstone formations in the canyon walls and mesas at Canyonlands have a distinctive red coloration (Fig. 1.8). Some mesas are capped by pyramid hills and spires. The maximum altitude of the park is 2,170 m. The deeply-dissected areas of badlands erosion are remarkably difficult to travel through, as described in 1869 by Major J.W. Powell. Powell also reported on the ancient culture of the indigenous people of the region, including rock paintings and artefacts (Harris et al. 2004).

1.4.1 Stratigraphy

The plateau in the southern part of Canyonlands is underlain by Upper Carboniferous and Lower Permian rocks, with

Fig. 1.8 The canyons and mesas at Canyonlands National Park reveal multiple beds of flat-lying, mostly red sandstones

Triassic and Jurassic strata dominating the northern part of the park (Fig. 1.4). These strata represent almost 150 Ma of the Earth's history (excluding the 40 Ma "gap" due to the absence of the Middle-Upper Permian (Fig. 1.5). The base of the canyon at Dead Horse Point is comprised of grey limestones of the Elephant Canyon Formation (Lower Permian); these rocks interfinger with red beds of the Cedar Mesa Sandstone (Lower Permian). The distinctive, light-coloured bench part way up the canyon is developed on the White Rim Sandstone (Lower Permian). This resistant stratum is exposed over large sections of the park adjacent to the Colorado River. The upper part of the canyon reveals cliff faces associated with the Moenkopi Sandstone and Chinle Formation (Triassic), as well as the Wingate Sandstone and Kayenta Sandstone (Lower Jurassic). The lower part of the Kayenta Sandstone is very resistant to erosion in the arid climate and forms the capping of mesas in this area. The hills and spires located on mesas are typically comprised of Wingate Sandstone and Kayenta Sandstone, although the overlying Navajo Sandstone occurs in some sections of the park.

1.4.2 Salt Domes

In parts of eastern Utah, including Canyonlands, thick evaporite or salt deposits of the Paradox Formation (Upper Carboniferous) have been squeezed by overburden pressure to form salt domes. Formation of salt domes is driven by the density contrast between evaporates and shales. The salt, which can flow plastically, is forced upward through zones of weakness in the overlying shales. Preferential weathering of evaporates in the core-parts of salt domes, mostly by chemical dissolution, can create sunken valleys. Upheaval Dome is a large salt dome (diameter of approximately 5 km) located in the Island in the Sky (Fig. 1.9). The chaotic nature of the strata in this locality is pronounced. The eroded, central part consists of Organ Rock Mudstone, located some 300 m lower than the resistant Wingate Sandstone found in the rim of the salt dome (Baars 1993). The concentric scarps between core and rim are separated by valleys illustrating differential erosion of the intervening formations. The presence of the salt-bearing Paradox Formation at depth in the Upheaval Dome has been ascertained from geophysical data.

Fig. 1.9 Upheaval Dome, Canyonlands. The pale grey rocks in the core are the Organ Rock Mudstone (Lower Permian) and the red-brown rocks in the rim are the Wingate Sandstone (Lower Jurassic)

1.4.3 Additional Features

The Grand View Point located above the confluence of the Colorado and Green Rivers, in the southernmost part of the Island in the Sky, is another area of interest at the Canyonlands National Park. The Maze is an almost inaccessible area of badlands associated with disjointed canyons located west of the Colorado River. The Maze includes the "Land of the Standing Rocks", an area of free-standing buttes and spires. The monoliths are comprised of Organ Rock Mudstone which overlies the Cedar Mesa Sandstone. The Needles District is located on the southeastern side of the Colorado River. This area contains a high density of natural sandstone arches together with groups of pedestals and small canyons. The arches are developed in the Cedar Mesa Sandstone. Pedestals are known as "Toadstools" as they include broad caps of resistant sandstone located above more fissile sandstone.

1.4.4 Trails

The Shafer Trail leads from the high plateau at Dead Horse Point into the Colorado Canyon. The precipitous descent with hairpin bends encourages stopping to examine the different strata exposed in the cliffs. The drive from the base of the Shafer Trail to Moab passes bright blue settling ponds where potassium salts are being recovered from evaporite deposits (Paradox Formation). This road also accesses dinosaur tracks preserved in the Navajo Sandstone. The sandstone containing the tracks reveals primary sedimentary textures, including chaotic cross bedding and mud cracks (Fig. 1.10). Parts of southern Canyonlands are accessed by the White Rim Trail, including many areas which historically could only be reached on foot. The trail is developed on the bench associated with the White Rim Sandstone.

1.5 Monument Valley

Monument Valley spans the border between southern Utah and Arizona and can be approached from a road linking the small towns of Kayenta and Mexican Hat (Fig. 1.2). This area of high-altitude desert offers an idealized view of the scenery for which the American West is universally known,

Fig. 1.10 a Chaotic cross-bedding indicative of wind-blown sands, Navajo Sandstone, Canyonlands; **b** Mud cracks consistent with a fluvial environment at the same locality

in part as it has featured in many films and television series. In the language of the indigenous Navajo people, the name of Monument Valley translates to "*Valley of the Rocks*", an accurate description of the isolated mesas and free-standing monoliths (Fig. 1.11). The average elevation of the pediment at Monument Valley is 1,500–1,800 m; the mesas and buttes are as much as 300 m higher.

1.5.1 Navajo Tribal Park

Some of the most well known buttes, spires, and pinnacles in the region are preserved in the Monument Valley Navajo Tribal Park (area of 13 km^2). The park is accessed by a 27 km-long gravel road which joins National Highway 163

Fig. 1.11 The desert landscape of the Monument Valley Navajo Tribal Park includes free-standing monoliths such as West Mitten Butte (left), East Mitten Butte, and Merrick Butte (right)

between the northern Arizonan town of Kayenta and the southern Utah town of Mexican Hat (Fig. 1.12). The monoliths at Monument Valley were initially named by the Navajo people, mostly with spiritual references. Additional names are accredited to early settlers, including prospectors, such as Merrick and Mitchell. A visitors centre located near the southwestern corner of the park includes a museum based on the historic Gouldings Trading Post. The trading post operated from the 1920s through to the 1960s. The first contacts with the film industry were made by the owner, Harry Goulding.

1.5.2 Stratigraphy

The Colorado Plateau in this part of southern Utah is dominated by Lower Permian and Triassic sediments. The stratigraphy used here is based on Baker (1936), including his Plate 4A (a panoramic section), but updated to the modern terminology (Fig. 1.5). The four principal formations exposed at Monument Valley are as follows: (i) Organ Rock Mudstone; (ii) De Chelly Sandstone; (iii) Moenkopi Sandstone; and (iv) Shinarump Conglomerate. They have a thickness of approximately 428 m. The stratification is illustrated in a

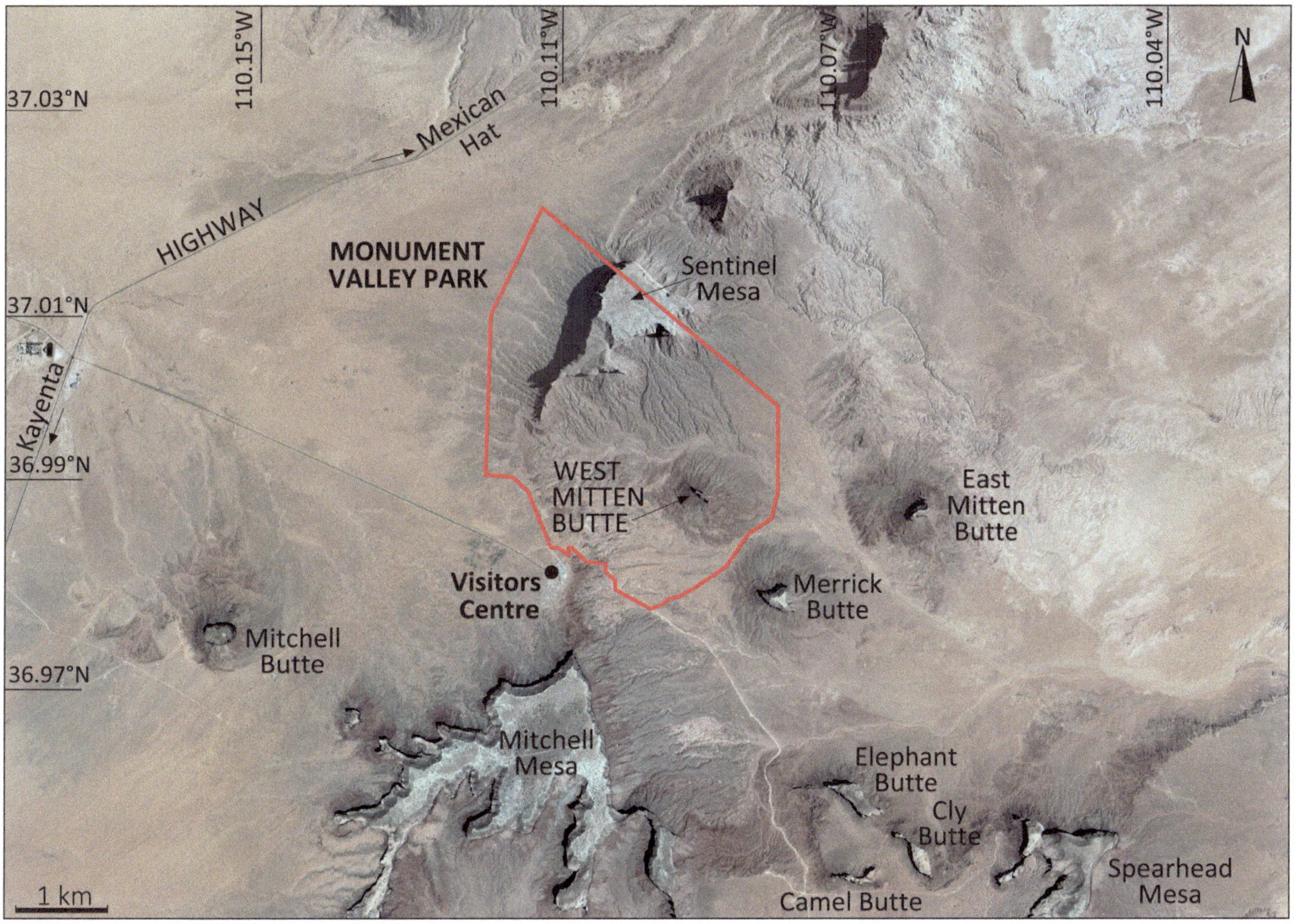

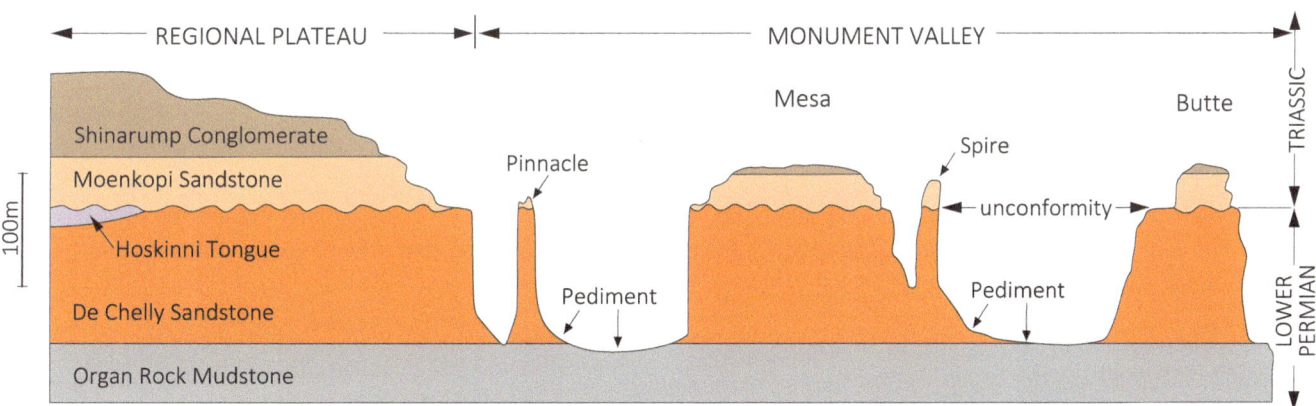

Fig. 1.12 Image showing location of the Monument Valley with some of the well known mesa and monoliths. Rain God Mesa is located just to the south of the image. The pediment is dominated by the Organ Rock Mudstone (light brown) and the highest mesas and monoliths are capped by the resistant Shinarump Conglomerate (pale green). *Source* Google Earth

Fig. 1.13 Idealized cross section of the regional plateau (left) and Monument Valley (right), the latter including mesas, buttes, pinnacles, and spires. The pediment is underlain by the relatively soft Organ Rock Mudstone. The near-vertical cliffs are associated with the resistant De Chelly Sandstone. The contact between the De Chelly Sandstone and the Moenkopi Sandstone is a major unconformity, although the Permian-Triassic boundary (the greatest mass extinction event known) is not found in Utah. The plateau, mesas and buttes are capped by the extraordinarily resistant Shinarump Conglomerate

Fig. 1.14 View looking south through the "North Window" at Monument Valley. The characteristic stratification of the Lower Permian and Triassic sediments can be observed in Cly Butte (left) and Elephant Butte (right). Pinnacles of the Rain God Mesa occur in the background with the Thunderbird Mesa in the far distance

schematic cross section (Fig. 1.13). The flat-lying nature of the strata is a pronounced feature. The pediment consists of relatively soft mudstones and siltstones; monoliths are associated with resistant sandstones and may be capped by the even more resistant conglomerate (Fig. 1.14).

Organ Rock Mudstone This formation is part of the Cutler Group (Lower Permian) and has an estimated age of 286-280 Ma. The Organ Rock Mudstone has a thickness of approximately 210 m and forms the majority of the pediment at Monument Valley. The red-coloured, finely bedded mudstones and siltstones are prominently stratified, a feature

which aids erosion. The red coloration is ascribed to the high content of iron oxides, although in some areas these rocks have a blue-grey colour associated with manganese oxides. The muds and silts were deposited in meandering rivers which were eroding the underlying strata.

De Chelly Sandstone This is the uppermost component of the Cutler Group (Lower Permian) and is associated with most of the near-vertical cliffs at Monument Valley. The average thickness is 113 m. The massive sandstones are extraordinarily resistant to erosion in the desert climate (Fig. 1.15a). Some rock faces are prominently jointed which

Fig. 1.15 a The massive nature of the De Chelly Sandstone can be observed in the near-vertical rock faces at Monument Valley; **b** Pinnacles of De Chelly Sandstone form due to closely-spaced vertical jointing

may result in formation of pinnacles and spires (Fig. 1.15b). The De Chelly Sandstone reveals both extensive and high angled cross-bedding, characteristic of aeolian sands.

Moenkopi Sandstone This is the lowermost formation of the Triassic with an estimated age of 254–247 Ma. The Moenkopi Sandstone is separated from the underlying De Chelly Sandstone by a major unconformity. The non-angular nature of such a pronounced unconformity (40 Ma of the geological record is missing) is unusual. The Moenkopi Sandstone has an average thickness of 40 m. The impure, fissile sandstones are more readily eroded than the De Chelly Sandstone. In parts of southern Utah, the Moenkopi Sandstone is underlain by the Hoskinni Tongue, although this member is not developed at Monument Valley.

Shinarump Conglomerate The stratigraphically-highest formation at Monument Valley is the Shinarump Conglomerate which has an estimated age of 225 Ma. There is some debate as to whether this constitutes a discrete formation (Upper Triassic) or should be classified as part of the Chinle Formation (Middle-Upper Triassic). The Shinarump Conglomerate at Monument Valley has an average thickness of 65 m. The coarse-grained sandstones and pebble beds (conglomerates) include siliceous cement (matrix) which is extremely resistant to erosion. The Shinarump Conglomerate is preserved over large parts of southern Utah and northern Arizona and constitutes the rimrock at the Canyon De Chelly National Monument.

1.5.3 Monuments

Possibly the two most well known sandstone monuments at Monument Valley are West Mitten Butte and East Mitten Butte; they each include distinctive pinnacles (Fig. 1.11). Rain God Mesa holds special significance for Navajo spiritual ceremonies (Fig. 1.14). Features of the West Mitten Butte can be observed from a 5 km hike, the only walking trail in the Tribal Park which doesn't require a local guide (Fig. 1.16). Other well known localities are Merrick Butte, Three Sisters, Camel Butte, and the Hub. Views from localities such as Artist's Point and North Window, the latter including the 137 m-high Totem Pole, are widely photographed (Fig. 1.17). Monument Pass and Hunt's Mesa are situated outside the Tribal Park; a permit is required to visit some of these remote areas. An additional area of interest in the region is Mystery Valley which includes the Honeymoon Arch, a natural sandstone structure.

Fig. 1.16 The West Mitten Butte at Monument Valley with its distinctive pinnacle

Fig. 1.17 The pinnacles and spires grouped to the south of Spearhead Mesa include the 137 m-high "Totem Pole" (right)

References

Baars, D. L. (1993). Canyonlands country: Geology of Canyonlands and Arches National Parks (pp. 138). University of Utah Press.

Baker, A. A. (1936). Geology of the Monument Valley-Navajo Mountain region, San Juan County, Utah. *United States Department of the Interior Geological Survey Bulletin, 865*, 98 p.

Blakey, R. & Ranney, W. (2008). Ancient landscapes of the Colorado Plateau: Chapter Five "Desert rivers and great dunes, Triassic and Jurassic (*251–145 Ma*)", 56–86.

Faulds, J. E., Howard, K. A., & Duebendorfer, E. M. (2008). Cenozoic evolution of the abrupt Colorado Plateau-Basin and Range boundary, Northwest Arizona: A tale of three basins, immense lacustrine-evaporate deposits, and the Nascent Colorado River. *Geological Society of America Field Guide, 11*, 119–151.

Flowers, R. M. (2010). The enigmatic rise of the Colorado Plateau. *Geology, 38*, 671–672.

Geological Survey of Utah. (2005). Map at scale of 1:2,500,000 original by G. C. Hintze (1974), modified by G. C. Willis.

Hamilton, W. (1987). Crustal extension in the Basin and Range Province, southwestern United States. *Geological Society London Special Publication, 28,* 155–176.

Harris, A.G., Tuttle, E., & Tuttle, S.D. (2004). Geology of National Parks. Kendal Hunt Publishing Company 6th Edition, 882 p.

Karlstrom, K. E., Lee, J. P., Kelley, S. A., Crow, R. S., Crossey, L. J., Young, R. A., et al. (2014). Formation of the Grand Canyon 5 to 6 million years ago through integration of older palaeocanyons. *Nature Geoscience, 7,* 239–244.

Abstract

The eastern part of the state of Utah, southwestern USA, is an area of considerable geological diversity. Desert landforms in the Colorado Plateau include free-standing monoliths at the Arches National Park with natural sandstone arches which have spans of several tens of metres. This part of the Colorado Plateau is separated from the Canyonlands National Park by the Moab Fault. The arches occur in a discrete structural block dominated by thick sequences of relatively flat-lying sandstone. The Dinosaur National Monument occurs in the western extremity of the Rocky Mountain terrain where an unsurpassed concentration of dinosaur fossils is preserved in impure sandstones, mudstones, and shales. The strata here are relatively steeply-dipping as they have been subjected to folding during the Laramide Orogeny. The Salt Valley anticline is a significant feature at the Arches National Park as the sandstone beds located on the flanks of the structure accumulated on irregular palaeo-surfaces. The anticline formed prior to deposition of the Middle Jurassic sandstones in response to salt domes associated with the Paradox Formation (Upper Carboniferous). The deposition of the sandstone beds on gently inclined palaeo-surfaces resulted in formation of regularly spaced joints. The jointed nature of the sandstone, together with differential erosion in the desert environment, is fundamental to formation of arches. Most arches and pinnacles are associated with the resistant Slickrock Member, the central component of the Middle Jurassic Entrada Sandstone. Examples include the well known Delicate Arch and Landscape Arch. The underlying and softer Dewey Bridge Member is readily eroded. The Moab Tongue Member, the uppermost component of the Entrada Sandstone, forms a resistant capping to some arches. Balanced rocks typically consist of resistant blocks of the Slickrock Member and Dewey Bridge Member which may rest on pedestals of the underlying Navajo Sandstone. The most well known feature of the Dinosaur National Monument is a museum constructed over a steeply-dipping quarry face. Numerous dinosaur skeletons, including multiple species of sauropods, can be observed in the Quarry Sandstone, a specific component of the Upper Jurassic Morrison Formation. The anomalous concentration of fossils is ascribed to flooding of a palaeo-river and trapping of the dinosaurs in a sand bar. The Morrison Formation extends over large parts of western USA and southern Canada.

Keywords

Arches • Balanced rocks • Differential erosion • Dinosaur fossils • Entrada Formation • Jurassic • Morrison Formation

2.1 Introduction

The dominant physiographic and geologic feature in the eastern part of Utah, southwestern USA, is the Colorado Plateau. This vast hinterland of remarkable landforms includes high-altitude deserts and localized mountain ranges. The Colorado Plateau contains some iconic landforms, including natural sandstone arches that may rise tens of metres above the pediment. The sandstone arches are so remarkable that the earliest travellers reported they were manmade, rather than natural phenomena (Fig. 2.1). The arches are, however, comprised of thick sequences of relatively flat-lying Middle Jurassic sandstones which are extremely resistant to erosion in the desert climate. Many of the arches are preserved in the Arches National Park, an area

Photographs not otherwise referenced are by the author.

© Springer Nature Switzerland AG 2021
R. N. Scoon, *The Geotraveller*,
https://doi.org/10.1007/978-3-030-54693-9_2

Fig. 2.1 Delicate Arch, Arches National Park, is comprised of gently-dipping Entrada Sandstone (Middle Jurassic). The snow-capped La Sal Mountains are visible in the distance

dominated by relatively flat-lying sandstone pavements. The occurrence of monoliths which rise directly from the pediment is an unusual feature. Other landforms include narrow canyons where the sedimentary successions can be examined in detail, as well as windows and balanced rocks. The dinosaur fossils at the Dinosaur National Monument—which transgresses the border of Utah and Colorado—occur in impure sandstones, clays and siltstones. The concentration of dinosaur fossils in the Upper Jurassic Morrison Formation is unsurpassed.

The Arches National Park is situated north of the regional town of Moab and the Dinosaur National Monument is accessed from the small town of Vernal (Fig. 2.2). The Colorado Plateau in eastern Utah is mostly underlain by flat-lying Upper Palaeozoic-Mesozoic sedimentary rocks which have undergone only modest deformation. At the Arches National Park, however, the sandstones accumulated on a gently-inclined palaeo-surface associated with the early-formation of salt domes. The Dinosaur National Monument is situated close to the western margin of the

Fig. 2.2 Map showing location of the Arches National Park and Dinosaur National Monument, eastern Utah

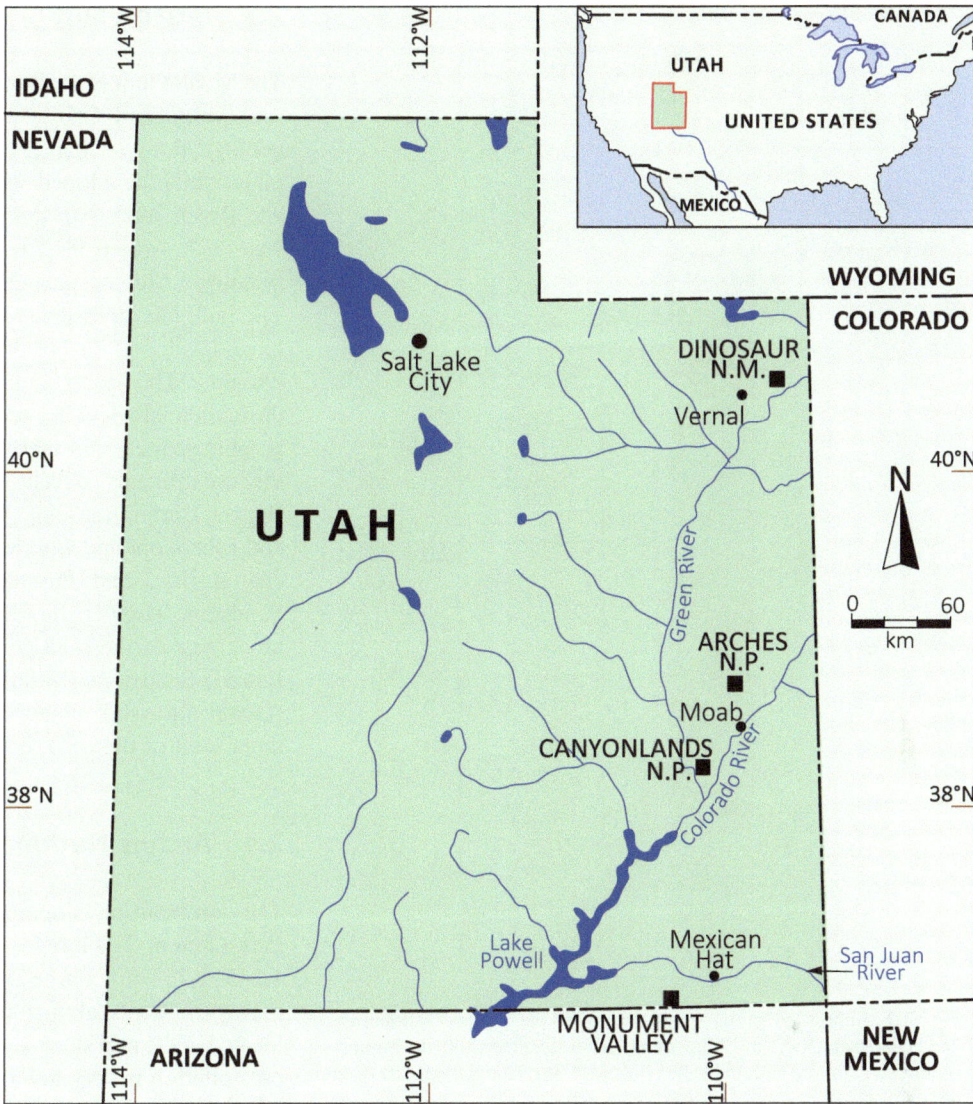

Rocky Mountain terrain. This discontinuous series of fold-and-thrust belts extends through much of western Canada and the USA. Folding associated with the Laramide Orogeny has resulted in relatively steep dips in the Upper Jurassic strata at the Dinosaur National Monument.

2.2 Regional Geology

The majority of the Colorado Plateau in eastern Utah is underlain by flat-lying Upper Palaeozoic-Mesozoic sedimentary rocks which reveal only modest levels of deformation (Chap. 1). Regional uplift during the Late Cenozoic initiated a period of intensive erosion. Arches National Park is

associated with a discrete structural block, separated from areas south of Moab by the Moab Fault. This block is dominated by thick successions of Middle Jurassic sandstone (Fig. 2.3). The resistant sandstone of the Navajo and Entrada forms isolated monoliths (Fig. 2.4). The Carmel Formation is absent from the sequence at Arches. Two Lower Jurassic formations, the Kayenta Sandstone and the Wingate Sandstone, occur in areas of low relief where outcrop is poor (Fig. 2.4). For details of the Upper Carboniferous-Triassic strata—these rocks are rarely observed in the national park, yet are an important component in formation of arches—the reader is referred to Chap. 1. The northern part of the Arches National Park is underlain by the Morrison Formation (Upper Jurassic). Despite the Dinosaur National Monument

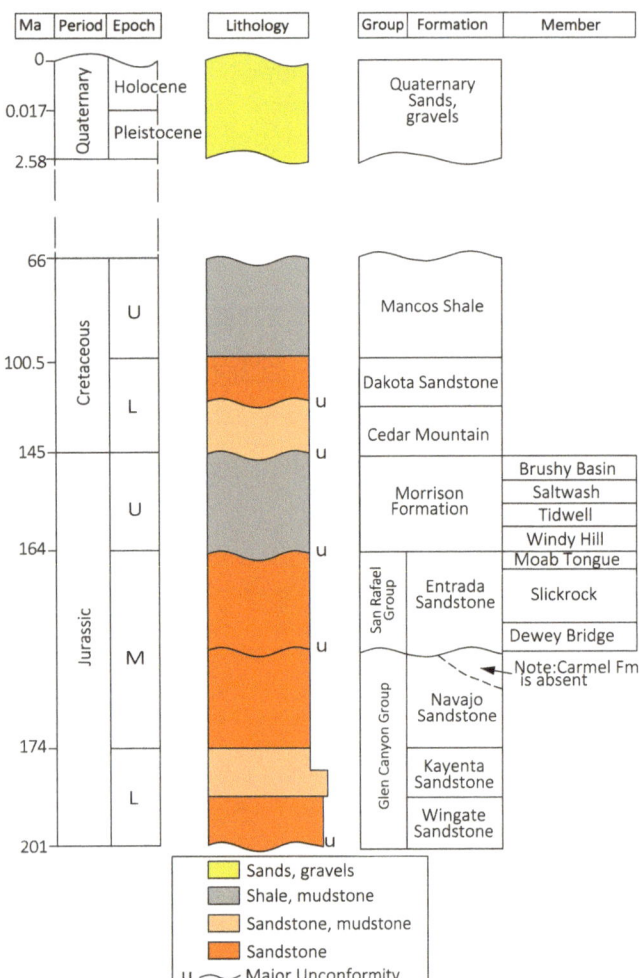

Fig. 2.3 Simplified stratigraphic section for the Arches National Park and Dinosaur National Monument. The thickness of the strata is schematic. *Source* Modified after Baker (1936), Harris et al. (2004) and others

2.3 Moab Fault

The Moab Fault is an important feature of the region between the Arches and Canyonlands National Parks (see regional geological map, Chap. 1). The town of Moab is located on a plain which developed where the fault intersects the Colorado River. Access to the Arches National Park is via Highway 191, which has been constructed in a north-south trending valley associated with the Moab Fault (Fig. 2.5a). The fault has an estimated age of 6 Ma, and is considered to have developed during the very late stages of the Laramide Orogeny. The Moab Fault has a vertical throw of 780 m. The up-faulted block west of the road reveals a similar stratigraphic sequence to that found at Canyonlands. The base of the escarpment consists of the Honaker Trail Formation (Upper Carboniferous), the Cutler Group (Lower Permian) and Moenkopi and Chinle Formations (Triassic) occur in the central part, and the capping is comprised of Wingate Sandstone (Lower Jurassic) (Fig. 2.5b). The down-faulted block east of the road consists primarily of the Navajo and Entrada Sandstone (Middle Jurassic); the Kayenta Sandstone (Lower Jurassic), which has been eroded from the escarpment west of the road, occurs in the floor of the valley.

2.4 Arches National Park

The dominant physiographic feature of the Arches National Park (area of 310 km^2) is a high-altitude desert (elevation of 1,245–1,723 m). Large sections of the pediment consist of rock pavements almost devoid of vegetation. The park is located in the rain shadow of the Rocky Mountains and annual precipitation is very low (13–23 cm). The paucity of vegetation means that even light showers can create sheet floods with runoff forming short-lived steams and waterfalls. The climate is unusually harsh. Temperatures may be as low as −29 °C in winter and are regularly over 40 °C in summer. The southern boundary of the park is formed by the Colorado River. The river is fed in spring by snow-melt derived from the Rocky Mountains and localized ranges such as the La Sal Mountains.

The location of the Visitor Centre at the main southern entrance gate to the Arches National Park, together with

occurring on the margins of the Rocky Mountain terrain, the stratigraphy is consistent with the northern part of the Arches National Park (Fig. 2.3). The majority of the dinosaur fossils occur in the Morrison Formation, although some of the overlying Cretaceous rocks also contain dinosaur fossils.

Fig. 2.4 Monoliths of the Entrada Sandstone (Middle Jurassic) project from the pediment at Arches National Park. Localized areas of badlands erosion on the flanks of the Salt Valley (centre) reveal pale buff coloured cliffs of the Wingate Sandstone (Lower Jurassic). The snow-capped La Sal Mountains in the background are associated with intrusive igneous rocks

selected geosites is shown in the map (Fig. 2.6). Sites where monoliths such as natural arches, columns, pinnacles, and balanced rocks are concentrated include Park Avenue, The Windows, and the Fiery Furnace. Arches are notably abundant in the Devils Garden. The most famous arch is Delicate Arch, a free-standing structure located on an extensive rock pavement (Fig. 2.1). A total of 700 arches may occur in the park and approximately 40 arches are known to have collapsed since 1979. The highest density of arches occurs on the fringes of the Salt Valley, an important physiographic feature readily observed on a Google Earth image (Fig. 2.7).

2.4.1 Geological Setting

The majority of the southern and central parts of the Arches National Park are underlain by the Navajo Sandstone and Entrada Sandstone (Fig. 2.8) (Doelling 1985; Baars 1993).

Most arches occur in the Entrada Sandstone. Exposure of the Middle Jurassic formations is related to the down-faulted block east of the Moab Fault. The Kayenta Sandstone and Wingate Sandstone (Lower Jurassic) are restricted to localized outcrops on the flanks of the Salt Valley. Upper Carboniferous strata occur in the centre of the Salt Valley, with Lower Permian and Triassic strata located on the flanks. Outcrop is poor in the relatively low-lying Salt Valley and this section of the park is structurally complex (there are numerous small-moderate sized faults). The northern and eastern parts of the park are underlain by the Morrison Formation (Upper Jurassic).

Proximity to the Uncompahgre Lineament, a major structural feature of the Colorado Plateau (Chap. 1), affected sedimentation patterns in the Arches National Park. The accumulation of relatively thick sequences of sediment, including of the evaporite deposits in the Paradox Formation (Upper Carboniferous), is a marked difference to areas distal

Fig. 2.5 a View looking south along Highway 191 from the entrance road to Arches National Park with Moab and the La Sal Mountains visible in the distance. The western escarpment is capped by the Wingate Sandstone and the eastern escarpment by the Entrada Sandstone; **b** Details of the Moab Fault in an information board at the Arches National Park with the up-faulted block to the left and the down-faulted block to the right (view looking north). Note the repetition of the Wingate Sandstone

Fig. 2.6 Map of the Arches National Park showing the location of the Visitor Centre and selected geosites. The southern boundary of the park abuts against the Colorado River

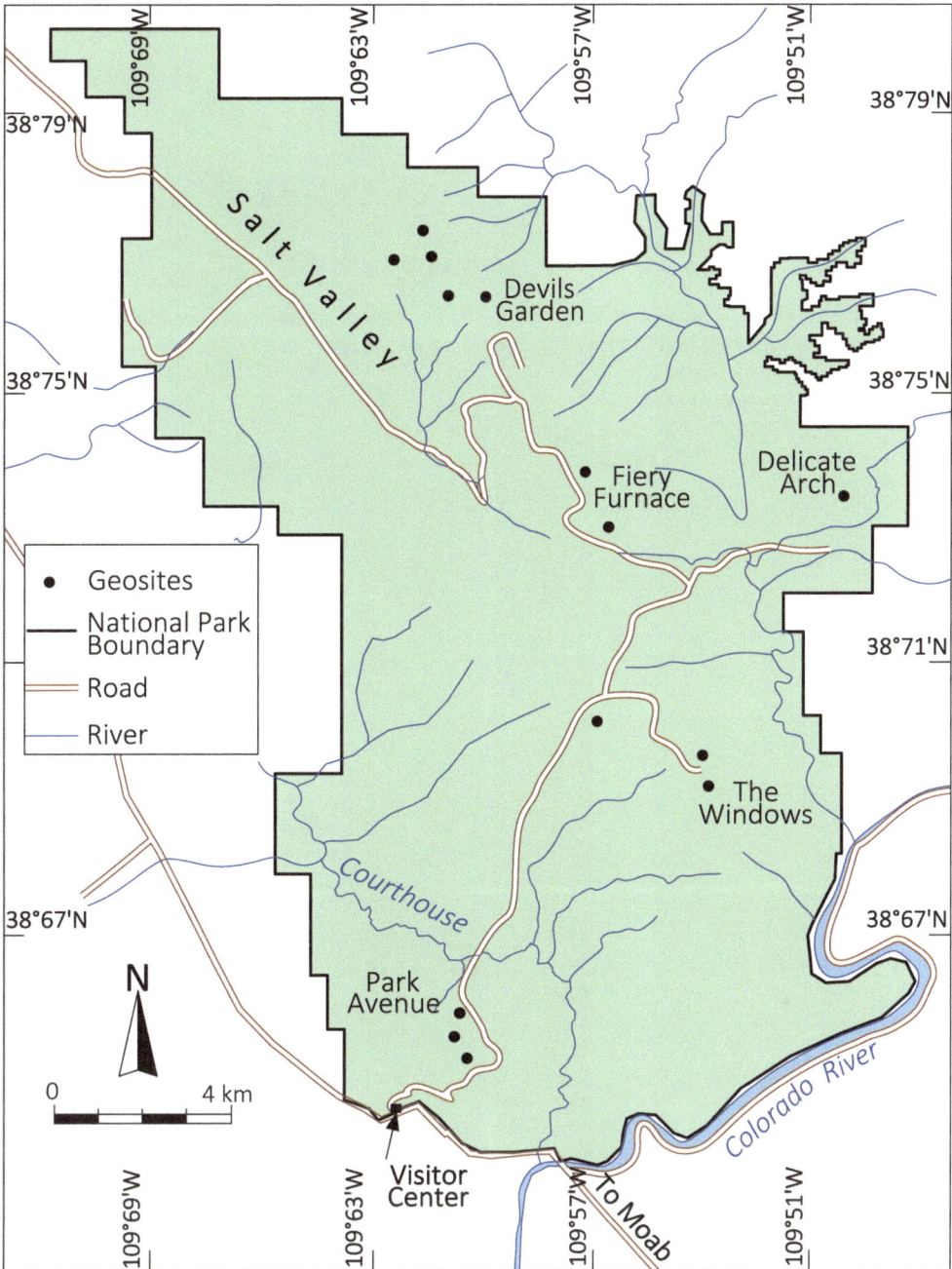

to the Uncompahgre Lineament. The relatively high overburden pressures in the sedimentary succession at Arches triggered formation of extensive salt domes and anticlines (Harris et al. 2004). (The mechanisms underlying formation of salt domes was described in Chap. 1). Dissolution of the evaporite beds in the core parts of domes or anticlines creates "salt valleys" (Fig. 2.4). The succession of rock units on the flanks of the Salt Valley is structurally complex (Fig. 2.8).

Fig. 2.7 The high-altitude desert (grey and pale mauve) of the Arches National Park contrasts with the narrow vegetated strip (green) associated with the sinuous Colorado River. Salt Valley (brown) is a prominent physiographic feature in the centre of the park. The majority of arches occur on the flanks of this structure (dark mauve). Highway 191 follows a prominent NW-SE aligned valley (red) associated with the Moab Fault: the national park is situated east of the fault. *Source* Google Earth

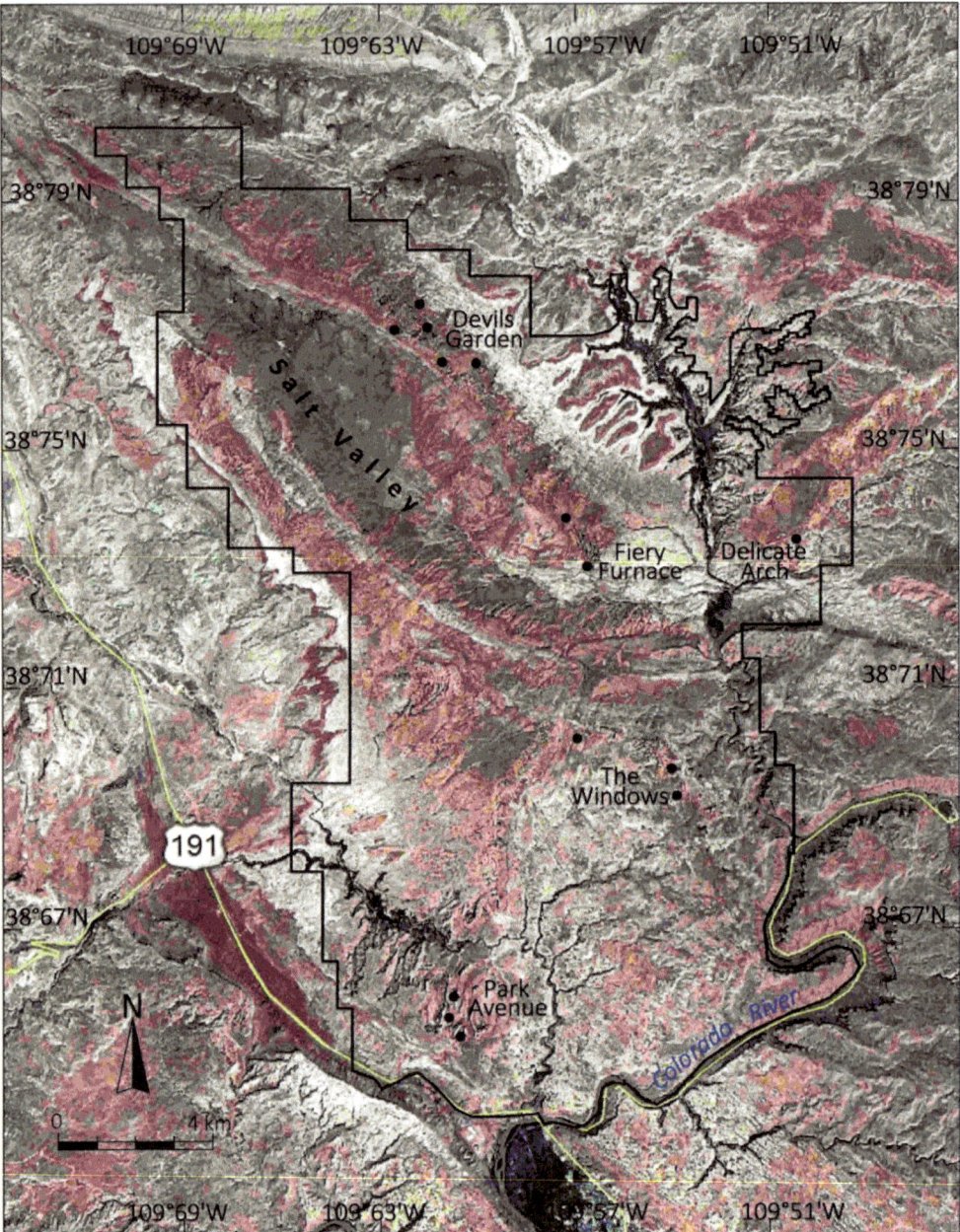

2.4.2   Middle Jurassic Sandstone

The Navajo Sandstone (estimated age of 180 Ma) consists of a thick succession of yellow-grey, massive sandstones. These rocks were deposited in coastal dunes and cliffs. The Entrada Sandstone has an estimated age of 180–160 Ma, suggesting this formation may overlap into the Upper Jurassic (Harris et al. 2004). The Entrada Sandstone is subdivided into three members, each of which can be observed at Park Avenue (Fig. 2.9). The lowermost Dewey Bridge Member consists of red-brown, silt-rich sandstones, deposited in a shallow-marine palaeo-environment. The impure and fissile nature of the sandstones—they may include contorted bedding—means they are relatively susceptible to erosion. The thickest component of the Entrada Sandstone is the Slickrock Member, a succession of massive, bright orange sandstones, mostly marine in origin. These rocks are particularly resistant to erosion in the desert

Fig. 2.8 Geological map and section of the Arches National Park. The area proximal to the Salt Valley contains numerous small faults which are not shown for reasons of scale. *Source* Simplified from maps of the Utah Geological Survey

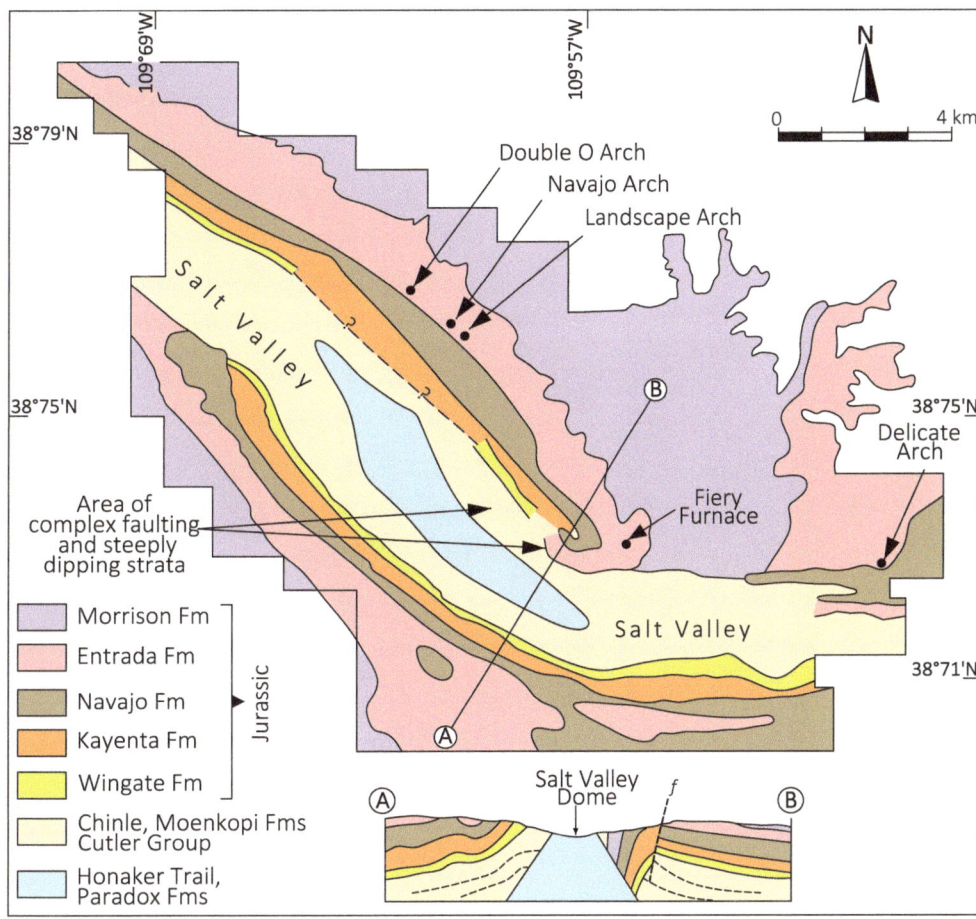

climate. The Slickrock Member locally reveals prominent cross-bedding, an indication the marine environment may have been displaced by coastal dune fields. The uppermost component of the Entrada Sandstone is the Moab Tongue Member. This sequence of resistant pale grey sandstones accumulated in coastal dune fields.

2.4.3 Formation of Arches

Arches in the jointed, Middle Jurassic sandstones developed as a consequence of differential erosion. There is no relationship with erosion by running water. (In comparison, natural bridges, e.g., Natural Bridges and Rainbow National Monuments, southern Utah, form from running water.) Arches may

take tens of thousands of years to form. Delicate Arch is estimated to have developed over a period of 70,000 years. Weathering by exfoliation (also known as onionskin weathering) is typical of massive rocks with little discernible layering or foliation, e.g., the Slickrock Member of the Entrada Formation. Exfoliation causes the massive rocks to peel off in concentric shells, skin-like, creating thin slabs. This process is assisted by frost, with water freezing in cracks during the winters. Enhanced erosion rates associated with the Colorado Plateau (which is estimated to have experienced as much as 600 m of uplift in the vicinity of the Arches National Park) are significant (uplift accelerates erosion).

An additional, equally significant factor in formation of arches is the occurrence of early-stage joints. Sandstones located on the flanks of anticlines accumulated on irregular

Fig. 2.9 The three members of the Entrada Formation are exposed in cliff faces at Park Avenue, a small canyon at Arches National Park

palaeo-surfaces. The gentle inclination of the sandstone beds caused development of flexures during the early stages of salt doming. Flexures are preserved in the sandstones as parallel northwest-trending fractures. Flexures induced minor compression and gentle warping, causing development of regularly-spaced joints in the sandstone (Cruikshank and Aydin 1994). The distribution of arches reflects this phenomenon: they are concentrated proximal to the Salt Valley (Fig. 2.7). The regularity of vertical joints, typically spaced every 3-6 m apart, is particularly significant at the Arches National Park.

Several stages in the formation of arches are recognized. The first stage is development of rock fins or buttresses. These features may be several hundreds of metres in length, tens of metres in width, and stand 10–20 m above the sandstone pavement (Fig. 2.10). Fins typically occur in

groups, the orientation of which is defined by fractures or joint patterns. Development of fins in the Slickrock Member is promoted by physical weathering and liberation of sand grains from the more fissile sandstones of the Dewey Bridge Member. In addition, some dissolution of calcium carbonate cement proximal to joints can occur. The caps of fins consist of the most massive sandstone. The bases weather more rapidly due to seepage.

The second stage in formation of arches is development of holes. Holes result from persistent erosion of fins, typically in beds which are less resistant to exfoliation. Holes may occur in beds interbedded with massive, cross-bedded units which are far more resistant to exfoliation weathering (Fig. 2.11). Holes may be sufficiently pervasive as to resemble the honeycomb weathering generally associated with carbonates.

Fig. 2.10 Formation of arches is initiated by erosion of broad sandstone fins (background) in the Slickrock Member, as seen in the Devils Garden, Arches National Park

The third stage in the process requires enlargement of holes to form small windows, principally by sand abrasion (Fig. 2.12). Continued erosion of windows may cause the unsupported sections of walls to partially collapse. The enlarged openings constitute a natural sandstone arch.

2.4.4 Delicate Arch

Delicate Arch is the most recognized of the arches, in part as it is depicted on Utah licence plates. The free-standing nature of the structure which rises from an open pavement to a height of 18 m is unusual (Harris et al. 2004) (Fig. 2.1). Most arches are associated with fins or buttresses, rather than open pavements. Delicate Arch occurs at an elevation of 1,472 m and is reached by a 5 km trail. The trail accesses extensive sandstone pavements, most of which are developed on the Slickrock Member. Delicate Arch is capped by the resistant Moab Tongue Member and can be predicted to endure longer than those developed wholly within the Slickrock Member. The arch was named by Frank Beckwith, leader of the Arches National Monument Scientific Expedition (1933–1934). Delicate Arch has been subjected to various unsavoury events, including lighting of a fire to demonstrate night-time photography (which discoloured a small section of the rock face) and various climbing stunts.

2.4.5 Landscape Arch

Landscape Arch is accessed by a trail at Devils Garden, which follows prominent fins and other, smaller arches. The span of over 87 m is reported to be the longest of any known

Fig. 2.11 Holes develop in slightly softer sandstone beds of the Slickrock Member (lower and top) which may be separated by more resistant cross-bedded units (centre), as seen on the trail to Delicate Arch

natural arch (Fig. 2.13). Large blocks of Landscape Arch broke off over the past several hundred years and the structure is notably slimmer than other arches. The current arch may, however, be better balanced than some of the older, thicker structures.

2.4.6 Balanced Rocks

Spectacular examples of balanced rocks occur at the Arches National Park. They are particularly concentrated at the Windows locality. The "Balanced Rock" consists of a resistant block of the Slickrock Member perched on an eroded relict of the softer Dewey Bridge Member, both of which rest on a pedestal of Navajo Sandstone (Fig. 2.14).

Balanced rocks form by a combination of wind erosion and seepage. Dissolution of interstitial cement in the sandstones assists with producing the spheroidal shape of some balanced rocks. The pedestals are subjected to relatively intense wind erosion (they occur closer to the ground). Abrasion by wind-driven sand particles is the most significant form of erosion and as the basal part of the pedestal preferentially thins, the balanced rock may tilt, causing compaction on one side due to gravity. Balanced rocks may tilt back-and-forth due to this process of differential erosion.

2.4.7 Trails

The short hike along the Park Avenue Canyon, named due to the resemblance of the cliffs to high-rise structures in New

Fig. 2.12 A window located in a sandstone fin comprised of the Slickrock Member next to the trail to Delicate Arch

York, reveals detailed sections of the Entrada Sandstone (Fig. 2.9). Features of interest include the 200 m-high Courthouse Towers, the Organ, and the Tower of Babel. The steep cliffs and monoliths are dominated by the Slickrock Member. Cliffs and monoliths may be capped by the Moab Tongue Member. Sedimentological textures, including cross-bedding and ripple marks, can be observed in rock pavements comprised of the Dewey Bridge Member at the foot of the cliffs. The relatively long hikes to Delicate Arch and the Devils Garden illustrate many of the geological features and erosional processes described here. The latter hosts a number of well-formed arches, including Wall Arch and the Double O Arch, as well as the Dark Angel Pinnacle.

2.4.8 Stone Tools

The presence of small amounts of chalcedony in some of the rock sequences at Arches National Park probably encouraged hunter-gather groups to migrate into the area, possibly as long ago as 10,000 BP. Chalcedony is a hard, cryptocrystalline form of silica, composed of very fine intergrowths of quartz and moganite, i.e., silica (SiO_2) with trigonal and monoclinic symmetry, respectively. Chalcedony can be sculptured into stone tools with knife edges almost as sharp as can be obtained from obsidian (i.e., volcanic glass). The groups of indigenous people who subsequently lived in the region left evidence of both tool making and pictographs. The latter are preserved on rock walls, some of which can be observed on the trails to Delicate and Landscape Arches.

2.5 Dinosaur National Monument

The Dinosaur National Monument is located approximately 100 km northeast of the Arches National Park (Fig. 2.2). The section located in Utah is a relatively small component of the Monument (total area of 856 km^2), accessed from

Fig. 2.13 Landscape Arch is reported to have the longest span of any natural arch. The arch is developed on a prominent fin or buttress comprised of massive sandstones of the Slickrock Member

Highway I-40 which links the towns of Vernal (Utah) and Dinosaur (Colorado). The arid or semi-arid landscape includes deeply eroded badlands. Rainfall is very low (less than 25 mm/annum), but the contrast of frosty nights and warm days accelerates erosion rates. The discovery of dinosaur fossils is aided by badlands erosion, a consequence of the uplift and arid climate, as well as the downward cutting of the Green and Yampa Rivers. Fossils were discovered in 1908, by Earl Douglass, a palaeontologist collecting for the Carnegie Museum of Natural History. Prior to declaration of the area in 1915 as a national monument, fossils were transported to Pittsburgh, Pennsylvania. The most complete skeleton of *brontosaurus* from the Monument can be viewed here. Over 800 palaeontological sites have been identified at the Dinosaur National Monument, where species including *Allosaurus*, *Deinonychus*, and *Abydosaurus* have been

identified (Foster 2003). The Monument also contains petroglypes accredited to some of the earliest inhabitants of the region. The most widely visited feature of the Monument, however, is the Dinosaur Museum (Fig. 2.15). This is a remarkable structure, built over part of a quarry face such that the dinosaur fossils can be viewed in the rock formation in situ. There are ten different species of dinosaur and more than 1,600 individual bones exposed in the Dinosaur Museum.

2.5.1 Geological Setting

The flat-lying stratum that is such a characteristic of the Colorado Plateau is displaced in the Rocky Mountain terrain by relatively steeply-dipping, folded strata (Fig. 2.16). The

Fig. 2.14 Balanced Rock, Arches National Park, consists of the Slickrock Member perched on a relict of the Dewey Bridge Member (both Entrada Sandstone) resting on a pedestal of Navajo Sandstone

Dinosaur National Monument includes a major W-E trending structural lineament in which Proterozoic rocks occur in the core; the Upper Palaeozoic and Mesozoic sedimentary strata are located on the southern flanks (Fig. 2.17). The Colorado section of the national monument is characterized by deeply-incised canyons which have exposed much older sequences, including the Uinta Mountain Group (Proterozoic), part of the crystalline basement (Hansen 1986). After lithification and burial, the sedimentary strata which contain the fossils were subjected to deformation during the Laramide Orogeny. The most severe deformation occurred in areas proximal to the Rocky Mountains and many sections of the Monument have been subjected to relatively modest upheaval. The majority of dinosaur fossils have been recovered from the Upper Jurassic Morrison Formation (Foster 2007) (Fig. 2.3). Both the Upper Jurassic and Cretaceous strata in this region

contain fossils of a range of fauna (not only dinosaurs) and flora (Foster and Lucas 2006).

2.5.2 Morrison Formation

The remarkable concentration of dinosaur fossils found in the Morrison Formation has resulted in this part of the Earth's stratigraphy being studied in great detail (e.g., Foster 2007). The Morrison Formation has an age of 156.3–146.8 Ma. The formation is dominated by mostly terrestrial rocks (sandstones, mudstones and siltstones) which reveal brown to purplish and green-grey colours. The sediments accumulated in rivers, streams, and lakes. The Morrison Formation has also been extensively prospected for uranium mineralization. The formation is subdivided into four

Fig. 2.15 The Quarry Museum, Dinosaur National Monument is constructed over the "Quarry Sandstone", a steeply-dipping and partially contorted rock face anomalously rich in dinosaur fossils

members (Fig. 2.3). The lowermost component is the Windy Hill Member, a sequence of shallow-marine and tidal-flat deposits. This is overlain by the Tidwell Member, a thin unit of dark red siltstones which accumulated in shallow lakes and mudflats. The Tidwell Member includes concretions of chalcedony (mostly found in the Arches National Park). The uppermost components are the Salt Wash and Brushy Basin Members. The latter two members are entirely terrestrial and consist of interbedded deposits of sandstone, siltstone, mudstone, and conglomerate. The siltstones and mudstones have a distinct green coloration. These rocks were probably deposited by meandering rivers on broad flood plains. Many of the most well known dinosaur fossils have been collected from the Brushy Basin Member (i.e., the uppermost part of the Morrison Formation).

2.5.3   Quarry Sandstone

The museum at the Dinosaur National Monument encloses part of a steeply-dipping (67°) rock face, known as the "Wall of Bones" (Fig. 2.18). More than half of all dinosaur species discovered in the Upper Jurassic of North America occur here. The most prominent are the giant, plant-eating sauropods, such as *Apatosaurus*, *Camarasaurus*, *Barosaurus*, and *Diplodocus*. Other, smaller plant-eaters and meat-eaters have been identified. The rock face is comprised of the Quarry Sandstone (age of approximately 145 Ma), part of the mudstone-dominated Brushy Basin Member. The Quarry Sandstone has an average thickness of 15 m, which represents only a small fraction of the Brushy Basin Member (total thickness of 143 m). The Quarry Sandstone has been exposed

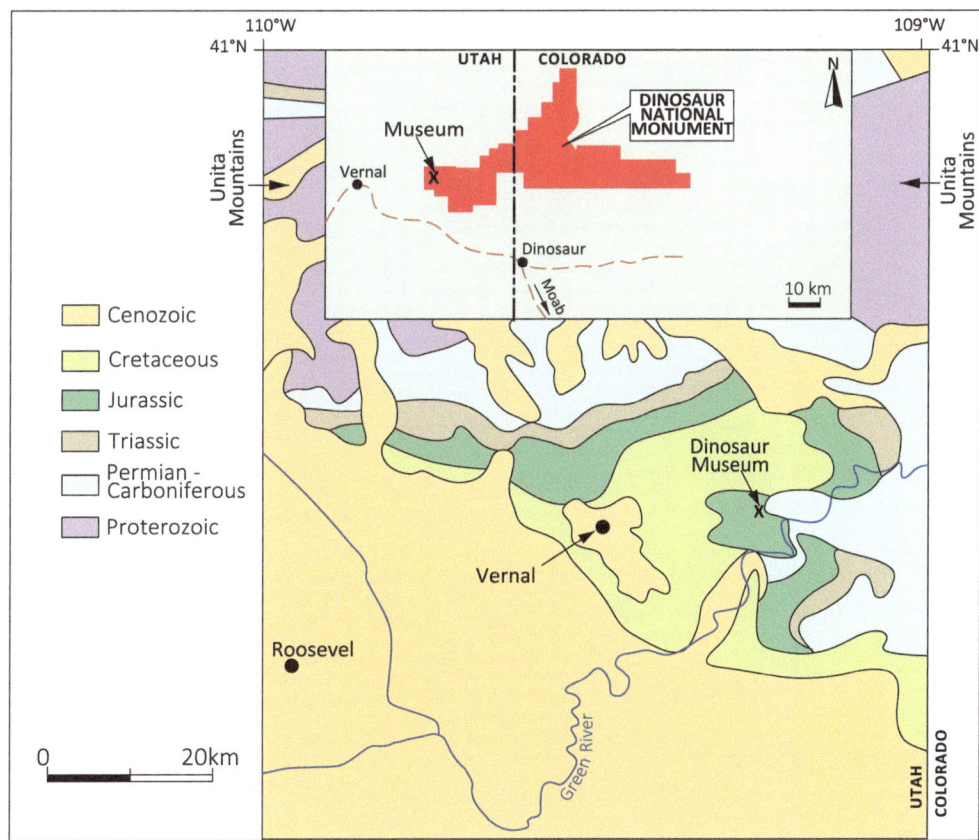

Fig. 2.16 The impure sandstone, shales and mudstones of the Morrison Formation are steeply-dipping in an exposure near the Dinosaur Museum

Fig. 2.17 Geological map of Dinosaur National Monument. *Source* simplified from maps of the Geological Survey of Utah (2005)

Fig. 2.18 **a** The Quarry Museum exposes a bedding plane of the Quarry Sandstone with more than 1,600 dinosaur fossils in situ; **b** Fossils of sauropods in the Quarry Sandstone

along a ridge for 1 km. The sandstone contains lens-shaped pebbles (they were deformed during the Laramide Orogeny), together with fragments of cherts and tuffs. The Quarry Sandstone is interpreted to have been part of a sand bar located in a meandering river and the mass burial of so many dinosaurs in one locality is ascribed to a devastating flood event.

References

Baars, D. L. (1993). Canyonlands Country: Geology of Canyonlands and Arches National Parks (pp. 138). University of Utah Press.

Cruikshank, K. M., & Aydin, A. (1994). Role of fracture localization in arch formation, Arches National Park, Utah. *Geological Society of America Bulletin, 106,* 879–891.

Doelling, H. H. (1985). Geology of Arches National Park, Utah. Geological and Mineral Survey, Department of Natural Resources, 15 p.

Foster, J. R. (2003). Paleoecological analysis of the vertebrate fauna of the Morrison Formation (Upper Jurassic), Rocky Mountain region, USA. *New Mexico Museum of Natural History and Science Bulletin, 23,* 1–95.

Foster, J. R. (2007). Jurassic West: The Dinosaurs of the Morrison Formation and their World (pp. 389). Indiana University Press.

Foster, J. R. & Lucas, S. G. R. M. (Eds.) (2006) Palaeontology and geology of the Upper Jurassic Morrison Formation. *New Mexico Museum of Natural History and Science Bulletin, 36,* 249 p.

Geological Survey of Utah. (2005). Map at scale of 1:2,500,000 original by G. C. Hintze (1974), modified by G. C. Willis.

Hansen, W. R. (1986). Neogene tectonics and geomorphology of the Eastern Uinta Mountains in Utah, Colorado, and Wyoming. *United States Geological Society Professional Paper, 1356,* 78 p.

Harris, A. G., Tuttle, E., & Tuttle, S. D. (2004). *Geology of National Parks.* Kendal Hunt Publishing Company (6th Edition, 882 p).

Yosemite National Park, California

Canyons, Granite Domes and Waterfalls

Abstract

Yosemite National Park and world heritage site is situated in a scenic section of the Sierra Nevada, California. Large sections of the national park are categorized as wilderness areas and occur at altitudes of almost 4,000 m. The near vertical faces of granite domes have resulted in the park being one of the world's premier climbing localities. Yosemite is famed for large canyons, or U-shaped glacial valleys, together with granite domes, Alpine lakes, and waterfalls. The park contains a diversity of wildlife and is particularly well known for bears. The broad range of ecosystems includes montane evergreen forest, moorland, snow-clad peaks, and deep valleys with lush pastures. Yosemite contains groves of giant sequoia trees. The geological setting is dominated by a granitic batholith which formed in the root zone of a volcanic-plutonic complex during the Mesozoic on the western margin of the North American Plate. The Sierra Nevada is a relatively young mountain range. The mountains are associated with uplift and tilting of the batholith in response to crustal thinning (they are not fold mountains). Crustal thinning was triggered by heat flow from the juxtaposed Basin and Range terrain. This process occurred in the Neogene at approximately 10 Ma. The granitic plutons exposed in Yosemite are subsidiary components of the granitic batholith at depth. The uplift and tilting of the batholith initiated a period of intense erosion. Exfoliation weathering of the granitic plutons, which typically have little or no layering or foliation, created large dome-shaped masses. Deep canyons formed as rivers cut aggressively into the granitic bedrock. A second phase of erosion at Yosemite occurred during the Late Pleistocene Ice Ages (0.68 Ma–12,000 BP). Canyon walls were steepened, the crests of granitic domes were further smoothed, and large waterfalls developed. Rock pavements were swept almost clean of soils and vegetation by ice sheets. Some glacial features at Yosemite are associated with ice advances during the Holocene, including the Little Ice Age (1300–1850 AD) when small ice sheets and glaciers advanced and retreated many times. The most widely visited section of the national park is Yosemite Canyon, an area that includes the El Capitan monolith (which has an exposed rock face of 1,095 m) and the Half Dome monolith (height of 1,443 m above the base of the canyon). Half Dome is an unusual feature as exfoliation weathering is in part controlled by vertical joints. Melting of the Late Pleistocene ice partially filled up hanging valleys to form Alpine lakes, such as Mirror Lake. Many of the high waterfalls at Yosemite, including the 739 m-high Yosemite Falls, are associated with hanging valleys. Waterfalls have also developed where glaciers plucked out jointed sections ("glacial steps") of the granitic basement.

Keywords

Canyons • Giant sequoia • Granite domes • Pleistocene Ice Ages • Sierra Nevada • Waterfalls

3.1 Introduction

Yosemite National Park (area of over 3,000 km^2) is a world heritage site situated in the state of California, western USA. Yosemite occurs in part of the Sierra Nevada ("the range of light") and over 95% of the national park is categorized as wilderness, which together with the invigorating climate encourages visitors to hike and camp. The park can be extremely crowded in summer with more than four million visitors estimated each year. The best times to visit are

Photographs not otherwise referenced are by the author.

Fig. 3.1 Half Dome is one of the most well known of the granitic monoliths at the Yosemite National Park. The 600 m-high, planar north face, which hangs above the upper sections of the Yosemite Canyon, is defined by a set of near-vertical joints. Exfoliated scree was removed by glacial activity and ice sheets scraped clean large sections of the granite plutons which dominate the Sierra Nevada in this region

probably spring and autumn. The mountain landscapes include deep canyons, or U-shaped glacial valleys, granitic monoliths and domes, Alpine lakes and waterfalls (Fig. 3.1). The domes are parts of large granitic plutons that protrude onto surface and have been eroded over long periods of geologic time. The domes contain unusually high and steep rock faces and Yosemite is one of the world's premier climbing localities.

Yosemite National Park is one of the least segmented sections of the North American coastal ranges which extend from Alaska to Mexico (Fig. 3.2). The Sierra Nevada is dominated by a granitic batholith (Mesozoic) which has been subjected to two discrete phases of erosion. The initial phase of erosion (Neogene) was triggered by uplift during formation of the mountain range. The second phase of erosion occurred during the Late Pleistocene when the Sierra Nevada was covered by thick ice sheets. The ice sheets

formed and receded innumerable times over a period of approximately 0.8 Ma. The relatively southerly-location of Yosemite, means that even with altitudes reaching almost 4,000 m, the permanent icefields are very small. The latter are probably relics of the Little Ice Age (1300–1850 AD), during which time small ice sheets and glaciers advanced and retreated many times. The Pleistocene ice almost certainly disappeared during the hotter regime of the Early Holocene.

3.2 History of the Park

Yosemite is the traditional land of the Ahwahneechee indigenous people (wikipedia.org 2018). One of their main encampments was at Wawona in the southwestern part of Yosemite. There are two possible derivations of the name

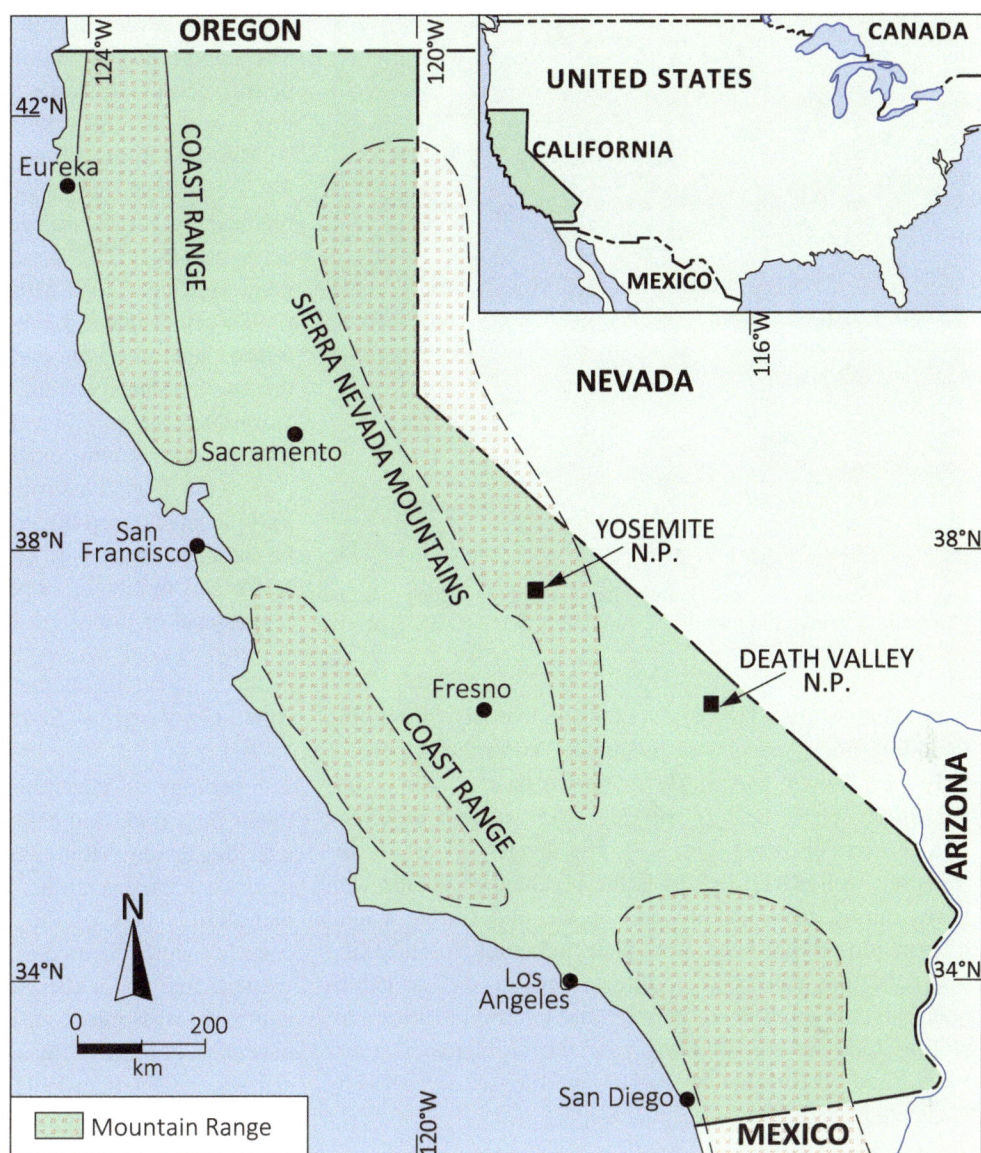

Fig. 3.2 Map showing location of the Yosemite National Park relative to coastal ranges in the state of California

Yosemite. The first explanation suggests the word means "some of them are killers" in the Miwok language. This was applied to a renegade tribe which the Mariposa Battalion was charged in 1851 with tracking down, and was mistakenly thought to refer to the main tribal group. An alternative explanation is derived from the word *Uzumati*, meaning grizzly bear. The protection of Yosemite is accredited to several early settlers, including Galen Clark and John Muir, who lobbied the government in the mid-19thC. The publication of articles in popular magazines by James Hutchings, illustrated by artist Thomas Ayres who also held an exhibition in New York which was widely supported, was also significant. The first tourists arrived in 1855, with Yosemite Canyon and the Mariposa Grove, Wawona as the main attractions. Yosemite attained national park status in 1890.

3.3 Ecosystems

Yosemite National Park contains a diverse range of ecosystems, including montane evergreen forest, Alpine moorland, snow-clad peaks, and deep canyons and valleys

Fig. 3.3 Yosemite is well known for black bear which can be observed on granite pavements

with lush pastures. The park contains a diversity of wildlife, including ninety species of mammals. Yosemite is particularly well known for grizzly or brown bear (*Ursus arctos horribilis*), black bear (*Ursus americanus*), and mountain lion (*Puma concolor*) (Fig. 3.3). The dangers of hiking the trails are well advertised, as is the potential problems with bears during berry season and at the end of the winter hibernation. Black bear were once infamous for breaking into cars (they can force the doors open), as well as cabins and tents, but controls on garbage and overnight lockers for storing food have solved many of the problems. Another issue faced by park management is wildfires exacerbated by alien plants.

3.4   Park Entrances

Yosemite is generally approached from the western, or Arch Rock entrance, either by road or train to the nearby town of Merced (Fig. 3.4). The principal attraction at Yosemite is Yosemite Canyon, an eleven km-long valley associated with the Merced River (Guide to the National Parks of the United States 2001). Cook's Meadow, proximal to Yosemite Village located near the Merced River, reveals the quiescence typical of valley floors, with the grassland, shrubs, and seasonal berries supporting wildlife including, in addition to bears, mule deer (*Odocoileus hemionus*) and a large number

of rodents, several species of ground squirrel and chipmunk. The opportunity for hiking at Yosemite from roads which permit access to some of the high plateaus and ridges downward into the valleys is particularly rewarding.

The only approach to Yosemite from the eastern side is the Tioga Pass (height of 3,031 m). This road is generally blocked by snow until early or mid-June. The Tioga Pass accesses Tioga Lake, a relatively large Alpine lake, and has good views of Tioga Mountain (3,509 m) and Mount Dana (3,980 m). Tenaya Lake is an Alpine lake located in a remote section of the park where the Tioga Road traverses the mountainous terrain between Tuolumne Canyon and Yosemite Canyon (Fig. 3.5). This region is dominated by granitic domes with extensive rock pavements. An advantage of the Tioga Pass route is that the Death Valley National Park, located approximately 150 km south of Yosemite, can be included in a visit. The contrast between the snow-clad peaks and mountain slopes of Yosemite and the arid landscapes east of the Sierra Nevada is marked.

3.5   Geological Setting

The geology of Yosemite is dominated by granitic plutons, rather than fold mountains associated with most cordilleras, e.g., the Rocky Mountains. The plutons are Mesozoic features associated with the giant Sierra Nevada Batholith. The batholith is part of the root zone of a volcanic-plutonic complex which formed in an ancient subduction zone on the western margin of the North American Plate. The granitic batholith was intruded into a Neoproterozoic-Palaeozoic basement dominated by metasedimentary and metavolcanic sequences (Bateman 1992). The western margin of the North American Plate changed to a major subduction zone during the Late Palaeozoic. One manifestation of this, over and above the Mesozoic granites, is the occurrence of a volcanic island arc assemblage (Late Devonian-Permian) in the basement. The oldest part of the basement preserved at Yosemite is Neoproterozoic quartzites and schists which occur on the margins of the granitic batholith.

Three ages of granitic pluton are indentified at Yosemite, each of which may be considered as a subsidiary feature of the Sierra Nevada Batholith. The three ages are as follows: Upper Triassic-Jurassic (210–150 Ma), Lower Cretaceous (120–100 Ma), and Upper Cretaceous (95–80 Ma) (Bateman 1992; Palmer 2004). The distribution of the plutons suggests major hiatuses in intrusive activity. Older batholiths have been intruded by younger granites, giving rise to a

Fig. 3.4 Image of the Yosemite National Park showing the mountainous terrain and entrance gates and geosites referred to in the text. *Source* Google Earth

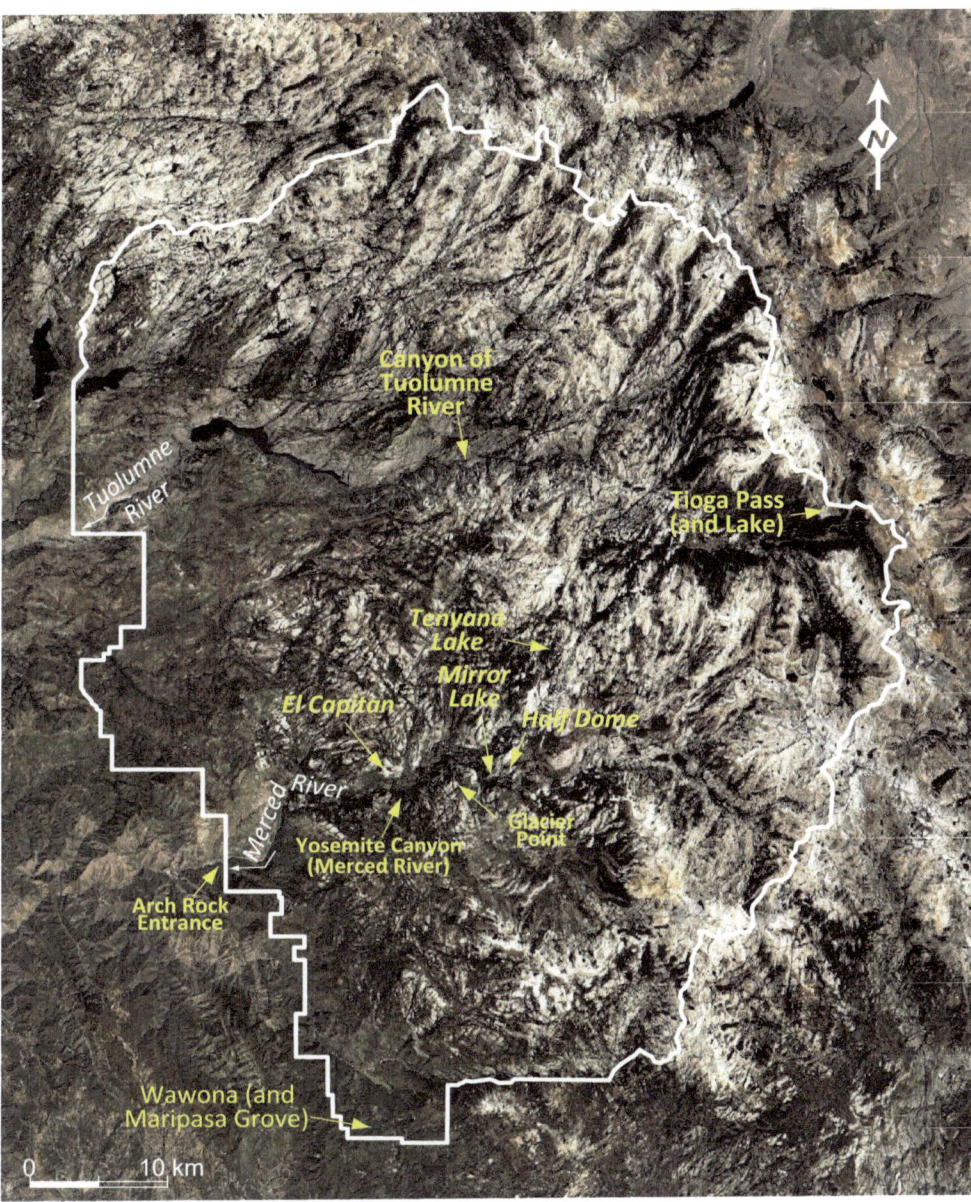

complex range of crosscutting relationships and xenoliths. The plutons consist of granite, granodiorite, quartz diorite, and monzonite. Compositional differences have had a pronounced influence on erosion and formation of features such as domes and pinnacles.

The Sierra Nevada Batholith was uplifted as a reaction to Late Cenozoic tectonic activity in the adjacent Basin and Range terrain (i.e., in Nevada and western Utah). The Sierra Nevada started to form in the Neogene, at approximately 10 Ma, when a section of the granitic crust located between the Basin and Range terrain and the coast was uplifted and tilted. Uplift and tilting was a reaction to crustal extension and heat transfer from the Basin and Range terrain. This thinned the crust in the eastern part of the Sierra Nevada. The thinned eastern section of the crust became more buoyant and was consequently uplifted. The prominent western tilt remains a feature of the Sierra Nevada. The relatively gentle and thickly forested western slopes at Yosemite contrast with the steeper and more arid eastern slopes.

The uplift at Yosemite during the Neogene initiated a period of intensive erosion. Deep canyons formed as rivers

Fig. 3.5 Tenaya Lake, located between the Tuolumne and Yosemite Canyons is an example of a high-altitude or Alpine lake located in a hanging valley at an elevation of 2,484 m

cut aggressively into the granitic bedrock. The tectonic activity associated with the Basin and Range terrain extended into the Sierra Nevada with extrusion of volcanic lavas. The youngest component of this volcanic activity, which is restricted to the eastern part of Yosemite, is dated at 3.5 Ma. Some of the lavas may have infilled the older canyons, but have been almost entirely removed by erosion.

3.6 Glaciations

A second phase of erosion at Yosemite was triggered by the Late Pleistocene Ice Ages (Box 3.1). Canyons which had formed prior to these events were steepened and rock monoliths were scraped clean of soils and vegetation. The glaciated valleys contain thick deposits of tills and moraines. The oldest glaciation recognized at Yosemite is probably the Illinois Glaciation. The effects of the Main Ice Age (110,000–12,000 BP), or Wisconsin Glaciation, are pronounced in the Sierra Nevada. The Tahoe, Tenaya, and Tioga Glaciations denote separate epochs of the Main Ice Age (Ehlers and Gibbard 2004; Guyton 1998; Palmer 2004).

The Tahoe Glaciation reached its maximum extent at 70,000 BP. The age of the intermediate Tenaya Glaciation is uncertain. The younger Tioga Glaciation was the least severe and occurred at 30,000–10,000 BP. This activity peaked at approximately 21,000 BP, which corresponds with the worldwide phenomenon of the Last Glacial Maximum.

Box 3.1: Late Pleistocene and Holocene Ice Ages

The Late Pleistocene Ice Ages included at least seven relatively long glacial epochs separated by shorter warmer periods. Changes in temperature and precipitation associated with the Ice Ages may be ascribed to the precession cycle of some 23,000 years. Precession is a regular change in the orientation of the axis of a rotating sphere, such as Earth. The half dozen or so **early glaciations** occurred at approximately 0.68–0.13 Ma. The most recent of the early glaciations is known as the Illinois Glaciation (238,000–126,000 BP) in North America. This was followed after a relative short interglacial period (of approximately

15,000 years) by the **Main Ice Age** (110,000–12,000 BP). The 100,000 year-long Main Ice Age, known as the Wisconsin Glaciation in North America and the Würm Glaciation in Central Europe, included several glacial peaks. The most recent peak of the Main Ice Age, the **Last Glacial Maximum**, occurred at approximately 20,000 BP.

There were several ice advances during the Holocene of which the most recent is the **Little Ice Age** (estimated duration of 1300–1850 AD). Mapping of glacial deposits and moraines enables geologists to identify the relative ages of glaciations, together with information on the advance and retreat of ice fields. The effects of the Late Pleistocene Ice Ages was particularly extreme in the northern latitudes, with large areas of Europe and North America being covered by ice caps and glaciers, but this was a global phenomenon which influenced the climate (and geology) of even the tropical regimes. The ages noted above are relatively inaccurate (precise radiometric methods such as those used for volcanic rocks are not obtainable), but comparing duration of the previous interglacial with the Holocene (the current interglacial) allows us to estimate (very crudely) that we are 3,000 years away from the next major ice advance.

Ice sheets at Yosemite may have reached a thickness of 1,200 m during the Tahoe Glaciation. Recession of the Tioga ice sheets from the Sierra Nevada occurred during the Early Holocene, i.e., several thousands of years later than in many other parts of the northern hemisphere (due to the enhanced elevation). Several large glaciers developed at Yosemite, of which the most prominent were the Yosemite and Tuolumne Glaciers. Branch glaciers were sufficiently well-developed as to cover large parts of the surrounding mountains. In some areas, only the highest peaks and domes were ice-free. Large granitic monoliths, such as El Capitan and Half Dome, projected above the ice as nunataks.

Several small slope glaciers remain at Yosemite, of which the two largest are the Lyell and Maclure Glaciers. They are not relicts of the Main Ice Ages, however, but are associated with ice which reformed during the historical period known as the Little Ice Age (1300–1850 AD). This activity included small ice sheets and glaciers which advanced and retreated many times (Fig. 3.6). Some estimates suggest the Little Ice Age peaked at Yosemite as recently as 1895 (Guyton 1998).

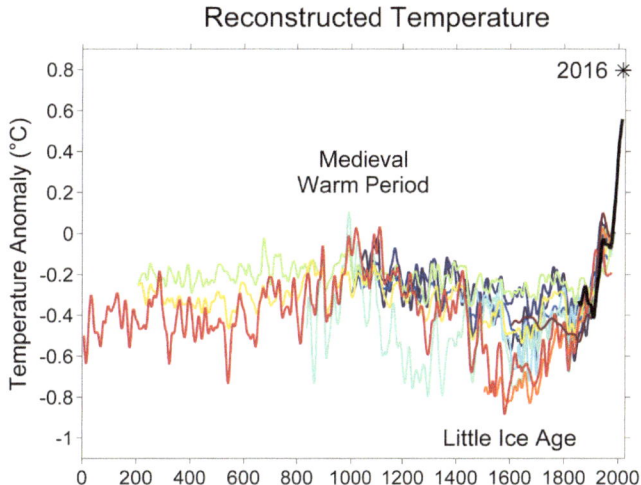

Fig. 3.6 Graph showing estimated temperatures during the previous 20 centuries. The most pronounced cycles (with estimated durations) are the Medieval Warm Period (950–1250 AD) and the Little Ice Age (1300–1850 AD). The relatively rapid temperature changes during the Little Ice Age promoted numerous advances and recessions of ice sheets at Yosemite. *Source* https://commons.wikimedia.org

The advances and retreats of ice sheets can be, in part, reconstructed from mapping of glacial deposits and moraines (or tills). Mapping of terminal moraines at Yosemite, for example, indicates the older Tahoe Glaciation was far more extensive than the younger Tioga Glaciation (the Tahoe moraines occur at elevations approximately 300 m lower than the Tioga moraines: Moore and Mack 2008). Some moraines at Yosemite attain considerable thickness in the deep canyons.

3.7 Geosites

Yosemite Canyon is the most widely visited area of the national park (Fig. 3.4). The U-shape is typical of ice-carved valleys, although the canyon initially developed due to erosion triggered by the Neogene uplift (Fig. 3.7a). The view of the base of the canyon from the Wawona Tunnel near Glacial Point is widely photographed (Fig. 3.7b). The canyon floor reveals areas of rich pastures which contrast markedly with the high rocky slopes. The El Capitan Dome, Half Dome, Cathedral Rocks, and the Bridal Veil Falls can also be observed from here. The El Capitan Dome or monolith has an exposed face with a height of 1,095 m (Fig. 3.8). This is one of the highest, near-vertical rock faces on Earth. Watching the climbers, some of whom sleep overnight in slings (due

Fig. 3.7 a The Yosemite Canyon is the principal feature of the national park. The U-shape is typical of ice-carved valleys although the canyon initially developed due to an older phase of erosion; **b** The view from Glacial Point includes pastures associated with nutrient-rich moraine deposits in the floor of the Yosemite Canyon

Fig. 3.8 El Capitan is the highest of the granitic domes with a near-vertical face of 1,095 m. Just visible are sections of younger granitic intrusives and cross-cutting granite pegmatite dykes and veins (darker)

to the verticality and absence of ledges) with binoculars from the base of the dome is a popular pastime. The monolith is comprised of granite associated with the Western Intrusive Suite (part of the youngest batholith). The monolith consists of massive granite with few fractures or joints.

The face of Half Dome soars 1,443 m above the base of Yosemite Canyon (Fig. 3.9). Half Dome is an unusual monolith as it is associated with both vertical jointing and exfoliation. It has been suggested that the monolith was sliced in two by a glacier, but this is probably incorrect: the "missing half" has never been identified. The orientation of the steep face of Half Dome is parallel to a set of prominent regional joints. Erosion is ascribed to proximity to the Merced River. Half Dome was modified by the Tenaya Glacier, which is estimated to have filled the canyon to within 150 m of the crest. Glaciation smoothed the faces of the dome, as well as other monoliths in the vicinity, and swept most of the exfoliation debris away.

3.7.1 Formation of Granitic Domes

The crests of many of the granitic plutons in the Sierra Nevada Batholith were exposed to extensive exfoliation weathering during the Neogene-Quaternary, i.e., after regional uplift but prior to onset of the Late Pleistocene Ice Ages. This style of erosion is a characteristic feature of massive rocks with little or no foliation. El Capitan is an example of a monolith consisting of granite with few fractures or joints. Erosion by exfoliation, with removal of concentric shells of granite, can be observed at Sentinel Dome and Turtle Dome. The Royal Arches are a consequence of exfoliation of granite in which the recessed arches are set back into a main arch, giving the illusion of a group of arches. Some granite domes and monoliths, however, are severely jointed, e.g., Half Dome. Jointing is a feature best observed on rock pavements on the crests of granite domes (Fig. 3.10). Joints developed in the granite as pressure was released by the regional uplift and erosion of the Sierra Nevada Batholith (Palmer 2004). The jointed domes of granite tend to erode to create castellated features. Granitic plutons with closely spaced joint sets erode to form columns, pinnacles, and spires (e.g., Watkins Pinnacle, Split Pinnacle, and Cathedral Spires).

3.7.2 Alpine Lakes

The landscape at Yosemite continued to evolve after cessation of the Late Pleistocene Ice Ages. Alpine lakes formed from meltwater trapped in hollowed-out sections of the canyons and hanging valleys, e.g., Mirror Lake (Fig. 3.11). The hiking trail around this 1,248 m-high lake, which is located on a tributary of the Merced River, includes Alpine meadows and views of some of the high peaks. Hanging valleys probably started to form in response to the Late Cenozoic uplift, with the main canyon glaciers accentuating the geomorphology during the Late Pleistocene. Landslips and rock debris, including extensive talus slopes, are in part related to seismic activity during the Holocene.

3.7.3 Waterfalls

Many of the highest waterfalls at Yosemite are fed from hanging valleys which drop into the main canyons. Yosemite Falls (740 m) is reputed to be the second highest in the world, although the upper and lower falls are divided by an intermediate section of cascades (Fig. 3.12a). Other falls of note in the Yosemite Canyon are Rippon Falls (491 m) and Bridal Veil Falls (189 m). Nevada Falls (181 m) and Vernal Falls (96 m) on the Merced River have a different geomorphic history: they are associated with giant glacial steps formed by the glaciers which plucked out jointed sections of the granitic basement (Fig. 3.12b).

3.8 Giant Sequoia

Yosemite is well known for groves of giant sequoia (*Sequoiadendron giganteum*), also known as giant redwood. The giant sequoia is the sole surviving species of the genus *Sequoiadendron* and one of three species of coniferous trees known collectively as redwoods. The Mariposa Grove in the southern part of Yosemite is the most well known locality (Fig. 3.13). This area was protected in 1864, prior to establishment of the national park and was a reaction to a time of undisciplined tree felling. The Mariposa Grove includes trees with estimated ages of over 2,000 years. This area has only recently been reopened to tourists. The largest tree in the grove has an estimated age of 1,900–2,400 years, and a height of 64 m. The Wawona Tree with a height of 89 m and diameter of 27 m is famous as a vehicle-sized tunnel was driven through it in 1881 (Fig. 3.14). The giant sequoia constitute (by volume) among the largest living organisms on Earth.

Fig. 3.9 The near-vertical-face of Half Dome includes dark vertical lines associated with water seepage

Fig. 3.10 Many of the granite domes proximal to the Yosemite Canyon reveal prominent joints. The rock slabs have been swept clean by glacial activity

Fig. 3.11 The idyllic scenery of Mirror Lake occurs in a hanging valley perched hundreds of metres above the Yosemite Canyon

Fig. 3.12 a The upper section of the Yosemite Falls, a tributary of the Merced River, plunges 436 m over granitic cliffs into the Yosemite Canyon; **b** The Vernal Falls on the Merced River is associated with a glacial step cut into the basement granite by the Yosemite Glacier

Fig. 3.13 The stands of giant sequoia in the Mariposa Grove, Yosemite encouraged protection of the area

Fig. 3.14 Historical photograph of the Wawona Tree, a giant sequoia in the Mariposa Grove, which fell in 1969 (internet)

References

Bateman, P. C. (1992). Plutonism in the central part of the Sierra Nevada batholith, California. *United States Geological Society Professional Paper, 1483,* 186 p.

Ehlers, J., & Gibbard, P. I. (2004). Quaternary glaciations – extent and chronology (p. 450). Part II: Elsevier Science.

Guide to the National Parks of the United States. (2001). National Geographic (3rd Edition, 448 p).

Guyton, B. (1998). Glaciers of California, modern glaciers, Ice Age glaciers, the origin of Yosemite Valley, and a glacier tour of the Sierra Nevada (p. 223). Berkley and Los Angeles: University of California press.

http://en.wikipedia.org. (2018). Yosemite National Park, 15 p.

Moore, J. G. & Mack, G. S. (2008). Map showing limits of the Tahoe Glaciation in Sequoia and Kings Canyon National Parks. California: United States Geological Society Scientific Investigations Map 2945, scale 1:125,000. http://pubs.usgs.gov/si,/2945.

Palmer, D. (2004). Yosemite National Park. In: A. G. Harris, E. Tuttle, & S. D. Tuttle (Eds.) Geology of National Parks. Kendal Hunt Publishing Company (6th Edition, pp. 385–402).

Abstract

The majority of the Yellowstone National Park is situated in the state of Wyoming, in the northwest USA. The park is particularly well known for large mammals, including bears and bison, and the opportunity to observe wildlife in association with an active geological system is unique. Yellowstone is named after the yellow and orange volcanic rocks associated with a volcanic system that contains the greatest density of geothermal features on Earth. The distinctive coloration of the rhyolitic ashes and pyroclastic rocks is the result of alteration by hydrothermal fluids and steam vents. Persistent supplies of geothermal water heated at depth and venting on surface, drive hot springs, mud pools, and fumaroles. Geysers develop where the heated groundwater is trapped in narrow underground plumbing systems. The regularity of some geysers, including the Old Faithful Geyser which erupts on average every 92 min, is remarkable. The Yellowstone Volcano reveals evidence of multiple and highly catastrophic, Plinian-style eruptions. Three major caldera events are recognized. The oldest occurred at approximately 2.1 Ma and the youngest has been dated at 0.6 Ma. The calderas are part of a 2,000 m-high volcanic plateau that forms a natural partition in the Rocky Mountains. The volcanic activity at Yellowstone is related to a continental hot spot driven by a deep-seated mantle plume. The hot spot triggered eruption of the Columbia River basalts (located west of Yellowstone, in Oregon and Idaho) during the Neogene. The westward drift of the North American Plate has caused the surface effect of the hot spot to migrate eastward and it is currently situated under Yellowstone. The caldera events at Yellowstone triggered extrusion of huge volumes of rhyolitic ash and destructive pyroclastic flows. The highly siliceous rhyolite magmas were probably derived from partial melting of continental crust. The rhyolitic magmas contain far higher concentrations of gases and volatiles in comparison to the Columbia River basalts. Yellowstone is gazetted as an extremely hazardous volcano, despite the most recent outpourings of rhyolitic ashes and pyroclastics having ceased in the Late Pleistocene (at approximately 60,000 BP). Most of the geothermal features are concentrated in the nested complex of calderas and are driven by heat associated with a shallow magma chamber. The magma chamber is part of the root system of an active volcano. Additional features of interest at Yellowstone are deposits of obsidian (volcanic glass), quarried by indigenous peoples for manufacturing stone tools, the travertine deposits at Mammoth Springs, and the waterfalls of the Yellowstone Canyon.

Keywords

Caldera • Geothermal activity • Geyser • Hot spot • Rhyolite • Supervolcano

4.1 Introduction

Yellowstone is the oldest and possibly most famous national park in the world. The name is derived from the distinctive yellow and orange volcanic rocks which outcrop prominently in the walls of the Yellowstone Canyon (Fig. 4.1). Yellowstone is one of the largest national parks in the USA (area of 8,982 km^2) and includes a diversity of landforms. Yellowstone Lake is one of the largest freshwater lakes in the country and the Yellowstone River includes two waterfalls in the deeply-incised canyon. The occurrence of abundant large fauna, including bison and bears, is well known, but the primary reason for establishing the park is the presence of the greatest concentration of geothermal features on Earth. The geothermal features are mostly found in a caldera. They include geysers, many of which erupt with great regularity, as well as hot springs, mud pools, and

Photographs not otherwise referenced are by the author.

Fig. 4.1 The multi-coloured rocks in the walls of the Yellowstone Canyon illustrate the derivation of the name of the Yellowstone National Park. The brightest patches of colour are due to hot springs and fumaroles that have hydrothermally altered the thick sequences of rhyolitic volcanic ashes and pyroclastics erupted from the Yellowstone Volcano during the most recent of the caldera events. Thin basaltic flows (a subordinate component of the volcanism) that display prominent columnar jointing are intercalated with the rhyolitic ashes and pyroclastics. The hydrothermal features, including the famous geysers, are driven by heat associated with a shallow magma chamber located beneath the Yellowstone Caldera

fumaroles. The average elevation of Yellowstone is 2,000 m, resulting in harsh winters and the park is closed until late spring due to snow fall.

The majority of the Yellowstone National Park is situated in the state of Wyoming, northwest USA, although parts of the park overlap into Idaho and Montana (Fig. 4.2). There is no evidence of a volcanic cone and the dominant physiographic feature is the Yellowstone Plateau. The Yellowstone Plateau is a natural partition in the Rocky Mountains that straddles the continental divide. The plateau is dominated by volcanic deposits associated with the Yellowstone Volcano. The geological record demonstrates the Yellowstone Volcano has experienced periodic catastrophic eruptions, including formation of calderas. The youngest of the calderas formed in the Late Pleistocene. This phase of activity included extrusion of huge volumes of volcanic ashes and pyroclastic flows. Yellowstone is gazetted as an extremely hazardous (active) volcano despite the absence of historical eruptions. The geothermal activity is driven by heat associated with a shallow magma chamber.

4.2 History and Physiography

The geology of the Yellowstone region was first documented in 1859. A government-sponsored expedition that visited the area in 1871 included several geologists (Langford 1905; Harris et al. 2004). An extensive report with paintings and photographs was submitted to the government in Washington extolling the importance of protecting the area. In 1872, Yellowstone was proclaimed as a *"public park or pleasuring ground for the benefit and enjoyment of the people"*.

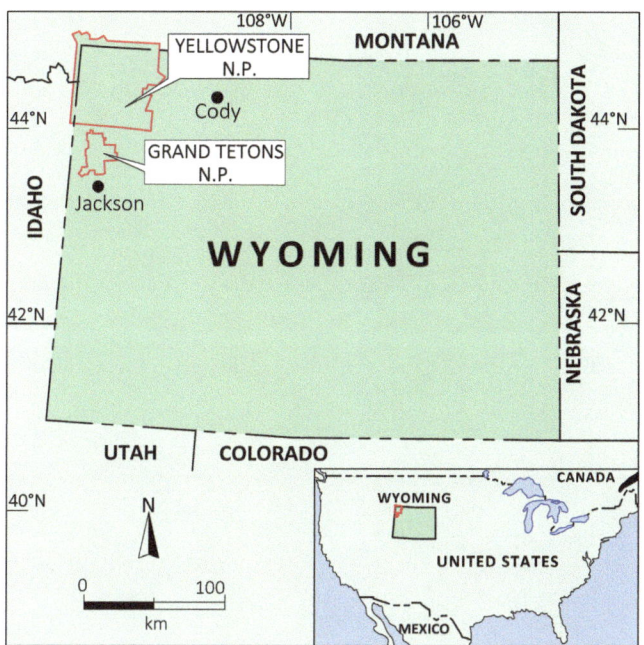

Fig. 4.2 Map showing location of the Yellowstone National Park, in the states of Wyoming, Idaho and Montana, northwest USA

Settlement and mining activities were expressively outlawed. The creation of Yellowstone during a time when there was little control on farming, forestry, and mining established a precedent, subsequently followed by other parts of the USA and other countries around the world.

The Yellowstone Plateau includes several mountain ranges, many of which are incorporated into the national park (Fig. 4.3). The highest peak in the caldera is Mount Washburn (3,111 m) which reveals some of the broader features of the park. The Red Mountains occur in the southern part of the park. The Absaroka Range includes several peaks over 3,000 m and overlaps the eastern boundary of the park. The Gallatin Range is part of the Rocky Mountain cordillera in the northwest of the park.

Several major rivers are sourced from the Yellowstone National Park. The Yellowstone River and its major tributary the Lamar flows northeast (they link with the Mississippi-Missouri system). The northwest and southern parts of park are drained by the northwest-flowing Madison River and headwaters of the westward-flowing Snake River, respectively.

There are multiple entrances to the national park. The most popular route is via the East Entrance, which links the park with the Grand Teton National Park and regional town of Jackson. The road linking the Northeast Entrance to Stillwater in Montana crosses the rugged Beartooth Mountains. The West Entrance on the western side of the continental divide provides access to the Snake River Plateau in Idaho.

4.3 Fauna

The Yellowstone National Park contains large herbivores such as bison (*B. Bison*), elk (*Cervus Canadensis),* moose (*Alces alces)*, and pronghorn antelope (*Antilocapra americanacan*), together with grizzly or brown bear (*Ursus arctos horribilis*), black bear (*Ursus americanus),* grey wolf (*Canis lupus)*, and coyote (*Canis latrans*). The opportunity to observe wildlife in association with an active geological system is unique (Fig. 4.4). Despite the possibility of relatively close encounters with bears and bison, visitors are encouraged by park authorities to view geosites by hiking unaccompanied on the trails.

4.4 Rocky Mountains

The Rocky Mountains includes several ranges in the vicinity of the Yellowstone National Park. The Beartooth and Madison Ranges occur north of the park and the iconic peaks of the Tetons are located to the south. The Rocky Mountain cordillera developed during the Laramide Orogeny, a long drawn out series of continental collisions (80–35 Ma), although the age limits are poorly defined. Tectonism may have persisted as recently as 10 Ma. The mountains in the cordilleras to the north of the national park are dominated by Palaeoproterozoic (2,000–1,600 Ma) granites and gneisses (Harris et al. 2004). The aptly named Granite Peak (3,901 m) in the Beartooth Range is the highest mountain in Montana. Palaeozoic strata with relatively high levels of metamorphism occur on the flanks of the ranges. The road through the Beartooth Mountains follows a deeply incised, U-shaped valley which formed in the Late Pleistocene Ice Ages (Fig. 4.5).

4.4.1 Grand Teton National Park

The Grand Teton National Park is located to the south of Yellowstone (Fig. 4.3). Trails using the shuttle boat at Jenny Lake provide views of the triangular-shaped, snow-capped peaks, including Grand Teton (4,197 m) (Fig. 4.6). The Tetons are dominated by Palaeozoic strata, principally marine sediments. Mesozoic strata which originally overlaid the Palaeozoic sediments in the Tetons have been eroded off most peaks. Unlike ranges north of Yellowstone, the Tetons reveal relatively modest levels of metamorphism. The Tetons are one of the youngest ranges in the Rocky Mountains cordillera and are related to block faulting, rather than orogenic collision. The release of regional stress following the compressional regime of the Laramide Orogeny was displaced in the Tetons by extensional tectonics.

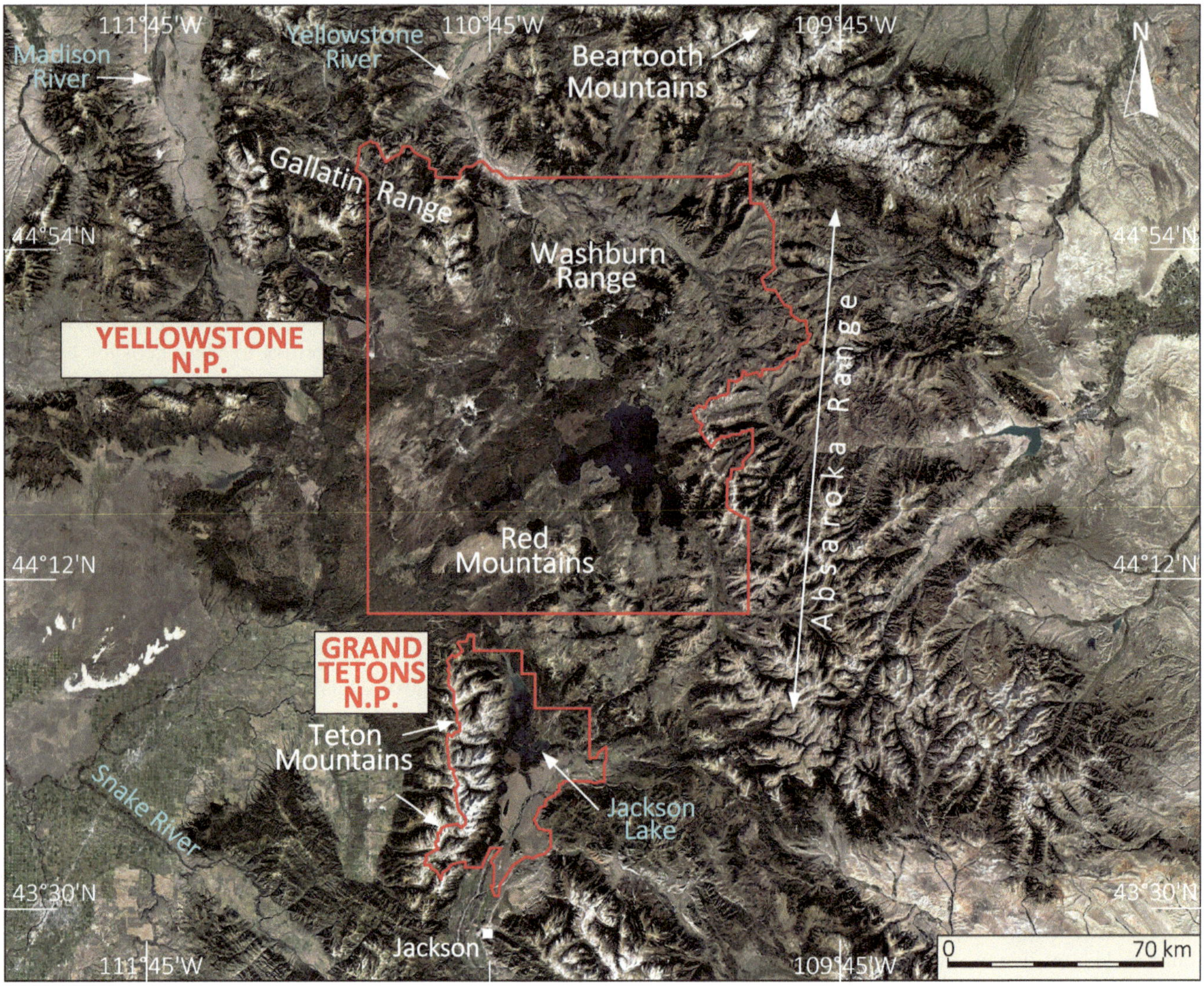

Fig. 4.3 Image showing the location of the Yellowstone National Park together with some of the mountain ranges, including the Tetons. *Source* Google Earth

Fig. 4.4 Views of wildlife, including bison, in association with active geothermal features is a unique feature of the Yellowstone National Park

Fig. 4.5 Deeply incised U-shaped valleys in the Beartooth Mountains formed during the Late Pleistocene Ice Ages

Fig. 4.6 View from Jenny Lake of the triangular-shaped and snow-capped peaks in the Grand Teton National Park

Fig. 4.7 Information board located close to the Teton Fault, looking west towards the uplifted Teton Range

Uplift of the Tetons is estimated to have occurred during the previous 9 Ma (Love et al. 1995; Harris et al. 2004). The Tetons Block is located on the western side of the N-S aligned Teton Fault (exposed on surface for 60 km), with the down-faulted Jackson Hole Block located on the eastern side (Fig. 4.7). The latter is a topographic low which includes the regional town of Jackson; the down-faulted block offers refuge from the harsh winters of the enclosing mountains. The Jackson Hole Block is dominated by Mesozoic-Cenozoic strata; the Palaeozoic strata associated with the Tetons are buried at depths >10 km. The Teton Fault is an active system and minor seismic events are experienced in the region. The Teton Block continues to rise and the Jackson Hole Block continues to sink.

4.5   Older Volcanism

The Older Volcanism of the region has a bearing on our understanding of the volcanism associated with Yellowstone. Eocene volcanics are exposed in the Absaroka Mountains. They are dominated by andesitic lavas and pyroclastics which were erupted during the early stages of the Laramide Orogeny. The Absaroka Volcanics have been severely eroded and occur in heavily dissected terrain in the

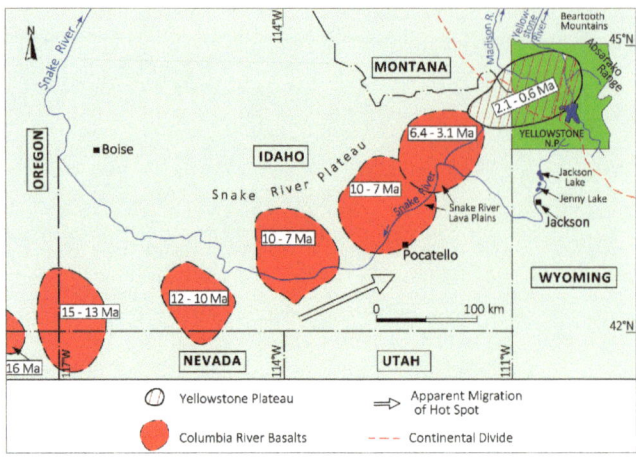

Fig. 4.8 The volcanism of the Columbia River basalts and the Yellowstone Volcano is related to a single deep-seated hot spot which appears to have "migrated" eastward due to the westward drift of the North Atlantic Plate

eastern part of the Yellowstone National Park. The Columbia River basalts, which formed after the peak of the Laramide Orogeny, cover large areas west of the Yellowstone Plateau. The oldest centres (ages of 16–13 Ma) overlap the Oregon/Idaho boundary; younger centres (ages of 12-3.1 Ma) are associated with the Snake River Plateau, Idaho (Fig. 4.8). The basalts are associated with a hot spot located deep in the lithosphere (e.g., Pierce and Morgan 2009). Volcanism reached northwestern Wyoming at approximately 2.1 Ma. The apparent eastward migration of the hot spot is ascribed to westward drift of the North American Plate (estimated at several cm/year). The Columbia River basalts are associated with fissures, rather than vents, whereby multiple eruptions of relatively fluid basaltic lava "flooded" the pediment (Reidel 2003). The basalt flows have smoothed out the older topography to form stepped plateaus, or cuesta landforms. The stacked lava flows attain thicknesses of more than 1,800 m (Reidel 1998). More than 300 individual lava flows have been recognized, representing a combined volume of 500–600 km^3 of basaltic magma (Bryan et al. 2010).

4.6 Yellowstone Volcano

The Yellowstone Volcano is a giant system and is popularly described as a supervolcano. The volcanism and thermal activity are driven by a continental hot spot, possibly the same feature associated with the Columbia River basalts. Volcanism at Yellowstone is characterized by long periods of quiescence broken by catastrophic, Plinian-style eruptions (Box 8.1). Plinian eruptions typically include major caldera events that generate sustained eruptions with high columns of ash. The collapsing ash columns form destructive ash-flows and/or extensive pyroclastic deposits. The Yellowstone Plateau was built up by repeated ash-flows and pyroclastic deposits. The volcanic deposits cover thousands of square kilometres. The volume of magma erupted at Yellowstone may have exceeded 6,000 km^3.

Three large caldera events have been identified at Yellowstone (Keefer 1975; Smith and Christiansen 1980; Harris et al. 2004). The oldest of the caldera events occurred at 2.1 Ma. The intermediate caldera event has an approximate age of 1.3 Ma. The youngest, or Yellowstone Caldera, has an age of 0.63 Ma. Each caldera is offset eastward, an observation consistent with movement of the North American Plate. The nested complex of calderas and associated volcanic deposits are the principal feature of the Yellowstone Plateau.

4.7 Late Pleistocene Ice Ages

The Rocky Mountains and Yellowstone Plateau were severely eroded during the Late Pleistocene Ice Ages (Box 3.1). In the Beartooth Mountains, the ice carved out U-shaped valleys more than 1,200 m deep (Fig. 4.5). The Tetons were eroded into the characteristic triangular-shaped peaks. Ice sheets (with thicknesses > 1,000 m) and glaciers transported rock and debris into valleys, such as Jackson Hole. Glacial moraines comprised of poorly sorted deposits of boulders, cobbles, and gravels may attain thicknesses of 6,000 m. The glacial deposits unconformably cover the

Mesozoic-Cenozoic strata in many valleys. Melting of glaciers and ice dams can release catastrophic floods and the Yellowstone Canyon, for example, may have been carved instantaneously, i.e. in a matter of days, or even hours (Bjornstad 2006). The high peaks together with the Yellowstone Plateau experienced additional ice advances and retreats during the Holocene, including during the Little Ice Age. The small glaciers currently observed in the Tetons are entirely Holocene. The occurrence of large glacial erratics at Yellowstone, as well as petrified trees, some of which were preserved by lahars, is of interest.

4.8 Yellowstone Caldera

The main features of the Yellowstone Caldera are shown on a simplified geological map (Fig. 4.9). The enormous thickness of volcanic rocks has meant basement rocks are only exposed in localities outside of the complex of nested calderas. The basement is comprised of Mesozoic sedimentary rocks together with the Absaroka Volcanics. Most of the volcanic deposits associated with the two older caldera events have not been mapped in detail (they mostly occur outside of the park) as they have been obscured by the younger activity. The volcanic deposits shown on the map, i.e. the Yellowstone Tuff, the Plateau Rhyolite, and the basalt flows are each associated with the Yellowstone Caldera. The latter forms a prominent scarp in some localities (Fig. 4.10). Most of the geothermal features are restricted to the Yellowstone Caldera where they cluster in well-defined geyser basins. The geothermal activity is driven by heat from a relatively shallow magma chamber (Fig. 4.11). The presence of the magma chamber has been established by electrical conductivity measurements and seismic surveys (Harris et al. 2004).

4.8.1 Yellowstone Tuff

Most calderas form due to volcanic collapse and near-instantaneous emptying of the magma chamber. This creates very fine-grained volcanic ash or tuff deposits. The Yellowstone Tuff was erupted explosively from ring fractures during volcanic collapse. The age is determined at 0.63 Ma. Most exposures occur in a circular pattern outside the perimeter of the scarp associated with the Yellowstone Caldera (Fig. 4.9). The Yellowstone Tuff is a generalized description of a sequence of rhyolitic ashes and ash-flows or pyroclastics. The pyroclastics included hot pumice and rock debris which cooled as extensive horizontal sheets. In some layers the particles were welded into resistant ignimbrite sheets that cover thousands of square kilometres. The palaeo-canyons and valleys which had eroded into the older calderas were rapidly covered by the fast-travelling pyroclastic flows. The pyroclastic flows also partially covered the Absaroka Volcanics east of the caldera. The Yellowstone Tuff obscures most of the palaeo-surfaces at Yellowstone and only localized areas of high relief remain intact.

4.8.2 Plateau Rhyolite

The Plateau Rhyolite is associated with a period of relatively quiescent activity, which followed the catastrophic caldera event at 0.63 Ma, and persisted until approximately 60,000 BP. The rhyolite fills most of the Yellowstone Caldera (Fig. 4.9). The eruptive style was quite different to the explosive activity associated with the Yellowstone Tuff. Magma forced upward onto surface, mostly within the caldera, was extruded as relatively slow-moving flows and lava domes. A prominent lava dome can be observed near the Lower Geyser Basin. The rhyolite magma was probably fed from the current magma chamber. The rhyolite can be identified as the primary colour is light grey or pink. Weathered surfaces include a very light-coloured crust. The rhyolite is altered by hydrothermal fluids to yellow and orange colours (Fig. 4.1). The multiple flows associated with the Plateau Rhyolite may form steep cliff-faces that are readily eroded and subjected to landslides due to seismic activity.

4.8.3 Obsidian

Most of the lava flows and domes associated with the Plateau Rhyolite were confined to the caldera, but in some areas the rhyolitic magma exploited external fissures. This promoted formation of obsidian, or volcanic glass; the occurrence at Obsidian Cliff is particularly well known (Fig. 4.9). The obsidian occurs in association with banded rhyolite, or

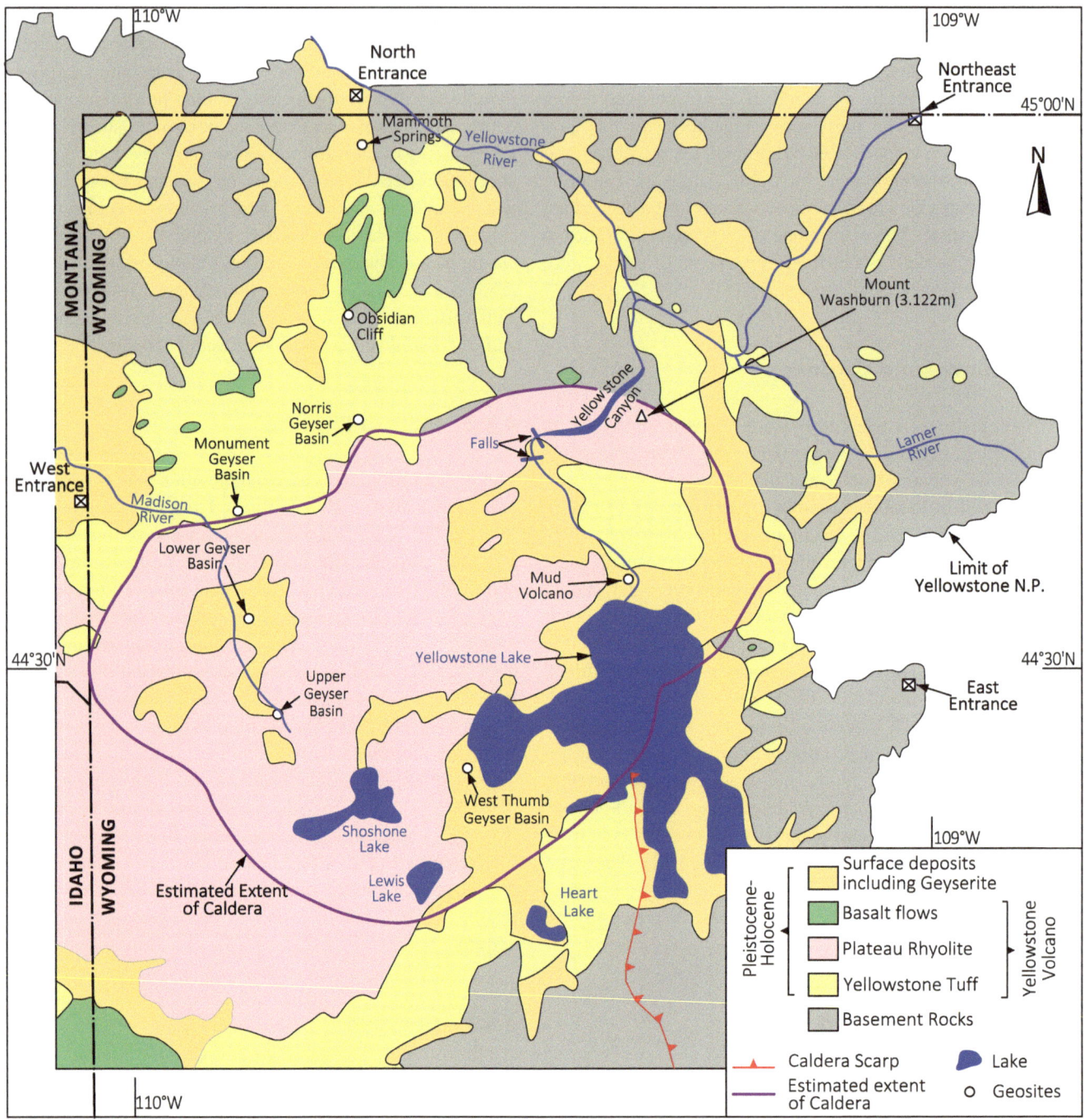

Fig. 4.9 Geological map of the Yellowstone National Park showing extent of the youngest caldera, a scarp related to an older caldera, and the larger geyser basins. *Source* Simplified after Keefer (1975)

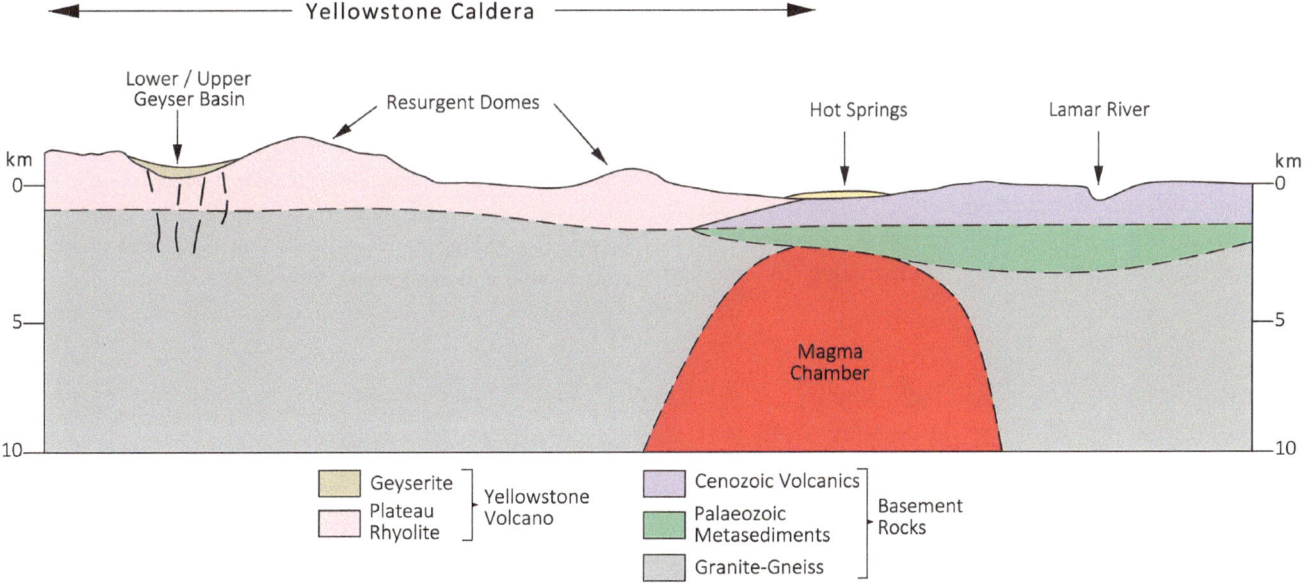

Fig. 4.10 View of the desolate landscapes typical of geyser basins at Yellowstone with the rim of the youngest caldera visible in the background

Fig. 4.11 Simplified cross-section of the Yellowstone Volcano illustrating a schematic magma chamber at depth. The position of the magma chamber relative to the caldera is offset to the east due to the apparent migration of the hot spot

Fig. 4.12 Obsidian (black volcanic glass) occurs in pink-coloured banded rhyolite at Obsidian Cliff, Yellowstone

Fig. 4.13 Columnar jointing typical of basaltic lava flows at Yellowstone

as discrete narrow veins (Fig. 4.12). The jet black appearance, glassy texture, and conchoidal fracture pattern is characteristic. Small pieces of obsidian readily shatter into knife-sharp shards. Erosion is constantly exposing new deposits and Obsidian Cliff was exploited by indigenous peoples for stone tools many thousands of years before the European explorers arrived. Obsidian forms when the mobility of ions within highly siliceous (rhyolitic) melts is prevented from achieving an ordered crystalline pattern; this is usually the result of very rapid cooling (Bakken 1977). The banded texture results from the lower viscosity of the obsidian flow in comparison to the rhyolite host. These processes are typically part of the final eruptive phases, i.e., after the majority of the gas and pumice has been vented. The temperature of formation is approximately 900–700 °C.

4.8.4 Basalt Flows

The Plateau Rhyolite is intercalated with flows of basaltic magma. Basalt is notably abundant in the area between Obsidian Cliff and Mammoth Springs. The basalt is darker then the rhyolite and typically displays columnar joints (Fig. 4.13). The cyclical eruption of alternating rhyolitic and basaltic magmas (bimodal volcanism) is typical of strato-volcanoes which periodically undergo catastrophic activity. At Yellowstone, however, the basalt constitutes < 5% by volume.

4.9 Geothermal Activity

Estimates of the number of individual geothermal features in the Yellowstone National Park range from 2,500 to 10,000 and is dependent on how many smaller features are included (Keefer 1975). The largest and most frequently visited sites occur in the Lower and Upper Geyser Basins and the Norris Geyser Basin (Fig. 4.14). Geyser basins are typically devoid of vegetation as trees are killed by the heat and eruptive activity. The geothermal activity is driven by the magma chamber at depth, as described above, and temperatures at Yellowstone have not cooled in the one hundred years that records have been kept. The high heat flux is estimated to have lasted since 40,000 BP. The amount of heat given off by the Upper Geyser Basin, for example, is estimated at 800 times that associated with a non-thermal area of the Crust.

Fig. 4.14 The Norris Geyser Basin contains numerous geothermal features including geysers, fumaroles and hot springs

4.9.1 Circulation of Groundwater

Groundwater heated by geothermal energy originates from rain and snow that circulates through fractures and faults to considerable depths (Fig. 4.15). The groundwater underlying geyser basins is heated to temperatures of over 250 °C (White 1967). Circulation of the less dense hot water, which rises back toward the surface, pushed upward by the colder and denser near-surface water, forms large convection cells. This ensures a continuous supply of hot water in thermally-active areas. The effect of pressure on the boiling temperature of water is also significant. The deepest water is subjected to the greatest pressures. Boiling is relatively quiescent if pressure is released gradually, but where pressure is released suddenly, or conduits become narrow or blocked, the water flashes explosively into steam. Differences in water flow and temperature produce a range of geothermal features. The plumbing systems of individual basins are probably interconnected at depth. The activity and frequency of geothermal features is affected by changes in the movement of groundwater, which in turn can be influenced by seismic activity. Geyser basins may reveal continuous displays of intense thermal activity, including plumes of "steam". Steam is condensed fog or water droplets, thus activity appears more intense on damp, humid days.

4.9.2 Hot Springs

Brightly coloured pools associated with hot springs at Yellowstone are as popular with visitors as the geysers. Hot springs occur where the geothermal system is connected to open feeder conduits and there is very little pressure build up. Most geothermal features that discharge water and steam consistently can be categorized as hot springs. Individual features, however, vary considerably, depending upon pressure, water temperature, rate of flow, heat supply, and size of underground passages. Some hot springs boil violently and emit dense clouds of vapour; others are relatively quiescent. The colour of the hot springs and pools is related to temperature which in turn controls the development of

Fig. 4.15 Section showing how geothermal features work The geyserite (grey) is underlain by rhyolite depicted schematically as getting hotter with depth (pink, orange, and red). *Source* based entirely on White (1967)

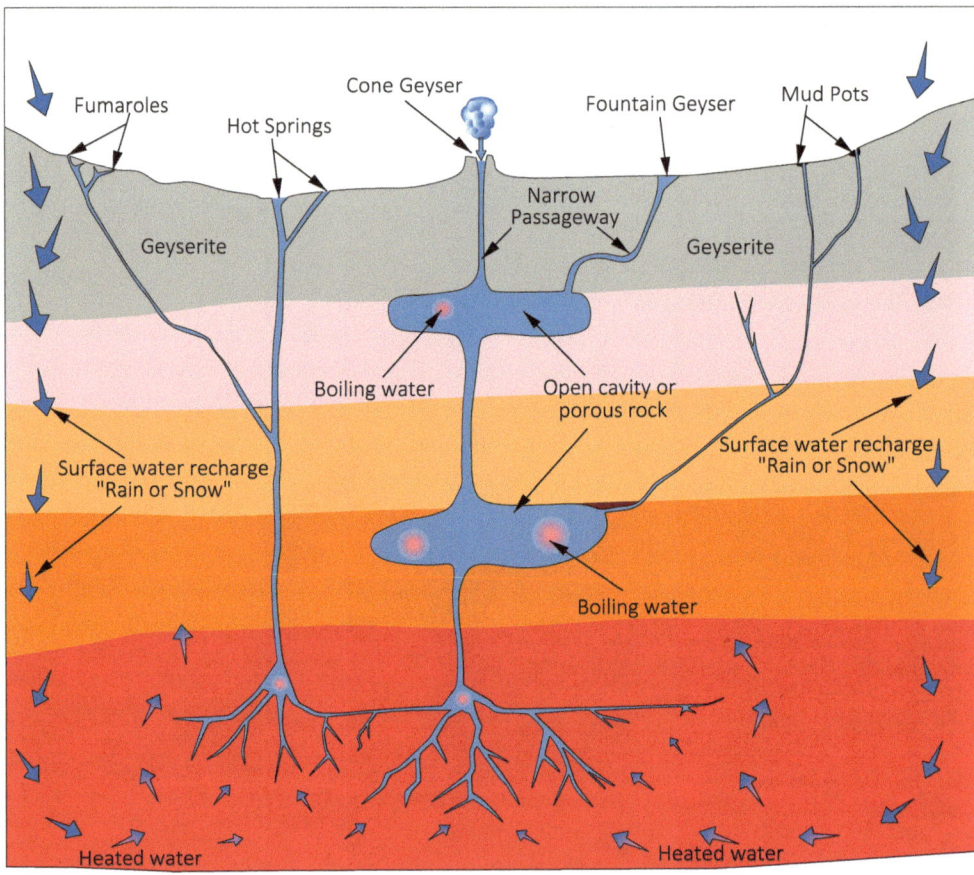

thermophiles (bacteria and algae). The very hot blue-coloured (or clear) springs are associated with cyanobacteria (Fig. 4.16). The temperature of the blue pools can be above boiling point due to the high elevation. The yellow (70–60 °C) and green (60–50 °C) pools contain different types of algae. The lower temperature brown springs (50–27 °C) host mostly mosses, crustaceans, and insects.

4.9.3   Mudpots and Fumaroles

Mudpots develop where the supply of hot water is restricted and water that is available is mixed with clay. This produces a black, grey or even cream coloured mud. Mudpots comprised of thick viscous mud may build up circular cones known as mud volcanoes. In early days of the national park, the Mud Volcano north of Yellowstone Lake was a prime attraction as it was continuously boiling and periodically shot clumps of mud some 17 m into the air. However, in the 20thC the plumbing system was blocked and the system is no

longer active. The Sulphur Cauldron at Yellowstone is an unusual example of a mudpot dominated by sulphuric acid (pH < 2) due to upwelling of sulphur gases. The pool contains bacteria known as thermoacidophiles which thrive in acidic solutions. They have a symbiotic relationship with the pools as the bacteria assist with converting the hydrogen sulphide gas into sulphuric acid. Fumaroles differ from hot springs in that they discharge only steam and other gases such as carbon dioxide and hydrogen sulphide. They are also called steam vents and are typically located on hill slopes above the height of the ground water.

4.9.4   Geysers

If underground channels are too narrow or the pressure is too great, periodic eruptions of water and steam are ejected as geysers. Four stages of geyser activity are identified (Box 4.1). The expansion of water into steam at depth, which may induce a volume change of several hundred times, provides the energy. Geysers require watertight

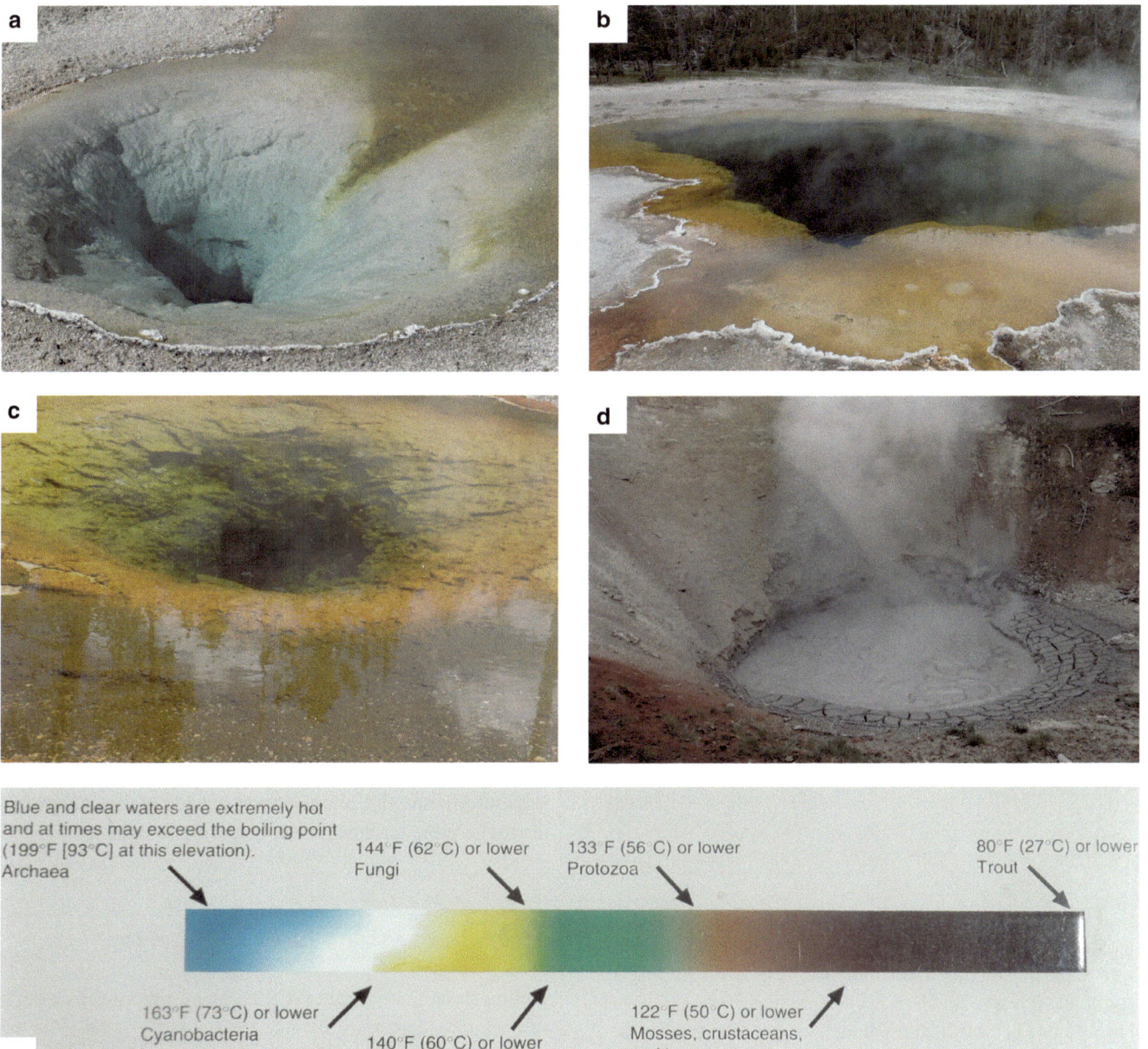

Blue and clear waters are extremely hot and at times may exceed the boiling point (199°F [93°C] at this elevation). Archaea

144°F (62°C) or lower Fungi

133°F (56°C) or lower Protozoa

80°F (27°C) or lower Trout

163°F (73°C) or lower Cyanobacteria

140°F (60°C) or lower Algae

122°F (50°C) or lower Mosses, crustaceans, and insects

Fig. 4.16 **a** Blue pools are relatively hot; **b** Brown pools are cooler; **c** Orange pools are rich in thermophiles; **d** Mud pools occur in areas where groundwater is sparse; and **e** A colour chart (from an information board in the park) illustrating the relationship between colour and temperature and/or the presence of thermophiles

plumbing systems and for this reason are generally associated with thick sequences of rhyolite that can withstand tremendous pressure. Passageways are sealed by silica leached from the rhyolite which is re-deposited as sinter. Water boiling in the plumbing system at depth produces steam bubbles, which expand and rise to the surface. The abrupt reduction of pressure causes water trapped in side chambers to flash into steam, forcing water and steam higher in the system to be ejected under intense pressure. At least 200 geysers occur in the Yellowstone National Park.

Box 4.1: Four Stages of Geysers

Stage 1: Recharge. After an eruption, the partly emptied plumbing system refills with water. Hot water enters through a feeder conduit from below, and cooler water percolates in from side channels near surface. Steam bubbles form in the up-flowing current as a decrease in pressure causes a corresponding decrease in boiling temperature. The bubbles condense in the cooler, near-surface water. Eventually water is heated sufficiently so that bubbles no longer condense.

Stage 2: Preliminary Eruption. As the rising gas bubbles grow in size and number, they concentrate in parts of the plumbing system, i.e. in constrictions. The expanding steam abruptly forces its way upward through the system and causes some of the water to discharge from the vent in preliminary spurts.

Stage 3: Full Eruption. After the preliminary spurt, the reduction in pressure at depth starts a chain reaction. Water in the side chambers and pore spaces begin to flash into steam, and the geyser rapidly surges into full eruption.

Stage 4: Steam. When most of the energy is spent, and the plumbing system is nearly empty, water remaining in pockets and pore spaces makes additional steam. This stage may be relatively long-lived.

After White (1967) and Keefer (1975)

Two broad types of geysers are identified, fountain geysers and cone geysers. Fountain geysers have wide openings and can be mistaken for hot springs. An example of a fountain geyser is Grand Geyser, possibly the worlds' highest predictable feature which erupts every 7–15 h. An eruption typically lasts 9-12 min and may attain a height of 60 m. Cone geysers occur where constricted passageways result in eruptions of narrow jets which can reach considerable heights. The regularity of the Old Faithful Geyser which erupts approximately every 92 min is remarkable (Fig. 4.17a). An eruption typically lasts 2–5 min. The regularity is a function of a stable plumbing system at depth (O'Hara and Esawi 2013). Some geysers can be extremely violent. The Seismic Geyser, which may have developed

from cracks caused by the Hebgen Lake Earthquake, has ejected rock rubble.

Geysers build mounds or terraces of mineral deposits known as geyserite; the pale grey colour is a distinctive feature of the geyser fields (Fig. 4.17b). The principal constituent of geyserite is silica which is leached from the rhyolite lavas and ashes by the hot water and steam. The sinter in the core of the mound associated with Castle Geyser has been dated using the Carbon-14 method and is approximately 1,000 years old (Duncan 2004). Each eruption of Castle Geyser (every 10–12 h) includes a hot water phase (which lasts for approximately 20 min and attains a height of 27 m) followed by a steam phase (which lasts for 30–40 min).

4.9.5 Travertine

The geothermal deposits at Mammoth Springs in the northern part of Yellowstone are comprised of travertine. The travertine consists of thin layers of calcium carbonate and is observed in shallow pools fed by streams of hot water (Fig. 4.18). Each layer or band of travertine has been precipitated over a constrained period of time from the hot water. The multi-coloured layers may reflect the presence of thermophiles (bacteria and algae). The occurrence of travertine, rather than silica-rich geyserite, reflects the lithology of the underlying rock formation: Mammoth Springs occurs outside of the caldera in an area where the basement rock is marine limestone.

4.10 Yellowstone Lake and Canyon

Yellowstone Lake located in the eastern part of the caldera is one of the largest freshwater lakes in the USA. The lake feeds the Yellowstone River which has cut an impressive canyon (known as the Grand Canyon prior to discovery of the Colorado River system) with two major waterfalls (Fig. 4.19). Canyon formation is ascribed to rapid (possibly near-instantaneous) erosion of the flat-lying, partially consolidated volcanic ashes. Catastrophic floods may have been released from the breaking up of ice dams at approximately 8,000 BP. This occurred several thousands of years after the main retreat of the ice sheets.

a

b

Fig. 4.17 a The column of hot water associated with Old Faithful, the most famous of the geysers at Yellowstone, erupts on average every 92 min; **b** Castle Geyser in the Upper Geyser Field has built up a substantial mound of geyserite and includes a relatively long steam phase at the end of each eruptive cycle

Fig. 4.18 Terraces of travertine at the Mammoth Hot Springs reveal constantly changing forms as they are shaped by subtle changes in the springs' plumbing system

Fig. 4.19 The Yellowstone River includes several large waterfalls

The unique landscape at Yellowstone is related, not only to volcanism and geothermal activity, but also earthquakes and erosion. Earthquakes have triggered significant landslides in parts of Yellowstone (e.g., at Hebgen Lake where a magnitude 7 event occurred in 1959).

References

Bakken, B. (1977). Obsidian and its formation. *North West Geology, 6,* 88–92.

Bjornstad, B. (2006). On the trail of the ice age floods: A geological guide to the Mid-Columbia Basin, Sand Point. Idaho: Keokee Books.

Bryan, S. E., Peate, I. U., Peate, D. W., Self, S., Jerram, D. A., Mawby, M. R., et al. (2010). The largest volcanic eruptions on Earth. *Earth-Science Reviews, 102,* 207–229.

Duncan, F. (2004). How does your geyser grow? 3-D laser scanning and preliminary 14C Dating of Castle Geyser, Upper Geyser Basin, Yellowstone National Park, Wyoming. Rocky Mountain (56th) and Cordilleran (100th) Joint Meeting. *Geological Society of America,* 11–11.

Harris, A. G., Tuttle, E., & Tuttle, S. D. (2004). Geology of National Parks. Kendal Hunt Publishing Company, 882 p.

Keefer, W. (1975). The geologic story of Yellowstone. *United States Geological Survey Bulletin, 1347.* 7th printing 1987, Yellowstone Library and Musuem Association.

Langford, N. P. (1905). The discovery of yellowstone park—Diary of the Washburn Expedition to the Yellowstone and Firehole Rivers in the year 1870 (p. 123). St. Paul, Minneapolis: Frank Jay Haynes.

Love, D. D., Reed, J. C., & Pierce, K. L. (1995). Creation of the Teton landscape. Grand Teton Natural History Association (ISBN 978–093189508 1).

O'Hara, K. D., & Esawi, E. K. (2013). Model for the eruption of the Old Faithful geyser, Yellowstone National Park. *GSA Today, 23*(6), 4–9.

Pierce, K. L., & Morgan, L. A. (2009). Is the track of the Yellowstone hotspot driven by a deep mantle plume?—Review of volcanism, faulting, and uplift in light of new data. *Journal of Volcanology and Geothermal Research, 188,* 1–25.

Reidel, S. P. (1998). Emplacement of Columbia River flood basalt. *Journal of Geophysical Research. Solid Earth, 103,* 27393–27410.

Reidel, S. P. (2003). The Columbia River flood basalts and the Yakima fold belt, *in* Swanson, T.W. ed., Western Cordillera and adjacent areas. *Geological Society of America Field Guide, 4,* 87–105.

Smith, R. B., & Christiansen, R. L. (1980). Yellowstone Park as a window on the Earth's interior. *Scientific American, 242,* 104–17.

White, D. E. (1967). Some principles of geyser activity, mainly from Steamboat Springs. *Nevada. American Journal of Science, 265,* 641–684.

National Parks of the Canadian Rocky Mountains

Fold Mountains and Glacial Landforms of the Banff, Jasper and Yoho National Parks

5

Abstract

The landforms and mountain scenery of the Banff, Jasper, and Yoho National Parks in the Canadian Rocky Mountains attract millions of tourists annually. The Banff and Jasper National Parks are located in the province of Alberta, east of the continental divide. The Yoho National Park occurs in British Columbia, on the more densely forested western slopes of the Rocky Mountains. Many visitors to the parks drive the 232 km-long Icefields Parkway, a major road, which exploits giant U-shaped valleys with views of the snow-clad peaks, icefields and glaciers, and Alpine lakes and waterfalls. The building of railways that crossed the continental divide brought the landforms and scenery of the Rocky Mountains to the attention of an international audience. The Yoho National Park includes exposures of marine fossils (discovered by railway surveyors) described as the "Cambrian Explosion of Life". This description emphasizes the rapid diversification of fossils over a period of approximately 40 Ma during the Middle Cambrian. Some of the unique lineages of the fossils are evolutionary experiments that became extinct, but the majority of the fossils are probably predecessors of modern life. The Rocky Mountains are a discontinuous series of fold-and-thrust belts that extend through western Canada and the USA. The cordilleras are in part arcuate and have a width of less than 100 km. An older, western section of the cordillera in Canada is associated with closure of an oceanic basin located off the west coast of the North American craton during the Mid Jurassic. A younger, eastern section of the Rocky Mountains developed from folding and thrusting associated with the Late Jurassic-Early Eocene Laramide Orogeny. The mountain peaks in the Front and Main Ranges are superbly exposed in the Banff and Jasper National Parks. They are comprised of thick sequences of metasedimentary rocks, part of the Neoproterozoic-Mesozoic cratonic platform. Erosion of huge thicknesses of strata during the Late Cenozoic and Late Pleistocene Ice Ages (i.e., after the peak of the Laramide Orogeny) exposed the older core of the orogenic belt. Textbook glacial landforms include deeply incised U-shaped valleys, triangular peaks, cirques, and hanging valleys with moraines and Alpine lakes. Large waterfalls occur on most of the major rivers. The Maligne Canyon and the Athabasca Glacier are highlights of the Jasper National Park. The picture book scenery of Lake Louise and Lake Moraine is the most widely visited section of the Banff National Park. The distinctive turquoise coloration of the Alpine lakes reflects the presence of fine-grained rock flour in the glacial meltwater. The clays and silts reflect the short wavelength blue and green components of the spectrum. The relative proportion of clay and silt in the meltwater causes seasonal variations, and some lakes vary from blue through turquoise to green. Icefields and glaciers in the Canadian Rockies reformed during the Holocene and were notably extensive in the Little Ice Age. The Athabasca Glacier has retreated several kilometres since 1844.

Keywords

Fold mountains • Glaciers • Icefields Parkway • Thrusts • Turquoise lakes • Waterfalls

Photographs not otherwise referenced are by the author.

The original version of this chapter was revised: Belated corrections have been incoporated. The corrections to this chapter are available at https://doi.org/10.1007/978-3-030-54693-9_18

Fig. 5.1 View looking south along the Icefields Parkway in the U-shaped valley of the North Saskatchewan River, Banff National Park. The ice-scoured mountains of the Front Ranges consist of well-bedded, eastward-dipping Palaeozoic strata

5.1 Introduction

The Banff, Jasper, and Yoho National Parks in the Canadian Rockies attract millions of tourists annually. The classic mountain scenery and the relatively easy access to icefields and glaciers are probably unsurpassed. The three national parks each obtained UNESCO heritage status in 1985. The Banff and Jasper National Parks are located in the province of Alberta, east of the continental divide. The Yoho National Park occurs in British Columbia, on the more densely forested western slopes of the Rocky Mountains. Many visitors to the parks drive the 232 km-long Icefields Parkway, a major road, which exploits giant U-shaped valleys with views of the snow-clad peaks, icefields and glaciers, and turquoise-coloured Alpine lakes and waterfalls

(Fig. 5.1). Over a hundred small icefields and glaciers are visible from this road, including the large Columbia Icefield. The building of railways that crossed the continental divide brought the landforms and scenery of the Rocky Mountains to the attention of an international audience. In 1886, the general manager of Canadian Pacific Rail commented, *"If we can't export the scenery, we'll import the tourists"*. Teams of surveyors led by Major Albert Bowman Rogers opened passes through the Rockies, and completion in 1885 of the Canadian Pacific Rail, the first transport link between eastern and western Canada, was a remarkable feat of engineering.

The Canadian Rocky Mountains are the source of major rivers, and most roads and railways follow the deeply incised valleys (Fig. 5.2). The railways originally followed a southern route through the Banff and Yoho National Parks; the Trans-Canada Highway (Route I-9) exploits the same

Fig. 5.2 Map showing location of the three national parks in the Canadian Rockies, the continental divide, major rivers, and access routes

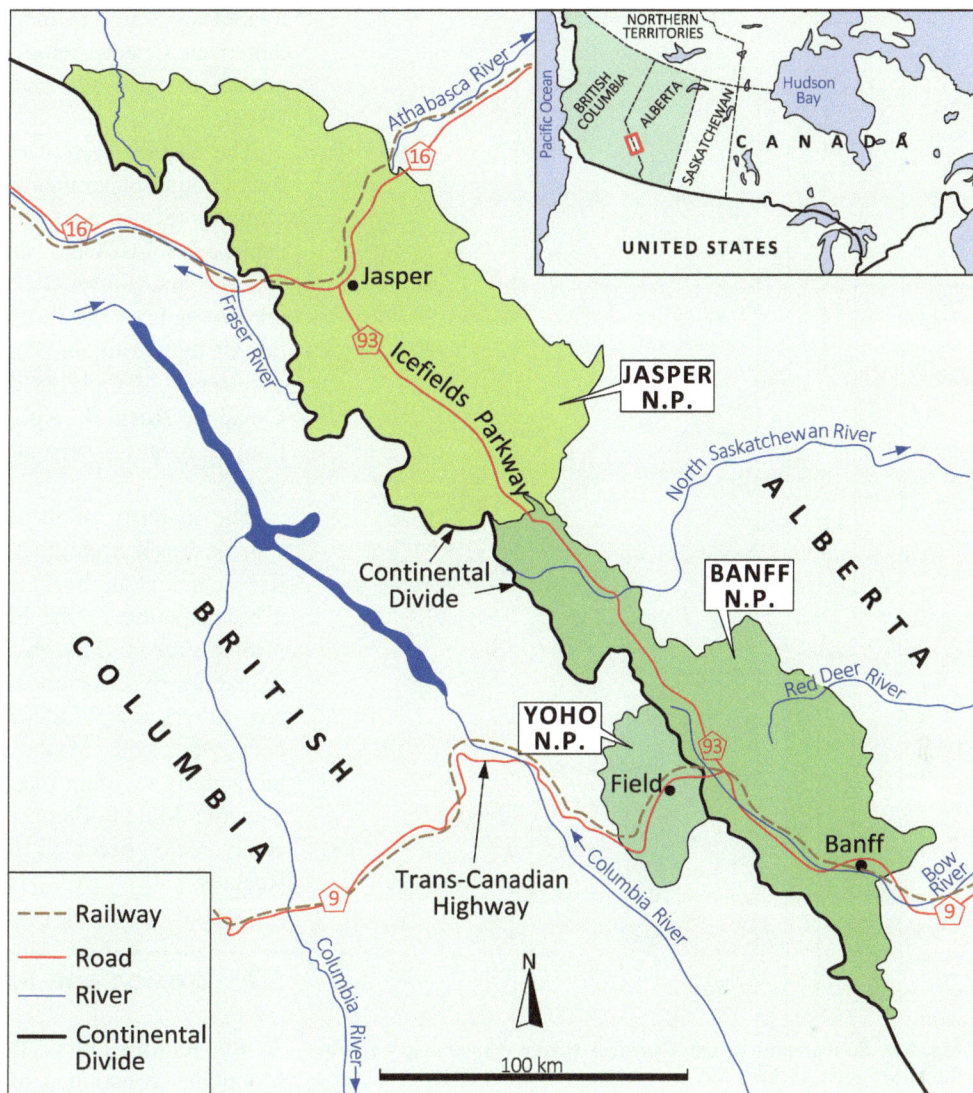

system of valleys. The northern route through the Jasper National Park also includes both rail and road access (Route I-16). The Icefields Parkway (Route I-93) connects the two major access roads.

The Rocky Mountains are a discontinuous series of fold-and-thrust belts that extend through western Canada and the USA. The majority of the tectonism is associated with the long-lived Laramide Orogeny. Fold mountains are typically associated with the collision of tectonic plates. Most strata are crumpled and folded. A significant tectonic element of the Canadian Rockies is the recognition of sub-horizontal thrust faults. Erosion during the Late Cenozoic, including during the Late Pleistocene Ice Ages, reshaped many of the landforms currently observed in the Canadian Rockies.

5.2 Rocky Mountain Fold-and-Thrust Belt

The Rocky Mountains are part of a discontinuous series of cordilleras that stretch from western Canada southward into the USA. The N-S aligned fold-and-thrust belts are in part arcuate and have a width of less than 100 km. The recognition of multiple ranges with distinctive geological features is an important feature. The Omineca Belt, an uplifted and eroded volcanic terrain with granitic plutons and high-grade metamorphics, flanks the Canadian Rockies to the west. The eastern extent of the cordillera abuts against the Western Canada Basin. The Canadian Rockies are associated with thick sequences of mostly metasedimentary

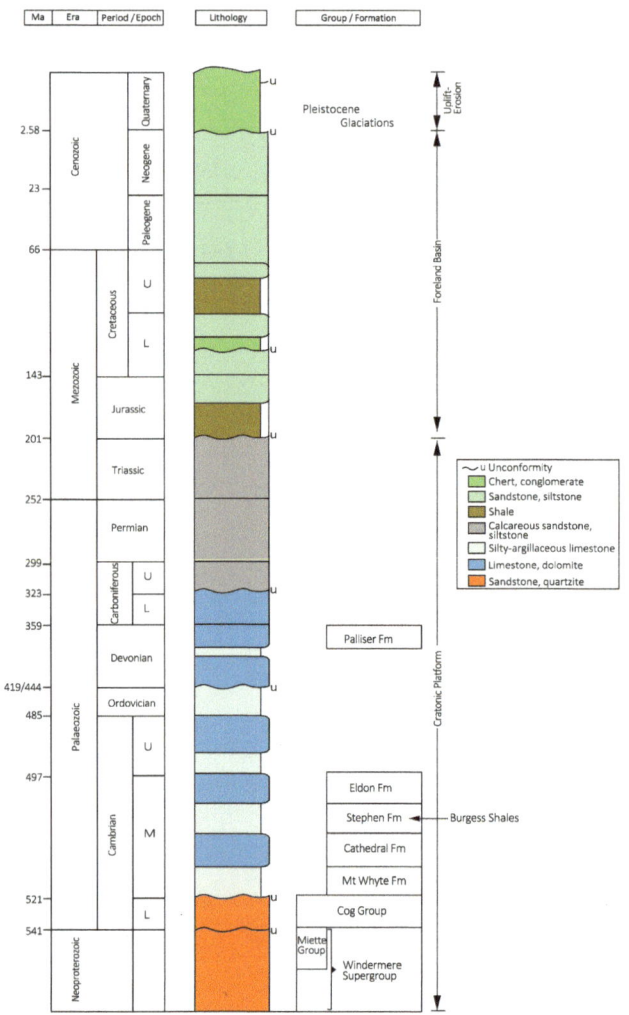

Fig. 5.3 Stratigraphy of the Canadian Rocky Mountains. *Source* Simplified after Leckie (2017)

rocks. In comparison, high-grade metamorphic rocks, including gneiss and granite dominate the cordilleras in the USA.

The western section of the Canadian Rockies developed in the ancient microcontinent of Cordilleria, located 1,500 km south of the current position (Chamberlain and Lambert 1985). During the Middle Jurassic, the microplate drifted northward and collided with the North American Craton, initiating formation of the fold-and-thrust belt (Bally et al., 1966; Engebretson et al. 1985). An alternative

hypothesis suggests the collision resulted from the North American Craton being dragged westward in response to closure of an ocean basin off the west coast (Chen et al. 2019).

The eastern section of the Canadian Rockies is a relatively young phenomenon ascribed entirely to the Laramide Orogeny (80–35 Ma). Sequences of overlapping sheets and detached folds were thrust eastward between the Late Jurassic and Early Eocene (Evenchick et al. 2007). The relative ages of the thrusts confirms the eastward propagation of the cordillera (Pana and Van der Pluijm 2015). The occurrence of multiple north-south trending ranges in the Canadian Rockies, e.g., the Front Range and the Main Ranges, relates to separate thrusts as the orogeny peaked and waned.

The majority of the sedimentary rocks in the Western Canada Basin accumulated prior to the Laramide Orogeny (Evenchick et al. 2007). The stratigraphic succession covers a broad period of the Earth's history, extending from the Neoproterozoic through to the Cretaceous (Fig. 5.3). Thick sequences of sandstone, grit, and shale of the Neoproterozoic Miette and Windermere Groups are the oldest components. Quartzite of the Lower Cambrian Cog Group is particularly resistant to erosion. Shale and limestone dominate the Middle Cambrian through Jurassic strata. Cretaceous rocks occur in the eastern part of the Canadian Rockies.

5.3 Erosion and Ice Ages

At the termination of the Laramide Orogeny, the Rocky Mountains constituted a high altitude plateau with an elevation of approximately 6,000 m. Erosion of huge thicknesses of strata during the Late Cenozoic exposed the cores of the older orogenic belt. Ice sheets and glaciers covered the entire Rocky Mountains during parts of the Late Pleistocene Glaciation (Box 3.1). The relatively northern latitude of the Canadian Rockies resulted in the ice being both extremely thick and long lasting. Glacial landforms include deep U-shaped valleys, cirques, triangular peaks, and hanging valleys with Alpine lakes. Erosion of the bedrock and glacial moraines is continuously reshaping the landforms. The icefields and slope glaciers partially melted and reformed during the Holocene. Ice advances and retreats during the Little Ice Age (1300–1850 AD) were significant.

The Canadian Rocky Mountains are the source of a number of major rivers (Fig. 5.2). The Athabasca River flows northward through the Jasper National Park towards the Arctic Ocean. The North Saskatchewan and Bow Rivers (the latter is the principal headwater of the South Saskatchewan River) flow through the Banff National Park, prior to draining into Hudson Bay. Rivers in the Yoho National Park are part of the Columbia River system, which flows westward into the Pacific Ocean. Most rivers are fast flowing, swollen by meltwater from ice sheets and glaciers in spring and early summer. Rivers include sections of rapids and waterfalls, typical of recently glaciated landscapes, with waterfalls developed at resistant knickpoints in the bedrock. The Icefields Parkway accesses many of the larger waterfalls in the Jasper and Banff National Parks (Fig. 5.4a).

5.4 Geological Setting

A simplified geological map shows the principal stratigraphy of the Main and Front Ranges proximal to the Icefields Parkway (Fig. 5.4b). The Front Ranges are the first of the mountains viewed when approached from the east. In a generalized way the strata become younger towards the east, albeit older sequences are thrust eastward over younger strata (Evenchick et al. 2007; Pana and Van der Pluijm 2015; Leckie 2017). Thrusts within the Main Ranges, e.g., the Pyramid Thrust (163.0 Ma) and the Simpson Pass Trust (161.7 Ma), are related to an early stage of deformation (Middle Jurassic). Thrusts located in the Front Range are considerably younger (Cretaceous-Early Eocene) and encompass a broad range of ages, e.g., the Sulfur Mountain Thrust (75 Ma) and the Bourgeau Thrust (52 Ma). Most fold mountains associated with collision zones form 300–600 km inland from subduction zones. The relatively interior position of the Canadian Rockies in the North American continent is an unusual feature.

5.5 Jasper National Park

The Jasper National Park (area of 11,000 km^2) contains numerous snow-capped ranges with peaks in excess of 3,000 m. The highest peak is Mount Robson (3,954 m). Two roads access many of the geosites described by Yorath

and Gadd (1995) and Leckie (2017). The road linking Jasper and Maligne Lake in the northeast of the park includes views of the Front Range, dominated by Devonian-Permian strata (Fig. 5.4b). Several major thrusts occur here, notably the Pipestone Thrust and the Sulphur Mountain Thrust. The Icefields Parkway parallels the Main Ranges in which the Neoproterozoic and Cambrian-Ordovician strata can be examined (Fig. 5.4b). This section of the Icefields Parkway follows the broad, flat-floored valley of the Athabasca River. The river has exploited a weakness associated with the Simpson Pass Thrust (Fig. 5.5).

5.5.1 Lake Maligne

The Maligne Valley parallels the Pyramid Thrust, a NW-SE trending structure in the Front Ranges (Yorath and Gadd 1995; Leckie 2017) (Fig. 5.4). A viewpoint overlooking Jasper is located close to the confluence of two rivers. The Maligne River occurs in a hanging valley that formed from a secondary glacier perched 90 m above the much larger (primary) Athabasca Glacier. The road to Maligne Lake includes views of moraines (or tills) comprised of sands and gravels deposited during the Late Pleistocene by ice sheets and glaciers (Roed 1964). Moraines reveal hummocky ground with depressions infilled by small lakes. A moraine that developed during the Last Glacial Maximum (at approximately 20,000 BP) dams the northern end of Maligne Lake. The finger shape is characteristic of lakes that develop in the U-shaped valleys created by glaciers. Maligne Lake is unusually deep (98 m), an indication of the immense size of the glacier. The eastern side of the lake is fringed by the saw-tooth Colin Range (Mount Brazeau has an elevation of 3,470 m), consisting of steeply dipping Devonian-Carboniferous limestones. On the western side of the lake is the relatively subdued Maligne Range, flat-lying Neoproterozoic-Cambrian quartzites and shales severely eroded during the Ice Ages.

5.5.2 Maligne Gorge

The second relatively large finger lake in the Maligne Valley is Medicine Lake. Rockslides consisting of large limestone blocks and boulders of the Palliser Formation (Devonian), together with mudstone of the Banff Formation (Carboniferous), block the outflow of Medicine Lake (Yorath and

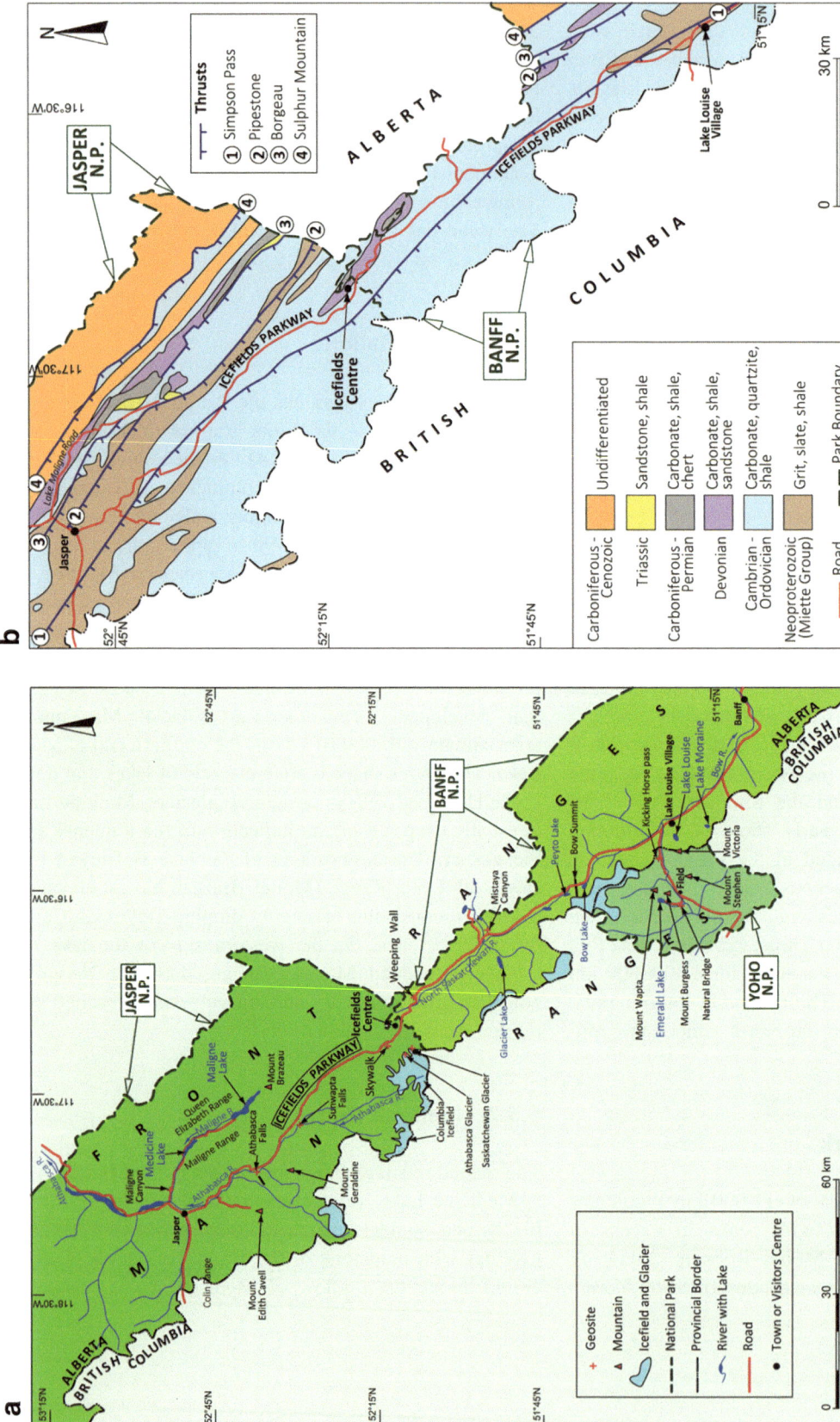

Fig. 5.4 **a** Selected geosites in the Banff and Jasper National Parks accessed by the Icefields Parkway and sites in the Yoho National Park close to the Trans-Canadian Highway. Only icefields in the national parks are shown (slope glaciers are too small to show). The provincial boundary corresponds to the "Great Divide"; **b** Geological map of the area proximal to the Icefields Parkway. *Source* Simplified after Yorath and Gadd (1995); Leckie (2017), and maps of the Geological Survey of Alberta

Fig. 5.5 The northern part of the Icefields Parkway, Jasper National Park, follows the course of the Athabasca River with views of the Main Ranges

Gadd 1995). Glacial undercutting of dip slopes near the base of the Colin Range triggers rockslides. The rockslides have caused the lake to drain into the Maligne cave system prior to re-emerging in the Maligne Gorge. Thick surficial deposits of glacial debris and moraine block access to the caves. The level of Medicine Lake shows considerable variability. In summer, the volume of meltwater causes the lake to rise by as much as 18 m (when the cave system floods), but in winter the reduced ingress exposes mudflats and isolated pools.

Access to the Maligne Gorge is by a hiking trail with six bridges (Fig. 5.6). The gorge is part of an exhumed cave system that formed prior to the Late Pleistocene Ice Ages. The gorge cuts into limestone of the Palliser Formation; the sinuous nature of the gorge relates to joints that developed during the Laramide Orogeny (Leckie 2017). In the upper section of the gorge, between First and Second Bridge, the gorge is narrow and the river drops 50 m over a succession of falls and potholes (Fig. 5.7a). Large chockstones occur wedged between the sidewalls. Between Third and Fourth Bridge, side streams enter a wider section of the gorge fringed by terrace gravels. Springs with small falls occur on

the walls of the gorge between Fourth and Fifth Bridge (Fig. 5.7b). The flow of the river varies seasonally due to changes in springs and filling of subterranean channels.

5.5.3 Mounts Edith Cavell and Geraldine

Views of the Main Ranges from the Icefields Parkway include two of the most well know peaks in the Jasper National Park (Yorath and Gadd 1995). Mount Edith Cavell (3,363 m) consists largely of the resistant Gog Group quartzite (Lower Cambrian) and contains three small glaciers (Fig. 5.8a). The Glacier Trail accesses the Cavell Glacier and includes views of the unusually steep Angel Glacier, perched 300 m above the valley, and the Ghost Glacier. The extent of the three glaciers is variable and the Angel Glacier is particularly unstable during the summer months. In the 1700s (during the Little Ice Age), Cavell Glacier extended to the trailhead. Until the mid-1940s, Angel Glacier merged with the Cavell Glacier. Part of the trail is no longer accessible as in August 2012 a large section of the Angel Glacier collapsed. Giant blocks of ice fell into Cavell Lake triggering shock waves that created a debris avalanche. The

Fig. 5.6 The narrow gorge of the Maligne River carved into flat-lying limestone of the Palliser Formation (Devonian)

a

b

Fig. 5.7 **a** The upper section of the Maligne Canyon reveals evidence of a cave system in the sidewalls; **b** The lower sections of the Maligne River includes springs exiting the cave system

avalanche cut a new channel through the moraines in the lower part of the Glacial Trail. The moraines contain large boulders partially covered by lichen (*Rhizocarpon geograpicum*), which can be used to obtain approximate ages. The crenulated face of Mount Geraldine (2,930 m) reveals cirques that contain small glaciers (Fig. 5.8b). Mount Geraldine is comprised of well-bedded Cambrian-Ordovician strata. The location proximal to the Simpson Pass Thrust may explain the eroded nature of the main face.

5.5.4   Waterfalls on the Athabasca River

The Athabasca River is one of Canada's most well known rivers and includes several large waterfalls in the upper reaches. The volume of water forced through an 18 m-wide gorge at the Athabasca Falls creates an impressive roar. The

gorge is unusual as rather than consisting of bedrock, it is associated with large blocks of Cog Formation quartzite deposited by a glacier (Leckie 2017). A walkway below the falls accesses an abandoned channel where the quartzite reveals prominent cross-bedding, indicative of the marine origin of the primary sandstone (Fig. 5.9a). The Sunwapta Falls occurs in a hanging valley eroded by tributary glaciers, perched above the primary Athabasca Glacier. The river has eroded a deep gorge into limestone of the Cathedral Formation (Middle Cambrian) (Leckie 2017). The gorge exhibits a zigzag pattern between the upper and lower falls due to orthogonal jointing (Fig. 5.9b). Large potholes located on the walls of gorges are relics of higher water stands. This is characteristic of rapidly eroded hanging valleys.

5.5.5   Columbia Icefield

The Colombia Icefield (area of 325 km^2), the largest icefield in the Jasper and Banff National Parks, transects the continental divide (Fig. 5.10). The icefield has an average thickness of 100–365 m and contains eleven peaks of over 3,000 m, including Mount Columbia (3,747 m) (Yorath and Gadd 1995; Gadd 2011).The icefield formed during the Illinois Glaciation (238,000–126,000 BP), with additional advances occurring during the Main Ice Age (110,000–12,000 BP) and Little Ice Age (1300–1850 AD). Prominent slope glaciers project finger-like from the core of the icefield down the slopes of the Main Ranges. The slope glaciers flow in valleys carved by larger palaeo-glaciers. Slope glaciers are an indication that the icefield is unstable; this typically occurs when the thickness exceeds 30 m. The slope glaciers are moving at approximately 15-125 m a year (Leckie 2017). Recession of the slope glaciers since the Little Ice Age has left extensive terminal moraines.

5.5.6   Athabasca Glacier

Tours to the Athabasca Glacier, a prominent feature of the Columbia Icefield, commence from the Icefields Centre. Prior to commencing the tour, a visit to the "Glacier Skywalk", a glass-floored structure that projects at a height of 280 m over the Athabasca Valley, is recommended. The Athabasca Glacier has a length of 6 km and despite forward motion of 10–25 m/annually—the icefield and glacier are fed by annual snowfall of as much as 7 m—is receding (Gadd 2011). During the final stages of the Little Ice Age, the glacier had a length of 8 km. Recession since 1844 has

a

b

Fig. 5.8 a Mount Edith Cavell is comprised almost entirely of resistant quartzite of the Gog Group (Lower Cambrian); **b** The crenulated face of Mount Geraldine includes cirques that contain small glaciers (foreground left and centre)

Fig. 5.9   a The gorge below the Athabasca Falls occurs in well-bedded quartzite of the Gog Group (Lower Cambrian); **b** Large potholes on the walls of the Sunwapta Gorge are relicts of higher water stands. The gorge cuts into limestone of the Cathedral Formation (Middle Cambrian)

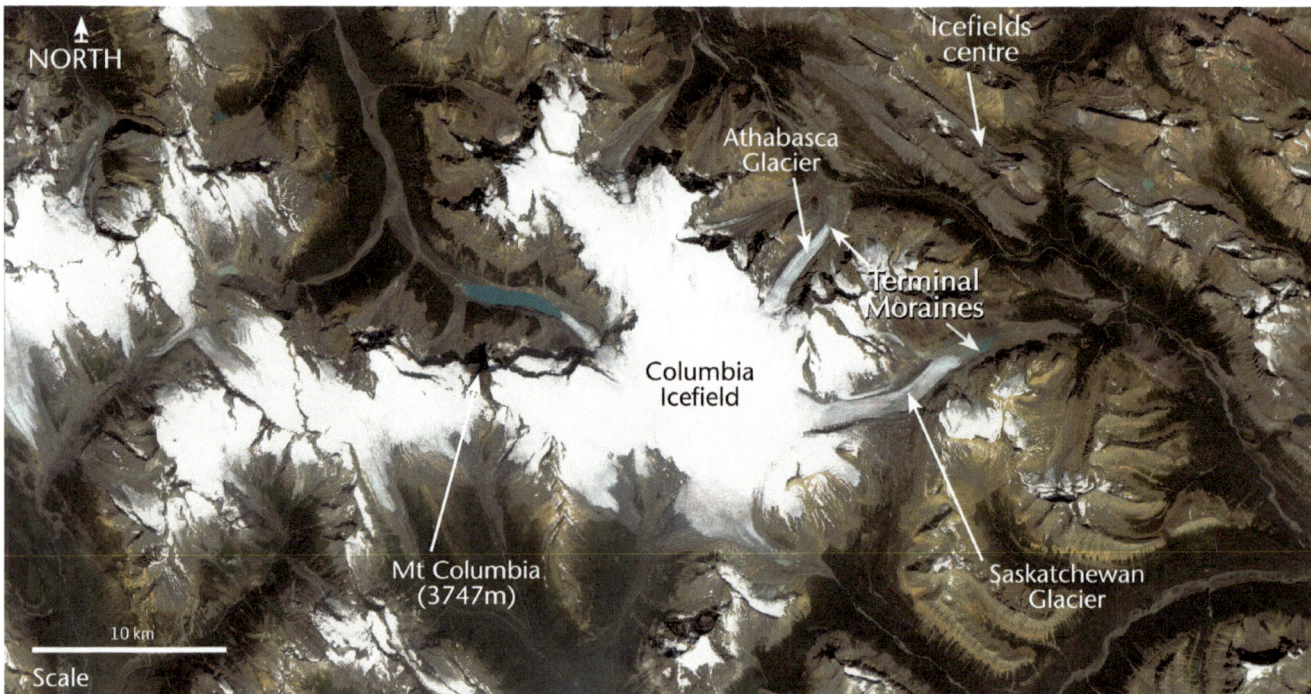

Fig. 5.10 Terminal moraines left by the receding Athabasca and Saskatchewan Glaciers, part of the irregular-shaped Columbia Icefield, are visible on this satellite image. *Source* Modified Copernicus Sentinel data of August 2018 processed by Philip Eales, Planetary Visons/DLR

exposed an extensive terminal moraine (Fig. 5.11). The glacier is also getting thinner as the estimated thickness is currently 90–320 m as compared with 2,000 m in 1844.

During the hotter and wetter periods of the Early-Mid Holocene, many of the icefields in the Canadian Rockies had disappeared entirely. At approximately 5,000 BP, the area where the Athabasca Glacier occurs was a forested valley (Leckie 2017). Icefields and slope glaciers either reformed, or grew considerably, during the Little Ice Age. The current recession of the icefields and glaciers in the Canadian Rockies is consistent with the warm period that we are currently experiencing, i.e. part of the natural cycle following the Medieval Warm period and the Little Ice Age. The anthropogenic effects of climate change are accelerating the rate of warming, but recession of the Athabasca Glacier is part of a natural cycle.

5.6   Banff National Park

The Banff National Park (area of 6,641 km^2) established in 1885 is the oldest national park in Canada. Access from the south and east is via the Trans-Canadian Highway and access from the north is via the Icefields Parkway (Fig. 5.2). The southern part of the Icefields Parkway follows the course of the North Saskatchewan River, parallel to the Front Ranges (Fig. 5.1). Neoproterozoic and Cambrian-Ordovician strata dominate the Front Ranges in this part of

the Rockies (Fig. 5.4b). The view from Bow Summit, a watershed at an elevation of 2,088 m, includes the high-altitude Bow Lake. Peyto Lake is an example of a turquoise-coloured Alpine lake fed by glacier meltwater (Box 5.1). The streams fed by the glacier have a braided pattern, typical of a new erosional surface, and deposits of coarse detritus form a broad delta. The delta acts as a sieve, separating the coarse sands and gravels, such that only the fine-grained rock flour (clays and silts) enters the lake. The rock flour feeds into the lake as turbid plumes.

Box 5.1: Turquoise Lakes

Artists and photographers prize the turquoise colour of Alpine lakes in the Canadian Rockies. Alpine lakes occur in hollows scooped out by branch glaciers in hanging valleys. Moraines and landslip deposits generally dam the lakes. Meltwater from the Late Pleistocene Ice initially filled the lakes, but the main supply is meltwater from current ice sheets and glaciers. The meltwater contains rock flour, i.e., fine-grained clay and silt. The coarser sands and gravels, which impart a white or pale grey discoloration to rivers and lakes, are generally absent. One reason the coarse sands and gravels do not enter the Alpine lakes is they may deposit upstream in small deltas. The relative fineness of the sediment load of meltwater is typical of glaciers

Fig. 5.11 a The Athabasca Glacier is located in a valley fed by the Columbia Icefield (vehicles on the glacier are specialized tour buses);
b Recession of the Athabasca Glacier reveals extensive terminal and lateral moraines

that grind ancient bedrock. The suspended particles of clay and silt reflect the shortwave length blue and green components of the spectrum. Red-orange colours are absorbed. Colour differences depend on the composition of the rock flour. The slightly coarser clay particles preferentially reflect green light and the very fine-grained silts reflect blue light.

Seasonal variations in the colour of Alpine lakes are a pronounced feature. During late spring and summer, for example, the green coloration of Emerald Lake, situated in the Yoho National Park is notably prominent. The green colour is due to an increase in the proportion of clay in the rock flour that enters the lake. In winter, the volume of sediment entering Emerald Lake is both lower and dominated by fine-grained silts. This imparts a deep blue colour.

5.6.1 Bow Valley

The main commercial centre of the Banff National Park is the regional town of Banff, situated on the Bow River. The river contains a well-known waterfall, Bow Falls, associated with resistant Triassic sandstones. The valley of the Bow River reveals vertical pinnacles, known as "hoodoos", which consist of cemented glacial tills and fluvial gravels. The unusual shapes of the hoodoos relates to preferential erosion of relatively poorly cemented tills deposited by ice advances towards the end of the Main Ice Age (at approximately 29,000–26,000 BP) (Leckie 2017). The Mount Norquay view site includes views of Bow Valley and the Front Ranges. The Front Ranges consist of multiple thrusts, stepped one upon the other, which have transported rock sheets by as much as 32 km eastward.

The thermal springs at Sulphur Mountain, Banff, are associated with sulphurous hot groundwater that circulates through Carboniferous limestones dissected by the Sulphur Mountain Thrust. The cave system associated with the springs is located in a thick deposit of travertine (Fig. 5.12). Indigenous peoples probably used the springs for thousands of years.

5.6.2 Lake Louise

The area surrounding Lake Louise is one of the most scenic areas in the Canadian Rockies. The proximity to the Canadian Pacific Railway encouraged development of a major tourist centre. An imposing, 19[th]C hotel located next to the lake is situated on an extensive glacial moraine. Lake Louise has a depth of 70 m and occupies a steep-sided valley located at the foot of several glaciers (Fig. 5.13a). A terminal moraine with

Fig. 5.12 Geothermal deposits, Cave Spring, Banff. *Source* https://www.google.com/ gonewiththewynns.com

an approximate age of 10,000 BP dams the lake. The Lake Louise Valley has views of the castellated peaks of Mount Victoria (3,464 m), Mount Whyte (2,983 m), and Mount Fairview (2,745 m), part of the Main Ranges. Erosion of flat-lying sedimentary strata. has produced a bench-and-cliff topography (Leckie 2017). The cliffs consist of the Miette Group quartzite (Neoproterozoic) in the lower part, Gog Group quartzite (Lower Cambrian) in the steepest middle sections, and quartzite and limestone of the Cathedral and Eldon Formations (Middle Cambrian) in the upper parts.

Lake Louise is one of the most well known of the turquoise coloured Alpine lakes and is fed by the Lower Victoria and Upper Victoria Glaciers. The glaciers are retreating despite the ice moving at approximately 30 m/year. The debris covering the glaciers is a pronounced feature. The hiking trail around the northern side of Lake Louise is suitable for most visitors. Trails that are more strenuous access several hanging lakes. The trail to Lake Agnes, with the Victorian tradition of a mountain teashop, includes views of the main valley (Fig. 5.13b).

5.6.3 Moraine Lake

The winding road that accesses Moraine Lake follows a hanging valley perched above and to the south of Lake Louise. Moraine Lake occurs in the Valley of Ten Peaks, or Wenkchemna Peaks, comprised of resistant Cog Group quartzite (Fig. 5.14a). The quartzite is steeply-dipping and forms prominent, cone-shaped scree (talus) deposits at the base of each peak (Fig. 5.14b). The Wenkchemna Glacier (area of 4 km^2) includes a terminal moraine associated with the Little Ice Age. The dam at the northeast end of Moraine Lake consists of a combination of a terminal moraine (which formed towards the end of the Main Ice Age) and a landslide deposit (Fig. 5.15). The latter contains large blocks of Cog Group quartzite. The view of Moraine Lake from the

a

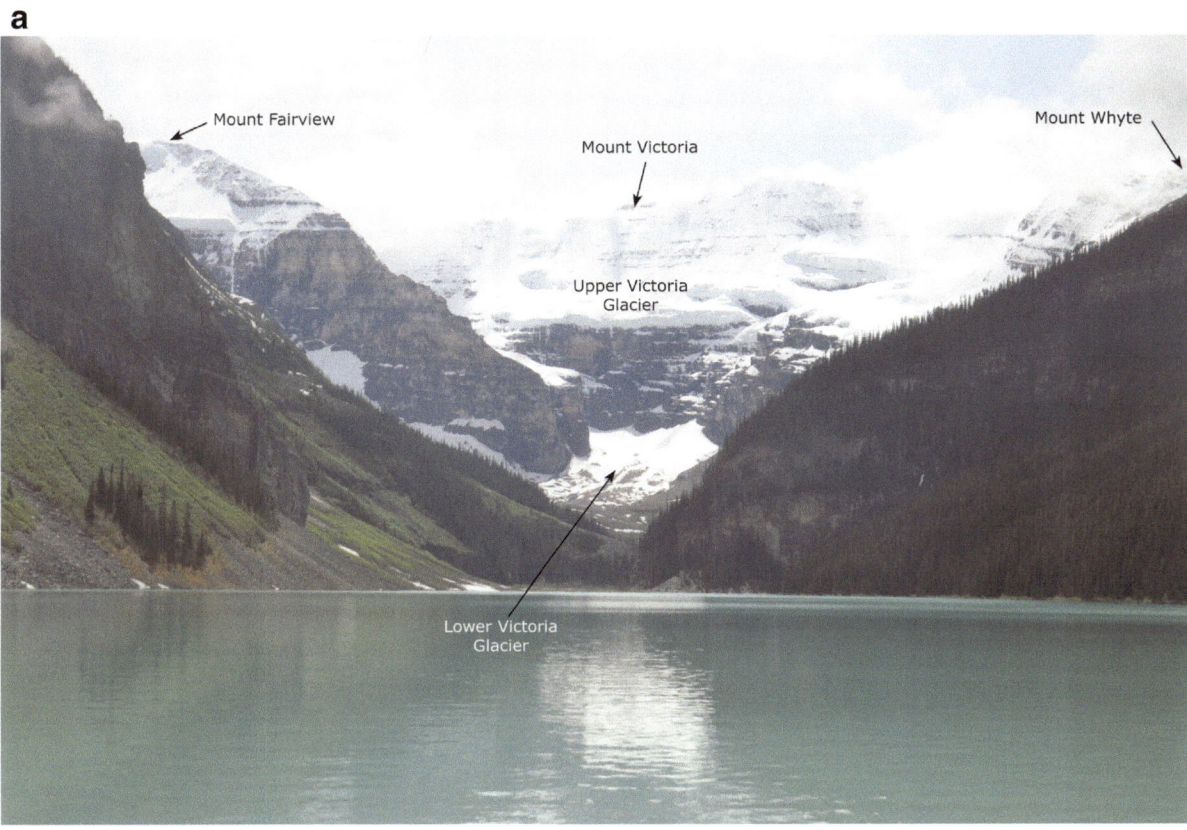

b

Fig. 5.13 a View of Lake Louise looking southwest towards the high peaks of Mount Victoria. A barren rock face comprised of Cog Quartzite separates the two glaciers on the lower flanks; **b** View of Lake Louise from the trail to Lake Agnes shows the renowned turquoise colour with gently dipping quartzites of the Cog Group exposed in the mountain face

a

b

Fig. 5.14 **a** The turquoise-coloured Moraine Lake is one of the most scenic of the glacial lakes in the Canadian Rockies; **b** Cone-shaped scree deposits occur at the base of the cliffs above Moraine Lake

Fig. 5.15 Terminal moraine and landside deposit with large blocks of Cog Group quartzite dams Moraine Lake

moraine/landslide deposit has been extolled for many years, but in comparison, Consolation Lakes to the southeast of Moraine Lake, occur in a remote and rather gloomy valley.

5.7 Yoho National Park

The Yoho National Park is considerably smaller than Banff and Jasper (area of 1,313 km^2). The indigenous Cree people used the name as an expression of amazement, i.e., a reaction to the remarkable landscape of jagged mountain peaks and glaciers. Access to the Yoho National Park from the east is via Kicking Horse Pass, the route pioneered by the Canadian Pacific Railway (Fig. 5.2). The steepness of the pass resulted in construction of two spiral tunnels for the railway. The spiral tunnels include large loops located within the mountainside of Cog Group quartzite. A pullout on the Trans-Canadian Highway provides views of the entrances to the spiral tunnels, together with a large debris fan associated

with an unstable face of Cathedral Mountain (Fig. 5.16). Several mine adits visible in the face of Cathedral Mountain access deposits of lead, zinc, and silver exploited between 1888 and 1952 from the Cathedral Formation. The deposits are examples of Mississippi Valley-type lead-zinc ores that have replaced the host carbonates (Paradis et al. 2007). The national park encompasses several ranges, including the Ottertail Range in which the highest peak is Mount Goodsir (3,567 m). The location on the western slopes of the Rockies results in considerably higher precipitation in comparison to Banff and Jasper, and the Yoho National Park includes extensive forests.

5.7.1 Natural Bridge and Emerald Lake

Natural Bridge is the site of a knickpoint on the Kicking Horse River. The "bridge" is associated with resistant calcareous slate of the Chancellor Formation (Middle

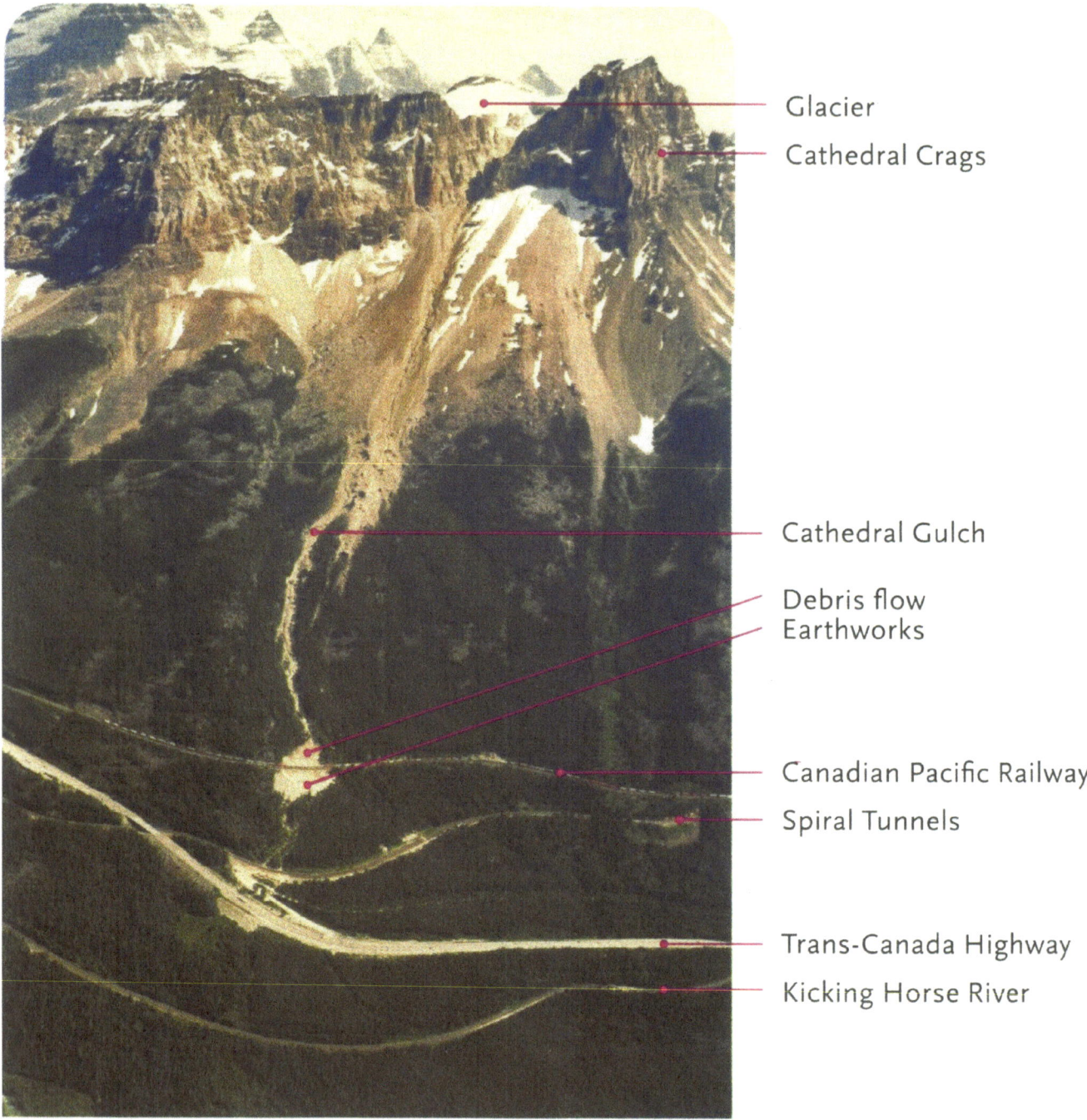

Glacier

Cathedral Crags

Cathedral Gulch

Debris flow
Earthworks

Canadian Pacific Railway

Spiral Tunnels

Trans-Canada Highway

Kicking Horse River

Fig. 5.16 The debris fan associated with the unstable face of Cathedral Mountain, on the northern side of Kicking Horse Pass, Yoho National Park, affects the Canadian Pacific Railway and Trans-Canada Highway. *Source* Original photograph by Lukas Arenson and annotation of Leckie (2017)

Cambrian) which displays a prominent cleavage (Leckie 2017). Upstream the river is several hundreds of metres wide but the gorge forms a narrow slot only a few metres wide (Fig. 5.17). Emerald Lake is an Alpine lake with a pronounced coloration and with views of a mountainous ridge that has featured in many paintings. Emerald Lake developed in a moraine associated with the Main Ice Age. The ridge, which occurs west of the lake, includes Wapta Mountain (2,778 m), which consists of massive carbonates of the Eldon Formation (Middle Cambrian). The face of the mountain is visible from the trail around the lake and reveals two dolomite pipes, lighter in appearance in comparison to

Fig. 5.17 Natural Bridge, Yoho National Park, is associated with a knickpoint on the Kicking Horse River. The calcareous slate of the Chancellor Formation (Middle Cambrian) reveals prominent cleavage

the host carbonate (Leckie 2017). The pipes are hydrothermal features, similar to those that host the lead-zinc ore deposits of Cathedral Mountain.

5.7.2 Cambrian Explosion of Life

The description *Cambrian Explosion of Life* emphasizes the rapid diversification of fossils that occurred over a period of approximately 40 Ma during the Middle Cambrian. The fossils at Mount Burgess (2,599 m), situated near Emerald Lake, were discovered by workers on the Canadian Pacific Railway (Royal Ontario Museum 2011), probably in 1886. The location, high above Kicking Horse Pass, suggests the surveyors were searching for timber required for railway sleepers (ties). Palaeontologist Charles Walcott described many of the fossils discovered here in detail. Walcott spent most field seasons between 1909 and 1924 at Fossil Ridge, excavating on the lower slopes of Mount Wapta, amassing over 65,000 specimens. Walcott Quarry is a historical site; subsequent quarries are located higher on the ridge. Imprints of the soft parts of marine organisms dominate the fossil beds. The interpretative centre for the Yoho National Park,

located at Field, includes detailed explanations and slabs of the Burgess Shales (Fig. 5.18). Fossil Ridge is a protected area and visits require organized groups with strict guidelines on collecting of fossils. The hike from Emerald Lake is 21 km (round trip) and involves some steep slopes.

5.7.3 Burgess Shale

The Burgess Shale contains the most complete record of the oldest marine fossils known. The shale is part of the Stephen Formation (Middle Cambrian) with an age of 510–505 Ma. This formation outcrops extensively at Mount Stephen (3,199 m), southeast of Kicking Horse Pass, as well as at the type localities of Mount Burgess and Mount Wapta near Emerald Lake. The Burgess Shale consists of calcareous mudstones deposited in a shallow tropical sea. Of fundamental importance to the occurrence and evolution of the marine life, as well as preservation of the fossils, was location of the palaeo-sea at the base of a 160 m-high palaeo-escarpment. The palaeo-escarpment formed the edge of a carbonate platform. The carbonate platform, which occurred on the western margins of the Laurentian Craton, is

Fig. 5.18 A slab of Burgess Shale in the interpretative centre at Field contains typical Middle Cambrian marine fossils, including a large trilobite, *Ogygopsis* (width of approximately 4 cm)

preserved as the Cathedral Formation. The Cathedral Formation is resistant to compression, resulting in extreme tectonic fracturing, such that fossils are almost impossible to find. Conversely, the Burgess Shale, which was located at the base of the palaeo-escarpment in the deeper ocean, contains almost pristine fossils. Moreover, the Cathedral Formation absorbed much of the deformation associated with the Laramide Orogeny.

The fossils of the Burgess Shale have been widely investigated since the pioneering work of Walcott, creating some controversy, as discussed, by for example, Collins (2009). Some fossils have proven to be unique species, entirely different from others in the geological record. Gould (1989) introduced the fossils of the Burgess Shales to a worldwide audience in a bestselling book entitled "Wonderful Life", interpreting the diversity to indicate that life forms were disparate in body form in comparison to those of today. An example is *Opabinia*, a fish with five eyes and a snout shaped like a vacuum cleaner. The unique lineages are evolutionary experiments that became extinct. These interpretations, based on palaeontologists such as Morris (1986), who was undertaking a detailed reinterpretation of Walcott's original collection, did not fit well with the primary authors. Classification of the fauna into modern day phyla is consistent with the original explanations by Walcott. Many of the fossils found in the Burgess Shale are predecessors of modern life.

References

Bally, A. W., Gordy, P. L., & Stewart, G. A. (1966). Structure, seismic data, and orogenic evolution of southern Canadian Rocky Mountains, British Columbia. *Bulletin of Canadian Rocky Mountains Petroleum Geology, 14*, 337–381.

Chamberlain, V. E., Lambert, R., & St, J. (1985). Cordilleria, a newly defined Canadian microcontinent. *Nature, 314*, 707–713.

Chen, Y., Gu, Y. J., Currie, C. A., Johnston, S. T., Hung, S. -H., Schaeffer, A. J., & Audet, P. (2019). Seismic evidence for a mantle suture and implications for the origin of the Canadian Cordillera. *Nature Communications, 10*. https://doi.org/10.1038/s41467-019-09804-8.

Collins, D. (2009). Misadventures in the Burgess Shale. *Nature, 460*, 952–953.

Engebretson, D. C., Cox, A., & Gordon, R. G. (1985). Relative motions between oceanic and continental plates in the Pacific Basin. *Geological Society of America Special Paper, 206*, 59 p.

Evenchick, C. A., McMechan, M. E., McNicoll, V. J., & Carr, S. D. (2007). A synthesis of the Jurassic–Cretaceous tectonic evolution of the central and southeastern Canadian Cordillera: Exploring links across the orogen. In L. W. Sears, T. A. Harms, & C. A. Evenchick (Eds.), Whence the mountains? Inquiries into the evolution of orogenic systems: A volume in honour of Raymond A. Price. *Geological Society of America Special Paper, 433*, 117–145.

Gadd, B. (2011). Athabasca Glacier and Columbia Icefield. Canadian Federation of Earth Sciences: GeoVistas Brochure.

Gould, S. J. (1989). Wonderful life (p. 236). New York: Norton.

Leckie, D. A. (2017). Rocks, ridges, and rivers: Geological wonders of Banff, Yoho, and Jasper National Parks (p. 216). Canada: A Roadside Tour Guide.

Morris, S. C. (1986). Community structure of the middle Cambrian Phyllopod Bed (Burgess Shale). *Palaeontology, 29*, 423–467.

Pana, D. I., & Van der Pluijm, B. A. (2015). Orogenic pulses in the Alberta Rocky Mountains: Radiometric dating of major faults and comparison with the regional tectono-stratigraphic record. *Geological Society of America Bulletin, 127*, 480–502.

Paradis, S., Hannigan, P., & Dewing, K. (2007). Mississippi Valley-type lead-zinc deposits. In W.D. Goodfellow, (Ed.), Mineral deposits of Canada: A synthesis of major deposit-types, district metallogeny, the evolution of geological provinces, and exploration methods. *Geological Association of Canada, Mineral Deposits Division, Special Publication, 5*, 185–203.

Roed, M.A. (1964). Geology of the Maligne Valley Jasper National Park Area. Open File Report 1964–01, Alberta Geological Survey, 61 p.

Royal Ontario Museum. (2011). The Burgess Shale: Historic and scientific explorations. Retrieved from http://www.virtualmuseum.ca/edu/ViewLoitCollection.do.

Yorath, C., & Gadd, B. (1995). Of Rocks, Mountains and Jasper: A Visitor's Guide to the Geology of Jasper National Park. (184 p). Toronto: Dundurn Press. ISBN-13: 978-1550022315.

National Parks, Rivers and Lakes of Southern Uganda

Albertine Rift, Giant Lakes, Rapid Speciation and River Capture

Abstract

The national parks of southern Uganda protect a remarkable diversity of fauna and flora, many species of which have evolved since the Miocene. A combination of factors is responsible for the unusually rapid Darwinian evolution during this period, including the influence of the East African Rift System. The East African plateau is sufficiently high as to result in a pleasantly mild climate, despite the equatorial setting, and large parts of southern Uganda are characterized by open savannahs with extensive areas of woody cover. The western border of Uganda is located within the Albertine Rift, the western branch of a rift system which extends from the Red Sea as far south as Mozambique. The Albertine Rift, which includes hilly regions associated with the uplifted shoulders, is a region of considerable climatic and botanic variability. The contrast between regional plateaus underlain by some of the oldest rocks on Earth and the youthful East African Rift System is an unusual feature of East Africa. The regional plateau formed due to uplift and erosion during the Late Mesozoic-Early Cenozoic, but the rifting is a Late Cenozoic phenomenon which remains active. The Albertine Rift contains a chain of ribbon-shaped lakes, including Lake Albert and Lake Edward, each of which is situated in a separate sedimentary basin. Basins are divided by basement horsts, of which the snow- and ice-capped Rwenzori Mountains is the most prominent example. The Albertine Rift is associated with several volcanic features. The Virunga Mountains are a chain of eight volcanic peaks which include two active cones. Four small volcanic crater fields occur in the vicinity of Lake Edward. Rifting severely impacted drainage patterns in southern Uganda, with major rivers, including the Victoria Nile, a relatively youthful channel which is aggressively eroding the regional plateau, captured. Large lakes, including Lake Victoria, formed in shallow basins. The Murchison Falls and Queen Elizabeth National Parks protect extensive grassy and wooded savannahs within the Albertine Rift. The Murchison Falls National Park is named after the waterfall where the Victoria Nile plunges into the Albertine Rift. The Rwenzori Mountain and Semliki National Parks contain a range of ecosystems including Alpine heathland and Afromontane forest. The Kibale National Park protects a forested area on the shoulders of the Albertine Rift known for studies on habituated groups of chimpanzee. The Lake Mburo National Park occurs on the regional plateau in an area of grassy savannah with swamps and small lakes. The Bwindi Impenetrable Forest and Mgahinga Gorilla National Parks (Virunga Mountains) contain Afromontane forests which support the endangered Mountain gorilla. The most spectacular feature of the Virunga Mountains is the Nyiragongo Volcano, located in the Democratic Republic of Congo . Nyiragongo is capped by a vertical-walled summit crater which contains possibly the largest, and most persistently active, lava lake on Earth.

Keywords

Albertine Rift • Crater fields • Lakes • Lava lake • Mountain gorilla • Speciation • Victoria Nile

Photographs not otherwise referenced are by the author.

The original version of this chapter was revised: Belated corrections have been incoporated. The corrections to this chapter are available at https://doi.org/10.1007/978-3-030-54693-9_18

6.1 Introduction

Southern Uganda is a landlocked region of East Africa with a broad range of ecosystems. Grassy and wooded savannahs, tropical and Afromontane forests, wetlands, and snow-capped peaks have produced scenic landscapes. The extensiveness of the woody cover is significant and some forested regions are associated with fertile volcanic soils (Fig. 6.1). The principal physiographic features of southern Uganda are the regional plateau, the broad linear valley and uplifted shoulders of the Albertine Rift, the Lake Victoria Basin, and two mountain ranges, the Rwenzori and Virunga. The Albertine Rift is associated with the western branch of the East African Rift System (EARS).

The national parks of southern Uganda reflect the unusual diversity of fauna and flora for which East Africa is renowned. Several national parks protect areas either within or proximal to the Albertine Rift, with others located on the uplifted rift shoulders, the active or dormant volcanic areas, and the regional plateau (Fig. 6.2). The Mount Elgon

National Park is associated with the Gregory Rift (eastern branch of the EARS) and is not described here. The two largest national parks, Murchison Falls and Queen Elizabeth, occur proximal to large finger-shaped lakes located in the Albertine Rift. The border between the Democratic Republic of Congo (DRC) and Uganda is aligned with the centre of the rift valley; Lake Albert and Lake Edward are shared between the two countries. Lake Victoria, the largest of the African Great Lakes, which is constrained to a shallow basin between the Albertine Rift and the Gregory Rift, is divided between Kenya, Tanzania, and Uganda. The Virunga Mountains are shared between the DRC, Rwanda and Uganda, and a brief mention of the Virunga National Park (DRC) and Volcanoes National Park (Rwanda) is included.

Large parts of southern Uganda are underlain by a regional plateau which formed due to uplift and erosion during the Late Mesozoic-Early Cenozoic. The EARS is primarily a Late Cenozoic phenomenon and remains active. River systems in southern Uganda are unusually complex as the physiography was disrupted by the rifting. Rivers were

Fig. 6.1 The Nyinambuga Crater viewed from Ndali Lodge in the Ndali-Kasenda Crater Field, southern Uganda

Fig. 6.2 Map showing location of national parks and major lakes in southern Uganda, together with the two components of the White Nile (Victoria Nile and Albert Nile) and parks in the DRC and Rwanda in the vicinity of the Virunga Mountains

6.2 National Parks and Speciation Patterns

Large parts of southern Uganda are extremely fertile and receive more than 2,000 mm of rain annually. The regional plateau experiences a pleasantly mild climate, as the heat and humidity expected of an equatorial setting is tempered by the elevation (1,200–1,800 m). The lower elevation of the Albertine Rift results in a region of considerable climatic and botanic variability. The elevation decreases northward, from Lake Edward (912 m) to Lake Albert (615 m). Lake Victoria is a relatively hot and humid region, despite being several hundreds of metres higher than the rift valley (elevation of 1,135 m). Most of the different ecosystems are protected in national parks and reserves (Fig. 6.3). The Murchison Falls and Queen Elizabeth National Parks protect extensive grassy and wooded savannahs within the rift valley. The Rwenzori Mountain and Semliki National Parks, although located in the rift valley, are associated with Alpine heathland and Afromontane forest. The Kibale National Park protects a forested area on the rift shoulders. The Lake Mburo National Park includes open savannahs together with swamps and small lakes located on the regional plateau. The Bwindi Impenetrable National Park contains extensive Afromontane forests, which together with the forested slopes of the Virunga Mountains (Mgahinga Gorilla National Park, Uganda; Virunga National Park, DRC; Volcanoes National Park, Rwanda) contain the only habitats of the endangered Mountain gorilla (*Gorilla beringei*). The Nyiragongo Volcano, which is located in the southern sector of the Virunga National Park, DRC, is an active volcano which poses a major hazard to the city of Goma and other localities northeast of Lake Kivu.

The text of the introduction continues:

captured by the Albertine Rift and Lake Victoria Basin. Two components of the White Nile occur in southern Uganda, the Victoria Nile which exits Lake Victoria at Jinja and plunges into the Albertine Rift at Murchison Falls, and the Albert Nile which flows out of Lake Albert.

Some species of large mammals, such as African elephant (*Loxodonta africana*) and Cape buffalo (*Syncerus caffer*) are found throughout East Africa, but others, of which the Uganda kob (*Adenota kob*), the primary grazer on the Ugandan savannahs, is a notable example, have a West African heritage (e.g., Williams et al. 1994) (Fig. 6.4). Southern Uganda is particularly well known for primates,

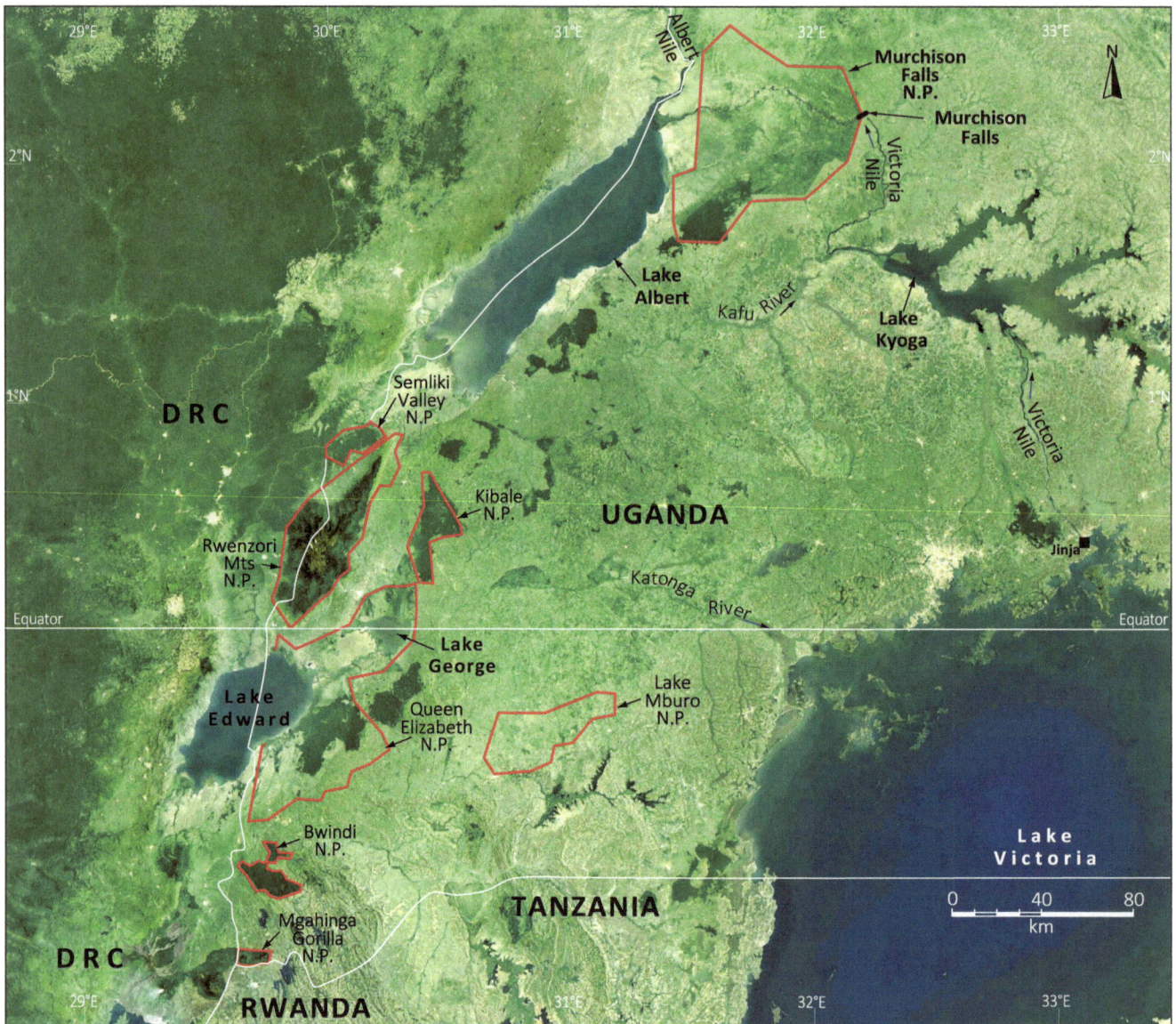

Fig. 6.3 Image of southern Uganda showing the extensiveness of the wooded savannahs on the regional plateau (light green), the forested rift shoulders and areas of central Africa west of the Albertine Rift (dark green), and the rift valley with its finger-shaped lakes. Lake Victoria and Lake Kyoga are associated with shallow warps on the regional plateau. The Victoria Nile exits Lake Victoria at Jinja, flows northwest via Lake Kyoga, and exits Lake Albert as the Albert Nile. *Source* Google Earth

including chimpanzee (*Pan troglodytes*) and Mountain gorilla. More than 1,000 species of birds, including the yellow-billed stork (*Mycteria ibis*) and little egret (*Egretta garzetta*), are found in Uganda. Rapid speciation is exemplified by cichlid fish of the African Great Lakes, with more than 2,000 species occurring just in Lake Victoria (Johnson 1996; Sturmbauer et al. 2011).

The rapidity of Darwinian evolution can be linked to a combination of factors, including impact of the relatively recent rifting and volcanism (e.g., Pickford et al., 1993; Danley et al. 2012; Ebisuzaki and Maruyama 2015). The evolutionary period coincides with the onset and persistence of rifting and volcanism. Speciation is following an island-style pattern, despite the continental setting, as regional plateaus are being dissected into smaller and smaller geological terrains (Scoon 2020). This is illustrated by restriction of the Mountain gorilla to regions where Afromontane forests developed in rift-related uplands isolated by savannah grasslands.

6.3 Regional Geology

The principal geological systems of southern Uganda are subdivided into basement complexes, which are dominated by crystalline metamorphic rocks, and the sedimentary basins

Fig. 6.4 **a** African elephant are a common site in the Queen Elizabeth National Park and can be approached to within a few metres in the Kazinga Channel; **b** Cape buffalo can be observed in swamps at the Murchison Falls National Park; **c** The Uganda kob is the most abundant grazer on grassy savannahs of the Queen Elizabeth National Park; **d** Chimpanzee are protected in forests of the Kibale National Park; **e** Mountain gorilla occur in Afromontane forests of the Virunga Mountains (Volcanoes NP, Rwanda); **f** Aquatic birds such as yellow-billed stork and little egret are a common site on the banks of the Kazinga Channel

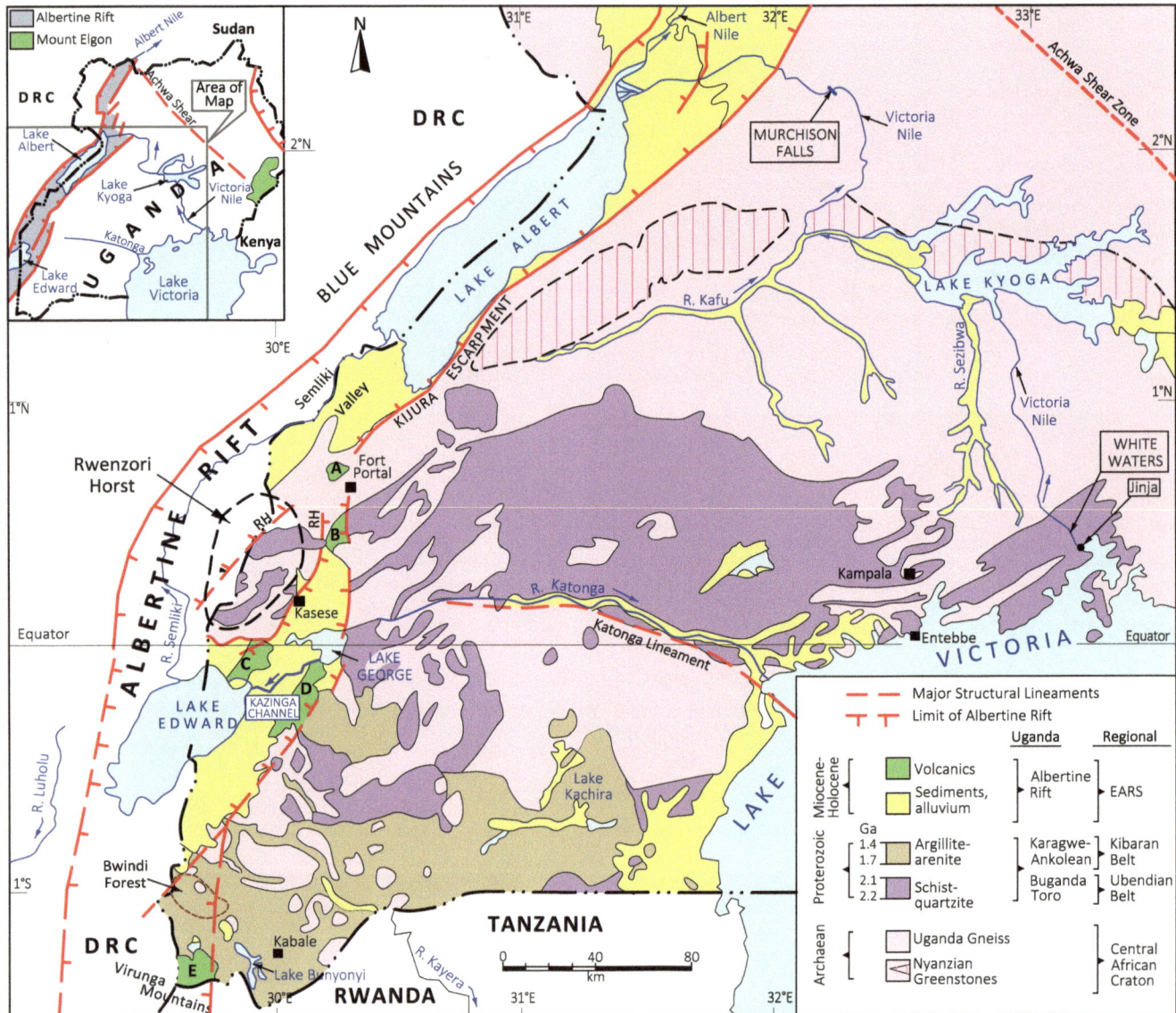

Fig. 6.5 Simplified geological map of southern Uganda. Each of the finger-shaped lakes of the Albertine Rift occurs in a separate sedimentary basin. Five rift-related volcanic terrains are recognized in the Albertine Rift: **a** Fort Portal-Kasekere; **b** Ndali-Kasenda; **c** Katwe-Kikorongo; **d** Bunyaraguru; **e** Virunga. The Mount Elgon volcanic field is part of the Gregory Rift (inset). *Source* Simplified from the 1:1,500,000 scale map compiled by R. MacDonald and published by the Department of Geological Survey and Mines, Uganda (1966)

and volcanic terrains of the EARS (Fig. 6.5). The basement complexes encompass more than a billion years of the geological record, prior to the appearance of hard-shelled animals. The basement complexes underlie the majority of the regional plateau and are also associated with the uplifted rift shoulders. The Rwenzori Mountains are an unusual example of a horst block consisting of basement rocks located within the Albertine Rift. The Albertine Rift is comprised of a

sequence of sedimentary basins, typically half-grabens, each of which contains a large finger-shaped lake. Shallow warps located on the regional plateau contain basins with irregular-shaped lakes, e.g., Lake Kyoga. Volcanic terrains are a subordinate feature of the Albertine Rift. The Virunga Mountains are a chain of extinct and active volcanic cones. Four crater fields occur in the Albertine Rift and the eastern shoulder.

6.3.1 Central African Craton

The oldest component of the basement complexes in southern Uganda is the Central African Craton (2,750–2,550 Ma). The craton formed from accretion of the Congo cratonic nuclei and the Tanzania cratonic nuclei during either the Archaean or Palaeoproterozoic (Dirks et al. 2015). Two components are recognized, the Ugandan Gneiss Complex (granitic gneiss) and the Nyanzian Greenstones (metavolcanics, chert, shale, and banded iron formation). The greenstones are possibly the oldest component, but the granitic gneiss is the most widespread. Greenstones are generally restricted to linear belts. Banded iron formations are a distinct component of the greenstones; these rocks constitute upland areas and may have impacted drainage patterns during rifting, notably of the River Kafu. A characteristic feature of the basement complexes throughout eastern and southern Africa is that outcrop is typically restricted to river channels. The greenstones are generally obscured by thick, black cotton soils or laterites, and the granitic gneiss is covered by sandy soils.

6.3.2 Proterozoic Belts

The Buganda-Toro System is a Palaeoproterozoic mobile belt (estimated age of 2,200–2,100 Ma) that underlies the regional plateau north of the River Katonga. The system consists of high-grade metamorphic rocks (argillite, schist, and quartzite) associated with accretion of the cratonic nuclei. The lowermost part of this system includes lenses and dykes of amphibolite which are exposed in the vicinity of Lake Victoria. The Karagwe-Ankolean System is a Mesoproterozoic belt (estimated age of 1,700–1,400 Ma) which covers large parts of southern Uganda near the border with Rwanda and northern Tanzania. The system consists of sandstone, conglomerate, and shale in the lowermost part, with molasse-like, coarse clastics in the uppermost part. From a regional perspective, the Buganda-Toro System is correlated with the high-grade Ubendian Belt and the Karagwe-Ankolean System with the low-grade Kibaran Belt (Dirks et al. 2015).

6.3.3 Regional Plateau

During the Palaeozoic Era, Gondwana was the largest landmass on Earth, covering an area of approximately 100 million km^2. The breakup of this ancient supercontinent was a long-drawn out process, which was initiated in the Jurassic (at approximately 180 Ma), and persisted into the Early Cenozoic (ending at approximately 35 Ma). The breakup isolated the African Plate as rifting opened up the Atlantic

Fig. 6.6 Large sections of the regional plateau in southern Uganda are covered by thick palaeosols and recent deposits of gravels and alluvium, as seen in this quarry near Mbarara

and Indian Oceans. The breakup and rifting was driven by a major thermal event, probably a mantle plume (e.g., Morley et al. 1999; Ebinger 2005; Merle 2011). The plume caused the upwelling mantle to invade and under-plate the lithosphere. As the asthenosphere cooled and expanded, almost the entire African Plate was uplifted. The uplift triggered intense cycles of erosion during the Cretaceous and Late Cenozoic. Regional plateaus in East Africa are generally associated with the 70 Ma-old African Erosion Surface (Saggerson and Baker 1965). This erosion surface is underlain by deeply-weathered basement terrains. Palaeozoic-Mesozoic strata are mostly absent. These rocks were either almost entirely stripped off by the erosion, or only developed sporadically in East Africa. Outcrop is sparse on the regional plateau, and large areas are covered by either thick palaeosols or recent deposits of sands and gravels (Fig. 6.6).

6.3.4 East Africa Rift System

The EARS incorporates three branches: Ethiopian, Albertine (western) and Gregory (eastern) (Box 6.1 and Fig. 6.7). The rift propagated from north to south; spreading rates decrease from 6.5 mm/yr in the north of the rift to 1.5 mm/yr in the south of the rift (e.g., Morley et al. 1999; Ebinger 2005; Chorowicz 2005; Ebinger 2005; Stamps et al. 2008; Merle 2011). An early phase of faulting and volcanism commenced in the Oligocene, but the peak of the rifting occurred in the Miocene, Pliocene, and Pleistocene. Continental rifting is associated with alkali volcanism (e.g., Bailey 1974; Woolley 2001). The Ethiopian and Gregory Rifts have a high output of volcanics. Volcanism is a subordinate feature of the Albertine Rift. The Kenyan-Tanzanian and Ethiopian Domes are areas of uplift (considerably higher than the regional plateau) associated with pre-rift volcanic activity. Thick

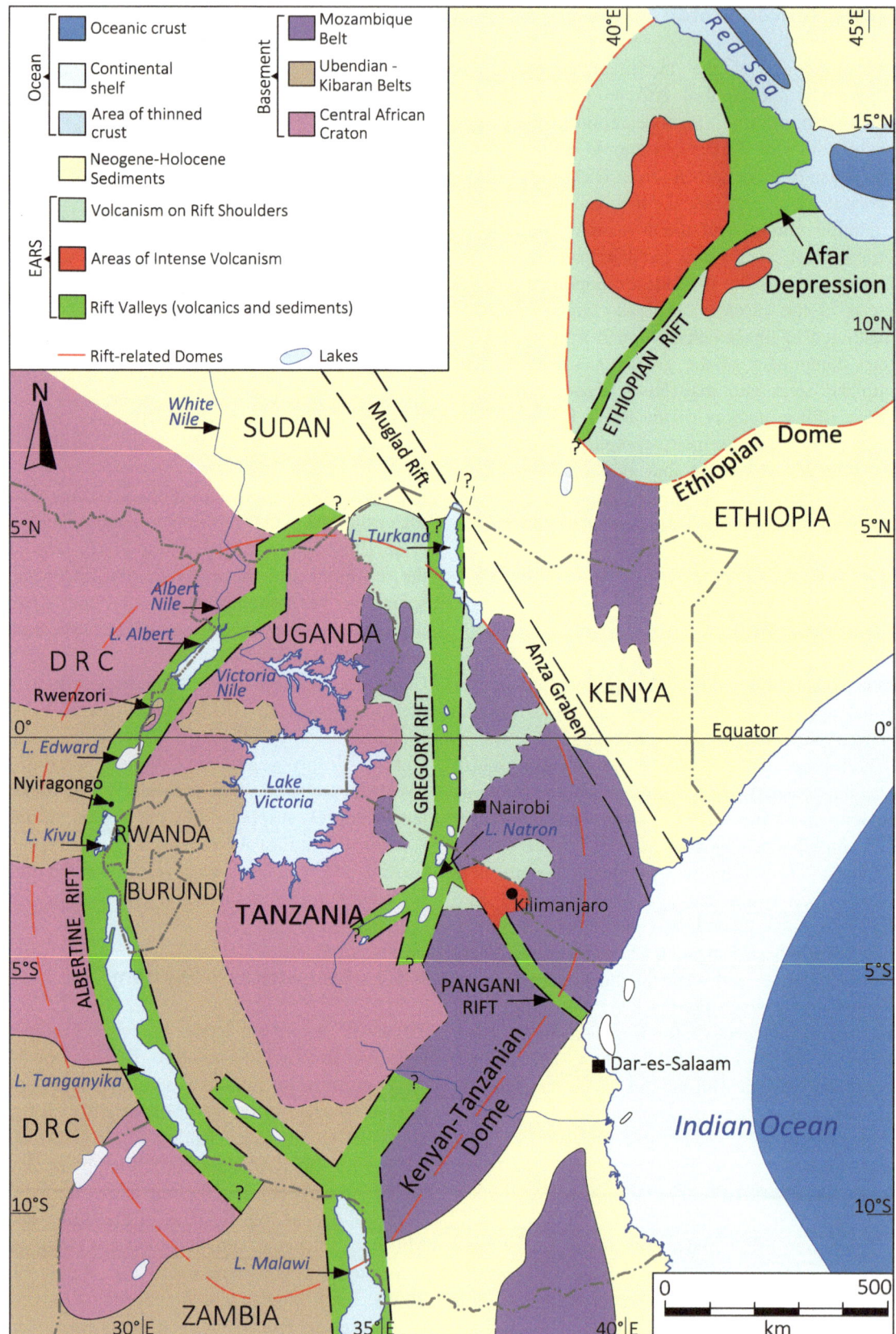

Fig. 6.7 Simplified geological map of East Africa depicting the three branches of the EARS. Some rifts are truncated by older, Cretaceous-age rifts (e.g., Anza Graben). Volcanism associated with the Ethiopian and Gregory Rifts is notably extensive. The Albertine Rift is dominated by sedimentary basins. *Source* Compiled from the 1:1,000,000 Geological Map of Kenya, the 1:2,000,000 Geological Map of Tanzania, and the 1:1,500,000 Geological Map of Uganda

sequences of rift-related volcanics dominate the Ethiopian and Gregory rift valleys and rift shoulders. The Albertine Rift, however, is comprised of sedimentary basins constrained by asymmetrical half-grabens (boundary faults alternate between the Ugandan and DRC side of the rift valley). This rather unusual geometry resulted from activation of horst blocks with the rift.

Box 6.1: East African Rift System

Rifting is one of the underlying principals of plate tectonics as it explains the dismantling of continents and development of oceans. The ability of lithospheric plates to drift, collide, and stretch is widely accepted (Wilson 1973; Windley 1977). The African Plate was isolated from the ancient supercontinent of Gondwana due to Mesozoic-Early Cenozoic rifting when the Atlantic and Indian Oceans formed. The EARS is a Late Cenozoic feature and constitutes one of the most significant physiographic and geologic features on Earth. The EARS is comprised of different segments that in their extremity extend for some 6,500 km from Lebanon in the Middle East, through the Red Sea and Afar region of Ethiopia, prior to petering out in southern Africa. The Albertine Rift is the most extensive branch and extends southward as far as Mozambique. Northwards, the rift extends into the Middle East. The joining of the Ethiopian Rift, the Red Sea Rift, and the Gulf of Aden Rift defines a triple junction. This feature has brought oceanic crust close to surface, an indication the African Plate could at some stage separate into two microplates (e.g., Ebinger 2005; Sarie et al. 2014). The EARS is an example of a continental rift (most rifts occur within oceanic basins) and has a high-volcanic output (e.g., Merle 2011). Large systems such as the EARS are related to major thermal events, although it is known that sedimentary basins can rift irrespective of volcanism (McKenzie 1978).

A key component of the hypothesis of continental drift is the concentration of seismic activity and volcanism on plate boundaries. The linear alignment of these phenomena in the interior is an anomalous feature of the African Plate. The EARS is stretching (thinning) and pulling-apart the continental crust. Rifting and volcanism commenced in the Oligocene (at approximately 30 Ma) and peaked in the Miocene, Pliocene, and Pleistocene (Chorowicz 2005; Baker and Wohlenberg 1971). Three discrete stages of faulting are recognized. Pre-rift activity is restricted to development of shallow warps in the plateau. This is followed by a half-graben stage with escarpments developing on one side of the down-faulted blocks. A full graben stage with linear valleys enclosed by two well-developed escarpments is the final stage. The half-grabens and full grabens are characterized by normal faulting. Faults typically propagate laterally for several tens of kilometres and can be observed on surface as escarpments with differential heights of as much as several thousands of metres.

Volcanism persisted throughout the EARS, including the early or pre-rift activity. The Ethiopian and Gregory Rifts are regions of intense volcanism, none more so than the Afar Triangle in Ethiopia. The rift-related volcanism may be driven by a long-lived mantle plume, possibly rejuvenation of the older Gondwana plume (Sarie et al. 2014). Linkage between the Ethiopian and Gregory Rifts has been widely debated (e.g., Ebinger et al. 2000). The Albertine Rift is a low volcanic output rift and is dominated by sedimentary basins.

Despite the EARS being geologically active (e.g., Sarie et al. 2014), the Holocene has been a relative quiescent epoch in East Africa. There is, however, modest seismic activity and some volcanism. Southern Uganda is subject to significant earthquakes (Maasha 1975). In 1966, a magnitude 7 event resulted in 157 fatalities in the Semliki Valley. In 1995, a magnitude 6 event caused a 20 km long crack to appear in the vicinity of Fort Portal. Earthquakes in the region are relatively deep-seated (27–40 km). The Albertine Rift has a higher than average heat flow. The Virunga Mountains include two active volcanoes, Nyiragongo and Nyamulagira, both of which are located in the DRC (Woolley 2001). The two volcanoes may account for almost half of the eruptions in Africa.

6.4 Lakes and Rivers

The EARS severely impacted drainage patterns in southern Uganda and resulted in extensive areas of internal drainage developing within the heart of the African continent (Doornkamp and Temple 1966; Beadle 1981; Pickford et al., 1993; Talbot and Williams 2008). Prior to activation of the EARS, southern Uganda was drained by rivers that flowed west towards the Atlantic Ocean, e.g., the Kafu, Katonga,

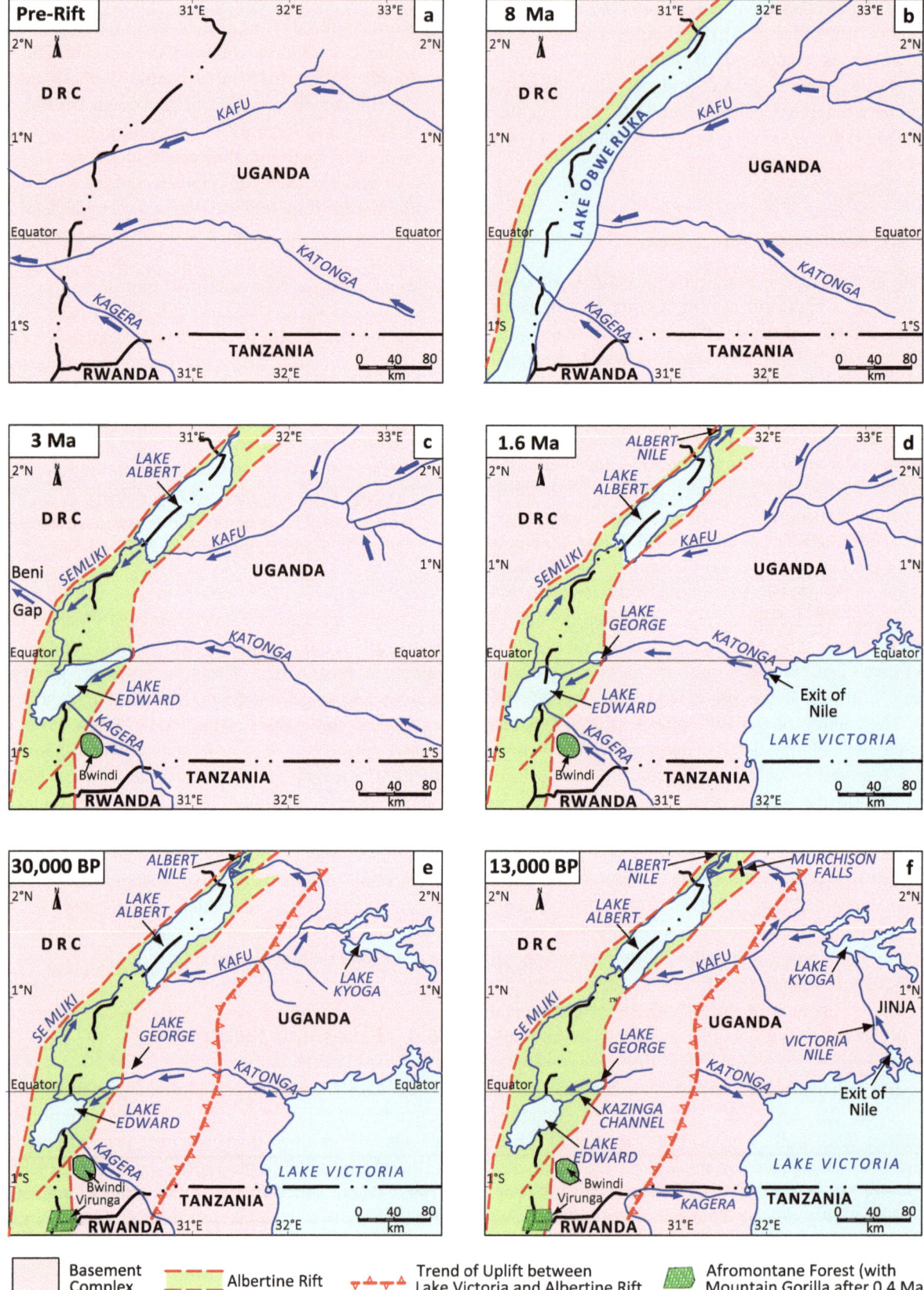

Fig. 6.8 The effects of the Late Cenozoic rifting on river systems and lakes in southern Uganda. **a** Pre-rifting; **b** Formation of a large palaeo-lake in the Albertine Rift; **c** Isolation of lake basins in the Albertine Rift; **d** Formation of the Albert Nile and Lake Victoria; **e** Upwarping between the Albertine Rift and Lake Victoria; **f** Development of the Victoria Nile. *Source* Compiled from Beadle (1981), Pickford et al. (1993), Talbot and Williams (2008) and others

and Kagera (Fig. 6.8a). These rivers were not part of the Congo system, as may be envisaged; prior to rifting the Congo flowed eastward towards the Indian Ocean (Stankiewicz and De Wit 2006). An initial phase of uplift occurred on the western escarpment of the Albertine Rift at approximately 8 Ma (Fig. 6.8b). This created a linear basin that trapped the westward-flowing rivers and culminated in formation of the 550 km-long Lake Obweruka.

6.4.1 Lake Albert and Lake Edward

The segmentation of the Albertine Rift into separate basins commenced at approximately 3 Ma due to uplift on the eastern shoulder (Fig. 6.8c). The activation of horst blocks within the rift valley created discrete basins, e.g., the Lake Albert Basin (fed by the Kafu River) and the Lake Edward Basin (fed by the Katonga and Kagera). The stratigraphy of the lake basins in southern Uganda has been described by Pickford et al. (1993) and Bishop (1969). Most lakes in the Albertine Rift are relatively deep and freshwater (Lake Tanganyika is the most pronounced example; Sander and Rosendahl 1989). This is a major difference with the Gregory Rift where lakes are shallow and most are strongly alkaline. A new westward-flowing outlet formed at approximately 3 Ma. The Semliki River exploited the Beni Gap. A new phase of uplift on the western shoulder at approximately 1.6 Ma triggered closure of the Beni Gap. This resulted in the Semliki River providing a link between Lake Edward and Lake Albert (Fig. 6.8d). The new outlet for the Albertine Rift was provided by the northward-flowing Albert Nile.

6.4.2 Lake Victoria

The closure of the Beni Gap coincided with formation of the Lake Victoria Basin (Fig. 6.8d). Lake Victoria occupies a shallow warp which formed as stress was released from the regional plateau in the area between the two branches of the southward-propagating EARS. The Lake Victoria Basin has a maximum age of 1.6 Ma, constrained by radiometric dating of sediments on the shores of the Kavirondo Gulf, Kenya (Kent 1944). At approximately 0.8 Ma, the basin tilted eastward to expose sediments in the Kagera Valley, southern Uganda (Doornkamp and Temple 1966). Lake Victoria was initially filled by rivers sourced in Kenya. The main outflow at this early stage was via the Katonga River, which drained into Lake Edward. The water was fed from Lake Edward into Lake Albert by the Semliki River, and exited the Albertine Rift as the Albert Nile.

6.4.3 Victoria Nile

Localized stress associated with depression of the Lake Victoria Basin resulted in a new zone of uplift developing in southeastern Uganda at approximately 30,000 BP (Fig. 6.8 e). This NNE-SSW trending ridge reversed the flow of the Katonga River (which now flows eastward into Lake Victoria), terminating the main connection between Lake Victoria and Lake Edward. This temporarily depleted the flow of the Albert Nile. The increase in water entering Lake Victoria caused a new outlet to develop, the Victoria Nile, at approximately 13,000 BP (Talbot and Williams 2008) (Fig. 6.8f). The Victoria Nile exits Lake Victoria at Jinja and flows into the sinuous Lake Kyoga, prior to entering the extreme northern part of Lake Albert. The Victoria Nile has captured the eastern part of the Kafu River, the flow of which was reversed earlier. The majority of the water which had previously entered the Albert Nile via the Katonga River is now provided by the Victoria Nile.

6.5 Quaternary Climatic Regimes

The Early Pleistocene in East Africa was characterized by hot and humid climates (Nicholson 1996), but onset of the Late Pleistocene Ice Ages (800,000–12,000 BP) saw a change to colder and drier conditions. Numerous desiccation events occurred in East Africa during the Late Pleistocene (Trauth et al. 2001). Lake Victoria may have dried entirely several times during this period. This is illustrated by the observation that Lake Victoria is underlain by 60 m of sediments, consistent with only 40,000 years of deposition, rather than a much thicker sequence which may be envisaged in light of the primary age of 1.6 Ma. Extensive icefields and glaciers developed in the Late Pleistocene on the higher peaks of East Africa (e.g., Kilimanjaro, Mount Kenya, Mount Elgon, and the Rwenzori) (Osmaston 2004; Pepin et al. 2014). The Late Pleistocene ice disappeared entirely from the Kibo component of the Kilimanjaro massif at approximately 12,000 BP. The ice reformed at the onset of the Holocene. The Early Holocene was dominated by the African Humid Period (11,700–5,000 BP). The Middle Holocene included a dry phase (4,000–3,700 BP), known to historians as the First Dark Age, as it impacted civilizations throughout North Africa/Middle East. The Late Holocene has probably experienced similar climatic patterns to those observed today (Olago et al. 2009). Climatic cycles, such as the Medieval Warm Period and the Little Ice Age, are also significant.

Fig. 6.9 **a** The Victoria Nile exits Lake Victoria at Jinja; **b** View of the Owens Dam on the Victoria Nile with hills (background) comprised of the Buganda-Toro System; **c** View from Wildwaters Lodge, Kangulumira Island, showing resistant outcrops of the Uganda Gneiss Complex in a fast-flowing, wide section of the Victoria Nile

6.6 Lake Victoria and the Victoria Nile

Lake Victoria and the Victoria Nile, although not part of national parks, illustrate the remarkable scale and diversity of the lakes and rivers in southern Uganda. The rectangular outline of Lake Victoria (surface area of 68,800 km^2) and shallow depth (<100 m) contrasts with the finger-lakes of the Albertine Rift. The exit of the Victoria Nile at Jinja has a memorial commemorating the "discovery" by John Hanning Speke (Speke 1863) (Fig. 6.9a). This was considered a major achievement in Europe, but the disputes and counterclaims by Richard Burton, who was far more widely travelled and a more successful explorer except in this one arena, make for interesting reading (Burton 1860). Jinja is not the primary source of the White Nile as Lake Victoria is fed by numerous rivers. Several sources have been identified, including the Kagera River and tributaries such as the Ruvyironza (Burundi) and Nyabaronga (Rwanda). The Mara River and tributaries such as the Grumeti River on the

Serengeti Plains, northern Tanzania, have a similar claim. Construction in 1954 of the Owens Dam on the Victoria Nile submerged the ravine and Ripon Falls, described by Speke and Burton as a feature of the river near Jinja (Fig. 6.9b). Downstream of the dam, the Victoria Nile is aggressively eroding the regional plateau. This is consistent with the youthfulness of the channel (i.e., maximum age of 13,000 BP). Rocks of the Buganda-Toro System are exposed in a shallow gorge (used for white-water rafting) which includes the Bujagali Falls and Kyabirwa Falls. The luxury resort of Wildwaters, located almost on the equator, is situated a few kilometres downstream from the gorge in the riverine forest of Kalagala Island, Kangulumira (Fig. 6.9c). Headward erosion at this locality has been slowed by the more resistant granite-gneiss of the Central African Craton (Fig. 6.10). The Victoria Nile is over a kilometre wide here and despite the fast-flow (and rapids) was used as a major transport route in the latter part of the 19thC and early 20thC.

Fig. 6.10 The Uganda Gneiss Complex is intruded by dark amphibolite dykes of the Buganda-Toro System, Kangulumira Island

6.7 Murchison Falls National Park

Murchison Falls National Park (area of 3,840 km^2) is located almost entirely within the Albertine Rift (Fig. 6.3). The park is dominated by extensive grassy and wooded savannahs located on the northern banks of the Victoria Nile. The most famous feature of the park is Murchison Falls. The height of the falls is only 43 m, but the volume of water is remarkable with approximately 300 m^3/second of water constricted from a wide channel into a narrow gorge (Fig. 6.11). Murchison Falls occurs close to the contact between the resistant Uganda Gneiss Complex and the younger and softer sediments of the Lake Albert Basin. The gorge has developed due to subparallel faults associated with the Bunyoro Escarpment, i.e., the eastern boundary of the rift valley. The extensiveness of the gorge suggests a rapid

a

b

Fig. 6.11 **a** The Victoria Nile at Murchison Falls is constricted from a wide channel to a narrow gorge as the river plunges into the Albertine Rift;
b The gorge is carved out of resistant rocks of the Uganda Gneiss Complex

a

b

Fig. 6.12 **a** In the Murchison Falls National Park, the Victoria Nile is fringed by riverine forest with hardwoods, including African mahogany and borassus palms; **b** The Victoria Nile enters Lake Albert by a delta that includes a myriad of channels and small islands. The water hyacinths have survived the passage from Lake Victoria

Fig. 6.13 View looking south along the Albertine Rift. Lake Albert is constrained between two escarpments. *Source of Image* Christian Hormann 2012 (http://earth.imagico.de)

rate of headward erosion (noting the youthfulness of the river). Murchison Falls was first reported by explorers Samuel and Florence Baker, but was named after Roderick Murchison, the then president of the Royal Geographical Society. There is some debate as to whether the Bakers observed the Murchison Falls or the Uhuru Falls, the latter associated with a secondary channel (that dries periodically) north of the main river. This channel flowed for the first time in over 50 years due to heavy rains in 1962. The southern entrance to the national park at Paraa is via a car ferry across the 1 km-wide Victoria Nile. The sluggish flow is typical of a deep, alluvium-filled channel. The banks of the river are rimmed by lush, riverine forest (Fig. 6.12a). A well known sight is the "African Queen", a tourist boat commemorating the usage of this area for scenes in the 1951 film of C.S. Forester's book (despite the book being based in Burundi). Boat trips can be taken upriver to view the gorge associated with Murchison Falls, or downstream to Lake Albert. The latter includes one of only a few localities in East Africa where the shoebill stork (*Balaeniceps rex*) can be readily observed.

6.7.1   Lake Albert

Lake Albert is the northernmost of the African Great Lakes (area of 53,000 km^2) and has a length of 150 km, width of 35 km, and maximum depth of 58 m (World Lake Database: Lake Albert 2014). The lake is primarily fed by the Semliki River. The Victoria Nile is characterized by less-saline water that enters so close to the outlet that even during floods the ingress of this river has little effect on the lake. The Victoria Nile has formed an extensive delta at the entrance to the lake that includes a myriad of channels with small islands built of sands and gravels (Fig. 6.12b). Some islands have formed due to papyrus grasses and floating masses of water hyacinth that captured river silt. (Water hyacinth is an alien weed that has infested parts of Lake Victoria; clumps of water hyacinth are a common sight at Jinja). The Lake Albert delta is rarely visited by tourists and remains a relatively pristine wilderness. The lake is constrained to the Albertine Rift by the Kijura Escarpment (Uganda) and the Blue Mountains (DRC) (Fig. 6.13). The location makes the lake susceptible to catastrophic storms as strong adiabatic winds can sweep down from the escarpments with little warning. In December 2016, a large number of people lost their lives when a ferry capsized, and there have been additional accidents in recent years. The Lake Albert Basin contains rift-related, lacustrine sediments (Plio-Pleistocene) with a thickness >1,200 m. Several hot springs and petroleum seepages have been documented, the latter having initiated extensive prospecting in recent years. The Kibiro Geothermal field is located on the eastern shores of the lake adjacent to the 300-m high Kijura Escarpment (Natukunda 2010). The escarpment includes outcrops of the Uganda Gneiss Complex.

6.8 Queen Elizabeth National Park

The Queen Elizabeth National Park (area of 1,978 km^2) is located mostly with the Albertine Rift and incorporates parts of Lake Edward and Lake George (Fig. 6.3). The half graben in this section of the rift valley is observed as a disjointed escarpment on the Ugandan side. A wide range of ecosystems

a

b

Fig. 6.14 a The tranquil setting of Lake Edward viewed from Uganda; **b** The sluggish, meandering Kazinga Channel which connects Lakes George and Edward is located in the Queen Elizabeth National Park

are identified and extensive grassy and wooded savannahs support large herds of grazers and browsers (Williams et al. 1994; Uganda Maps: Queen Elizabeth National Park). The northern, southern and western parts of the park constitute discrete regions. The northern area includes the Kazinga Channel, the link between Lake George and Lake Edward, and several crater fields, as described below. The southern area includes the open savannahs of the Ishasha Plains. This area extends to the Ishasha River, the border with the DRC, which flows northward into Lake Edward. The rift shoulders in the western part of the park include wooded uplands, e.g., the Maramagambo Forest where several species of primate can be viewed.

6.8.1 Lake Edward and Lake George

Lake Edward (area of 2,325 km^2) has a length of 77 km, width of 40 km, and maximum depth of 112 m (World Lake Database: Lake Edward, 2014) (Fig. 6.14a). Lake Edward is fed by several rivers, but the only outlet is the Semliki River. Lake George is considerably smaller (area of 250 km^2) and shallower (average depth of 2.4 m) (World Lake Database: Lake George, 2014). The two lakes occur at similar elevations, 614 m and 612 m, respectively. A highlight of a visit to the Queen Elizabeth National Park is an excursion on the 32 km-long Kazinga Channel, a RAMSAR locality famous for numerous species of birds. The sluggish flow is a reflection of the similar elevation of Lake George and Lake Edward (Fig. 6.14b). The Kazinga Channel was formerly part of the westward-flowing Katonga River. The Lake Edward Basin is constrained by NE-SW fault zones which underwent significant extension within the last 130,000 years (Nicholas et al. 2016). Shallow-penetration seismic surveys have enabled correlation of sequences (swamp clays and clastics) that reflect alternating arid and wet intervals during the past 20,000 years.

6.9 Kibale National Park

The Kibale National Park (area of 766 km^2) is located on the eastern shoulders of the Albertine Rift (Fig. 6.3). The escarpment is poorly defined in this area and the elevation of the park varies between 1,110 m and 1,590 m. Kibale is one of the largest and ecologically most important areas in southern Uganda. The region includes remarkable biodiversity and the thickly forested national park is surrounded by additional forest reserves. Three forest variants occur: moist evergreen forest, lowland forest, and Afromontane forest. The majority of Kibale is underlain by the Uganda Gneiss Complex, but small fault-bounded basins contain outcrops of the Buganda-Toro System. Several wetlands occur in areas of internal drainage which are associated with Late Cenozoic rifting. The principal attraction of the Kibale National Park is the opportunity to view family groups of chimpanzees that have been habituated for both scientific study and tourism. The relationships established by trackers enable some close encounters (Fig. 6.4d). The chimpanzees have been widely studied. Since 1987 a team led by Richard Wrangham with members from Harvard University, Massachusetts, and Makerere University, Kampala, have presented findings that have complemented and provided additional information to the pioneering work of Jane Goodall and colleagues at Gombe Stream National Reserve, Tanzania. There are an additional twelve species of primate at Kibale. They include the distinctive Redtail monkey (*Cercopithecus ascanius*), the endangered Red colobus (*Procolobus badius*) and Black-and-white colobus (*Colobus guereza*), together with several species of Central African monkey including the Ugandan mangebey (*Lophocebus ugandae*).

6.10 Crater Fields

The Bunyaraguru and Katwe-Kikorongo Crater Fields are located in the Queen Elizabeth National Park (Fig. 6.5). Both crater fields are probably Pleistocene (Woolley 2001). The Bunyaraguru volcanism occurs in a region of open savannahs proximal to the Kazinga Channel and may have initiated subdivision of Lake Edward and Lake George. The Katwe-Kikorongo Crater Field (area of 450 km^2) occurs in the northern part of the park. This scenic area is accessed by the circular "Crater Drive". Dozens of explosion craters with diameters of up to 4 km and depths of as much as 100 m can be viewed. Some craters contain small lakes (e.g., Kitagata Crater) and others contain salt deposits (e.g., Kikeronga Salt Pan) (Fig. 6.15). The Ndali-Kasenda Crater Field is located south of Fort Portal (area of 200 km^2), in a subsidiary graben of the Albertine Rift. The landscape of lush, conical hills dotted by deep, thickly forested craters is very scenic (Fig. 6.1). Thirty of the Ndali-Kasenda craters contain lakes (Uganda Maps: Fort Portal and Ndali Crater Area). Some of the smaller craters constitute blowholes with little ejected material. During a visit in 2012, a local guide reported that villagers had witnessed a small eruption. The Fort Portal-Kasekere Crater Field (area of 142 km^2) is located west of Fort Portal. This field includes prominent cones (comprised of lapilli tuff) up to 150 m in height. Both the Fort Portal-Kasekere and Ndali-Kasenda Crater Fields are thought to be Holocene features. Some volcanic rocks in the Fort Portal area have been radiometrically dated at 6,000–4,000 BP (Woolley 2001).

All four crater fields described above are dominated by tuffs and pyroclastic rocks; lava flows are of restricted occurrence (Woolley, 2001). Amygdaloidal lavas and agglomerate outcrop on the rims of some craters (Fig. 6.16). Pyroclastic rocks may contain basement xenoliths, indicative of their explosive origin. Most of the volcanic rocks in the crater fields are rich in the mineral leucite, indicative of potassium-rich alkaline magmas. In comparison, the Gregory Rift is characterized by sodium-rich alkaline magmas (e.g., nephelinite). Some of the potassium-rich alkaline lavas in the Albertine Rift have unusual compositions, prompting local names such as "*Katungite*" and "*ugandite*", as discussed by Yoder (1986).

6.11 Rwenzori Mountain and Semliki National Parks

The 120-km long, north-south trending rampart of the Rwenzori Mountains, although generally obscured by mist and clouds, can sometimes be observed from Fort Portal or tourist lodges such as next to the Nyinambuga Crater. Mount Stanley is the third highest mountain in Africa with several serrated peaks over 5,000 m (the highest is the Margherita Peak at 5,109 m). Icefields and glaciers are currently found on only the three highest peaks in the Rwenzori. The Rwenzori Mountain National Park (area of 996 km^2) protects the forested lower slopes, Alpine heathland, and snow-covered peaks (Fig. 6.17). The botanical zones constitute annular rings defined by altitude, similar to those on other East African mountains, such as Kilimanjaro and Mount Kenya. The slopes of the Rwenzori reveal a tremendous diversity of flora and fauna, including endemic giant groundsels and lobelia. Whereas all of the other high peaks in East Africa are extinct or active volcanoes, the Rwenzori is associated with basement rocks (Fig. 6.5). Rocks of the Uganda Gneiss Complex and Buganda-Toro System occur in a horst located wholly within the Albertine Rift.

The Semliki National Park is one of Uganda's smallest protected areas (194 km^2) and is a low-lying segment of the Albertine Rift (elevation of 670–760 m) at the foot of the Rwenzori (Fig. 6.3). The national park connects with a much larger reserve that extends to the southern shores of Lake Albert. The Semliki Valley is associated with lowland rain forest of Central African heritage. The national park includes a hot spring and small geyser, indicative of geothermal activity associated with the active rifting. Significant archaeological finds have been identified in the Semliki Forest, including evidence of Plio-Pleistocene hominins.

6.12 Lake Mburo National Park

The Lake Mburo National Park is the smallest of Uganda's savannah parks (area of 260 km^2). The area was formerly far larger, but reduced after land was reclaimed by local

a

b

Fig. 6.15 **a** Kitagata Crater Lake and **b** Kikeronga Salt Pan. Both localities occur in the Katwe-Kikorongo Crater Field, Queen Elizabeth National Park

Fig. 6.16 Outcrops of amygdaloidal alkali lavas and agglomerate cap the walls of the Kyemengo Crater, Katwe-Kikorongo Crater Field

pastoralists in the early 1980s. The national park is situated on the regional plateau, at an elevation of 1,220–1,828 m. Approximately 20% of the area is wetlands and a group of fourteen lakes are fed by the Ruizi River. The wetlands and lakes occur in depressions associated with rift-related warping of the plateau. The basins are probably related to

uplift at 30,000 BP, the same event which caused the flow of the Katonga River to reverse (Fig. 6.8e). The region is mostly underlain by basement rocks and large monoliths of the Uganda Gneiss Complex are a pronounced feature (Fig. 6.18). Quaternary sands and gravels occur proximal to the lakes.

6.13 Bwindi Impenetrable National Park

The main feature of the Bwindi Impenetrable National Park (area of 331 km^2) is a dense forest located at an elevation of 1,160-2,607 m. The park is also a world heritage site. Bwindi is part of the Kigezi Highlands, an uplifted shoulder of the Albertine Rift located between Lake Edward and the southern border. The forest consists of a combination of lowland and Afromontane variants (Fig. 6.19). Large sections are so densely vegetated it is only accessible on foot. The Bwindi Impenetrable Forest is one of the oldest forests in East Africa, dating from at least 25,000 BP (i.e., prior to the Last Glacial Maximum)—but is probably considerably older. The relative age has resulted in extraordinary diversity of fauna and flora. The resident population of approximately

Fig. 6.17 Rwenzori Mountains. *Source* en.Wikipedia.org

Fig. 6.18 Monoliths of the Uganda Gneiss Complex are a feature of the Lake Mburo National Park

400 Mountain gorillas at Bwindi accounts for half of the population of this endangered species (Fig. 6.8e). Bwindi is underlain by rocks of the Karagwe-Ankolean System (1.7–1.4 Ga) (Fig. 6.5). An inlier of the Uganda Gneiss Complex occurs in the eastern part of the national park.

6.14 Mgahinga National Park and the Virunga Mountains

The Mgahinga Gorilla National Park is located in one of the most scenic areas of East Africa. Mountain gorillas do not reside permanently at Mgahinga, but migrate freely between the three national parks that encompass the Virunga Mountains, i.e. including the Virunga National Park (DRC) and the Volcanoes National Park (Rwanda) (Fig. 6.2). The habituation of family groups within the parks, which has taken more than 25 years, has been a great success, despite times of civil unrest which the gorilla may react to by crossing the borders.

6.14.1 Virunga Volcanic Province

The Virunga volcanic province, the principal manifestation of volcanism in the Albertine Rift, is unusual as the chain of eight volcanoes are aligned W-E, i.e., at right angles to the rift. The eastern group of peaks (they are probably dormant) are located on the regional plateau, and only the active volcanoes of Nyiragongo and Nyamulagira occur within the rift. There are few reliable age dates, but most of the volcanism is younger than 100,000 BP (reviewed by Woolley, 2001). Mount Muhabura (4,127 m) and Mount Mgahinga (3,474 m) occur on the border of Rwanda and Uganda (Fig. 6.20a). Mount Sabinyo (3,634 m) defines the position where the DRC, Rwanda, and Uganda meet. Mount Visoke (3,711 m) and Mount Karisimbi (4,507 m) are located on the border of Rwanda and the DRC. The deeply eroded peak of Mikeno (4,437 m), which is situated entirely within the DRC, is extinct (the main activity occurred at 2.5–0.8 Ma) and may be an unrelated feature. The volcanic province is related to a localized hot spot, rather than the rift-related

Fig. 6.19 The Bwindi Impenetrable National Park is situated in an area of uplands (on the shoulders of the Albertine Rift) covered by dense Afromontane and lowland forest

lineaments (Tazieff 1977). The hot spot is currently located beneath Nyiragongo and Nyamulagira. The volcanism is dominated by potassium-rich foidite, or nephelinite, lavas of which are extremely fluid with anomalously low viscosity.

6.14.2   Nyiragongo

The Nyiragongo Volcano is located in the Virunga National Park (area of 7,844 km²), Africa's oldest park having been established by the Belgian government in 1925 (Fig. 6.21). The park is also a world heritage site as it protects an unusually diverse range of ecosystems with more than 2,000 plant species (Languy and de Merode 2009). Nyiragongo is a stratovolcano characterized by a steep-sided cone, an indication of the fluid nature of the potassic-rich lava (Fig. 6.22). The volcano has been more-or-less continuously active since the earliest reports were submitted in 1882. The cone is capped by a vertical-walled summit crater (diameter of 1.2 km) which contains an active lava lake. Two catastrophic eruptions have occurred in recent years. In 1977, the lava lake drained entirely (in less than an hour). In 2002,

streams of lava emitted from fissures on the southern flanks of the cone entered the city of Goma and partially destroyed the airport (Tedesco et al. 2007). The lava reached as far as Lake Kivu (Fig. 6.23).

The 2002 eruption did not drain the lava lake, but the crater was lowered by approximately 250 m. The 2002 eruption was the most destructive effusive volcanic eruption in recent history and piles of lava from this event can still be observed in the city. Eruptions are unusually hazardous as they include anomalously fast-flowing lava flows with large volumes of toxic gas. The problem of asphyxiating gases associated with the Virunga volcanism is well known and natural gas chambers (**"makuzu"**) occur in surface depressions in the Virunga National Park (Languy and de Merode, 2009).

Since 2002, the lava lake in the summit crater of Nyiragongo has remained intact (Fig. 6.24a). A spatter cone which remains active formed in 2016. The hike to the summit of Nyiragongo (3,470 m) ascends the southeastern flanks of the cone and starts at the Kibati gate (altitude of 1,984 m), with overnight huts located on the rim of the crater. The sidewalls of the crater are built up from multiple

Fig. 6.20 The volcanic peak of the Sabinyo Volcano in the Virunga Mountains, viewed from the Volcanoes National Park, Rwanda

Fig. 6.21 Image of the Nyiragongo Volcano showing location in the southern part of the Virunga National Park (DRC) and proximity to Goma and Lake Kivu. The ascent of the cone starts at the Kibati gate and includes sections located on the 2002 lava. *Source* Google Earth

layers of lava and tephra; the rim is capped by lava from the 1977 eruption (Fig. 6.24b). The view of the lava lake and spatter cone in the twilight is spectacular (Fig. 6.25a). The surface of the lava lake has a temperature of 980 °C, with the liquidus temperature of the magma fed into the lake estimated at 1100 °C (Sahama 1978). The surface of the lava lake reveals cracks and flares which constantly disappear and reform (Fig. 6.25b).

The lava lake currently has a diameter of approximately 400 m, area of 46,500 m^2, and is modelled as having a funnel-shape (Burgi et al. 2014). The spatter cone is fed from a dyke or conduit from the shallow magma chamber (rather than the juxtaposed lava lake), which contains an estimated volume of 10 km^3 of magma (Burgi et al. 2018). The shallow magma chamber occurs at an estimated depth of 1–4 km, with a deeper reservoir occurring at 10–14 km

Fig. 6.22 The steep-sided and rugged Nyiragongo cone (foreground) as compared with the smoother and gentler slopes of Nyamulagira (background). Both volcanoes are more-or-less continuously active

(Wauthier, et al. 2012). The semi-permanent nature of the lava lake remains a hazard to the inhabitants of Goma and surrounding villages (Tedesco 1995). The active nature of the crater is, however, a major attraction to tourists. The volcano is actively monitored from a research observatory based in Goma.

6.15 Evolution of Mountain Gorilla

Ancestors to chimpanzees, gorillas, and hominins, i.e., monkeys and apes, first appeared in the Oligocene Period (33.9–23.03 Ma). The fossil record of the genus *gorilla* is poor, but the evolutionary history may be traced to an extinct ape *Proconsul africanus*, discovered in 1948 by Mary Leaky at Rusinga Island, Lake Victoria. The gorilla is thought to have split from the primates at approximately 9 Ma (i.e., during the Miocene). Two subspecies are recognized,

including the Mountain gorilla (Stanford 2001). Note that the Western gorilla and Eastern Lowland gorilla were formerly thought as separate subspecies, but are now classified as one species (*Gorilla gorilla*). The Mountain gorilla separated from the Eastern Lowland gorilla at approximately 400,000 BP. Only two populations of Mountain gorilla are recognized. They occur in the Bwindi Forest (Uganda) and the Virunga Mountains (DRC, Rwanda, and Uganda). The geological evidence can be construed to suggest the oldest population is Bwindi, as this locality was uplifted to elevations sufficient for Afromontane forest to develop in the Pliocene (Fig. 6.8c). The Mountain gorilla could only have colonised the Virunga Mountains in more recent times, i.e. after 100,000 BP (Fig. 6.8e). The active nature of the Virunga volcanism, which characteristically includes fast-moving lava flows, would have made this a risky venture, possibly a reaction to the isolated habitat of the Bwindi forests.

Fig. 6.23 View of Goma airport, partly covered by the 2002 lava flow, looking south over the city which is situated next to Lake Kivu. *Source* Guido Potters (http://www.gnu.org/copyleft/ fdl.html)

a

b

Fig. 6.24 a The active summit crater of the Nyiragongo Volcano contains a lava lake (left) and an active spatter cone (right). Lava fountains associated with the spatter cone attain a height of approximately 50 m; **b** The rim of the near vertical-walled summit crater is capped by lava from the major eruption of 1977

a

b

Fig. 6.25 **a** The lava lake (left) and active vent (right) in the summit crater of the Nyiragongo Volcano are a spectacular sight in the twilight; **b** A night time view of the lava lake shows the occurrence of cracks and segments with magma flares

References

Bailey, D.K. (1974). Continental Rifting and Alkaline Magmatism. In: Sorensen, H. (Ed.). The Alkaline Rocks, Wiley p. 148–159.

Baker, B. H., & Wohlenberg, J. (1971). Structure and evolution of the Kenyan Rift Valley. *Nature, 229,* 538–542.

Beadle, L.C. (1981). The inland waters of tropical Africa-an introduction to tropical limnology. Longman, London 2nd Edition, 475 p.

Bishop, W.W. (1969). Pleistocene stratigraphy of Uganda. *Geological Survey of Uganda Memoir, X,* 1–128.

Burgi, P.-Y., Darrah, T. H., Tedesco, D., & Eymold, W. K. (2014). Dynamics of the Mount Nyiragongo lava. *Journal of Geophysical Research Solid Earth, 119,* 4106–4122.

Burgi, P.-Y., Minissale, S., Melluso, L., Mahinda, C. K., Cuoco, E., & Tedesco, D. (2018). Models of the Formation of the 29 February 2016 New Spatter Cone inside Mount Nyiragongo (abstract). *Journal of Geophysical Research Solid Earth, 1239,* 9469–9485.

Burton, R. (1860). Lake regions of Central Equatorial Africa, with notices of The Lunar Mountains and the sources of the White Nile being the results of an expedition undertaken under the patronage of Her Majesty's Government and the Royal Geographical Society of London in the years 1857–1859. London: Clowes.

Chorowicz, J. (2005). The East African Rift System. *Journal of African Earth Sciences, 43,* 379–410.

Danley, P.D., Husemann, M., Ding, B., DiPietro, L.M., Beverly, E.J. & Peppe, D. (2012). The impact of the geologic history and paleoclimate on diversification of East African cichlids. *International Journal of Evolutionary Biology,* Article ID 574851, 20 p.

Dirks, P.H.G.M., Blenkinsop, T.G. & Jelsma, H.A. (2015). The geological evolution of Africa. In: Geology volume IV, Encyclopaedia of Life Support Systems (EOLSS).

Doornkamp, J. C., & Temple, P. H. (1966). Surface, drainage and tectonic instability in part of the southern Uganda. *The Geographical Journal, 132,* 238–252.

Ebinger, C. (2005). Continental break-up: The East African perspective. *Astronomy and Geophysics, 46,* 16–21.

Ebinger, C. J., Yemane, T., Harding, D. J., Tesfaye, S., Kelley, S., & Rex, D. C. (2000). Rift deflection, migration, and propagation: linkage of the Ethiopian and Eastern rifts, Africa. *Geological Society of America Bulletin, 112,* 163–176.

Ebisuzaki, T., & Maruyama, S. (2015). United theory of biological evolution: disaster-forced evolution through Supernova, radioactive ash fall-outs, genome instability, and mass extinctions. *Geoscience Frontiers, 6,* 103–119.

Johnson, T. C., Scholz, C. A., & Talbot, M. R. (1996). Late Pleistocene desiccation of Lake Victoria and rapid evolution of cichlid fishes. *Science, 273,* 1091–1093.

Kent, P. E. (1944). The Miocene Beds of Kavirondo, Kenya. *Quarterly Journal of the Geological Society London, 100,* 85–118.

Languy, M. & de Merode, E. (2009). Virunga: The survival of Africa's First National Park. Uitgeverij Lannoo, 350 p. 978–90-209-6562-9.

Maasha, N. (1975). The seismicity and tectonics of Uganda. *Tectonophysics, 27,* 381–393.

McKenzie, D. (1978). Some remarks on the development of sedimentary basins. *Earth Planetary Science Letters, 40,* 25–32.

Merle, O. (2011). A simple continental rift classification. *Tectonophysics, 513,* 88–95.

Morley, C.K., Ngenoh, D.K., and Ego, J.K. (1999). Introduction to the East African Rift System. In: Morley, C.K. (Ed.) Geoscience of Rift Systems – Evolution of East Africa. *AAPG Studies in Geology, 44,* 1–18.

Natukunda, J.F. (2010). Geology of the Kibiro, Katwe and Buranga geothermal prospects of Uganda. Proceedings World Geothermal Congress, Indonesia, 11p.

Nicholas, C.J., Newth, I.R., Abeinomugisha, D., Tumushabe, W.M., & Twinomujuni, L. (2016). Geology and stratigraphy of the southeastern Lake Edward basin (Petroleum Exploration Area 4B), Albertine Rift Valley, Uganda. *Journal of Maps, 12,* 237–248.

Nicholson, S.E. (1996). Sedimentary processes and signals of past climatic changes in the large lakes of the African Rift Valley. In: (Ed.) Johnson, T.C., and Odada, E.O. The Limnology, Climatology and Paleoclimatology of the East Africa Lakes. Gordon and Breach Amsterdam, 367–412.

Olago, D., Opere, A. & Barongo, J. (2009). Holocene palaeohydrology, groundwater and climate change in the lake basins of the Central Kenya Rift. *Hydrological Sciences Journal, 54,* 765–780.

Osmaston, H. (2004). Quaternary glaciations in the East Africa Mountains. In: Ehlers, J. and Gibbard, P.I. (Eds.) Developments in Quaternary Sciences, Quaternary Glaciations Extent and Chronology Part III: South America, Asia, Africa. Australasia, Antarctica, 2C, Amsterdam, Elsevier, 139–150.

Pepin, N.C., Duane, W.J., Schaefert, M., Pike, G. & Hardy, D. (2014). Measuring and remodelling the retreat of the summit icefields on Kilimanjaro, East Africa. *Arctic, Antarctic, and Alpine Research, 46,* 905–917.

Pickford, M, Senut, B. & Hadoto, D. (1993). Geology and Palaeobiology of the Albertine Rift Valley Uganda-Zaire. Volume I Geology. Occasional Publication CIFEG BP5517, Orleans, France, 190p.

Saggerson, E.P., & Baker, B.H. (1965). Post-Jurassic erosion surfaces of East Africa and their deformation in relation to rift structure. *Journal of the Geological Society of London, 121,* 51.72.

Sahama, T.G. (1978). The Nyiragongo main cone. Musée Royale de l'Afrique Centrale, Tervuren, Belgique. *Annales Série In-8 Sciences Geologiques, 81,* 1–88.

Sander, S., & Rosendahl, B. R. (1989). The geometry of rifting in Lake Tanganyika, East Africa. *Journal of African Earth Sciences, 8,* 323–354.

Sarie, E., Calais, E., Stamps, D. S., Delvaux, D., & Hartnady, C. (2014). Present day kinematics of the East African Rift System. *Journal of Geophysical Research, 119,* 3584–3600.

Scoon, R.N. (2020). Geotourism, Iconic Landforms and Island-style Speciation Patterns in National Parks of East Africa. *Geoheritage, 12*:66. https://doi.org/10.1007/s12371-020-00486-z.

Speke, J.H. (1863). Journal of the Discovery of the Source of the Nile. William Blackwood, Edinburgh.

Stamps, S. D., Calais, E., Saria, E., Hartnady, C., Nocquet, J.-M., Ebinger, C. J., et al. (2008). A kinematic model for the East African Rift. *Geophysical Research Letters, 35*(L05304), 4. https://doi.org/10.1029/2007gl032781.

Stanford, C. (2001). The Subspecies Concept in Primatology: The Case of Mountain Gorillas. *Primates, 42,* 309–318.

Sturmbauer, C., Husemann, M. & Danley, P.D. (2011). Explosive speciation and adaptive radiation of East African cichlid fishes. In: (Eds.) Zachos, F. E., & Habel, J.C., Biodiversity Hotspots-distribution and protection of conservation priority areas, Springer, pp. 333–362.

Stankiewicz, J., & De Wit, M. J. (2006). A proposed drainage model for central Africa – did the Congo flow eastward. *Journal of African Earth Sciences, 44,* 75–84.

Talbot, M.R., & Williams, M.A.J. (2008). Cenozoic Evolution of the Nile Basin. In Dumont. H.J. (Ed.) The Nile: Origin, Environment, Limnology and Human Use. Springer Science-Business Media B. V., p. 37–60.

Tazieff, H. (1977). An exceptional eruption: Mt Nyiragongo, Jan 10th 1977. *Bulletin Volcanology, 30,* 189–200.

Tedesco, D. (1995). Report on the Virunga volcanic region and on the related volcanic risks (p. 33). Department of Human Affairs, Geneva: United Nations.

Tedesco, D., Vaselli, O., Papale, P., Carn, S. A., Voltaggio, M., Sawyer, G. M., et al. (2007). January 2002 volcano-tectonic eruption of Nyiragongo volcano, Democratic Republic of Congo. *Journal of Geophysical Research, 112,* B09202. https://doi.org/10.1029/2006jb004762.

Trauth, M. H., Deino, A., & Strecker, M. R. (2001). Response of the East African climate to orbital forcing during the last interglacial (130-117 ka) and the early last glacial (117-60 ka). *Geology, 29,* 499–502.

Wauthier, C., Cayol, V., Kervyn, F. & d'Oreye, N. (2012). Magma sources involved in the 2002 Nyiragongo eruption, as inferred from an InSAR analysis. *Journal of Geophysical Research, 117,* https://doi.org/10.1029/2011JB008257.

Williams, J.G., Arlott, N. & Fennessy, R. (1994). Collins Field Guide National Parks of East Africa. Harper Collins Hong Kong, 336 p.

Wilson, J. T. (1973). Continental drift, transcurrent and transform faulting. In A. E. Maxwell (Ed.), The Sea (Vol. 4, pp. 623–644). New York: Wiley.

Windley, B. (1977). The evolving continents. Wiley and Sons, 399 p.

Woolley, A. (2001). Alkaline Rocks and Carbonatites of the World. Part 3: Africa. Geological Society of London, 372 p.

World Lake Database: Lake Albert (2018). *wldb.ilec.or.jp/data/databook_html/afr/afr-11.html.*

World Lake Database: Lake Edward (2018). *wldb.ilec.or.jp/Details/Lake/AFR-12.*

World Lake Database: Lake George (2018). *wldb.ilec.or.jp/data/databook_html/afr/afr-10.html*

Yoder, H. S. (1986). Potassium-rich Rocks: Phase Analysis and Heteromorphic Relations. *Journal of Petrology, 27,* 1215–1228.

Lake Natron and the Ngorongoro Conservation Area, Northern Tanzania

7

Alkaline Lakes, Gregory Rift, Ngorongoro Caldera, Oldoinyo Lengai Volcano, Oldupai Gorge, and Serengeti Migration

Abstract

Northern Tanzania is a region of spectacular landforms, with some of the largest and most diverse concentrations of wildlife remaining on Earth. The fauna and flora are protected in world-famous national parks and conservation areas. The grassy savannas of the Serengeti Plains and the self-contained ecosystems of the Ngorongoro Caldera are gazetted as natural wonders of the world. The Serengeti Plains are underlain by some of the oldest rocks on Earth, granite-gneiss and greenstone of the Central African Craton, but the most significant geologic feature of the region is the East African Rift System. This Late Cenozoic phenomenon is still active. The Gregory Rift (eastern branch) is associated with intensive volcanism. The northern Tanzania divergence is an area of the Gregory Rift where three branches are recognized. The Natron-Manyara branch contains large alkaline lakes, including Lake Natron (pH > 12) which is a major breeding ground for flamingo. Several volcanic cones occur in this section of the Gregory Rift, including Oldoinyo Lengai which erupts on average every 15–20 years. It is here that the existence of two immiscible magma-types, nephelinite and the anomalous natrocarbonatite, were first recognized in nature. Calcareous ashes derived from eruption of the natrocarbonatite at Oldoinyo Lengai support nutrient-rich grasses which have influenced the migration of grazers on the Eastern Serengeti Plains. The Ngorongoro Conservation Area occurs in a discrete structural step of the Gregory Rift, located between Lake Natron and the Eastern Serengeti Plains. The principal physiographic feature of this area is the Ngorongoro Highlands, an extinct volcanic complex (Pliocene) which contains multiple volcanic cones and calderas. The Ngorongoro Caldera is a near-circular feature protected by largely intact internal walls and which contains a diversity of ecosystems. The caldera formed from a catastrophic volcanic eruption at approximately 2 Ma. The palaeoanthropological sites of Oldupai Gorge and Laetoli occur on the Eastern Serengeti Plains. Hominin fossils and tracks have been discovered in lacustrine sediments and volcanic ashes which accumulated in small basins associated with warping of the regional plateau. The fossil-rich beds are exposed in shallow ravines carved by ephemeral rivers on the arid plains. The co-existence of multiple species of hominids is an intriguing feature of Oldupai Gorge. Discoveries of *Zinjanthropus boisei* (fossil OH-5 dated at 1.848 Ma) and *Homo habilis* (fossil OH-7 dated at 1.848–1.832 Ma), together with various fossils of *Homo erectus*, greatly influenced our understanding of human evolution. The footprints at Laetoli are considerably older (dated at 3.6 Ma). They are associated with the earliest hominids, *Australopithecus afarensis*, which may have roamed over large parts of East Africa.

Keywords

Caldera • Gregory Rift • Hominins • Migration • Palaeoanthropology • Volcanism

7.1 Introduction

The eastern branch of the East African Rift Valley extends from Kenya into northern Tanzania, and is named after John Walter Gregory, a geologist and intrepid explorer (Gregory 1894a). During his pioneering travels in Kenya, Gregory reported on many features of the rift valley, including the alkaline nature of lakes and the icefields of Mount Kenya (Gregory 1894b). Most of the spectacular landforms in northern Tanzania are associated with the Gregory Rift. The rifting is accompanied by intensive volcanism and in the vicinity of Lake Natron, the escarpment that constrains the

Photographs not otherwise referenced are by the author.

© Springer Nature Switzerland AG 2021
R. N. Scoon, *The Geotraveller*,
https://doi.org/10.1007/978-3-030-54693-9_7

rift valley on the western side, is built of layer-upon-layer of volcanic lava and tephra (Fig. 7.1). This part of the Gregory Rift includes a number of free-standing volcanic cones, including Oldoinyo Lengai ("Mountain of God" in the Maasai language), one of the most persistently active volcanoes in East Africa. The Ngorongoro Conservation Area is located in a discrete structural step between the Gregory Rift and the Eastern Serengeti Plains. The conservation area includes the Ngorongoro Highlands, an extinct volcanic complex, which includes the Ngorongoro Caldera. The caldera forms a self-contained sanctuary and is designated one of the natural wonders of the world. The world-famous palaeoanthropological sites of Oldupai Gorge and Laetoli are located in the Eastern Serengeti Plains.

The Ngorongoro Conservation Area and the adjoining Lake Manyara and Serengeti National Parks contain some of the largest and most diverse concentrations of wildlife remaining on Earth (Fig. 7.2). Safaris generally commence from the regional town of Arusha and the addition of a visit to the wilderness area in the vicinity of Lake Natron is recommended. The geological heritage of this region is recognized in the newly-formed Ngorongoro-Lengai Geopark, a region dominated by rifting and volcanism associated with the East African Rift System. Several of the Quaternary volcanoes in northern Tanzania are gazetted as active or quiescent. These include Oldoinyo Lengai, as noted above, but the most potentially catastrophic volcano in the region is Mount Meru (4,565 m), located in the Arusha National Park. The highest peak of Kibo (5,895 m) in the Kilimanjaro massif, also protected in a national park, includes a quiescent summit crater. The rift valley in the vicinity of Lake Natron is relatively low-lying (elevation of 600 m) and is a hot, arid environment, but the Ngorongoro Highlands (where most areas lie above 2,000 m) have a temperate climate. Parts of the Ngorongoro Highlands are densely forested, but the northern and western parts of the conservation area merge imperceptibly with the grassy savannas of the Eastern Serengeti Plains. This part of the Serengeti is an arid and inhospitable region in the dry seasons, but the biannual rains can produce a remarkable transformation.

Fig. 7.1 The Western Escarpment of the Gregory Rift near Lake Natron is constructed of multiple, near-horizontal layers of lava and tephra

Fig. 7.2 Map showing location of selected national parks and wilderness areas in northern Tanzania

7.2 Regional Geology

The East African Rift System (EARS) is a Late Cenozoic phenomenon that is still active, as described in Chap. 6 (Box 6.1). Rifting has dissected the older basement terrains, of which two broad divisions are recognized. The Lake Victoria Terrain (granite-gneiss and greenstone) is part of the Archaean-age Central African Craton. The craton has an average age of 2.6–2.5 Ga in northern Tanzania (Dirks et al. 2015). The Mozambique Belt (quartzite, schist, and granite) is a Neoproterozoic mobile belt (estimated age of 800–550 Ma) which extends through much of East Africa (Shackleton 1986; Mosley 1993). Preservation and exposure of these ancient, crystalline rocks relates to repeated cycles of uplift and erosion which culminated in formation of the regional plateau. The uplift is ascribed to a major thermal event associated with break-up of the supercontinent of Gondwana. The uplift and erosion commenced in the Early Jurassic and persisted into the Early Cenozoic. Intense periods of erosion produced several surfaces, including the

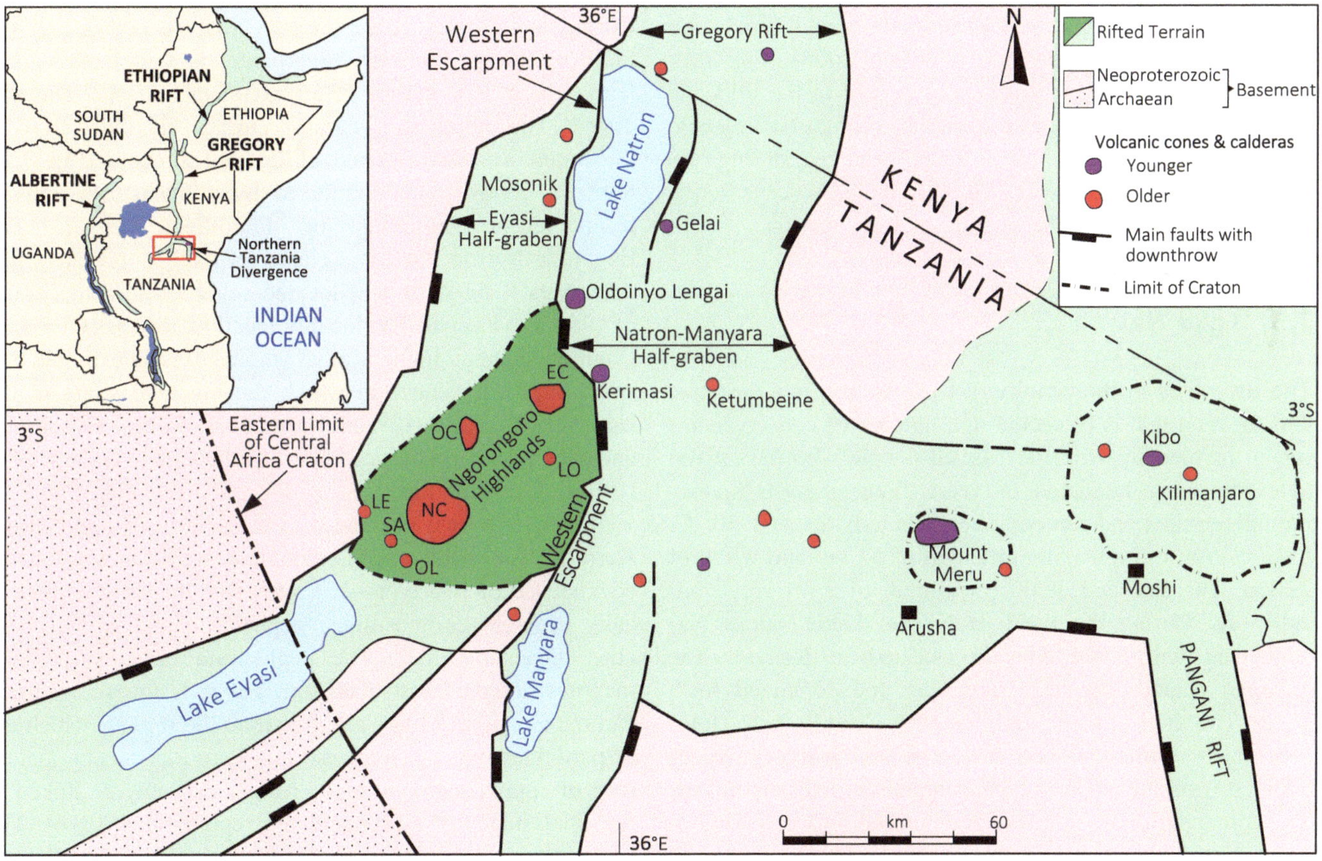

Fig. 7.3 Geological map of the northern Tanzanian divergence. The rifted terrains are dominated by volcanics (light green) and include the extinct volcanic complex associated with the Ngorongoro Highlands (dark green). Some of the larger cones and calderas are shown, but for reasons of scale the subordinate sedimentary basins are omitted. *Source* Simplified after Dawson (2008)

70 Ma-old Africa Erosion Surface (Saggerson and Baker 1965).

The Gregory Rift in Kenya is characterized by a relatively narrow graben (width of 50 km or less) constrained by major faults on either side. In northern Tanzania, however, the rift is associated with a structurally complex region: the "northern Tanzania divergence" (Fig. 7.3). This feature has a width of 200 km and incorporates regions of intense volcanism within three branches (Ebinger et al. 1997; Dawson 2008; Le Gall et al. 2008). The southward-propagating rift appears to have been refracted eastward by the Central African Craton, a deep-rooted block of crystalline rocks massively resistant to deformation; the Mozambique Belt is more readily stretched and thinned. The two branches of the divergence of interest here—the Gregory Rift (or Natron-Manyara Half-graben) and the Eyasi Half-graben—are both half-grabens in which faults are largely restricted to the western sides.

Continental rifting is generally associated with extensive alkali volcanism (e.g., Bailey 1974) and the northern Tanzania divergence is a textbook example where volcanism has produced both stepped plateaus (fissure eruptions) and giant, free-standing cones (discrete centres). The occurrence of giant calderas is characteristic of the volcanism in the Gregory Rift. Two groups of volcanoes are recognized in the northern Tanzania divergence (Dawson 2008) (Fig. 7.3). The Older Volcanism (extinct) commenced in the Pliocene and Early Pleistocene, e.g., the Ngorongoro Volcanic Complex. The Younger Volcanism occurred during the Upper Pleistocene and Holocene and includes some active features, e.g., cones in the vicinity of Lake Natron.

7.3 Lake Natron

The rift valley in the vicinity of Lake Natron is a desolate, arid location that is protected in a number of conservancies run in partnership with the Maasai people who have lived here for many hundreds of years. Vegetation is sparse, rainfall erratic, and temperatures regularly exceed 40 °C. Lake Natron has a maximum length of 57 km and width of 22 km. The size and depth (maximum of 3 m) vary considerably. During the Early Holocene, Lake Natron was connected with Lake Magadi (located in Kenya). The enlarged palaeo-lake was far deeper and contained fresh water. The drier climatic regimes of the Middle-Late Holocene have resulted in both lakes becoming exceptionally alkaline (average pH of 12) as evaporation exceeds inflow. The composition of the brines is enhanced by weathering of the sodium-rich volcanic rocks in the catchment. The concentrated alkaline brines, particularly in shallow pools, are sufficiently caustic to burn the skin. An unusual feature is the occurrence of freshwater springs and pools in the southwest corner of the lake. The fresh and saline waters do not mix readily.

Lake Natron and Lake Magadi contain extensive salt deposits. The salt is a mixture of two natural minerals, natron and trona (sodium carbonate and sodium bicarbonate is precipitated from the sodic brines). The salt can be intercalated with micro-crystalline silica, or Magadi chert (Eugster 1980). The deposits at Lake Magadi are quarried for production of soda ash, and there is a potential scheme to exploit the salts at Lake Natron. The extreme alkalinity and remote setting have resulted in Lake Natron being a major breeding ground for both Lesser Flamingo (*Phoeniconaias minor*) and Greater Flamingo (*Phoenicopterus ruber*). The flamingo has evolved remarkably efficient filtration systems; lake brines are poisonous to most other species. The principal food of the flamingo is the cyanobacteria which flourish in the highly alkaline waters (Fig. 7.4).

7.3.1 Geological Setting of Lake Natron

Rifting in this part of the Gregory Rift peaked during the Pleistocene. The Western Escarpment is associated with a fault which was active at approximately 1 Ma (Dawson 2008). The relatively youthful surface of the valley is illustrated by the presence of braided streams (Fig. 7.1). Surficial deposits are dominated by volcanic ash derived from recent eruptions of Oldoinyo Lengai. The Engaruka-Natron Field is a cluster of explosion craters and tuff cones (0.57–0.14 Ma) located several kilometres south of Oldoinyo Lengai. The locality locally known as God's Pit (Shimo la Mungu in the Maasai language) is an example of a small explosion crater in which the volcanic strata in the sidewalls are distinctly deformed. Lake Natron is partially enclosed by two giant volcanoes, to the east is Gelai (0.99–0.96 Ma) and to the west is Mosonik (estimated age of 1.24 Ma). The two volcanoes to the south of Lake Natron, Kerimasi and Oldoinyo Lengai, are well known for the coexistence of two types of immiscible magmas with radically different compositions, nephelinite and natrocarbonatite (Dawson 1962). The nephelinite (alkaline silicate) magmas are the most abundant, but it is the sodium-rich carbonate magmas (they are unique to the region) which are of particular interest. Kerimasi is a Late Pleistocene feature (age of approximately 0.6–0.4 Ma) which may be linked to the final stages of faulting in the region (Hay 1976). The cone was severely eroded during the hot and humid climates of the Late Pleistocene-Early Holocene.

Fig. 7.4 The unusual colour of the northern and central parts of Lake Natron is due to the red photosynthesizing pigment in the cyanobacteria. The rugged terrain on either sides of the lake is associated with deeply-eroded Pleistocene volcanics. View looking north; width of lake in southern part of image is approximately 10 km. *Source* NASA Terra-ASTER image for 2003, processed by Philip Eales, Planetary Visions/DLR

7.4   Oldoinyo Lengai

Oldoinyo Lengai has been studied in great detail as this is the only active volcano where natrocarbonatite lava flows have been observed in nature. The symmetrical cone (3,188 m), indicative of minimal erosion, rises some two and a half kilometres above the base of the valley to the south-west of Lake Natron (Fig. 7.5). Light-coloured natrocarbonatite flows and ashes partially cap the cone, giving the illusion of an icecap when viewed from a distance. The natrocarbonatite magma has an unusually low viscosity and low eruption temperatures (Dawson 2008). The magma can erupt as fast-travelling lavas that freeze as extraordinarily thin flows. On eruption, the magma is black, but the sodium reacts rapidly with meteoric water to form light-coloured secondary minerals within a few days. Ash from Oldoinyo Lengai is dispersed westward due to the prevailing easterly winds. They form extensive deposits on the Eastern Serengeti Plains and dominate the uppermost formations of the Oldupai Group.

Oldoinyo Lengai has been active since the Late Pleistocene (oldest activity is estimated at 0.37 Ma) with significant eruptions having occurred throughout the Holocene. Recent eruptions have been documented in some detail. Eruptions from the early 1990s were restricted to a shallow summit crater which could formerly be entered to inspect the flows and other features such as hornitos, spatter cones, and vents (Dawson 2008). Unfortunately, violent eruptions that commenced in 2007 destroyed this crater and the new crater is too deep and unstable to enter. This activity marked a change, from the quiescent effusive activity that had characterized the eruption of natrocarbonatite flows since 1960, to explosive activity associated with nephelinite magmas (Mattson and Vuorinen 2009). In 2010, two lava flows were erupted and a small lava lake was reported to be present in the crater. In 2013, a sequence of particularly violent eruptions occurred. Oldoinyo Lengai erupts on average every 15–20 years and presents a potential hazard to the local Maasai who live in the proximity of Lake Natron, as well as to air travel, as ash columns may attain heights of tens of kilometres.

Fig. 7.5 The symmetrical cone of the Oldoinyo Lengai Volcano rises abruptly from the floor of the rift valley near Lake Natron

7.5 Ngorongoro Conservation Area (NCA)

A satellite image shows the principal physiographic features of the NCA (area of 8,200 km^2) and diversity of ecosystems (Fig. 7.6). The eastern and southern slopes of the Ngorongoro Highlands are covered by afromontane forests. The forest is displaced on the higher peaks by heathland and moorland. The internal walls of calderas may also be forested, but the floors are dominated by grassy savannahs. The northern and western parts of the NCA, e.g., the Salei Plains, the Gol Mountains, and the Eastern Serengeti Plains, are remote wilderness areas.

The Ngorongoro Highlands is a watershed, with most rivers flowing eastward towards the rift valley. The calderas include areas of internal drainage. Most rivers on the Eastern Serengeti Plains flow westward toward Lake Victoria. The exception is the ephemeral Oldupai River which terminates in the Olbalbal Depression. The increased rainfall resulting from the enhanced elevation results in the Ngorongoro Highlands being associated with thick red soils on the eastern and southern slopes. This area is intensively farmed and large parts of the drier northern and western areas of the NCA are traditional lands of the Maasai people who have ancient rites to inhabit and graze cattle.

The NCA is approached from Arusha by a road that ascends the Western Escarpment near Lake Manyara and crosses a relatively densely populated part of the Ngorongoro Highlands prior to reaching the Loadere Gate (Fig. 7.7). The eastern entrance gate to the NCA is located

near the base of the thickly forested outer slopes of the Ngorongoro Volcano. A view point on the rim of the Ngorongoro Caldera reveals the immensity of the inner part of this famous sanctuary (Fig. 7.8). The approach from the Eastern Serengeti Plains involves a steep ascent as after the Naabi Gate, the road ascends the western slopes of the Ngorongoro Highlands. The paradox of crossing into a graben and yet gaining altitude is reconciled with the great thickness of lavas and ashes associated with the volcanism.

7.5.1 Geological Setting

Five principal geological formations are recognized in the NCA (Fig. 7.9): (i) Parts of the Eastern Serengeti Plains are associated with granite gneiss and greenstone of the Lake Victoria Terrain (Archaean); (ii) The Gol Mountains and small hills near Oldupai Gorge are erosional relicts of the Mozambique Belt (Neoproterozoic), also known as inselbergs ("Island Mountains"); (iii) The Ngorongoro Volcanic Complex is a group of seven or eight extinct volcanoes (Pliocene age of 4.9–1.8 Ma), three of which are comprised of calderas partially infilled by younger sediments; (iv) Rift-related sedimentary basins (Neogene-Quaternary) occur in association with the large alkaline lakes; (v) Deposits of calcareous ash derived from the Oldoinyo Lengai Volcano cover large areas, including part of the Ngorongoro Caldera. A simplified cross section shows the relationship of the rift-related volcanics and sedimentary rocks to the ancient basement complexes (Fig. 7.10).

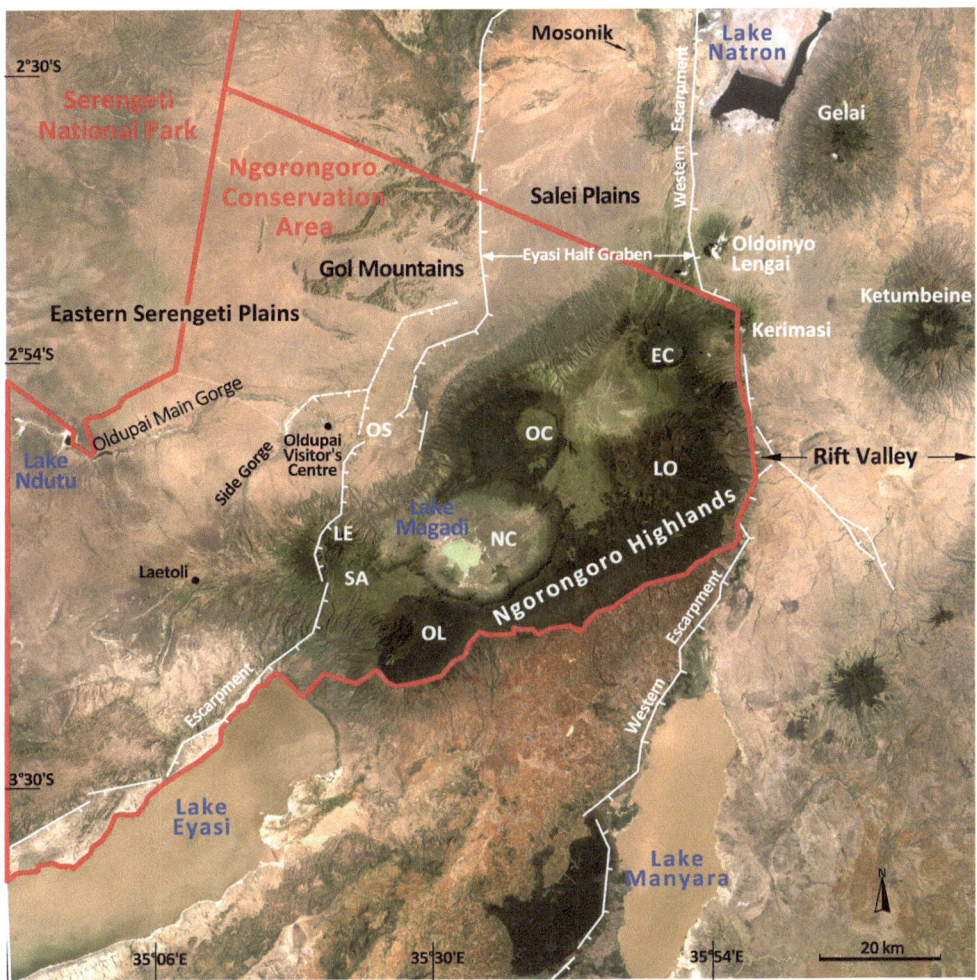

Fig. 7.6 The eastern part of the Ngorongoro Conservation Area (NCA) is constrained to the Eyasi Half-graben, a discrete structural block (located between the Rift Valley and Eastern Serengeti Plains) defined by prominent escarpments. The Rift Valley contains Lake Natron and Lake Manyara; Lake Eyasi is situated in the Eyasi Half-graben. The contrast between the lush Ngorongoro Highlands (dark green) and arid terrains of the Salei Plains, Gol Mountains, Rift Valley and Eastern Serengeti Plains (beige) is pronounced. The Ngorongoro Highlands is associated with the extinct Ngorongoro Volcanic Complex (EC: Empakaai Caldera; LE: Lemagrut; LO: Loolmalasin; NC: Ngorongoro Caldera; OC: Olmoti Caldera; OL: Oldeani; SA: Sadiman). Other components of the NCA are dominated by wind-blown ash from the active Oldoinyo Lengai Volcano. The palaeoanthropological sites of Oldupai and Laetoli are located in the Eastern Serengeti Plains. The Oldupai River peters out in the Olbalbal Swamps (OS). Approximate width of view is 150 km. *Source* NASA Landsat 7 ETM + image mosaic for the year 2000 sourced from the University of Maryland Global Land Cover Facility, processed by Philip Eales, Planetary Visions/DLR

7.5.2   Ngorongoro Caldera

Ngorongoro is a giant shield volcano with a base diameter of 35 km and caldera shoulders that rise to a maximum height of 2,417 m. The immense size of the volcano is illustrated by estimates the cone may have had a height of 5,000 m prior to caldera collapse (Pickering 1965). The volcano has an age of between 2.5 and 1.9 Ma, with the caldera event having occurred at 2.0 Ma (Dawson 2008). The caldera is a near-circular feature with an internal diameter of 22 km by 18 km (Fig. 7.11). The caldera floor occurs at an elevation of approximately 1,700 m. The internal walls have a maximum height of 350 m. The two main components of the volcano are dark-coloured lava flows and lighter, tephra-dominated units. These features can be observed in the south walls of the caldera. Lava flows and small scoria cones (e.g., Engitati Hill) form low hills within the centre of the caldera. The average composition of the lava and scoria is basaltic trachyandesite (the reader is referred to Chap. 8 for details of the chemical compositions of magmas). Magma was erupted from ring fractures close to the base of the caldera walls. Two flows of rhyolite ignimbrite associated with the caldera event can be observed in the Lerai ascent road (Mollel et al. 2009). The ignimbrite has yielded geochemical trends consistent with eruption of a stratified magma chamber, with the silicic top and basaltic base inverted by sequential eruptions (Mollel et al. 2008).

Fig. 7.7 The eastern approach to the NCA ascends the Western Escarpment near Lake Manyara (background, left). The escarpment is composed of multiple, near-horizontal volcanic layers of lavas and tephra associated with the Ngorongoro Volcanic Complex

Fig. 7.8 Ngorongoro Caldera and Lake Magadi from the viewpoint on the southeast of the caldera during the wet season

The caldera is an internally-drained basin fed by two external streams, the Munge River and the Oljoro Nyuki stream. The freshwater Ngoitokitok hot springs support a small lake and wetlands (Fig. 7.12). Lake Magadi is a large yet shallow alkaline lake which contains deposits of sodium and magnesian carbonates (Deocampo 2005). The extent and alkalinity of the lake varies considerably between seasons. These features have also changed markedly since the lake formed in the Pleistocene. Carbon dating of ostracod-bearing clays has yielded dates for a high-stand at 24,000 BP. Parts of the caldera floor are covered by lacustrine deposits (clays and silts). The southwestern quadrant is underlain by calcareous ash derived from the Oldoinyo Lengai Volcano. The ash provides a localized area of nutrient-rich short grasses, similar to those on the Eastern Serengeti Plains. These areas are favoured by grazers, and the concentration of herds to the southwest of Lake Magadi in the Ngorongoro Caldera is notable in the dry seasons.

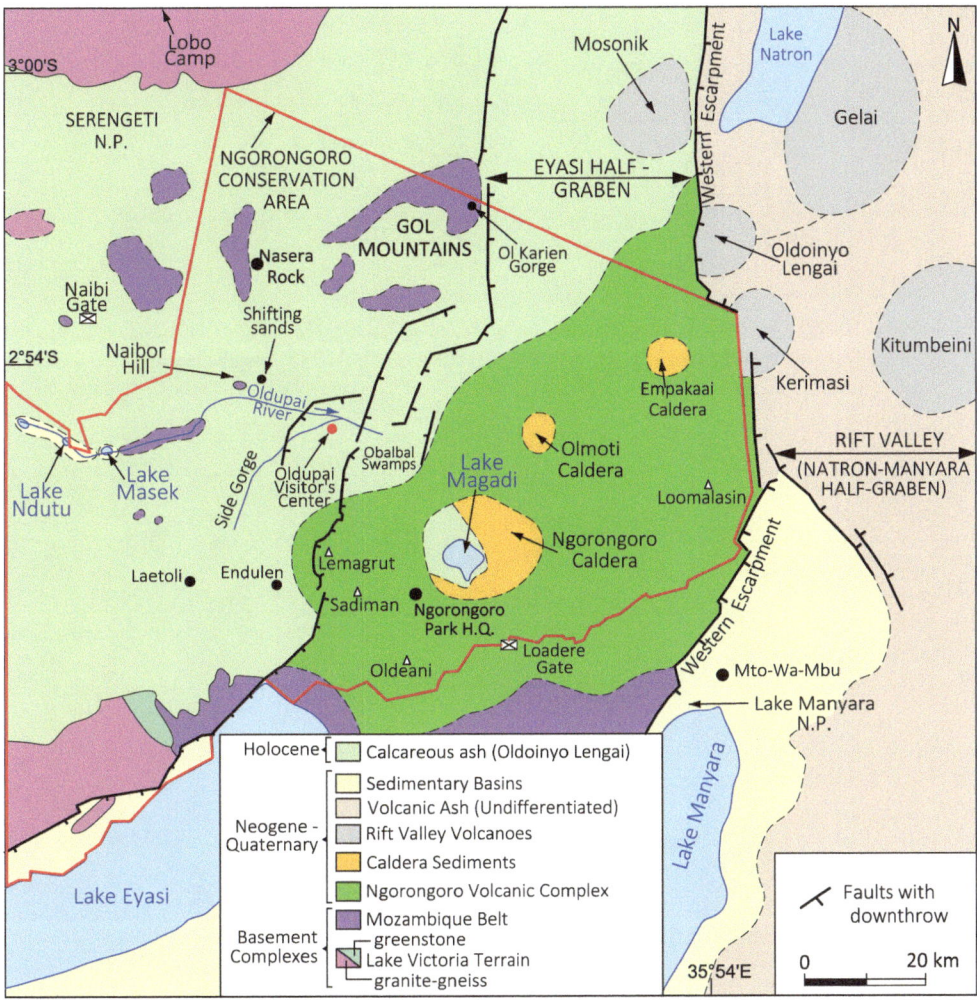

Fig. 7.9 Geological map of the Ngorongoro Conservation Area. Outcrop of the Oldupai and Laetoli Basins are not shown for reasons of scale. *Source* Simplified after Pickering (1958; 1964; 1965), Orridge (1965) and Dawson (2008)

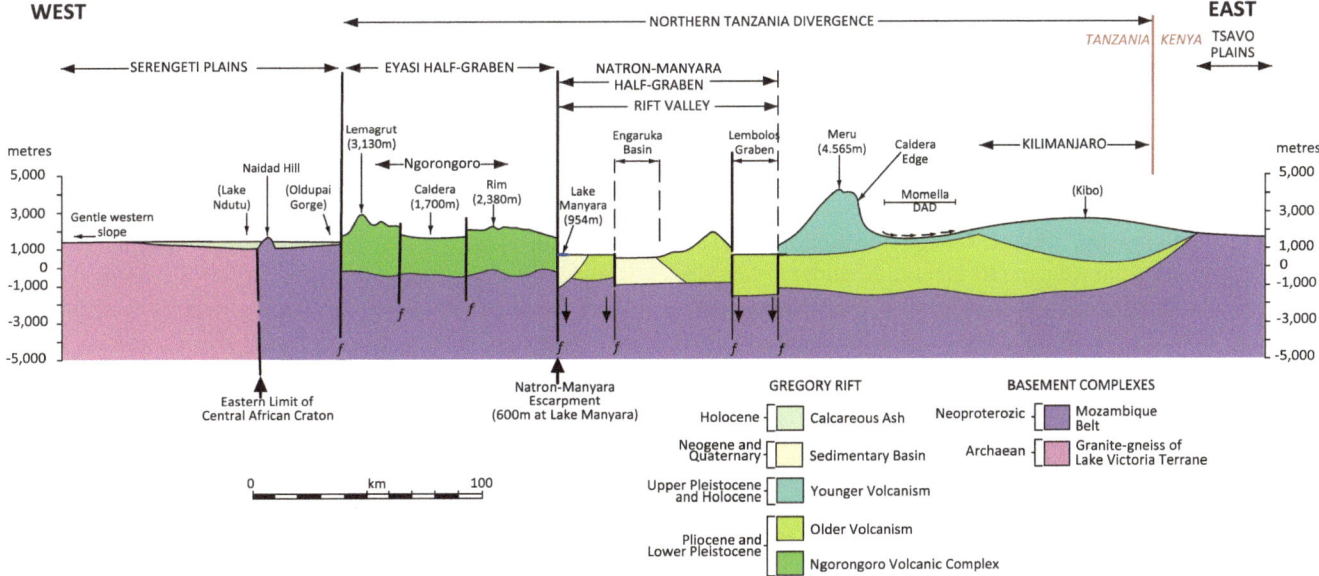

Fig. 7.10 Simplified geological section of the NCA and adjacent terrains. Section centred on the Ngorongoro Caldera (latitude 3° 10' South). Localities in brackets (e.g., Lake Ndutu, Oldupai Gorge, Kilimanjaro etc.) are located north of the section line. The thickness of the rift-related volcanics is schematic

Fig. 7.11 Geological map of the Ngorongoro Caldera and part of the Olmoti Caldera. *Source* simplified from the Geological Survey of Tanzania 1:125,000 quarter degree sheet 53 by Pickering (1965)

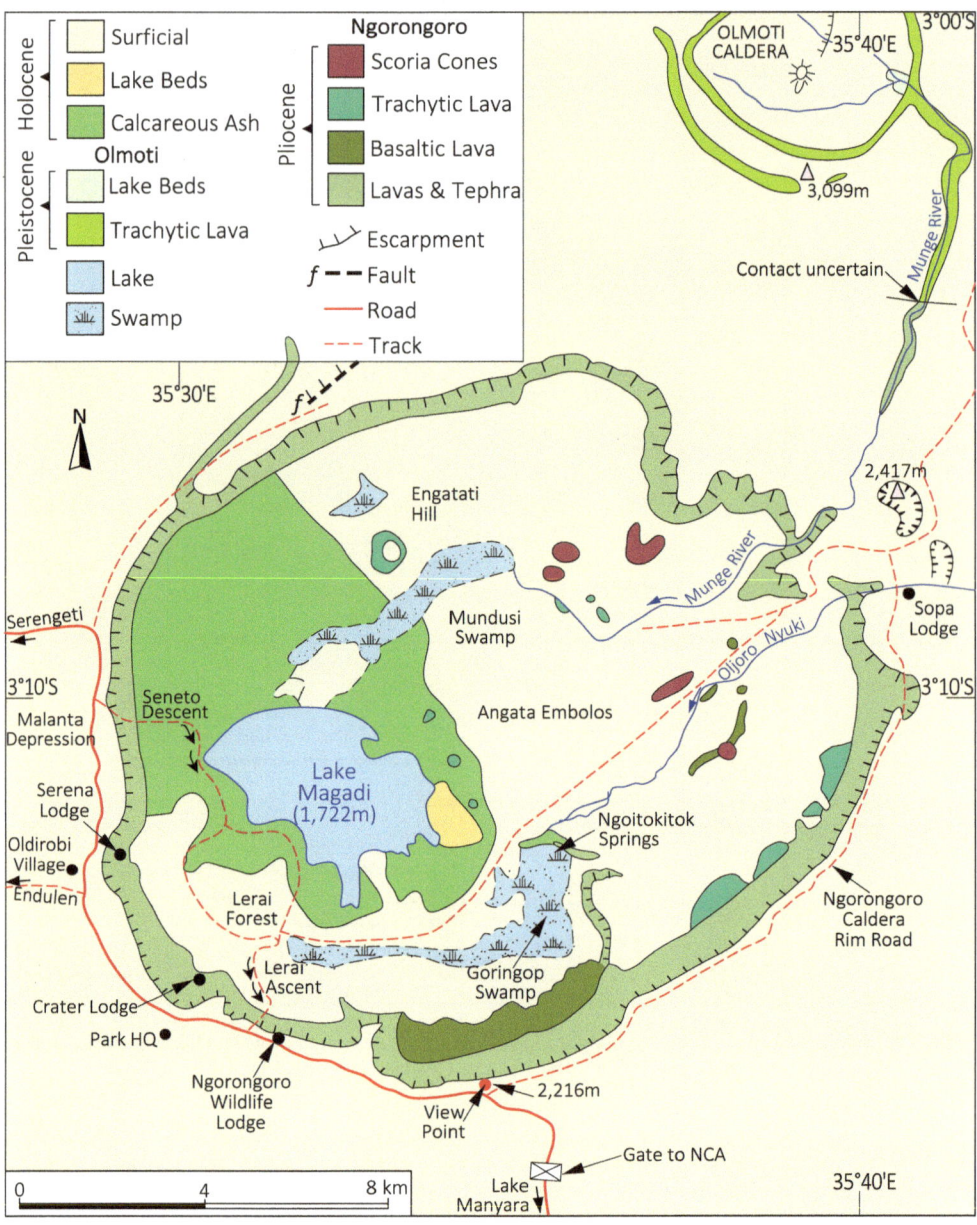

7.5.3   Other Volcanoes of the Ngorongoro Volcanic Complex

Olmoti ("cooking pot" in the Maasai language) is a large shield volcano (diameter of 30 km) with caldera shoulders that rise to a height of 3,101 m. The short-lived volcanism of Olmoti (2.01–1.79 Ma) was contemporaneous with Ngorongoro (Dawson 2008). The Olmoti Caldera has a diameter of 6.5 km and depth of 100–200 m. The southeastern flanks of the caldera are breached by the Munge River, which plunges by a series of waterfalls and ravines into the Ngorongoro Caldera (Fig. 7.11). Four small blow holes (diameters of a few tens of cm) that emit strong

draughts of cold air near the base of the Olmoti Caldera may be fossilized fumaroles.

Empakaai is a large shield volcano (diameter of 30 km) in the northeastern part of the Ngorongoro Highlands with a well-preserved caldera. The caldera has a diameter of 8 by 6 km and the thickly forested internal walls have a height of approximately 1,000 m (Fig. 7.13). The Empakaai Caldera contains a relatively deep alkaline lake. The Elanairobi Ridge (3,235 m) on the northern side of the caldera affords views of Oldoinyo Lengai and Lake Natron in the rift valley. The Loolmalasin Cone, located to the southeast of Empakaai, is the highest peak in the Ngorongoro Highlands (3,648 m). The three large volcanoes located in the

Fig. 7.12 The lush paradise of the Ngoitokitok Springs, Ngorongoro Caldera. The northern wall of the caldera is dwarfed by the flanks of the Olmoti Volcano

Fig. 7.13 A Maasai herdsman on a trail leading to the Elanairobi Ridge. The background shows the thickly forested inner slopes and lake of the Empakaai Caldera

southwestern part of the Ngorongoro Highlands are as follows: Sadiman (2,680 m)—also known as Satiman—Lemagrut (3,130 m)—or Makarot—and Oldeani (3,216 m). Sadiman has been radiometrically dated and has an age of 3.7 Ma. The composition varies from nephelinite to phonolite (Zaitsev et al. 2012). The Ogol Lavas (2.41 Ma) is a group of basaltic lavas erupted from parasitic cones that occur on the Eastern Serengeti Plains at the base of the Lemagrut Volcano.

7.5.4 Shifting Sands

The "Shifting Sands" is a group of isolated black dunes located near Oldupai Gorge (Fig. 7.9). The dunes consist of volcanic ash derived from the Oldoinyo Lengai Volcano (possibly from the 1940/1 eruption). The ash is related to eruptions of phonolite and nephelinite magmas (ashes are dispersed by the prevailing easterly winds). The largest of the dunes has a height of 9 m and length of 100 m; the crescent-shape is typical of a barchan (Fig. 7.14). The dunes consist of coarse-grained ash; the finer ash has been removed by winnowing. Dispersion of the coarse-grained ash is suppressed by a negative electric charge which is created by movement of the particles by wind and bouncing of individual grains off the ground. The electrical field created by the process of saltation assists with binding the grains together. The individual dunes are migrating west at the remarkable average rate of approximately 17 m/year. The tracks left by the dunes are clearly visible. Species of fossilized (extinct) dune beetles have been collected from the tracks.

7.5.5 Serengeti Migration

The Serengeti migration is currently estimated to include more than 1.5 million wildebeest or white-bearded gnu (*Connochaetes taurinus*), 0.5 million Burchell's zebra (*Equus quagga* or *Equus burchellii*), 0.2 million Thomson's gazelle (Eudorcas thomsonii) and 18,000 eland (*Taurotragus oryx*). Migration patterns are not as well established as may be envisaged, and the larger grazers are more-or-less continuously moving in vast numbers. Witnessing the herds crossing the major river channels and collecting in huge, disjointed groups is an unforgettable sight (Fig. 7.15). In recent times, the migration patterns have become more erratic, possibly due to climate change. In the early part of the 20[th]C, the eastward migration, instead of stopping in the northern and western parts of the NCA as is currently the case, continued as far as the Tarangiri National Park, located to the southeast of Ngorongoro. Calving, the most static time for the wildebeest, usually occurs during January-March on the Eastern Serengeti Plains and in the northern and western parts of the NCA. The alternation of the wet and dry seasons has a pronounced effect on the ecology, and the onset of the bi-annual rainy seasons (April-May; September-November) triggers remarkable changes. Arid landscapes convert rapidly into lush pastures.

The established pattern of the Serengeti migration appears to have been influenced by the distribution of the volcanic ash associated with Holocene eruptions of Oldoinyo Lengai (Scoon 2018). The ash is associated with a distinctive ecological region of the Serengeti: the "short grass plains". Growth of the nutrient-rich grasses in this region persists for a considerable period due to the shallowness of the water table. The water table is controlled by a shallow layer of hard pan located between volcanic ash and underlying rocks. The volcanic ash covers the underlying rocks as a thin blanket, albeit minor topographic features cause the basement to be exposed, e.g., quartzite (Mozambique Belt) at Oldupai Gorge (Figs. 7.9 and 7.10). The volcanic ash was originally thought to be a sedimentary deposit (calcrete), but was reinterpreted by Dawson (1964) who linked the calcareous nature to the unusual composition of some of the parental magmas associated with Oldoinyo Lengai.

7.6 Oldupai Gorge

The Oldupai Gorge and its largest tributary the Naisuri River (popularly known as the "Side Gorge") are ephemeral rivers located on the Eastern Serengeti Plains (Fig. 7.6). The spelling of Oldupai has been changed in recent years from "Olduvai" to record the meaning as "place of wild sisal" accurately in the Maasai language. The Oldupai River has a length of 48 km and is fed by two seasonal lakes, Lakes Masek and Ndutu. The river has carved a steep-sided, albeit relatively shallow ravine, the walls of which contain fossils that have greatly advanced our understanding of human evolution (Fig. 7.16a). Oldupai Gorge occurs at the foot of the Ngorongoro Highlands and many of the most important hominin fossils are preserved in volcanic ash derived from the Pliocene volcanism (Fig. 7.16b).

The initial discoveries of *Zinjanthropus boisei* (cranium OH-5) and *Homo habilis* (cranium OH-7), made by Mary Leakey in 1959 and 1960, are supported by more than sixty additional significant discoveries, yielding the most continuous record of human evolution known (Leakey 1984). The Visitors Centre at Oldupai Gorge has a museum with replicas of most of the findings. Many of the original fossils are located in the National Museum of Tanzania, Dar-es-Salaam, and the British Museum, London.

7.6.1 Geological Setting

In the vicinity of Lake Ndutu, the Oldupai River is underlain by granite-gneiss of the Lake Victoria Terrain. The river cuts progressively upward into the stratigraphy towards the east. Thus, quartzite of the Mozambique Belt is exposed in the river bed near the road which leads to the Visitors Centre and one of the youngest components of the Ngorongoro Volcanic Complex is exposed near the confluence of the river and the Side Gorge (Fig. 7.16a). The Naabi Ignimbrite is a distinctive black and lustrous pyroclastic rock linked to the catastrophic caldera event of the Ngorongoro Volcano (McHenry et al. 2008). The ignimbrite has been dated very accurately (2.038 ± 0.005 Ma), as this places a lower age limit on fossils that may occur here (Deino 2012). The Naabi Ignimbrite is unconformably overlain by the Oldupai Group, a sequence of mostly lacustrine sediments with subordinate volcanic ashes derived from the Ngorongoro Volcanics (Fig. 7.17). Seven formations—colloquially known as "Beds"—are recognized (Hay 1976; Hay 1994). The lateral dimensions and thickness of the Oldupai Group is poorly

a

b

Fig. 7.14 a Isolated, barchan-style dune of black volcanic ash at Shifting Sands with the western rampart of the Ngorongoro Highlands visible in the background; **b** The front of the dune (which is migrating towards the right of the photograph) reveals a steep slope and distinctive outer horns

Fig. 7.15 Migration of wildebeest and zebra in the vicinity of Lake Ndutu on the Eastern Serengeti Plains

constrained as exposures are restricted to linear sections within the main and side gorges (Stollhofen and Stanistreet 2012). The maximum thickness is estimated as 100 m, but a recent drilling programme suggests that this may have to be increased. The Oldupai Group was deposited in a shallow basin during the Late Pliocene. Sediments and ashes in the lowermost beds were deposited in a palaeo-lake; the uppermost beds are dominated by windblown volcanic ash. The basin developed due to warping of the basement associated with release of stress on the westward side of the southward-propagating Gregory Rift.

7.6.2   Oldupai Group

Beds I and II are dominated by pale-coloured lacustrine clays and sands (Fig. 7.17). Relatively hard and well-defined tuff layers are a subordinate feature, yet constitute important markers. The tuff layers yield accurate radiometric dates. Tuff layers 1a through 1e in Bed I are dated at 2.015–1.803 Ma (Deino 2012). The tuffs in Bed I are related to the waning activity of the Ngorongoro Volcano (McHenry et al. 2008). Bed I also contains clasts eroded from the Sadiman Volcano. The Marker Tuff (1f) at the base of Bed II records a change in the source of the tephra, which switched from Ngorongoro to the Olmoti Volcano (McHenry 2012). The palaeo-environment of Beds I and II has been interpreted as a wooded peninsula which hominins visited (and hunted in), rather than settled semi-permanently as was envisaged for many years (Blumenschine et al. 2012). The new interpretation is supported by discovery of a fossilised tree in Bed I (Habermann et al. 2016).

Bed III (which has yielded few fossils) consists of red-brown coloured sandstone with minor conglomerate and clay. The colour is associated with periodic changes from wet to dry environments along the fringes of small alluvial fans. Bed IV (which contains abundant hominin fossils) consists of fluvial clays, conglomerates, and sands. The uppermost formations are dominated by volcanic ash derived from Kerimasi (Bed V) and from Oldoinyo Lengai (Beds VI and VII). The volcanic ashes contain few fossils, although Bed VII contains evidence of *Homo sapiens* (age of 17,000 BP).

7.6.3   Discovery of the Hominin Fossils

The first scientist to visit Oldupai was probably Wilhelm Kattwinkel, a German entomologist, who in 1911 discovered fossils of extinct mammals. The first hominin remains were discovered in 1913 by Hans Reck, a German palaeontologist and volcanologist. The hominin fossils included the "Reck Skeleton" (probably *Homo erectus*), which is surrounded by mystery as it disappeared from the Munich Museum during World War II. The discovery of hominin fossils at Oldupai Gorge is, however, largely related to the British husband and wife team of Mary and Louis Leakey. The Leakey's invited Reck to join their first expedition to Oldupai, in 1931, during which time the team discovered numerous stone tools. The potential for finding additional hominin fossils was sufficiently encouraging as to result in the Leakey's establishing a semi-permanent camp (Leakey et al. 1931).

The partial cranium of a hominin (OH-5) was discovered by Mary Leakey in 1959, at level 22 in a trench located at

Fig. 7.16 **a** The stratigraphy at Oldupai Gorge includes distinctive, near-horizontal layers of lacustrine sediments and volcanic lavas and ashes (view close to the Visitors Centre). The principal hominin discoveries were made in Beds I and II, which are separated by the Marker Tuff; **b** A view of the red-brown buttes associated with Bed III shows the proximity of the extinct volcanoes associated with the Ngorongoro Highlands (background)

Fig. 7.17 Stratigraphy of the Oldupai Gorge with column on right showing details of Bed I and the lowermost part of Bed II. *Source* Modified after Hay (1976) and Stollhofen and Stanistreet (2012) with radiometric data of Deino (2012)

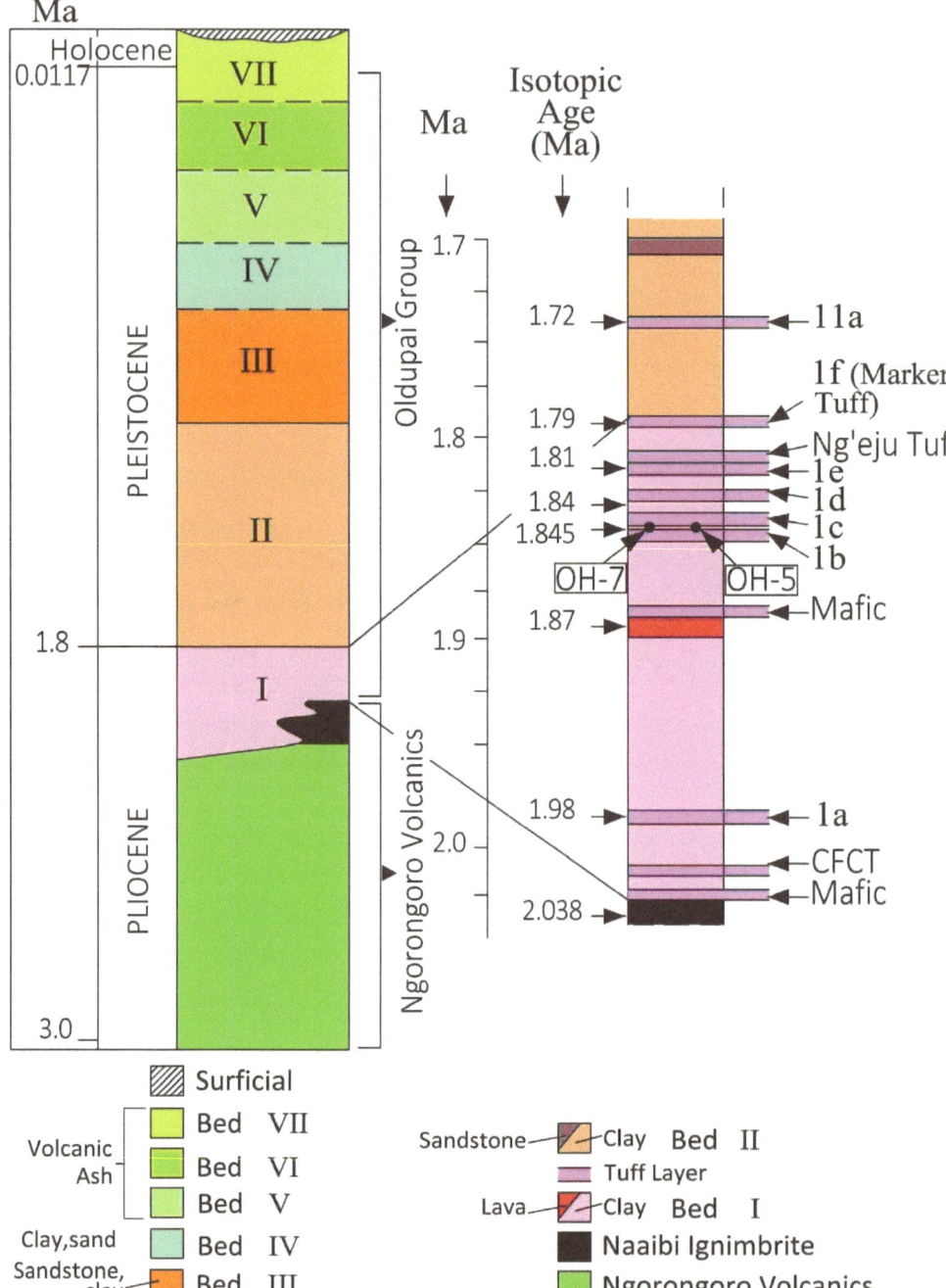

the FLK site (named for Louis's first wife, Frida Leakey Korongo) (Fig. 7.18). OH-5 was named by Louis Leakey as *Zinjanthropus boisei* in part to commemorate the benefactor of the research, Charles Boise (Leakey 1981). The stratigraphic height at which OH-5 was discovered is known as the FLK-*zinj* level: this surface occurs between Tuff Ib and Tuff Ic (Fig. 7.17). The age of OH-5 is estimated at 1.848 Ma, based on the position relative to the two tuff layers. OH-5 is known colloquially as "Nutcracker Man", in recognition of the powerful jaws and teeth, and some researchers use the description *Paranthropus* (others have reverted to the original name). A cast of the OH-5 cranium is displayed together with the Peninj mandible, the latter having been discovered by Kamoya Kimeu at Peninj, near Lake Natron (Fig. 7.19a). *Zinjanthropus boisei* has distinctive, robust features in comparison to the younger hominins (e.g.,

Fig. 7.18 The FLK site where the cranium OH-5 (*Zinjanthropus boisei*) was discovered in Bed I at Oldupai Gorge is commemorated by a stone cairn and plaque. The section in the cliff face includes resistant tuff layers (pale grey) interbedded with clays (brown) in the upper part of Bed I. The uppermost tuff layer in the section is the Marker Tuff which is overlain by clays (dark brown) in the lower part of Bed II

H. habilis; H. erectus), and is classified with the relatively primitive *Australopithecine. Zinjanthropus boisei* may have occurred over large parts of East Africa.

The second major discovery made at Oldupai Gorge by Mary Leakey was of OH-7, located at the FLK-South site (at a similar height to OH-5) and with an estimated age of 1.848–1.832 Ma. OH-7 is comprised of the lower jaw and parts of a hand and was named as *Homo habilis* ("Handy Man"). *Homo habilis* differs from the *Australopithecine* on the basis of larger brain capacity and decreased teeth size. Complete skulls of *Homo habilis* were subsequently

discovered by Kamoya Kimeu and Richard Leakey at Koobi Fora, on the eastern shores of Lake Turkana, Kenya (Fig. 7.19b). Other notable finds at Oldupai include OH-8 (foot of *H. habilis*) and OH-9 (flattened cranium of *H. habilis*; also known as "Twiggy"). Numerous fossils of *Homo erectus* (1.2–0.70 Ma) have been located at Oldupai Gorge in Beds II and IV.

7.7 Laetoli

Laetoli occurs in a remote location on the Eastern Serengeti Plains west of the village of Endulen (Fig. 7.6). Visitors are few and far between as the access road from Ngorongoro is not on the main tourist routes. Laetoli occurs close to a watershed between two ephemeral river systems, the Naisuri and Gadjingero Rivers, which flow northeast and southwest, respectively (Andrews and Bamford 2008). Laetoli is subjected to extensive badlands erosion, and the footprints discovered by Paul Adell in 1978 occur on the banks of a wide river bed (Fig. 7.20). New gullies develop during the annual floods. The footprints were made by *Australopithecus afarensis*, as described and interpreted by Mary Leakey (Leakey 1981). The type locality for this genus is, however, in Ethiopia (the "Lucy" skeleton dated at 3.2 Ma) (e.g., Stringer and McKie 1996). A full-size cast of the footprints can be viewed in the on-site museum (Fig. 7.21). The original footprints have been covered by a rock pavement for protection.

The footprints occur in volcanic ash dated at 3.6 Ma (Leakey 1981). The volcanic ash is part of the Laetoli Group, a succession of lacustrine sediments and volcanic ashes subdivided into a Lower Unit (4.3–3.8 Ma), which contains few fossils, and an Upper Unit (3.8–3.5 Ma) rich in both fossils and tracks (Hay 1987). Exposures of the Laetoli Group are mostly restricted to gullies and areas of badlands erosion. The plateau enclosing the gullies is covered by younger formations, including calcareous ash derived from Oldoinyo Lengai. The ash in the Upper Unit of the Laetoli Group was originally assigned to the Lemagrut Volcano (Leakey and Hay 1979), but recent evidence suggests the source is the Sadiman Volcano (Zaitsev et al. 2011). A track made by two individuals approximately 150 m from the original site, but travelling in a similar direction as the early footprints, is a recent discovery (Masao et al. 2016). This supports the theory that the body size of *Australopithecus afarensis* varied considerably amongst individuals. Footprints of large antelope,

Fig. 7.19 a Reconstruction of the cast of the 1.85 Ma old cranium of *Zinjanthropus boisei* (OH-5) discovered in Bed I at Oldupai Gorge, together with the lower jaw (the Peninj mandible) found near Lake Natron. Photograph by Lillyundfreya of the Westfälisches Musuem; **b** Replica of the 1.9 Ma old cranium of *Homo habilis* (KNM-ER1813) from Koobi Fora, Kenya. Photograph by Locutus Borg

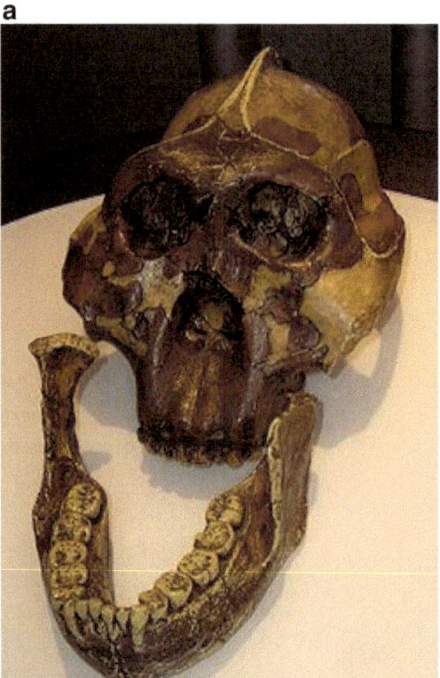
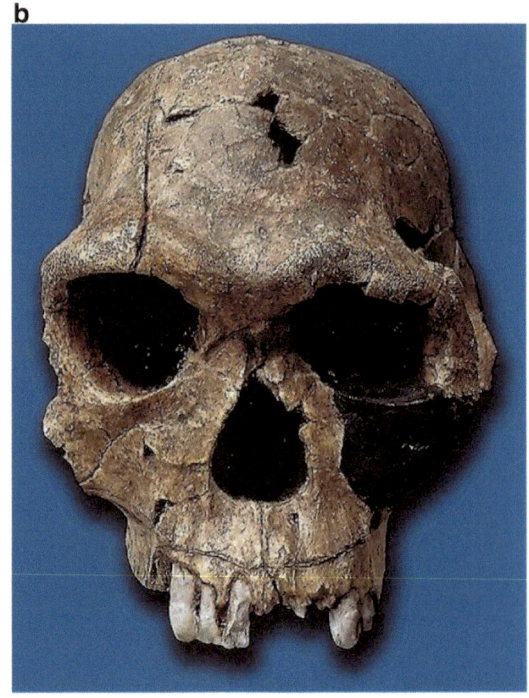

Fig. 7.20 Laetoli is located in an area of badlands erosion on the Eastern Serengeti Plains. The museum is just visible in background

elephant, and horses, as well as raindrops, have also been discovered, suggesting the hominins were part of Pliocene-age migrations during rainy seasons.

7.8 Evolution of Hominins

The identification of multiple species of hominins that co-existed over long periods of geological time is an intriguing feature of the Oldupai Gorge anthropological site. *Zinjanthropus boisei* (2.3–1.2 Ma) and *Homo habilis* (1.8–1.2 Ma) co-existed during the Upper Pleistocene. *Homo habilis* and *Homo erectus* (1.2–0.7 Ma) co-existed during the Middle Pleistocene. Evidence of increased development of hominins at Oldupai Gorge is supported by studies of stone tools. The Leakey's excavated more than 2,000 "Oldowan" tools, covering the period 1.8–0.6 Ma. Tools obtained from Bed I consist of pebbles with edges that have been sharpened. The younger beds reveal more advanced tools, including hand axes. Mary Leakey observed that *Homo habilis* used several different types of tools and achieved a high level of social development (Leakey 1984). Some tools were carved from lava clasts associated with the Sadiman Volcano. The clasts consist of dense homogeneous rocks with an attractive greenish colour.

Evidence suggests *Homo sapiens* arrived comparatively recently at Oldupai. Stone artifacts (dated at 30,000 BP) linked to *Homo sapiens* have been discovered at Nasera Rock in the Gol Mountains. The artifacts are constructed of quartzite (Mozambique Belt). The Nasera Rock site includes rock paintings. Anthropological sites in southern Africa suggest the oldest known age of *Homo sapiens* is approximately 200,000 BP, e.g., Pinnacle Point near Mossel Bay, Western Cape (Marean 2010).

The influence of geological catastrophes and climate change in hominin evolution is significant. The retreat of forests that formerly covered much of East Africa during the Early Pliocene, and the subsequent development of extensive grassy savannahs has been widely described (e.g.,

Fig. 7.21 Cast of part of the footprints made by *Australopithecus afarensis*, Laetoli Museum

Cerling et al. 2011). Hominins at Oldupai and Laetoli evolved during times of intense geological upheavals, including rifting and volcanism. The catastrophic caldera event associated with the Ngorongoro Volcano, for example, may have initiated extinction of the *Australopithecine* and evolution of more advanced hominins. The abrupt appearance of *Homo sapiens* during the Late Pleistocene Ice Ages (in the Western Cape) may also reflect climate change.

The possibility of our human ancestors having evolved in Africa, particularly in areas proximal to the East African Rift Valleys is widely accepted (Stringer and McKie 1996). Categorization of hominin fossils is, however, contentious and the story of how the discoveries at Oldupai were received into the scientific community is full of intrigue. The evolutionary relationship between different species of hominins, and particularly the link between hominins and *Homo sapiens*, is unlikely to be ever fully proven (due to the paucity of fossil evidence). Traditional arguments that evolution occurred in a regular, tree-like pattern with each subsequent jump being more evolved is probably best discarded. The evolutionary patterns are far more complex, and in the words of geneticist Adam Rutherford "there was no beginning, and there are no missing links, just the ebb and flow and ebb again of living through epochs" (Rutherford 2017).

7.9 Treks

Several treks can be undertaken in the NCA where the interrelationship between fauna, flora, and geology is apparent. The salt flats of Lake Natron reveal the extensiveness of the sodium carbonate salt deposits. A longer trek involves hiking from Empakaai down into the rift valley near Lake Natron. This includes the scenic waterfalls of the Engare Sero Gorge, accessed by a path which includes scrambling on the rock walls of the gorge and wading in the river (Fig. 7.22a). The gorge reveals exposures of the Ngorongoro Volcanic Complex, including lava flows, ash layers, debris deposits, and breccia dykes (Fig. 7.22b). Hikes around the forested rim of the Ngorongoro Caldera and to the summit of the Olmoti Cone are also recommended, as is the trek to the base of the Empakaai Caldera to examine the brackish lake and salt deposits. A highlight of a visit to the Gol Mountains is a hike in the Ol Karien Gorge, a narrow slot in the quartzite only a few metres wide in some places. The valleys between the quartzite hills are dry and inhospitable in the dry season, but are lush and verdant after the rains when large herds of grazers spill over from the Eastern Serengeti Plains. Many of the treks in the NCA and surrounding areas are guided by the Maasai from local communities, who are extremely knowledgeable as regards the natural and human history of the region.

Fig. 7.22 **a** Hiking in the Engare Sero Gorge near Lake Natron reveals exposures of volcanic debris avalanche deposits (foreground, right) as well as thick sequences of lavas and tephra; **b** Nephelinitic breccia dyke cuts a basaltic lava flow, Engare Sero Gorge

References

Andrews, P., & Bamford, M. (2008). Past and present vegetation ecology of Laetoli, Tanzania. *Journal of Human Evolution, 54,* 78–98.

Bailey, D.K. (1974). Continental Rifting and Alkaline Magmatism. In: Sorensen, H. (Ed.). *The Alkaline Rocks,* Wiley p. 148–159.

Blumenschine, R. J., Stanistreet, I. G., Njau, J. K., Bamford, M. D., Masao, F. S., Albert, R. M., et al. (2012). Environments and hominin activities across the FLK peninsula during Zinjanthropus times (1.84 Ma), Olduvai Gorge, Tanzania. *Journal of Human Evolution, 63,* 364–383.

Cerling, T. E., Wynn, J. G., & Andanje, S. A. (2011). Woody cover and hominin environments in the past 6-million years. *Nature, 476,* 51–56.

Dawson, J. B. (1962). Sodium carbonatite lavas from Oldoinyo Lengai, northern Tanganyika. *Nature, 196,* 1065–1066.

Dawson, J. B. (1964). Carbonatatic volcanic ashes in northern Tanganyika. *Bulletin of Volcanologique, 27,* 81–92.

Dawson, J. B. (2008). The Gregory Rift Valley and Neogene-recent Volcanoes of Northern Tanzania. *Geological Society Memoir, 33,* 102p.

Deino, A. L. (2012). $^{40}Ar/^{39}Ar$ dating of Bed I Olduvai Gorge, Tanzania and the chronology of early Pleistocene climate change. In: R. J. Blumenschine, E. T. Masao, I. G. Stanistreet, & C. C. Swisher (Eds.), Five decades after Zinjanthropus and Homo habilis: Landscape paleoanthropology of Plio-Pleistocene Olduvai Gorge, Tanzania. *Journal of Human Evolution, 63,* 252–273.

Deocampo, D. M. (2005). Evaporative evolution of surface waters and the role of aqueous CO_2 in magnesian silicate precipitation: Lake Eyasi and Ngorongoro Crater, northern Tanzania. *South African Journal of Geology, 108,* 493–504.

Dirks, P. H. G. M., Blenkinsop, T. G., & Jelsma, H.A. (2015). The geological evolution of Africa. In: *Geology* (Vol. IV). Encyclopaedia of Life Support Systems (EOLSS).

Ebinger, C. (2005). Continental break-up: The East African perspective. *Astronomy and Geophysics,* 46, 16–21.

Eugster, H. P. (1980). Lake Magadi, Kenya, and its Pleistocene precursors. In A. Nissenbaum (Ed.), Hypersaline brines and evaporitic environments (pp. 195–232). Amsterdam: Elsevier.

Gregory, J. W. (1894a). Contributions to the physical geography of British East Africa. *Geographical Journal, 4,* 290–315, 408–424, 505–514.

Gregory, J. W. (1894b). Contributions to the geology of British East Africa: Glacial geology of Mount Kenya. *Quarterly Journal Geological Society of London, 50,* 515–530.

Habermann, J. M., Stanistreet, I. G., Stollhofen, H., Albert, R. M., Bamford, M. K., Pante, M. C., et al. (2016). In situ 2.0 Ma trees discovered as fossil rooted stumps, lowermost Bed I, Olduvai Gorge, Tanzania. *Journal of Human Evolution, 90,* 74–97.

Hay, R. L. (1976). Geology of the Olduvai Gorge: A Study of sedimentation in a Semiarid Basin (p. 203). Berkeley: University of California Press.

Hay, R. L. (1987). Geology of the Laetoli Beds. In M. D. Leakey & J. M. Harris (Eds.), Laetoli, a Pliocene Site in Northern Tanzania (pp. 23–47). Oxford: Clarendon Press.

Hay, R. L. (1994). Geology and dating of Beds III, IV and the Masek Beds. In M. D. Leakey & D. A. Roe (Eds.), Olduvai Gorge (Vol. 5, pp. 8–14)., Excavations in Beds III, IV and the Masek Beds, 1968–1971 Cambridge: Cambridge University Press.

Leakey, M. D. (1981). Discoveries at Laetoli in Northern Tanzania. *Proceedings of the Geologists' Association, 92,* 81–86.

Leakey, M. D. (1984). Disclosing the Past. New York: Doubleday and Co.

Leakey, M. D., & Hay, R. L. (1979). Pliocene footprints in the Laetoli Beds at Laetoli, northern Tanzania. *Nature, 278,* 317–323.

Leakey, L. S. B., Hopwood, A. T., & Reck, H. (1931). Age of the Oldowan bone beds, Tanganyika. *Nature, 128,* 724.

Le Gall, B., Nonnotte, P., Rolet, J., Benoit, M., Guillou, H., Mousseau-Nonotte, M., Albaric, J. & Deverchere, J. (2008). Rift propagation at craton margin: distribution of faulting and volcanism in the north Tanzanian divergence (East Africa) during Neogene times. *Tectonophysics* 448, 1–19.

Marean, C. (2010). Pinnacle Point Cave 13B (Western Cape Province, South Africa) in context: The Cape Floral Kingdom, shellfish, and modern human origins. *Journal of Human Evolution, 59,* 425–443.

Masao, F. T., Ichumbaki, B .T., Cherin, M., Barili, A., Boschian, G., Lurino, D. A., Menconero, S., Moggi-Cecchi, J. & Manzi, G. (2016). New footprints from Laetoli (Tanzania) provide evidence for marked body size variation in early hominins. *eLife, 5,* 1–29.

Mattson, H. B., & Vuorinen, J. (2009). Emplacement and inflation of natrocarbonatite lava flows during the March-April 2006 eruption of Oldoinyo Lengai, Tanzania. *Bulletin of Volcanology, 71,* 301–311.

McHenry, L. J. (2012). A revised stratigraphic framework for Olduvai Gorge Bed I based on tuff geochemistry. In: R. J. Blumenschine, E. T. Masao, I. G. Stanistreet, & C. C. Swisher (Eds.). Five decades after Zinjanthropus and Homo habilis: Landscape Paleoanthropology of Plio-Pleistocene Olduvai Gorge, Tanzania. *Journal of Human Evolution, 63,* 284–299.

McHenry, L. J., Mollel, G. F., & Swisher, C. C. (2008). Compositional and textural correlations between Olduvai Gorge Bed I tephra and volcanic sources in the Ngorongoro Volcanic Highlands, Tanzania. *Quaternary International, 178,* 306–319.

Mollel, G. F., Swisher, C. C., Feigenson, M. D., & Carr, M. J. (2008). Geochemical evolution of Ngorongoro Caldera, Northern Tanzania: implications for crust-magma interaction. *Earth and Planetary Science Letters, 271,* 337–347.

Mollel, G. F., Swisher, C. C., McHenry, L. J., Feigenson, M. D. & Carr, M. J. (2009). Petrogenesis of basalt-trachyte from Olmoti, Tanzania. *Journal of African Earth Sciences, 54,* 127-143.

Mosley, P. P. (1993). Geological evolution of the late Proterozoic "Mozambique Belt" of Kenya. *Tectonophysics, 221,* 223–250.

Orridge, G.R. (1965). Mbulu. Tanzania Mineral Resources Division, Quarter Degree Sheet 69.

Pickering, R. (1958). Quarter Degree Sheet 38: Oldoinyo Ogol. Geological Survey of Tanganyika (1:125,000).

Pickering, R. (1964). Quarter Degree Sheet 52: Endulen. Geological Survey of Tanzania (1:125,000).

Pickering, R. (1965). Quarter Degree Sheet 53: Ngorongoro. Geological Survey of Tanzania (1:125,000).

Rutherford, A. (2017). A brief history of everyone who ever lived (p. 419). London: Weidenfeld and Nicolson.

Saggerson, E. P., & Baker, B. H. (1965). Post-Jurassic erosion surfaces of East Africa and their deformation in relation to rift structure. *Journal of the Geological Society of London, 121,* 51–72.

Scoon, R. N. (2018). Geology of National Parks of Central/southern Kenya and Northern Tanzania (220p.). Springer International. ISBN: 978-3-319-73784-3.

Shackleton, R. M. (1986). Precambrian collision tectonics in Africa. In: M. P. Coward & A. C. Ries (Ed.). Collision Tectonics. *Geological Society London Special Publication, 19,* 324–326.

Stringer, C., & McKie, R. (1996). African Exodus (p. 267). London: Jonathan Cape.

Stollhofen, H., & Stanistreet, I. G. (2012). Plio-Pleistocene synsedimentary fault compartments, foundation for the eastern Olduvai Basin paleoenvironment, Tanzania. *Journal of Human Evolution, 63,* 309–327.

Zaitsev, A. N., Wenzel, T., Spratt, J., Williams, T. C., Strekopytov, S., Sharygin, V. V., et al. (2011). Was Sadiman Volcano a source for the Laetoli Footprint Tuff? *Journal of Human Evolution, 61,* 121–124.

Zaitsev, A. N., Marks, M. A. W., Wenzel, T., Spratt, J., Sharygin, V. V., Strekopytov, S., et al. (2012). Mineralogy, geochemistry and petrology of phonolitic to nephelinitic Sadiman volcano, Crater Highlands, Tanzania. *Lithos, 152,* 66–83.

Mediterranean Basins and Italian Island Volcanoes

8

Active Volcanoes, Earthquakes, Microplates and Greek Mythology

Abstract

The Ancient Greeks settled many of the coastal areas in southern Italy and the Italian Islands. Well known archaeological sites of *Magna Graecia* include Paestum, near Naples, which includes three well-preserved temples, and Syracuse in Sicily. Mythology was an important part of the Ancient World and the difficulties that *Odysseus* experienced on his return voyage from Troy (western Turkey) to his home island of Ithaca (northwest Greece), as recounted by the Greek poet Homer, contain descriptions of the perils of navigating the Mediterranean Sea. The historical (and mythological) record includes geological catastrophes such as earthquakes, volcanic eruptions, tsunamis, and tidal whirlpools. The Stromboli Volcano has been described as the world's oldest lighthouse as the glow of the vent can be seen from many kilometres at night. The Mediterranean is comprised of multiple interlocking basins, each of which contains a sea identified by the ancient geographers. The Adriatic Sea, the Aegean Sea, the Ionian Sea, and the Tyrrhenian Sea each have different physical and chemical characteristics. Large parts of the central and eastern Mediterranean remain tectonically active. The Alpine Orogeny is a long drawn out process of continental collision associated with the northward-migrating African Plate and the Eurasian Plate. Tectonism peaked in the Oligocene-Miocene with formation of multiple chains of fold mountains. The collision is ongoing and includes subduction of oceanic crust associated with the ancient Tethys Ocean. The complexity of the tectonic setting is illustrated by the recognition of microplates, as well as the curvilinear nature of the Hellenic Trench, the current location of the plate boundary. The compressional tectonism was displaced in the Pliocene by localized regions of crustal extension. Crustal extension associated with development of fore-arc and back-arc basins has resulted in some regions being subjected to frequent and relatively shallow, earthquakes. Catastrophic events between 1169 and 1908 resulted in the destruction of the cities of Catania and Messina. The Italian Island volcanoes are related to the convergent plate boundaries. There are several potentially hazardous volcanoes in the Aeolian Islands, including Stromboli and the Fossa cone, associated with a volcanic island arc. The Strombolian style of volcanism is characterized by relatively small eruptions constrained to discrete craters. The Fossa cone on the island of Vulcano (named after *Vulcan*, the Roman god of fire) is characterized by short-lived, yet violent eruptions. The Etna Volcano in northeast Sicily is one of the largest stratovolcanoes on Earth and also one of the most active. Historical activity includes the 1669 eruption during which lava flowed into the city of Catania. Etna is a major tourist attraction and the volcanic cone is protected in a national park. Recent eruptions of Etna are generally restricted to the upper parts of the cone, but can be sufficiently hazardous as to restrict tourist visits. The small island of Pantelleria is part of an active volcanic system associated with a transform fault on the plate boundary between Sicily and North Africa.

Keywords

Etna • Mediterranean basins • Odyssey • Plinian eruptions • Straits of Messina • Stromboli • Vulcano

Photographs not otherwise referenced are by the author.

8.1   Introduction

Southern Italy and the Italian Islands are part of one of the most tectonically active areas on Earth. There is a historical record of geological catastrophes including earthquakes, volcanic eruptions, tsunamis, and tidal whirlpools. A sequence of catastrophic earthquakes occurred in the Calabria region of southern Italy and eastern Sicily between 1169 and 1908, resulting in the destruction of the cities of Catania and Messina. The Italian Island volcanoes include active cones, some of which are potentially hazardous. The volcanoes described here include Stromboli and the Fossa cone in the Aeolian Islands, Etna in northeast Sicily, and the island of Pantelleria located between Sicily and North Africa. Stromboli is characterized by relatively small-scale eruptions which are constrained to summit craters. The activity includes showers of scoria, or glowing lumps of lava. The Fossa Cone on the island of Vulcano is characterized by short-lived, yet violent eruptions. The Etna Volcano is a major tourist attraction, and the mostly moderate eruptions that have occurred in recent years have been widely reported (Fig. 8.1). Pantelleria is part of a large volcanic centre, the majority of which is submerged beneath the waters of the Straits of Sicily. Italian geologists can claim to have been among the founders of the science of volcanology, and many of the volcanoes, have long-established observatories.

The Mediterranean is comprised of multiple interlocking basins, each of which contains a sea identified by the ancient geographers, e.g., the Adriatic Sea, the Aegean Sea, the Ionian Sea, and the Tyrrhenian Sea (Fig. 8.2). Some of the seas are associated with discrete microplates and/or back-arc basins. The majority of the volcanoes, both in southern Italy and the Italian islands, occur in well-defined regions associated with either active subduction zones or regional

Fig. 8.1 Lava fountains and small lava flows associated with strombolian activity, eruption of the southeast summit crater Etna (24th December 2018). *Source* Emanuela Carone, VolcanoDiscovery Italy (https://images.app.goo.gl/)

sutures. The tectonic framework of the Mediterranean is unusually complex. Earthquakes and volcanic activity is driven by ongoing collision of the African Plate and Eurasian Plate. The complexity of the tectonic framework is explained in terms of a complex array of microplates, with development of basins associated with localized areas of crustal extension.

8.2   Ancient Greeks and Romans

The Mediterranean Sea was an important trade route for ancient communities based in parts of the Middle East and southeast Europe. Many of these communities played fundamental roles in shaping European history. The Ancient Greeks settled extensively in coastal areas of southern Italy and Sicily. Some historians refer to this period of colonization as the "Greek Mediterranean Adventure" and the region is known as *Magna Graecia* (Great Greece). The migration commenced in the 8[th]C BC and probably followed routes established by older civilizations, including the Mycenaeans and Phoenicians. Settlements established during this period originated from the disjointed city-states of Ancient Greece and western Turkey. Groups included the Athenians (the largest of the city-states in southeast Greece) and communities from the Peloponnese, such as the Corinthians, Messenians, and Spartans (Chap. 11). The colonizers arrived in different waves and mixed and integrated with the original inhabitants. Two of the most well known archaeological sites of *Magna Graecia*, the combined Paestum and Velia site, as well as Syracuse, are described here.

A well known example of Greek mythology is the epic poem the *Odyssey*, probably written by Homer in the 8[th]C BC, but describing events from the 13[th]C BC. The difficulties that *Odysseus* experienced on his return voyage from Troy to his home island of Ithaca, including the twin perils of *Scylla* and *Charybdis*, which are today used as a metaphor for a situation with difficult choices, are in part descriptions of the perils of navigating the Mediterranean Sea. The Aeolian Islands, an archipelago located off the north coast of Sicily, is named after the Greek god of the wind, *Aeolus*. The Stromboli Volcano has been described as the world's oldest lighthouse as the glow of the vent can be seen from many kilometres at night.

Historical records of geological catastrophes, from both the Ancient Greek and Roman times, include details of earthquakes, volcanic eruptions, tsunamis, and whirlpools. Some of the descriptions have been incorporated into modern geological science. A well known example is the 79 AD eruption of Vesuvius, which devastated Roman settlements in the vicinity of Naples, and was witnessed and described by Pliny the Younger (Chap. 9). Catastrophic volcanic eruptions are known to volcanologists as *Plinian* events. The

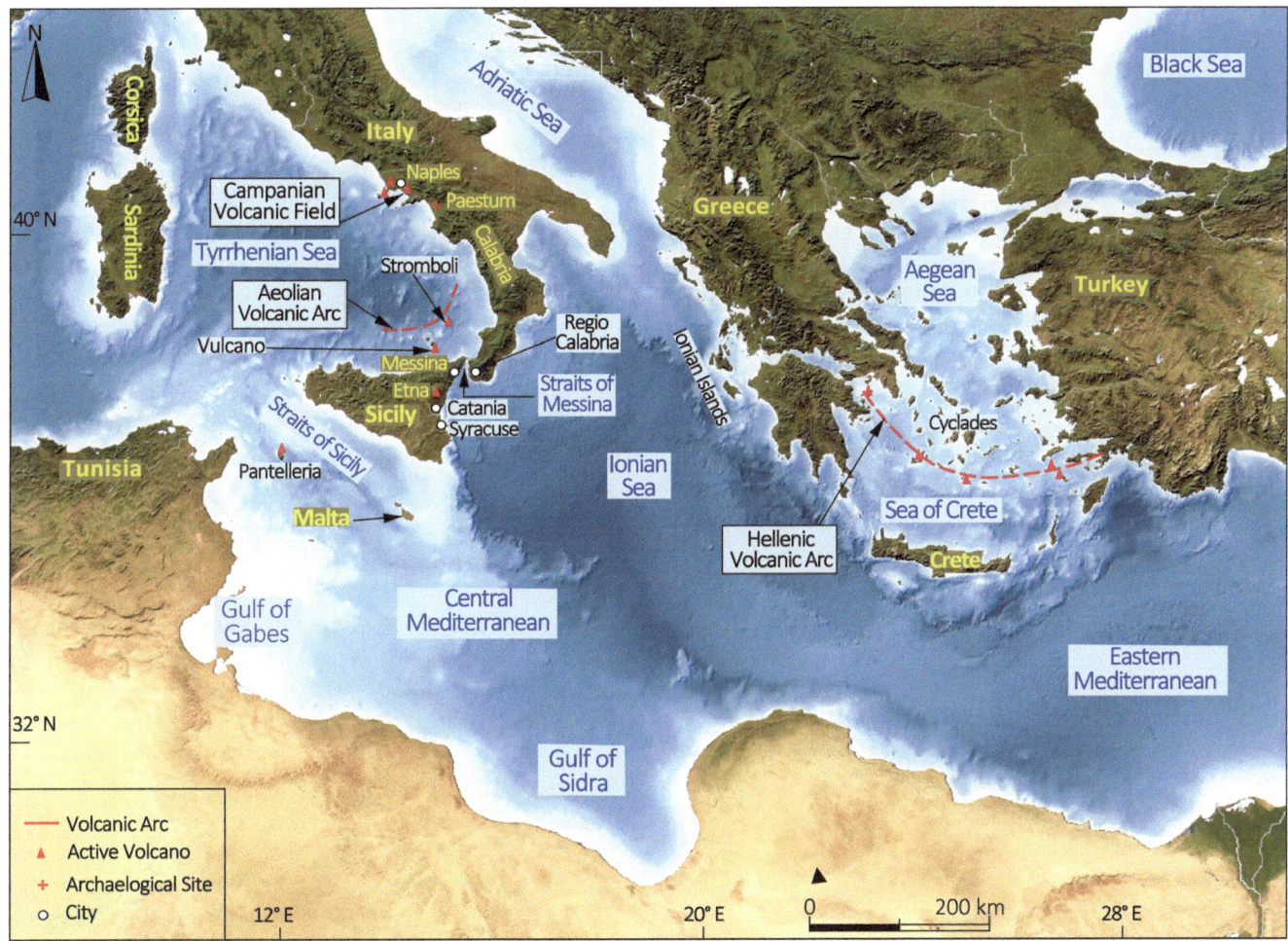

Fig. 8.2 Image of the central and eastern Mediterranean, including bathymetry, showing the seas and volcanoes described here. *Source* Satellite Image of Europe based on NASA MODIS data processed by Philip Eales, Planetary Visions/DLR

name of the island of Vulcano is derived from *Vulcan*, the Roman god of fire. In Roman mythology, the island was thought to comprise a chimney for Vulcan's underground workshops.

8.3 Regional Geology

The regional geology of the Mediterranean region is relatively complex as remnants of the ancient Tethys Ocean occur in the surrounding terrains (Laubscher and Bernoulli 1977). The Tethys Ocean initially developed between the ancient supercontinents of Gondwana (southern land mass) and Laurentia (northern land mass) during the Palaeozoic (Hsü 1977; Windley 1977; Moores and Fairbridge 1998). Repeated opening and closure of the ocean occurred due to convergence and collision of the supercontinents as part of the Lower Palaeozoic Caledonian Orogeny. The Tethys Ocean reformed during the Triassic-Jurassic i.e., after the final dismantling of the two supercontinents and after the

Upper Palaeozoic Variscan Orogeny. The Neo-Tethys was a relatively shallow, subtropical ocean or sea in which thick sequences of evaporates and carbonates accumulated. The closure of the Neo-Tethys in the Mid Jurassic (at approximately 170 Ma) triggered onset of the long drawn out Alpine Orogeny. Jurassic-age ophiolite complexes (i.e., obducted slabs of oceanic crust) are a prominent feature of the Alpine fold belts surrounding the Mediterranean (Dilek et al. 2000). The Mid-Jurassic convergence involved collision of the African Plate (located to the south) and the Eurasian Plate (located to the north).

Box 8.1: Ocean Trenches and Back-arc Basins

Oceanic trenches, back-arc and fore-arc basins, and volcanic island arcs are distinct physiographic features associated with the convergence of oceanic and continental plates. In the more typical subduction-dominated convergent plate boundaries, the dominant process is the subduction of relatively dense oceanic crust beneath lighter continental crust. The surface location of the subducted plate is demarcated by an oceanic trench, i.e., relatively narrow, linear features associated with the deepest parts of oceans. In accretionary convergent margins, however, subduction is a relatively minor process. The key process here is extensional tectonism and formation of back-arc and fore-arc basins. These features are located between the ocean trench and the continental plate, as shown in the attached cross-section. Back-arc and fore-arc basins do not always develop in convergent plate boundaries; they tend to be restricted to situations where the ocean crust is relatively old. The forearc lithosphere is subjected to extension caused by the subducted slab. Two processes are identified in development of the basins, rollback of the trench as it migrates away from the subduction zone and collapse of the edge of the continental plate. The lithosphere overriding the subducted oceanic plate, or slab, typically contains a volcanic island arc. The volcanic arc develops parallel to the ocean trench. The volcanism is associated with physical-chemical interaction between the subducted plate and the underlying (convecting) asthenospheric mantle. The forearc lithosphere has an anomalously low heat flow as the asthenospheric convection is blocked by the cold subducting slab.

A key difference between the two types of convergent plate boundaries is that in the subduction-dominated margins the lithosphere is thickened, whereas in accretionary margins the lithosphere is drastically thinned. Typically, back-arc and fore-arc basins are relatively long (several hundreds to thousands of kilometers) and relatively narrow (a few hundred kilometers). The seafloor in the basins is spreading asymmetrically. Accretion of sediment associated with the subducting slab causes the forearc lithosphere to grow as an accretionary prism. The thickness and age of the sediment in the accretionary prism is related to the age of the underlying oceanic crust. Volcanic island arcs may also extend for hundreds of kilometres.

The plate boundary associated with the central and eastern Mediterranean is a rather complex accretionary margin. The plate boundary is demarcated by the Hellenic Trench, which is located south of Crete and east of Sicily. The curvilinear nature of the trench and the complexity of the collision, which includes the northward pushing Arabian Microplate, triggered formation of microplates located between the African Plate and the Eurasian Plate, e.g., the Anatolian

Microplate and the Hellenic Microplate. An unusual feature of the Mediterranean component of the collision is the subduction of thick sequences of evaporites, rather than basaltic ocean crust. The evaporites developed as part of the Messinian salinity crisis. The Mediterranean Ridge is associated with a prominent accretionary prism. The shallow taper of the prism may be related to the different thermal-mechanical properties of evaporites in comparison to basaltic crust. The Aegean Sea and the Tyrrhenian Sea are back-arc basins with unusual ellipsoidal or rounded shapes. The back-arc basins contain volcanic island arcs, the Aeolian Volcanic Arc (Tyrrhenian Basin) and the Hellenic Volcanic Arc (Aeolian Basin).

8.3.1 Alpine Orogeny

The Alpine Orogeny peaked in the Oligocene-Miocene (34–5.3 Ma). The principal expression is the occurrence of fold mountains in terrains surrounding the Mediterranean, e.g., in North Africa (Atlas), in Southern Europe (Alps, Apennines, Balkan Peninsula, and Pyrenees) and in the Middle East (ranges in Turkey and Iran). The complexity of the Alpine convergence has been widely discussed (e.g., Hsü, 1977; Windley 1977; Panza and Suhadolc 1990; Moores and Fairbridge 1998; Okay and Tüysüz 1999; Bozkurt et al. 2000; Cavazza and Wezel 2003; Flower and Dilek 2003; Robertson and Mountrakis 2006; Jolivet et al. 2013). The orogeny resulted in formation of microplates with complex boundaries, e.g., the Adriatic, Anatolian, and Hellenic Microplates (Fig. 8.3). The Mediterranean microplates may have started to form in the Early Cretaceous, but the main development is thought to have been in the Oligocene-Miocene (i.e., concurrent with the peak of the Alpine Orogeny). An additional complexity involved accretion of the Arabian Microplate onto the northward-migrating African Plate (probably during the Oligocene-Miocene).

The anomalous nature of the geomorphology of the Mediterranean was discussed by Hsü (1977). Three features were emphasized: (i) the opposite sides of the Mediterranean Basin have a disparate form (unlike the Atlantic Ocean where continental margins can be fitted together); (ii) the Alpine fold mountains are not aligned parallel to coastlines and in many cases constitute peninsulas almost at right angles to the basin axis (e.g., the Apennines); and (iii) the sinuous nature of the current plate boundaries. The Alpine Orogeny is ongoing and oceanic crust associated with the ancient Tethys Ocean, and the Neo-Tethys Ocean, continues to be subducted beneath the Eurasian Plate.

Fig. 8.3 Image of the central and eastern Mediterranean showing the tectonic framework of plates, microplates and sutures, together with the bathymetry. *Source* Geological boundaries and faults simplified from articles referenced in the text; Satellite Image of Europe based on NASA MODIS data, processed by Philip Eales, Planetary Visions/DLR

8.3.2   Accretionary Convergent Margin

The central and eastern Mediterranean constitute an accretionary convergent margin, rather than a subduction-related collision (Box 8.1). A simplified cross-section shows the principal features of an accretionary margin, including occurrence of an ocean trench, fore-arc and back-arc basins, and a magmatic or volcanic island arc (Fig. 8.4). The geomorphology of the Mediterranean is illustrated by a bathymetric map (Fig. 8.2). Each sea constitutes a discrete basin or microplate, (Fig. 8.3). The Mediterranean Ridge correlates with the accretionary prism located north of the oceanic trench. The plate boundary in the eastern Mediterranean is aligned with the sinuous Hellenic Trench. In the central Mediterranean, the boundary is associated with the Apennine-Mahgreb Thrust.

8.3.3   Apennine-Mahgreb Thrust

The Apennine-Mahgreb Thrust demarcates one of the principal inter-continental collision zones in southern Europe (Panza and Suhadolc 1990; Cavazza and Wezel 2003). The northward-migrating African Plate and the westward-pushing Adriatic Microplate are being subducted beneath the Eurasian Plate (Fig. 8.5). The Apennine-Mahgreb Thrust is primarily a Pliocene-Quaternary feature and may be envisaged as a rejuvenation of the Alpine Orogeny. In southern Italy and eastern Sicily, areas of crustal extension have developed where stress is being released from the Eurasian Plate affected by the subduction (Panza and Suhadolc 1990). The axis of the Apennine Mountains, a prominent NW-SE trending Alpine fold belt is similarly subjected to ongoing extension (Panza and Suhadolc 1990).

Fig. 8.4 Cross-section showing the position of active fore-arc and back-arc basins in an accretionary convergent margin. *Source* By Zyzzy2 at the English language Wikipedia, CC BY-SA 3.0, https://commons.wikimedia.org/w/index.php?curid=54512237

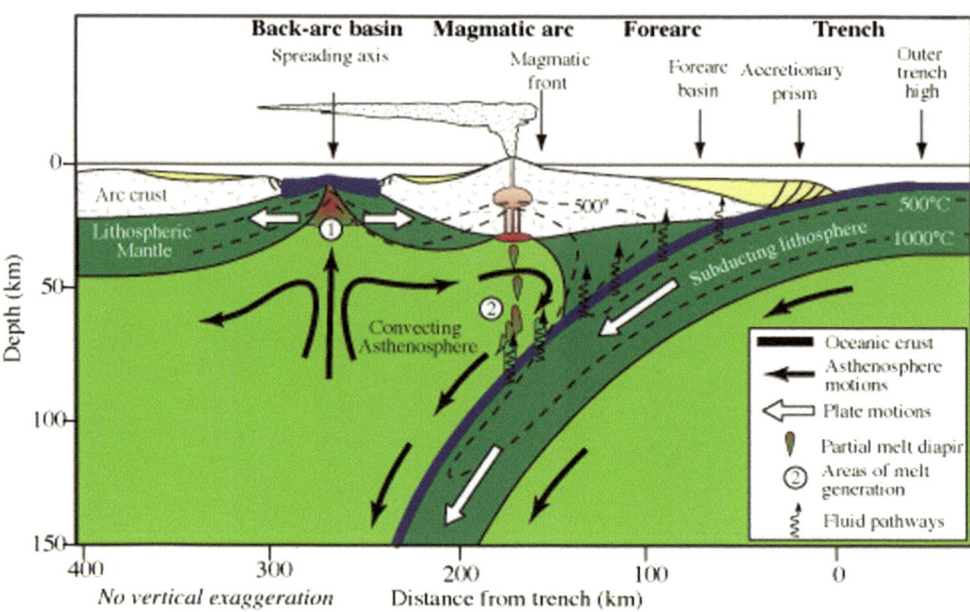

Fig. 8.5 Simplified tectonic map of southern Italy and the Italian Islands showing location of the Quaternary volcanoes. The Apennine-Maghreb Thrust is an active subduction zone. The Tyrrhenian Sea is a back-arc basin, the Aeolian Islands are associated with a volcanic island arc, and the Calabrian Wedge is a fore-arc basin. The Etna Volcano is a discrete stratovolcano located proximal to the plate boundary. The Pantelleria Volcano is situated on a transform fault on the convergent plate boundary. The fault zone located between the west coast of Calabria and eastern Sicily is associated with the Siculo-Calabrian Rift. *Source* After Panza and Suhadolc (1990)

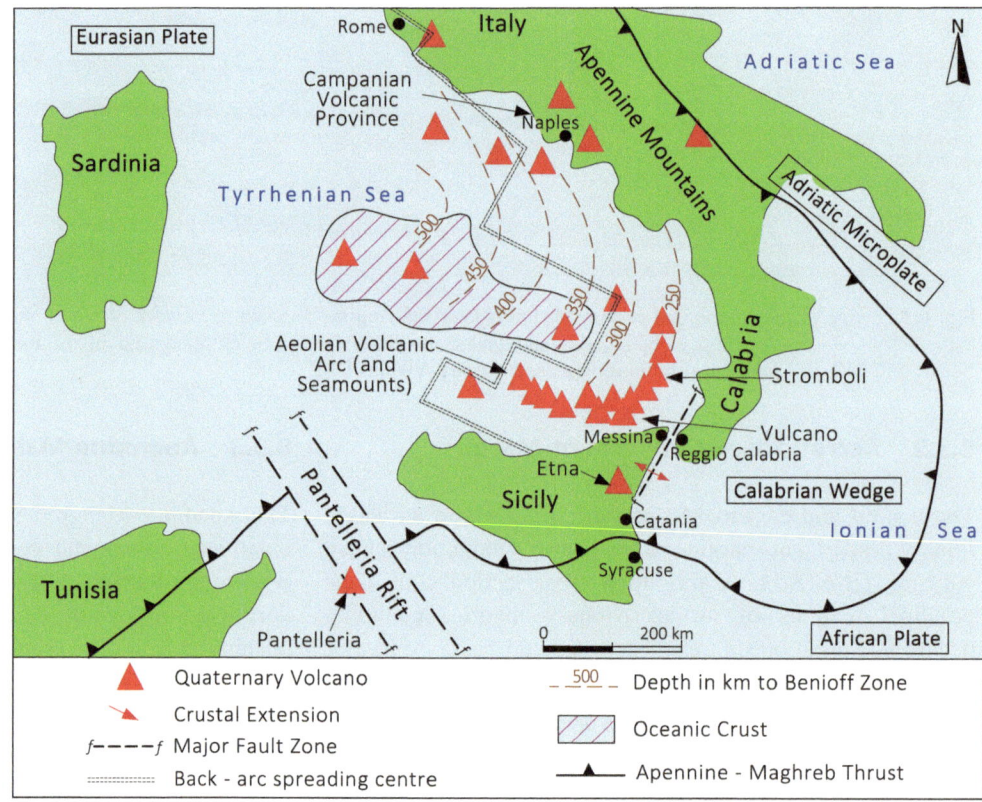

The Apennine Mountains are comprised of thick sequences of Mesozoic rocks, primarily carbonates (limestone, dolomite, and marble). Possibly the most significant feature of the crustal extension is development of the Siculo-Calabrian Rift, a 350 km long feature associated with NE-SW trending faults located between west coast of Calabria and eastern Sicily. The faults are primarily Quaternary features.

8.4 Mediterranean Basins

The Mediterranean Sea has a total surface area of 2,500,000 km^2. The depth is very variable and evaporation generally exceeds runoff from rivers and precipitation. In the eastern Mediterranean, the level is consistently dropping and the salinity increasing. The deepest sections have an average water temperature of 13.2 °C, but shallower areas are over 30°C in the summer months. In detail the Mediterranean consists of multiple, interlocking basins (Fig. 8.2). Each basin reveals significant geological and hydrological differences. The sea in each of the basins has different physical and chemical characteristics. Transient effects include changing circulation patterns driven by climatic events.

Formation of the Mediterranean Basins can be linked to the tectonism associated with the collision. Most basins developed after the peak of the Alpine Orogeny. The Messinian salinity crisis (5.9–5.3 Ma) records a period of intense desiccation in the Late Miocene, at which time the Mediterranean Basins were separated from the Atlantic Ocean. During this period great thicknesses of evaporites accumulated (Cita 2006). The Zanclean flood event (at approximately 5.3 Ma i.e., defining the Miocene-Pliocene boundary) records the refilling of the Mediterranean Basins, due to breaching of the Straits of Gibraltar.

The **Adriatic Sea** is a relatively shallow, elongate body of water associated with a discrete microplate. The Adriatic Microplate may have started to form in the Mesozoic. Subsequent movements have resulted in formation of parallel fold mountains on both the southwest side (Italy) and the northeast side (Balkan Peninsula). The Adriatic Microplate is migrating west at approximately 0.4 cm/year. Earthquakes are common in the region and more than 15 tsunamis have been documented in the last 600 years. The last major earthquake (measuring 7.0 on the Richter scale) occurred in Montenegro in 1979.

The **Aegean Sea** is associated with a large back-arc basin that is experiencing active extension (Jolivet et al. 2013; Meier et al. 2007). Details of this basin and the Hellenic Volcanic Arc are presented in Chap. 12. In the proximity of Cape Malae, where the Aegean Sea and Ionian Sea join, there is often an unusually turbulent wave pattern caused by the differences in density and salinity (Chap. 11).

The **Ionian Sea** is associated with a relatively old basin (in comparison with the Adriatic Sea and Aegean Sea) that includes relicts of the ancient Tethys Ocean. The deepest section is the Calypso Trench (depth of 5,267 m), part of the active Hellenic Trench. The western extremity of the Ionian Sea is clearly demarcated on the bathymetric map (Fig. 8.2). The Ionian Basin is part of a seismically active region where the African Plate is being subducted either beneath the Eurasian Plate (central Mediterranean) or beneath the Hellenic Microplate (eastern Mediterranean) (Fig. 8.3). The Ionian Basin contains a number of volcanic seamounts.

The **Tyrrhenian Sea** occurs in a large, and relatively deep, back-arc basin (maximum depth of 3,785 m), located between the west coast of the Italian Peninsula and the islands of Corsica and Sardinia (Fig. 8.2). The Tyrrhenian Sea is notably warmer and less salty than the Ionian Sea. These differences are manifested in the Straits of Messina (the narrows between southern Italy and Sicily) where counter currents, bores, and whirlpools develop. Currents typically flow southward on and near surface, but below a depth of approximately 30 m they flow northward (Heikell 2002). The counter currents create eddies and whirlpools. At each turn of the tide there is a brief stand, followed by one or more bores (waves that travel against the tide or current). Immediately after the bores, whirlpools well up and then disappear. This is caused by the sinking of denser water with surrounding less dense water rising. The combination of wind against tide creates short breaking seas which, together with the bores and whirlpools, can endanger small craft.

8.5 Earthquakes

Catastrophic earthquakes occurred in southern Italy and Sicily in both the Ancient Greek and Roman times, as well as in the more recent history. The regions of Calabria (southern Italy) and eastern Sicily have been particularly affected. Five historical earthquakes had devastating consequences: the Sicilian earthquakes of 1169 AD and 1693 AD, the Calabrian earthquakes of 1783 AD and 1905 AD, and the Messinian earthquake of 1908 AD (Mowbray 1909; Piatanesi and Tinti 1998). All of the above earthquakes are related to the fault zone associated with the Siculo-Calabrian Rift (Fig. 8.5). The shallow nature of the active extensional faults, particularly in comparison to the deep-seated faulting associated with the subduction zones, is a key parameter in the destructive nature of the earthquakes in this region.

The Sicilian earthquake of 1693 AD measured 7.4 on the Richter scale and affected a total area of 5,600 km^2 extending as far as Malta (the 1169 AD event was probably of a similar intensity). An estimated 60,000 fatalities occurred in the 1693 AD event and the cities of Catania and Syracuse were almost entirely destroyed. Many fatalities were associated with a tsunami (maximum height estimated at 8 m) that struck the Ionian coasts of Sicily and Calabria. The tsunami was particularly significant in the Straits of Messina and historical reports are consistent with a strong sea withdrawal, followed by a violent sea return and coastal flooding.

The Calabrian earthquake of 1783 AD included a sequence of five seismic events (measuring between 5.9 and 7.0 on the Richter scale) which caused estimated 32,000–

50,000 fatalities. The city of Messina was destroyed, together with many small towns and villages in Calabria. The seismicity was focused on faults aligned parallel with the Straits of Messina. The first two seismic events triggered large tsunamis which affected regions bordering the Tyrrhenian Sea as far north as Naples. Fatalities associated with the Calabrian earthquake of 1905 AD (which measured 7.2 on the Richter scale) were relatively few in comparison to the previous event, although Messina was again affected and many villages in Calabria were destroyed.

The Messinian earthquake of 1908 AD (measuring 7.1 on the Richter scale) caused estimated 75,000–82,000 fatalities. The cities of Messina and Reggio Calabria were destroyed. This event was caused by movement on a low angle, SE-dipping and relatively shallow fault located in the Straits of Messina. Many of the fatalities were ascribed to 12 m-high tsunamis resulting from subsidence of the seabed. In some areas the coastline dropped by almost a metre.

8.6 Volcanism

Subduction of the African Plate and Adriatic Microplate beneath the Eurasian Plate has resulted in chains of volcanoes developing in southern Italy and the Italian Islands. Subduction is the main driver of volcanism on Earth and results in the concentration and alignment of volcanoes in well-defined belts (Fig. 8.3). As the subducted slab is heated and subjected to partial melting, magmas are created that rise upward through the overlying crust. The volcanism of the Aeolian Islands, which includes Stromboli and the island of Vulcano, is typical of volcanic island arcs associated with subduction zones (Box 8.1). Etna is a giant stratovolcano located proximal to the plate boundary. The Pantelleria Volcano is associated with a transform fault on the convergent boundary between Sicily and North Africa.

The volcanism in this part of the Mediterranean commenced in the Pliocene, although the majority of volcanoes are Quaternary (Kilburn and McGuire 2001). There are a handful of active and potentially hazardous volcanoes, as gazetted by the Smithsonian Global Volcanism Programme (2016). Etna is one of the world's most active volcanoes and, together with Stromboli, is almost continuously erupting. Other active volcanoes in the central and eastern Mediterranean include Campi Flegrei and Vesuvius (Campanian Volcanic Field: Chap. 9), and Methana, Milos, and Santorini (Hellenic Volcanic Arc: Chap. 12).

The simplistic, near-linear distribution of the volcanic trends in the central Mediterranean masks the complexity of the collision zone (Panza and Suhadolc 1990) (Fig. 8.5). The curvilinear nature of the Apennine-Mahgreb Thrust is a prominent feature. The Aeolian Volcanic Arc encompasses a chain of islands, dominated by accreted volcanic cones and

calderas located on the seabed, as well as numerous seamounts. The distribution of the volcanic arc corresponds to the depth of the Benioff Zone, which in turn is related to the angle of the subducted slab (Fig. 8.4). The Benioff Zone is a planar zone of seismicity which correlates with a region of partial melting of the subducted slab. Differential motion along the Benioff Zone (i.e., associated with the subduction process) generates deep-seated earthquakes.

A feature of both the Aeolian and Campanian volcanism is the occurrence of catastrophic eruptions. Magma is generally fed from relatively shallow staging chambers where immense pressure may build up prior to eruptions. This is typical of island or continental volcanic arcs found on the back-arc side of subduction zones (Kilburn and McGuire 2001; Peccerillo 2003). Recent eruptions of Etna and Pantelleria have been relatively quiescent, although catastrophic events, including formation of calderas, occurred in the earlier activity. The non-explosive nature of some eruptions is typical of magmas fed from depth which do not get blocked in complex plumbing systems. The location of contrasting tectonic regimes within close proximity is an unusual feature of the southern Italian and Italian Island volcanoes. This has led to remarkably differing styles of volcanic activity. The volcanoes described here form the basis of classification schemes that define the explosive nature of eruptions (e.g., Cas and Wright 1988).

8.6.1 Explosiveness of Eruptions

The relative explosiveness of volcanic eruptions is calculated from details of the fragmentation index and area of dispersal of the volcanic ash (Walker 1973; Wright et al. 1980; Cas and Wright 1988). Three of the eruptive groups generally recognized are based on the southern Italian and Italian Island volcanoes, namely Stromboli, Vesuvius ("Plinian"), and Vulcano (Fig. 8.6). Plinian eruptions are the most catastrophic of the volcanic eruptions: they have the highest fragmentation indices (ash is very fine-grained) and the ash is dispersed over the largest areas (Box 8.2). Plinian eruptions are extremely hazardous and regularly include caldera events which generate extensive pyroclastic flows. Vulcanian activity is a relatively uncommon variant of Plinian activity, characterized by explosive, but short-lived eruptions. Strombolian eruptions define a less explosive style of eruption. The least explosive systems plot in the Hawaiian field, named after the group of volcanic islands in the Pacific Ocean.

> **Box 8.2: Plinian Eruptions** The observations and letters sent by Pliny the Younger to the historian Tacitus describing the 79 AD eruption of Vesuvius have been incorporated into geological science. Pliny

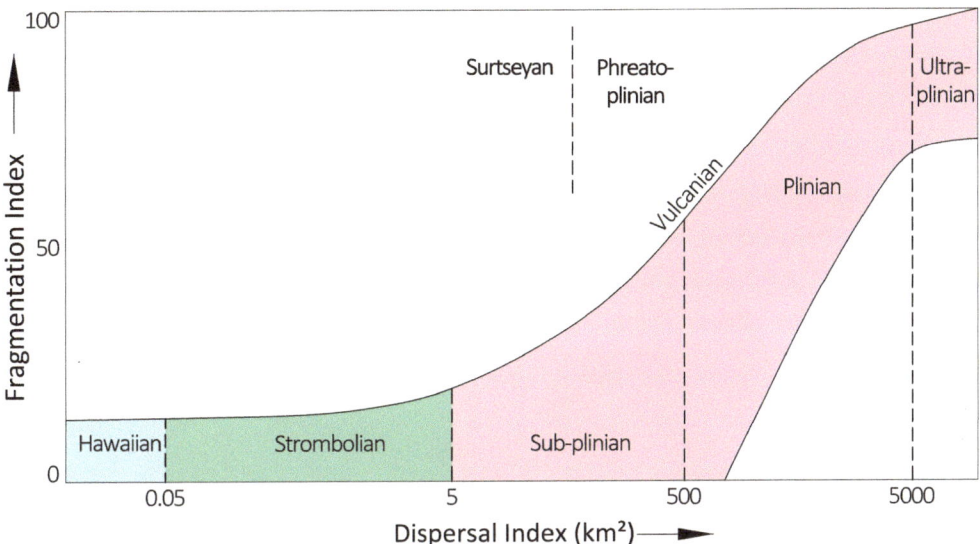

Fig. 8.6 Volcanoes that generate pyroclastic fall deposits can be divided into well-defined fields on a plot of fragmentation index (percentage of deposit with grains finer than 1 mm) against the area of dispersal. Coloured areas include the majority of eruptions on Earth. Ash-fall deposits associated with the relatively quiescent Hawaiian eruptive style have restricted distribution. Most eruptions of Etna and Pantelleria plot in this field. The catastrophic Plinian eruptions (which are divided into three separate fields) have larger components of pyroclastics that disperse over great distances. Many of the eruptions associated with the Neapolitan volcanoes plot in these fields. Strombolian eruptions define an intermediate field between Hawaiian and Sub-plinian. Eruptions located above the shaded area are associated with either short-lived explosive activity (Vulcanian) or systems that have reacted with water (Surtseyan or Phreatoplinian). *Source* After Walker (1973) and Wright et al. (1980)

observed the initial stages of the eruption from a relatively safe location at Misenum, to the north of Naples, from where he sketched the enormous ash column. The column reached a height of approximately 32 km and Pliny compared the shape to that of an "umbrella pine" (a common Mediterranean tree). Plinian eruptions are catastrophic events generally associated with large stratovolcanoes. They usually commence with sustained ash columns and include destructive pyroclastic flows. Ash columns may achieve great heights and generate extensive fallout deposits which cover large areas. Roof collapse due to the enormous volumes of ash erupted is an additional hazard. Plinian eruptions may also include basal surges, an incipient stage of pyroclastic flows, which involve explosive magma-water reactions. Basal surges generate centrifugal winds that are extremely hazardous and may result in cross laminated ash beds. The wind may be so extreme as to cause asphyxiation (air is simply blasted out of the way) and yet strong buildings may remain intact. The basal surges are considered as near-instantaneous events and can be considered as a dilute type of pyroclastic event (the volume of lava and ash is relatively low). Basal surges often precede pyroclastic flows, and the two events are characteristically interrelated. Basal surges and pyroclastic flows may extend as far as 20 kilometres from vents. These events often cause near-instantaneous catastrophes (e.g., in the 79 AD eruption of Vesuvius).

Plinian eruptions are in some cases followed by volcano-tectonic collapse and formation of calderas. Thick breccia deposits, also known as debris avalanche deposits (DAD's), may extend for tens of kilometres from cones subjected to even partial sector collapse. Plinian eruptions also trigger earthquakes and tsunamis which may have a regional affect. Plinian eruptions are restricted to specific types of volcanoes, including those located in tectonic settings where magmas are driven by subduction zones and continental rifting. Melting of continental crust (i.e., located above subducted slabs or associated with crust thinned and stretched by rifting) produces volatile-rich alkaline magmas which may include siliceous end members driven by differentiation in crustal staging chambers.

8.6.2 Magma Compositions

Volcanic rocks can be classified by use of geochemical variation plots, of which the TAS diagram is particularly useful for alkaline–rich magmas (Le Maitre et al. 2002; Gill 2010). In most cases, alkaline volcanic rocks crystallize or

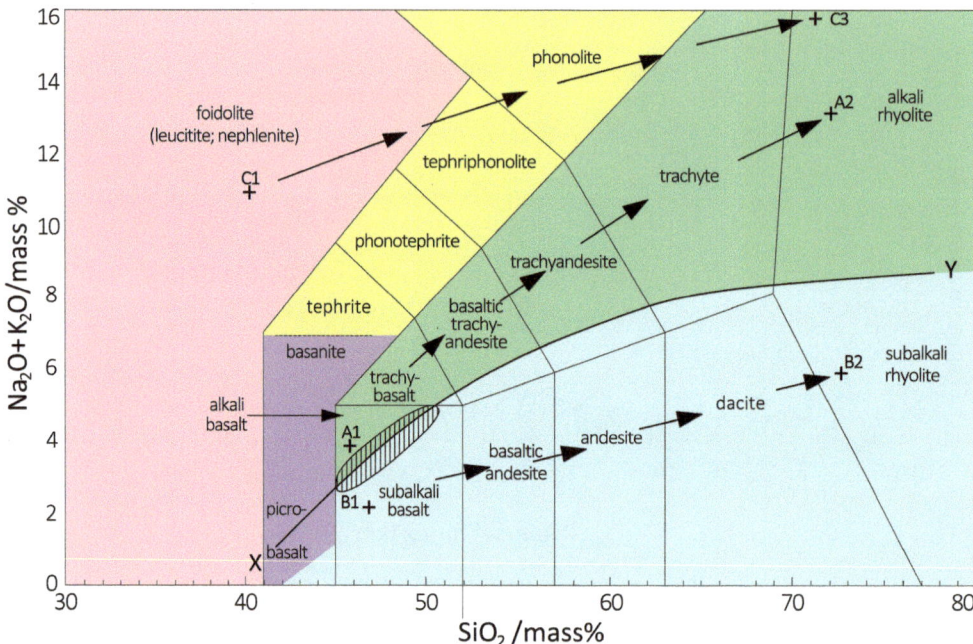

Fig. 8.7 Plot of total alkali (sodium and potassium oxides) versus silica ("TAS diagram") is a popular method of categorizing alkaline volcanic rocks. The line X–Y is used to separate the alkali and sub-alkali groups. The basalt subdivision can be divided into three groups: alkali basalt, transitional basalt (shaded area straddling the division), and subalkali basalt. The alkali-rich foidolite group includes two principal variants, nephelinite (potassium-rich) and leucitite (sodium-rich). Also shown are three fractional crystallization sequences, A1–A2 (potassium or sodium alkali basalt trend), B1–B2 (calc-alkaline basalt trend) and C1–C2 (foidolite trend)

accumulate from magmas erupted as part of differentiation series identified by an increase in SiO_2 and alkalis. Two examples of differentiation series are portrayed here (Fig. 8.7). Trend A1–A2 is representative of an alkali basalt-alkali rhyolite series. Trend B1–B2 is representative of a subalkali basalt-subalkali rhyolite series, also known as the calc-alkaline basalt trend. The broad field of extremely alkali-rich magmas (which are undersaturated in terms of silica) are described as foidolite. The two major variants of foidolite, potassium-rich magmas (which form nephelinite) and sodium-rich magmas (which form leucitite) cannot be separated on the TAS diagram. A possible differentiation series is portrayed on the diagram as C1–C2, i.e., a foidolite-alkali rhyolite series.

Historical eruptions in the Aeolian Volcanic Arc fit the calc-alkaline basalt trend, although recent eruptions of Stromboli and Vulcano include similar potassium-rich alkali magmas to those of the Campanian volcanoes (Kilburn and McGuire 2001; Bertagnini et al. 2003; Orsi et al. 2004; Peccerillo et al. 2013). Etna and Pantelleria are associated with a broad range of magmas, including tholeiitic basalt and alkali basalt (Corsaro and Cristofolini 1997, 2000). (The tholeiitic trend cannot be distinguished on the TAS diagram as tholeiitic basalts are reduced, rather than oxidized; most

tholeiites are subalkali, including those in flood basalt provinces, but they can plot in all three of the primary basalt fields.) A key difference between Etna and Pantelleria and the volcanism of the Aeolian Islands and the Campanian Volcanic Field is the sodium-rich nature of the alkali basalt. Pantelleria is the type locality for pantellerite, a peralkaline variety of sodic-rich rhyolite.

The broad range of magma compositions associated with the southern Italian and Italian Island volcanoes reflects the complexity of the tectonic setting. Position relative to the plate boundaries is particularly significant. In summary, the following broad differentiation trends (shown primarily by increasing SiO_2) are recognized in the southern Italian/Italian Island volcanoes:

a. **Potassium-rich alkali basalt trend** (alkali basalt-trachybasalt-trachyandesite-trachyte-alkali rhyolite) and **potassium-rich foidolite trend** (nephelinite-tephriphonolite-phonolite-alkali rhyolite). Examples: Campanian Volcanic Field and recent volcanism of Stromboli and Vulcano;

b. **Calc-alkaline basalt trend** (subalkali basalt-basaltic andesite-andesite-dacite-subalkali rhyolite). Example: older volcanism of the Aeolian Volcanic Arc;

c. **Tholeiitic basalt trend** (tholeiitic basalt-sodic trachy-basalt (or hawaiite), basaltic trachyandesite (or mugearite), trachyandesite (or benmoreite), and trachyte). Example: older eruptions of Etna;

d. **Sodium-rich alkali basalt trend** (alkali basalt-trachybasalt-trachyandesite-trachyte-alkali rhyolite). Examples: younger eruptions of Etna and Pantelleria.

8.7 Aeolian Islands

The Aeolian Islands are part of a volcanic island arc located in the Tyrrhenian Sea, approximately 25 km from the north coast of Sicily (Fig. 8.2). The arc has a length of more than 140 km. There are eight islands, of which the largest, Lipari, can be reached by a regular ferry service from Milazzo, northern Sicily (Fig. 8.8). The islands have a relatively small permanent population (approximately 15,000), but summer visitors are attracted by the relaxed Mediterranean lifestyle. There are two active volcanoes, Stromboli and the Fossa cone (on the island of Vulcano). Most islands reveal evidence of geothermal heat and include active fumaroles. The archipelago is surrounded by numerous seamounts (submerged volcanic cones), some of which are active. Located to the west of the Aeolian Islands is the small island of Ustica, the upper section of a volcanic cone dominated by black basaltic lava and cinder (the name translates from Latin as "burnt").

8.7.1 Volcanic Island Arc

Each island in the Aeolian archipelago is typically a composite structure formed from accretion of multiple volcanic cones and calderas (Peccerillo et al. 2013). Individual volcanoes have experienced complex histories of caldera and sector collapses. The subaerial components of the volcanic islands are all younger than 270,000 BP. The submerged parts (and the seamounts) may be considerably older. The island arc is associated with a tectonic block which detached from Sardinia-Corsica and drifted southwest as the Tyrrhenian Basin opened in the Neogene (Panza and Suhadolc 1990; Kilburn and McGuire 2001). The island arc is cut by a complex pattern of faulting, and some islands are located on a major NW-SE lineament. The western islands (Alicudi; Filicudi) occur in an area of compression where volcanism is extinct. The eastern islands (Salina; Lipari, Vulcano; Panarea; Stromboli) occur in an area of crustal extension where volcanism is quiescent or active.

The older volcanism of the Aeolian Islands is dominated by calc-alkaline magmas. A moderately potassium-rich variety of basaltic trachyandesite known as shoshonite occurs in some of the older volcanic centres. As a general rule, the younger volcanism is associated with potassium-rich alkali basalt magmas. The silica-rich components of this trend (including alkali rhyolite) are abundant on the islands of Lipari and Vulcano. The eruption of potassium-rich alkali basaltic magma in the Aeolian volcanoes is indicative of crustal signatures superimposed upon

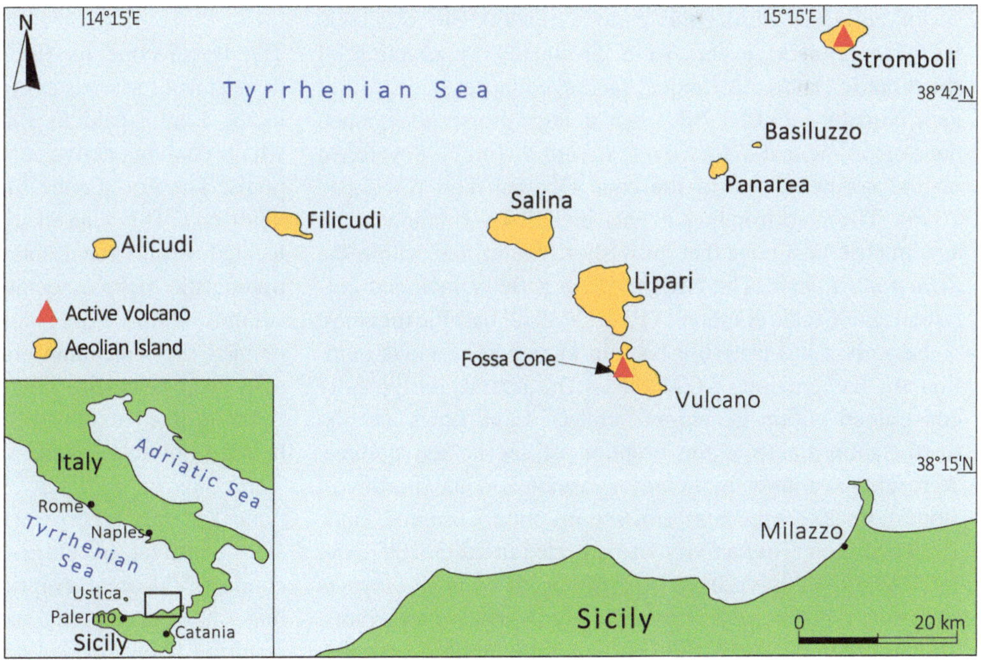

Fig. 8.8 Map of the Aeolian Islands showing location of the two active volcanoes

melted mantle material (Bertagnini et al. 2003). This may be ascribed to the relative maturity of the island arc.

8.7.2 Stromboli

Stromboli is the northernmost and most well known of the Aeolian Islands. The island has an area of 12.6 km^2 and is dominated by a volcanic cone that rises to a height of 926 m above sea level (Fig. 8.9a). The total height relative to the sea floor is approximately 2,700 m. The volcano is persistently active, although eruptions are typically short-lived and restricted to three summit craters. The Strombolian-style of volcanic activity is characterized by relatively small eruptions with periodic explosions and eruption of ash and lapilli up to heights of several hundreds of metres. Showers of glowing scoria are observed by numerous visitors every year (Fig. 8.9b). Small pieces of pumice can be observed in the sea some distance off the island. The glow of the vent can be seen from several kilometres at night (but is not usually apparent during the day). The historical evidence reveals that eruptions were previously larger and included sizeable ash clouds and lava flows.

Stromboli has been described as an "an open window on the deep-feeding system of a steady state basaltic volcano" (Bertagnini et al. 2003). Four main periods of activity have been identified from the geological record: Palaeostromboli (older than 26,000 BP), Vancori (26,000–13,000 BP), Neostromboli (13,000–6,000 BP), and Recent Stromboli. The oldest subaerial activity is dated at approximately 100,000 BP. Palaeostromboli experienced several caldera events and sector collapses during the period 100,000 BP to 26,000 BP. The Vancori events were dominated by eruption of shoshonitic lavas. A major sector collapse occurred at approximately 13,000 BP when a large horseshoe-shaped depression, *Sciara del Fuoco* ("Stream of Fire"), developed on the northeast side of the cone (Vezzoli and Corazzato 2016). The Neostromboli events included formation of an asymmetric lava cone that grew shortly after and within the *Sciara del Fuoco*. The Neostromboli activity included persistent small-scale eruption of potassic alkali basaltic magmas.

Stromboli has probably been in almost continuous eruption for the previous 5,000 years. The activity is typically constrained within the summit craters. Lava flows are distinctly subordinate to the eruption of scoria and pumice. A recent lava flow is, however, visible on the northwest slopes. In 2002, the first effusive eruption observed since 1985 occurred. This activity was repeated in 2003, 2007, and 2013–2014. In July 2019, two significant explosive events occurred, together with approximately twenty minor events. This resulted in some injuries and one fatality among a group of hikers. In August 2019, a pyroclastic flow erupted on the northern flanks. The flow spread across the surface of the sea for several hundreds of metres prior to collapsing. This eruption produced an ash column with a height of 2 km.

8.7.3 Vulcano

Vulcano is the southernmost of the Aeolian Islands with an area of 21 km^2. The mountainous landscape is associated with several accreted cones and multiple calderas (Fig. 8.10). The highest point on the island is Mount Aria (501 m). During Roman times, the island supported mining of alum and sulphur from the volcanic rocks. The southern end of the island consists of three extinct stratovolcanoes, Monte Aria, Monte Saraceno and Monte Luccia. They are each associated with the older style of calc-alkali volcanism. The three extinct cones have partially collapsed into the large Il Piano caldera. The central and northern part of the island is dominated by the Lentia caldera (approximate age of 6,000 BP). The Lentia caldera has experienced at least nine major eruptions since formation (Traglia et al. 2013). At the northern extremity of the island is the Vulcanello Islet, connected to the island by an isthmus that floods in poor weather conditions. The islet is part of a small cone which emerged from the sea in 183 BC. The historical activity, which persisted until 1550, included pyroclastic flows and lavas, some of which built up the isthmus.

8.7.4 Fossa Cone

The Fossa cone on the island of Vulcano is climbed by thousands of tourists every year (Fig. 8.11). The cone occurs in the Lentia caldera and is capped by the Gran Cratere, which contains active fumaroles, sulphur deposits, and mud pools. The Fossa cone formed during an eruption in 1888–1890 AD. This caused many fatalities in the sulphur mines located within the Lentia caldera. A 5 m-thick deposit of pyroclastic rocks associated with the eruption occurs on the summit of the Fossa cone. The 1888–1890 AD eruption defined the Vulcanian eruptive style. Eye-witness accounts refer to a "clearing of the volcanic throat" with irregular, "cannon-like" explosions and large blocks of chilled lava being thrown into the sea between Vulcano and Lipari. Variations in both the rate and violence of eruptions are typical of this type of activity. The loud cannonades result from removal of plugs of chilled lava which block the conduit. Vulcanian eruptions eject a high proportion of lithic material and tephra, but are typically short-lived. The resulting finely-bedded deposits are often characterized by a bread-crust texture.

a

b

Fig. 8.9 **a** The near-symmetrical cone of the island of Stromboli (September 2004). *Source* By Steven W. Dengler, CC BY-SA 3.0, https://commons.wikimedia.org/; **b** Erupting incandescent molten lava fragments, Stromboli (December 1969). *Source* By B Chouet from the website of the United States Geological Survey

Fig. 8.10 The northern side of the island of Vulcano viewed from Lipari. The island is comprised of several accreted cones and calderas. Extinct features are located on the left and the active Fossa cone with the Grand Cratere occur in the centre above the town. *Source* By Brisk g—Own work, Public Domain, https://commons.wikimedia.org/

a

b

Fig. 8.11 a Blocks and veins of sulphur crystals located on the rim of the Fossa cone, Vulcano, with Stromboli visible in the background (right); **b** The Gran Cratere includes veins and pockets of sulphur crystals

8.8   Etna

Etna is one of the largest stratovolcanoes on Earth with an area of approximately 1,190 km^2 and diameter of 40 km (Fig. 8.12). The main cone reaches a height of 3,350 m and dominates the northern skyline of the city of Catania (Fig. 8.13a). Etna is a major tourist attraction and is visited by an increasingly large number of tourists every year. The mountain may be snow-covered in winter and can be cloud-covered in summer when searing heat and violent winds can create difficult conditions for climbers. The Parco dell 'Etna protects most of the central and upper parts of the massif. Many visitors approach the volcano from the village of Nicolosi, located on the southern slopes, and which can be reached by a regular bus service from Catania. Nicolosi is the lower base of the cableway and funicular which save many hours of hiking the central slopes. Some of the upper slopes are reached by four-wheel drive tracks. Footpaths to the summit craters are well-defined. The lower, relatively gentle slopes of the Etna massif are either well wooded or intensely farmed, but the central and upper parts reveal a barren, near lunar landscape. Recent eruptions of volcanic ashes and lava flows are particularly well exposed on the southern slopes (Fig. 8.13b). The upper part of the cone includes hiking on recent lava flows (Fig. 8.14a). Craters are partially masked by steam and include evidence of active geothermal vents with fumarolic activity (Fig. 8.14b). Subsidiary cones of cinder and agglomerate can also be observed (Fig. 8.15a). The active geothermal vents form deposits of silica and sulphur (Fig. 8.15b).

Fig. 8.12 The Etna Volcano constitutes a large massif in northeast Sicily. Also visible are the Aeolian Islands. *Source* 3D Image based on NASA MODIS data, processed by Philip Eales, Planetary Visions/DLR

8.8.1 Early Volcanism

Etna is located at the edge of a major crustal discontinuity, a fault system that separates the continental crust of western and central Sicily from the oceanic crust of the Ionian Basin (Monaco et al. 1997; Branca et al. 2008). Since the Late Pleistocene, eastern Sicily has been affected by an extensional tectonic regime, rather than collision and subduction. The maximum age of the volcanism is estimated at 0.5 Ma. The early part of Etna's history (0.5–0.2 Ma) included eruption of calc-alkali basalts (Corsaro and Cristofolini 1997, 2000; Branca et al. 2008). Magma was probably derived from melting of a deep crustal slab. At approximately 0.2 Ma, a change to eruption of sodium-rich alkali basalt occurred, which can be ascribed to melting of different proportions of hydrous phases (amphibole and phlogopite) from a deeper part of the lithosphere (Casseta et al. 2019).

Four large caldera events occurred at approximately 15,000 BP. Since the caldera events, Etna has experienced persistent, relatively quiescent, activity i.e., of the Hawaiian or Strombolian styles ascribed to stabilization of the plumbing system (Coltelli et al. 2000). At approximately 6,000 BP, a catastrophic debris avalanche occurred on the eastern flank, which generated a tsunami that possibly affected ancient civilizations throughout the Mediterranean. The gentle lower slopes of Etna (with the exception of the highly dissected eastern flank) attest to the relatively quiescent nature of the recent activity.

The central cone of Etna is capped by four summit craters and there are reported to be some 300 subsidiary vents on the flanks. Eruptions from the summit craters typically include strombolian activity with spectacular lava fountains. Lava tubes feed flows on the lower slopes at a distance of more than 10 km from the craters. The magma is fed by a relatively complex plumbing system and each crater and vent can erupt a differing array of magma differentiates (within the sodium-rich alkali basalt trend).

8.8.2 Historical Activity

The eruption of Etna in 122 BC may have been the volcanic activity described by the Roman poet Virgil in the *Aeneid*. This event is remembered for the reaction by the Roman government in granting of a 10 year tax amnesty to the affected citizens of Catania. A major eruption in 1169 AD destroyed much of the city of Catania and caused estimated 15,000 fatalities. The historical eruption in 1669 AD included a 1.5 km-wide lava flow that travelled several tens of kilometres southward. The city of Catania was again almost entirely destroyed. After this and the ravages of the earthquake of 1693 AD, the city was rebuilt with wide, straight streets leading off a central square designed to enable rapid evacuation (Fig. 8.16). A lava flow from an eruption in 1928 led to the destruction of several villages for the first time since the eruption of 1669 AD. The 1928 eruption started

a

b

Fig. 8.13   a The Etna massif towers threateningly above the city of Catania, northeast Sicily; **b** Southern flank of Mount Etna showing lateral cones and a lava flow from the 2001 eruption. *Source* Wikipedia (Wilson 44691 own work)

Fig. 8.14 **a** The ascent of Etna includes hiking on recent (unvegetated) lava flows with cinder and agglomerate; **b** The summit craters include active geothermal vents and steam is more or less continuously being emitted

a

b

Fig. 8.15 a Subsidiary cones of cinder and ash occur near Etna's four summit craters; **b** Active geothermal vents have deposited silica and sulphur

Fig. 8.16 The city of Catania was rebuilt after the 1669 eruption of Etna and the 1693 earthquake with classical buildings, squares, and wide streets to enable rapid evacuation. Parts of the city were developed on foundations constructed on Roman buildings (including part of an amphitheatre built of blocks of Etna lava) destroyed by earlier eruptions and earthquakes

high on Etna's northeast flank, with subsequent fissures opening at decreasing elevations. One fissure eruption at the relatively low elevation of 1,200 m destroyed the village of Mascali and part of the Messina-Catania railway line. The rebuilding of the village and railway was used by Mussolini for propaganda purposes and is a well known event in Italian history.

8.8.3 Recent Activity

Etna is one of the most active volcanoes in Europe, and reports of new eruptions rarely create much reaction from Sicilians. One reason for this is that recent activity is generally restricted to the upper part of the cone. This can cause tourist visits to be temporarily restricted. There is a good record of predictions by the Etna Observatory, based on an increase of seismic tremors and build-up of gas. Major eruptions in the 20thC occurred in 1949, 1971, 1979, 1981, 1983 and 1991–1993. The 1971 event destroyed the Observatory. The 1991–1993 activity included relatively fast-moving lava flows that threatened several villages and the small town of Zafferana. Artificial barriers were surmounted by one of the lava flows and a 7 km-long lava tube (which was the primary feed) was blocked by use of explosives. A period of unusually intense activity in 1995–2001, when all four summit craters erupted, was followed by a flank eruption. This occurred near the cable car station, one of the main tourist areas, and was widely covered in the media. In 2003, a subsidiary crater located approximately 450 m below the summit of Etna erupted and ash was dispersed as far as Libya. This eruption destroyed the majority of the tourist infrastructure on the slopes above Nicolosi, together with some houses on the lower flanks. Eruptions in 2004–2005 were followed in 2006–2007 by intense activity focused on one of the summit craters and which included lava fountains. In 2008–2009, a series of eruptions triggered a swarm of more than 200 earthquakes east of the summit craters. The 2006–2007 and 2008–2009 events were accompanied by lava flows that extended as far as 4.5 km and 6.5 km down the eastern slopes of the cone, respectively.

Between January 2011 and February 2012, the summit craters recorded persistent activity including columns of ash. Catania airport was closed several times. In July 2011, a lava flow was successfully diverted away from the main tourist area. In 2014–2015, flank eruptions included strombolian activity and small lava flows. Lava fountains and ash plumes reached heights of 1 km and 7 km, respectively. In March 2017, 10 people in a television crew who were filming an eruption were injured by an explosion caused by lava reacting with snow. In December 2018, strombolian activity occurred in the southeast summit crater (Fig. 8.1). The eruption was triggered by the intrusion of a dyke at a relatively shallow depth. This event caused a magnitude 4.9 earthquake which affected Catania. The most recent events have included intermittent eruption of lava and columns of ash from near the summit, between May and September 2019.

8.9 Pantelleria

Pantelleria is a small satellite island of Sicily (area of 83 km^2) located in the Straits of Sicily, and is closer to Africa than to Europe, being only 60 km from the coast of Tunisia (Fig. 8.2). The Straits of Sicily are relatively shallow and constitute a segment of the Mediterranean which separates the deep basins associated with the Ionian Sea (to the northeast) and the Tyrrhenian Sea (to the northwest). Pantelleria has a long history of settlement. Archaeological sites include evidence of Bronze Age cultures with large tombs, known as "*sesi*". The largest of the tombs has an elliptical shape and measures 20 m by 18 m. Tombs are constructed of volcanic rocks. Pantelleria was an important part of the Roman Empire, noting the location between Sicily and the colonies in North Africa. The *Speccio di Venere* (Venus' Mirror) is a popular nature reserve in which the main attraction is a circular lake situated in an extinct volcanic crater. The lake is fed by hot springs and mud pools.

The island of Pantelleria is part of two large Pleistocene-age calderas, the main masses of which are submerged (Civetta et al. 1984). The calderas have ages of approximately 114,000 BP and 45,000 BP. The subaerial section is restricted to a 15 km-long shoulder of the calderas, which includes the highest peak on the island of Montagna Grande (836 m). The exposed section of the volcanic complex includes extensive ignimbrites and the "Green Tuff", a distinctive deposit related to the younger of the two calderas. The Green Tuff was associated with a major event and has been observed as far away as the island of Lesbos, in the Aegean Sea. Pantelleria is gazetted as an active volcano. Eruptions during the Holocene formed pumice and lava domes, as well short flows of blocky lava. The only historical activity recorded is associated with a submarine vent located off the northwest coast (in 1891). The recent volcanism is dominated by sodic alkali basalt (similar to the Etna Volcano).

8.10 Archaeological Sites of Magna Graecia

The Cilento and Vallo di Diano National Park, located approximately 80 km south of Naples, contains the archaeological sites of Paestum and Velia. The Ancient Greek cities were constructed on the Sele Plain, a broad coastal strip located between the Apennines and the Gulf of Salerno. The Apennines are largely comprised of resistant Mesozoic carbonates, but the Sele Plain is underlain by poorly consolidated Quaternary sediments and volcanic ash. Detailed investigations by means of borehole cores and radiometric dating of sediments have enabled a reconstruction of the recent sedimentation patterns (Pescatore and Cinque 2004). Flooding and deposition of alluvium meant the plain became uninhabitable in Roman times. Paestum was a major Greek city with three well-preserved temples built in the Doric style (Fig. 8.17). The temples date from the period 600–450 BC. In addition to the temples, Paestum includes city walls, an amphitheatre, and other monumental buildings. Buildings are constructed of Mesozoic limestone. Investigations of the Hera Argiva site (which includes a sanctuary dedicated to the Greek god *Hera*), formerly located at the mouth of the River Sele, has shown the shoreline has retreated westward since building of the sanctuary.

The city of Velia (located 30 km south of Paestum) was founded by the Greeks in 540–535 BC and subsequently incorporated into the Roman Empire The city experienced several periods of decline, in part due to silting up of the harbour and in part due to earthquake damage (Fig. 8.18a). Persistent sedimentation after the Roman period resulted in deposition of considerable thicknesses of alluvium in some parts of Velia (Pescatore and Cinque 2004) (Fig. 8.18b).

The island of Ortigia is part of the city of Syracuse, which is situated on a prominent indentation of the Ionian Sea in southeast Sicily. Ortigia was settled in 734 BC by the Corinthians, one of the seven city-states located in the Peloponnese region of southern Greece. The Corinthians defeated an Athenian expedition to Ortigia in 413 BC (part of the Athenian-Peloponnese war). The 396 BC invasion of Syracuse by the Carthaginians is reported to have been held up by an eruption of Etna. The Piazza Archimedes is dedicated to the mathematician Archimedes who was killed in the Second Punic War (estimated at 212 BC), when Roman forces captured the city from the Greeks after a two year long siege. The Neapolis Archaeological Park is part of a UNESCO world heritage site which protects both the Ancient Greek and Roman buildings in an eroded section of the Syracuse Plateau.

The Syracuse Plateau is underlain by the Syracuse Limestone (Late Oligocene-Pliocene) which has eroded to a typical karstic landscape with dry valleys, bare rock pavements, and caves (Fig. 8.19a). Building stones for the monumental buildings were derived from nearby quarries. The Syracuse Limestone was used for construction of many of the classical buildings in the city, and unfortunately the building stones are relatively soft and are deteriorating rapidly (Giuffrida and Ciliberto 2013). The most well known structure at the archaeological park is the *Teatro-Greco*, a 5thC BC theatre that has been partially rebuilt (and was

a

b

Fig. 8.17 a The archaeological site of Paestum is situated on a broad coastal plain at the base of the Apennines. The Sele Plain is underlain by Quaternary sediments. Hills in the background consist of Mesozoic carbonates; **b** Paestum includes three well-preserved temples, including the Temple of Athena built in the Doric style (possible age of 500 BC)

Fig. 8.18 a The Velia archaeological site includes Roman walls damaged by earthquakes; **b** Excavations at Velia have exposed an alluvial sequence which includes historical remains (including terracotta Roman tiles) from the 2ndC AD until present

previously reconstructed by the Romans), known for presenting plays by the Greek poet Aeschylus (including "The Persians" at approximately 455 BC). The theatre is cut into the limestone bedrock (Fig. 8.19b). The *Grotta del Ninfeo* is an excavation located behind the theatre that is reached by a path known as the "Street of Tombs". The cave formerly contained a large pot made of *opus signinum* (a building material consisting of broken pottery and mixed with mortar used in Roman times) that fed underground water via an aqueduct into the theatre. The flat-lying nature of the marine limestones in the Syracuse Plateau can be observed at the ancient quarries, collectively known as *Latomie* (Fig. 8.20a). The most famous of the quarries contains a cave known as the "Ear of Dionysus" (Fig. 8.20b).

8.11   The Odyssey

Mythology was an important part of the Ancient World and provided a way of recording unusual events, including epic voyages of exploration and discovery. History and mythology are interwoven. The epic poems the *Iliad* and the *Odyssey*, probably written by the Greek poet Homer in the 8thC BC, are thought to describe historical events from the 13thC BC, i.e., associated with the Mycenaean civilization. Classical scholars have attempted to reconstruct the poems, although there is little consensus in this regard (e.g., Fox, 2009). The *Odyssey* can, in a broad sense, be envisaged as an early record of the perils of navigation in the Mediterranean, as described in detailed pilot books for the waters around Greece and Italy (Heikell, 2002; Heikell and Heikell, 2014). The Trojan Wars described in the *Iliad* probably refer to the ancient city of Troy (Homer's *Ilium*), which is thought to be located on the southern shores of the Dardanelles, western Turkey (Chap. 13). Most classical scholars (although there is no consensus) suggest that *Odysseus* lived on the rocky island of Ithaca, one of the seven Ionian Islands located off the northwest coast of Greece (Fig. 8.21).

A map located in the village of Stavrós, in the northwest of Ithaca, shows a highly speculative route that *Odysseus* may have endured on his way home after the Trojan Wars (Fig. 8.22). The Stavrós Museum has the only physical evidence of the existence of *Odysseus*, fragments of pottery derived from a sea-cave, at Port Pólis, which has subsequently been closed due to earthquake damage. The Cave of Pólis yielded Late Bronze Age or Mycenaean relics. There is an archaeological site dating to this period at Pelikata Hill, near Port Pólis, which some historians accept as a possible location of *Odysseus' Palace*. Other historical sites on Ithaca include Port Frikes, which may be linked to the *Reithron* of the *Odyssey*, the ravens rock (the *Korax*) and the Cave of the Nymphs. Ithaca includes several sheltered harbours from where *Odysseus* may have commenced his journey (Fig. 8.23a). Historical walks on the island include these localities as well as a rocky valley (Homer's School), near Stavrós, where Homer was alleged to teach from a pulpit-like rock (Fig. 8.23b).

Odysseus inadvertently opened the contrary winds, which *Aeolus* the Greek god of winds had tied up in a bag, near Ithaca. These winds blew *Odysseus* and his ship as far as the Aeolian Islands. This part of the Tyrrhenian Sea is well known for sudden, violent winds and constitutes possibly the

Fig. 8.19 **a** The Neapolis Archaeological Park at Syracuse is situated on a limestone pavement (Syracuse Limestone) that erodes to form a karstic landscape; **b** The *Teatro-Greco* is constructed into the limestone bedrock. Natural caves and quarries in the limestone are visible in the background

Fig. 8.20 **a** The flat-lying nature of the marine limestones can be observed at the *Latomie* quarries; **b** The "Ear of Dionysus" is located in the Syracuse Limestone

Fig. 8.21 Image of the Eastern Mediterranean showing some of the possible locations associated with Homer's *Odyssey*. *Source* Satellite Image of Europe based on NASA MODIS data, processed by Philip Eales, Planetary Visions/DLR

Fig. 8.22 A map located in the village of Stavrós, Ithaca, shows a highly speculative route that *Odysseus* may have endured on his voyage from Troy (location 1) to Ithaca (location 14)

stormiest part of the Mediterranean (Heikell, 2002). The island of Ustica has been identified by some authorities to be *Odysseus*' home, although this is not generally accepted. The Ancient Greeks described active volcanic islands in the central and eastern Mediterranean, including Etna and Stromboli, as the private foundries of the Olympian god, *Hephaestus*. The Stromboli Volcano has been linked to the "wandering rocks" of *Circe* and the persistently glowing cone was used as a natural lighthouse by ancient mariners, long before the concept of a lighted tower was invented.

The *Ciclopi* at Acitrezza, a small town north of Catania, are black rock pillars that rise from the sea close to the harbour (Fig. 8.24). Legend has it that these are the rocks that the blinded *Polyphemus*, one of the *Cyclopes*, hurled after *Odysseus* and his men. The fine-grained dolerite is eroded from near-surface sills associated with the Etna Volcano. The basaltic magma was injected into the Syracuse Limestone.

Despite the whereabouts of localities in the *Odyssey* being debated by historians, the twin perils of *Scylla* and *Charybdis* have never been in doubt in the minds of seamen: they occur in the **Straits of Messina** (Fig. 8.2). Passage through the narrow waters between Calabria and Sicily was far more perilous in historical times than the relatively minor problems now encountered. The dangers have diminished since the catastrophic earthquakes described above. The 1908 AD earthquake, in particular, appears to have smoothed out the seabed and reduced the effectiveness of the

a

b

Fig. 8.23 **a** The village and sheltered harbour of Kioni is one of the many idyllic locations on the island of Ithaca; **b** "Homer's School", located in the limestone (Mesozoic) hills near Stavrós, Ithaca, includes a pulpit-like rock

a

b

Fig. 8.24 a The *Ciclopi* are black pillars of fine-grained dolerite that rise from the sea close to the harbour at Acitrezza. The dolerite reveals prominent columnar jointing; **b** The dolerite occurs in near-surface sills injected into the Syracuse Limestone

Fig. 8.25 Waterspouts are common phenomena in the Straits of Messina and other parts of the Mediterranean (example here photographed in the Adriatic Sea, between Albania and southern Italy). They typically occur as clusters with multiple centres that drop instantaneously over several square kilometres

whirlpools. Heikell (2002) reports that during the period associated with the Ancient Greek legends "there is every reason to suppose a whirlpool did exist off the town of Scylla (Sicily) and that both it and the whirlpool off the town of Charybdis (Calabria) were rather more impressive than the latter is today". As late as 1824, the British Admiralty Pilot noted the dangers of this passage and Admiral Smith wrote that "to the undecked boats of the Ancient Greeks, it must have been formidable, for even in the present day small craft are sometimes endangered by it, and I have seen a 74-gun ship whirled around on the surface" (quoted in Heikell, 2002). The supposition is that it was a large waterspout, or cluster of waterspouts, which was Homer's Scylla (a monster with numerous tentacles which sucked seaman from above). Waterspouts are common phenomena today in the Straits of Messina and other parts of the Mediterranean (Fig. 8.25).

References

Bertagnini, A., Métrich, N., Landi, P., & Rosi, M. (2003). Stromboli volcano (Aeolian Archipelago, Italy): An open window on the deep-feeding system of a steady state basaltic volcano. *Journal of Geophysical Research, 108,* 2336. https://doi.org/10.1029/2002jb00214.

Bozkurt, E., Winchester, J.A. & Piper, J.D.A. (2000). Tectonics and magmatism in Turkey and surrounding areas. *Geological Society of London Special Publication, 173,* 521 p.

Branca, S., Coltelli, M., De Beni, E., & Wijbran, J. (2008). Geological evolution of Mount Etna volcano (Italy) from earliest products until the first central volcanism (between 500 and 100 ka ago) inferred from geochronological and stratigraphic data. *International Journal of Earth Sciences, 97,* 135–152.

Cas, R.A.F. & Wright, J.V. (1988). Volcanic Successions Modern and Ancient: A geological approach to processes, products, and successions. Springer Netherlands. ISBN: 978-0-412-44640-5.

Casseta, F., Giacomoni, P. P., Ferlite, C., Bonadiman, C., & Coltorti, M. (2019). The evolution of the mantle source beneath Mt. Etna (Sicily, Italy): from the 600 ka tholeiites to the recent trachybasaltic magmas. *International Geology Reviews, 62,* 338–359.

Cavazza, W., & Wezel, F. C. (2003). The Mediterranean region – a geological primer. *Episodes, 26,* 160–168.

Cita, M. B. (2006). Exhumation of Messinian evaporites in the deep-sea and creation of deep anoxic brine-filled collapsed basins. *Sedimentary Geology, 188–189,* 357–378.

Civetta, L., Comette, Y., Crisci, G., & Gillot, P. Y. (1984). Geology, geochronology and chemical evolution of the island of Pantelleria. *Geological Magazine, 121,* 541–562.

Coltelli, M., Del Carlo, P., & Vezzoli, L. (2000). Stratigraphic constraints for explosive activity in the last 100 ka at Etna volcano, Italy. *International Journal of Earth Sciences, 89,* 665–677.

Corsaro, R. A., & Cristofolini, R. (1997). Geology, geochemistry and mineral chemistry of tholeiitic to transitional Etnean magmas. *Acta Vulcanologica, 9,* 55–66.

Corsaro, R. A. & Cristofolini, R. (2000). Subaqueous volcanism in the Etnean area: evidence for hydromagmatic activity and regional uplift inferred from the Castle Rock of Acicastello. *Journal of Volcanology and Geothermal Research, 95,* 209–225.

Dilek, Y., Moores, E.M., Elthon, D. & Nicolas, A. (2000). Ophiolites and Oceanic Crust: new insights from field studies and the ocean drilling program. *Geological Society of America Special Paper, 349,* 552 p.

Flower, M.F.J. & Dilek, Y. (2003). Arc–trench Rollback and Forearc Accretion: 1. A Collision–Induced Mantle Flow Model for Tethyan Ophiolites. In: Dilek, Y. and Robinson, P.T. (Eds.) Ophiolites in Earth History. *Geological Society London Special Publications, 218,* 21–41

Fox, R. L. (2009). Travelling Heroes: Greeks and their myths in the epic age of Homer. Penguin, 514 p.

Giacomelli, L., Perrotta, A., Scandone, R. & Scarpati, C. (2003). The eruption of Vesuvius 79AD and its impact on human environment of Pompeii. *Episodes, 26,* 235–238.

Gill, R. (2010). Igneous Rocks and processes: a practical guide. Wiley-Blackwell, London, 429 p.

Giuffrida, A. & Ciliberto, E. (2013). Syracuse Limestone: from the past a prospect for contemporary buildings. *Geosciences, 3,* 159–175.

Heikell, R. (2002). Italian Waters Pilot. Imray, Laurie, Norie and Wilson Ltd., 6th edition, 430 p.

Heikell, R. & Heikell, L. (2014). Greek Waters Pilot. Imray, Laurie, Norie and Wilson Ltd., 12th edition, 568 p.

Hsü, K. J. (1977). Tectonic evolution of the Mediterranean Basins. In A. E. M. Nairn, W. H. Kanes, & F. G. Stehli (Eds.), *The Ocean Basins and Margins* (Vol. 4A, pp. 29–76)., The Eastern Mediterranean New York and London: Plenum Press.

Jolivet, L., et al. (2013). Aegean tectonics: Strain localisation, slab tearing and trench retreat. *Tectonophysics, 597–598,* 1–33.

Kilburn, C. J., & McGuire, W. J. (2001). Italian Volcanoes (p. 174). Harpenden: Terra Publishing.

Laubscher, H., & Bernoulli, D. (1977). Mediterranean and Tethys. In A. E. M. Nairn, W. H. Kanes, & F. G. Stehli (Eds.), *The Ocean Basins and Margins* (Vol. 4A, pp. 1–28)., The Eastern Mediterranean New York and London: Plenum Press.

Le Maitre, R. W., et al. (Eds.). (2002). Igneous Rocks: A Classification and Glossary of Terms, Recommendations of the International Union of Geological Sciences. Subcommission of the Systematics of Igneous Rocks: Cambridge University Press. ISBN 0-521-66215-X.

Meier, T., Becker, D., Endrun, B., Rische, M., Bohnhoff, M., Stockhert, B., & Harpjes, H-P. (2007). A model for the Hellenic subduction zone in the area of Crete based on seismological investigations. In: Taymaz, T., Yilmaz, Y., and Dilek, Y. (Eds.) The Geodynamics of the Aegean and Anatolia. *Geological Society London Special Publications, 291,* 183–199.

Monaco, C., Tapponnier, P., Tortorici, L., & Gillot, P. Y. (1997). Late Quaternary slip rates on the Acireale-Piedimonte normal faults and tectonic origin of Mt. Etna (Sicily). *Earth and Planetary Science Letters, 147,* 125–139.

Moores, E. M., & Fairbridge, R. W. (Eds.). (1998). Encyclopaedia of European and Asian Regional Geology (p. 825). London: Encyclopaedia of Earth Sciences Series.

Mowbray, J.H. (1909). Italy's Great Horror of Earthquake and Tidal Wave. National Publishing, Philadelphia, 358 p (urn: colic: record: 1047490596).

Okay, A. I., & Tüysüz, O. (1999). Tethyan sutures of northern Turkey. *Geological Society London Special Publication, 156,* 475–515.

Orsi, G., de Vita, S, di Vito, M.A. & Isaia, R. (2004). The Neapolitan active volcanoes (Vesuvio, Campi Flegrei, Ischia): Science and impact on Human Life. Field Trip Guidebook B28, 32nd International Geological Congress, 44 p.

Panza, G. F., & Suhadolc, P. (1990). Properties of the lithosphere in collisional belts in the Mediterranean – a review. *Tectonophysics, 182,* 39–46.

Peccerillo, A. (2003). Plio-quaternary magmatism in Italy. *Episodes, 26,* 222–226.

Peccerillo, A., Astis de, G., Faraone, D., Forni, F. & Frezzotti, M.L. (2013). Compositional variations of magmas in the Aeolian arc: implications for petrogenesis and geodynamics. In: Lucchi, F., Peccerillo, A., Keller, J., Tranne, C. A. and Rossi, P. L. (Eds.) The Aeolian Islands Volcanoes. *Geological Society London Memoirs, 37,* 491–510.

Pescatore, T.S. & Cinque, A. (2004). Historical-geological events and their impact on Man. Field Trip Guidebook P14, 32nd International Geological Congress, 44 p.

Piatanesi, A., & Tinti, S. (1998). A revision of the 1693 eastern Sicily earthquake and tsunami. *Journal of Geophysical Research, 103,* 2749–2758.

Robertson, A. H. F., & Mountrakis, D. (2006). Tectonic development of the Eastern Mediterranean region: an introduction. *Geological Society of London Special Publication, 260,* 1–9.

Santacroce, R., Christofolini, R., La Volpe, L., Orsi, G., & Rosi, M. (2003). Italian active volcanoes. *Episodes, 26,* 227–234.

di Traglia, F., Pistolesi, M., Rosi, M., Bonadonna, C., Fussilo, R., & Roverato, M. (2013). Growth and erosion: The volcanic geology and morphological evolution of La Fossa (Island of Vulcano, Southern Italy) in the last 1,000 years. *Geomorphology, 194,* 94–107.

Vezzoli, L., & Corazzato, C. (2016). Geological constraints of a structural model of sector collapse at Stromboli volcano, Italy. *Tectonics, 35,* 2070–2081.

Walker, G. P. L. (1973). Explosive volcanic eruptions - a new classification scheme. *Geologische Rundschau, 62,* 431–446.

Windley, B. (1977). The Evolving Continents. Wiley and Sons, 399 p.

Wright, J. V., Smith, A. L., & Self, S. (1980). A working terminology of pyroclastic deposits. *Journal of Volcanology and Geothermal Research, 8,* 315–336.

Neapolitan Volcanoes, Southern Italy

Active Volcanoes, 79 AD Eruption of Vesuvius, and Roman Archaeology

Abstract

The Neapolitan volcanoes are located on the margins of the Campanian Plain adjacent to the Gulf of Naples, southern Italy. The volcanoes are closely monitored from the Vesuvius Volcanological Observatory as they pose a major threat to the cities of Naples and Pozzuoli. Long periods of quiescence have encouraged extensive settlement of the region and yet at least two of the volcanoes are active features that periodically experience Plinian-style eruptions. Plinian eruptions are catastrophic events named after Pliny the Younger who witnessed the 79 AD eruption of Vesuvius that devastated Roman settlements at the base of the volcano. Plinian eruptions typically include pyroclastic flows and partial or total collapse of cones. Two volcanic districts are recognized, Somma-Vesuvius and Phlegraean. The Somma-Vesuvius volcanic edifice towers threateningly above the city of Naples. The Phlegraean district includes the Campi Flegrei Volcano, a nested complex of calderas in which the city of Pozzouli is situated, and the volcanic island of Ischia located in the Gulf of Naples. The Campi Flegrei and Vesuvius Volcanoes are possibly the most hazardous volcanoes in Europe. Archaeological evidence and historical descriptions of the Neapolitan volcanoes has been incorporated into the modern discipline of volcanology. The Campanian Plain is a graben situated between the western slopes of the Southern Apennine Mountains and the Gulf of Naples. The graben commenced forming in the Neogene due to stress released from subduction of the Adriatic Microplate beneath the Eurasian Plate. The Neapolitan volcanoes are Quaternary features. The region remains tectonically active. The Adriatic Microplate continues to be subducted on the Apennine-Maghreb Thrust. Melting of the subducted slab occurs at depth. Extensional faults associated with the Campanian Graben promote the ascent of magmas. The volatile- and alkali-rich basaltic magmas invariably trigger catastrophic eruptions. The Ischia Volcano reports the oldest activity (150,000 BP). Two caldera events associated with the Campi Flegrei Volcano, the Campanian Ignimbrite (38,000 BP) and the Neapolitan Yellow Tuff (15,000 BP), were sufficiently extreme as to trigger regional climatic changes. Multiple Plinian and sub-Plinian events have been identified in the Somma-Vesuvius massif. The oldest event occurred at 22,000 BP and the youngest is the historical event of 79 AD. No systematic periodicity has been identified. Geological and archaeological tourism is a significant component of the local economy and the Roman settlements excavated from the deposits of volcanic ash and pumice attract huge numbers of tourists. The city of Herculaneum was entirely buried, but Pompeii suffered considerably less damage due to its more distal location. Pliny the Younger documented several important features of the eruption including the giant, umbrella-shaped ash column and the destructive pyroclastic flows. The pyroclastics flows were preceded by basal surges which are estimated to have travelled down the slopes of the cone at speeds > 100 kph. Temperatures are calculated at 350–400 °C. The Vesuvius National Park protects a crater on the summit of the Vesuvius cone. The crater developed in 1906. The last significant eruption of Vesuvius occurred in 1944.

Keywords

Campi Flegrei • Gulf of Naples • Herculaneum • Ischia • Plinian eruptions • Pompeii • Somma-Vesuvius

Photographs not otherwise referenced are by the author.

9.1   Introduction

The Neapolitan volcanoes are located on the margins of the Campanian Plain, a prominent physiographic feature situated between the Southern Apennines and the Gulf of Naples in southern Italy. The Gulf of Naples is a 15 km-wide indentation in the Tyrrhenian Sea that contains three islands, Capri, Ischia, and Procida. The Neapolitan or Campanian Volcanic Field is separated into the Somma-Vesuvius and Phlegraean districts. The Somma-Vesuvius massif includes the twin peaks of the Somma caldera and the Vesuvius cone (Fig. 9.1). The Phlegraean District incorporates the active volcanic complex of Campi Flegrei and the quiescent Ischia Volcano. Ischia is the largest of the volcanic islands in the Gulf of Naples. The Campanian Plain has been inhabited since Neolithic times and has an extensive historical record. The popular legend suggesting that the Roman emperor Caligula drove a chariot across the Gulf of Naples on a bridge of boats, whilst wearing the armour of Alexander the Great, may have arisen because of historical records associated with the 79 AD eruption of Vesuvius. This eruption devastated the Roman settlements at the base of the Somma-Vesuvius massif. The 79 AD eruption is a famous historical event and has been widely investigated by geologists, archaeologists, and historians.

The Gulf of Naples is constrained to the south by the Sorrento Peninsula and to the north by the Miseno Peninsula (Fig. 9.2a). The cities of Naples and Pozzuoli constitute a more-or-less contiguous urban area. Naples is situated at the base of the Somma-Vesuvius massif and Pozzuoli occurs within the nested calderas of the Campi Flegrei Volcano (Fig. 9.2b). The interior mountains and some of the rugged coastal landforms are dominated by Mesozoic carbonates, but the Campanian Plain is underlain by Neogene and Quaternary sediments. The Neapolitan volcanoes are Quaternary features.

Geological and archaeological tourism is a significant component of the local economy and Mount Vesuvius includes a national park that protects part of the volcanic cone. Visitors are also attracted to the regions' coastal resorts, including the Sorrento Peninsula, the scenic towns of Amalfi and Positano, and the islands of Capri and Ischia.

Fig. 9.1 The two peaks of the Somma-Vesuvius Volcano (Vesuvius on the right) tower above the city of Naples with mountains of the Southern Apennines (left) and Sorrento Peninsula (far right) visible in the background

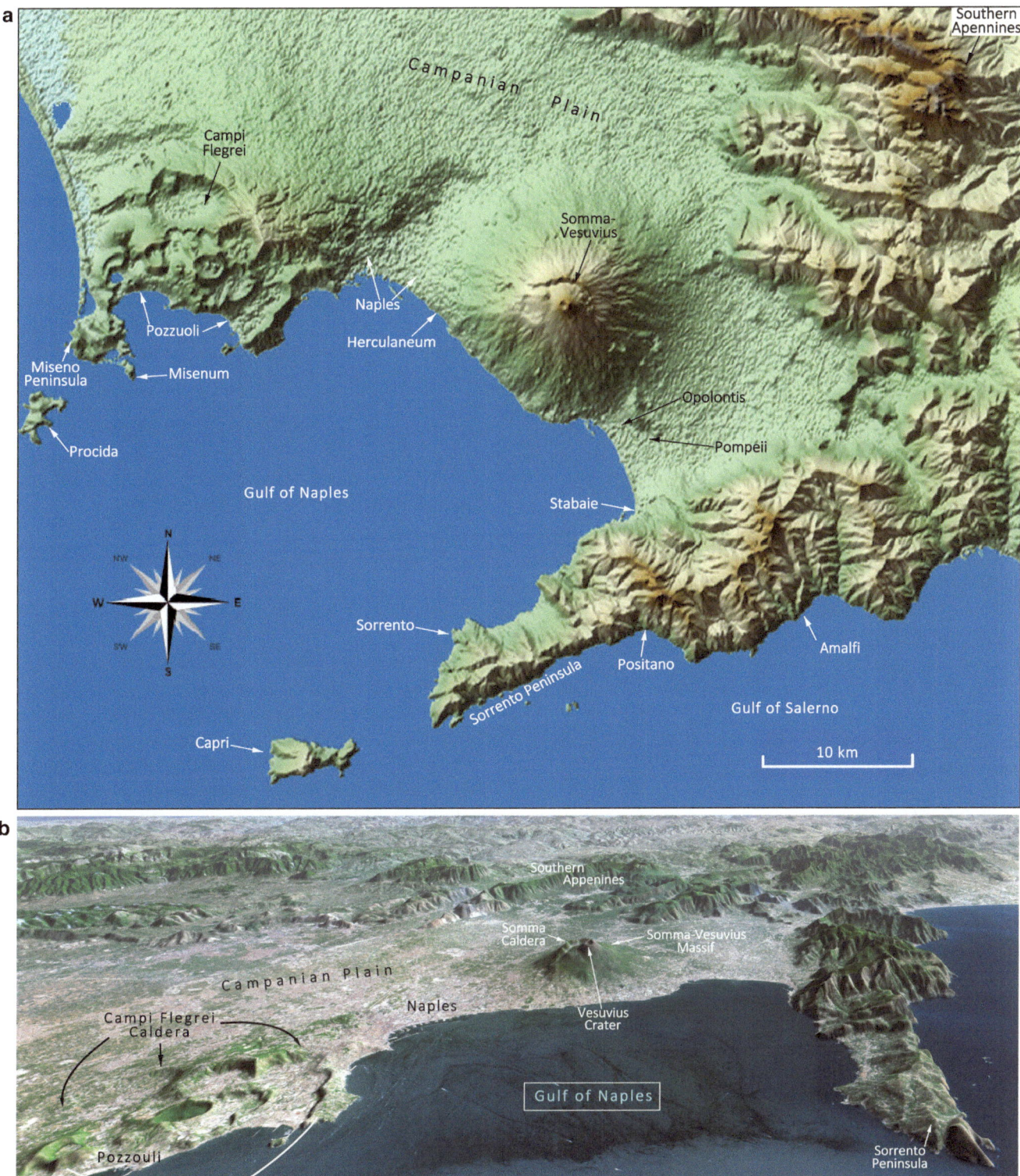

Fig. 9.2 **a** The Neapolitan volcanoes are located on the margins of the broad Campanian Plain between the Southern Apennines and the Gulf of Naples. The islands of Capri and Procida are included in the image, but Ischia is located to the west of the area shown. Pliny the Younger observed the 79 AD eruption from Misenum. Pliny the Elder died at Stabaie; **b** Three-dimensional image looking east showing the more-or-less continuous urban conurbation on the shores of the Gulf of Naples. The dome-shaped Somma-Vesuvius massif and the nested calderas of the Campi Flegrei Volcano are prominent features. *Source* 3D Image based on NASA MODIS data, processed by Philip Eales, Planetary Visions/DLR

9.2 Volcanology

The modern discipline of volcanology i.e., the study of volcanoes including the origin of magmas and their physical and chemical properties is named after *Vulcan*, the Roman god of fire. The history of volcanology may be traced to the earliest civilizations, including references in Homer's poem the *Odyssey* (Chap. 8). The Greek philosopher Plato (428–348 BC) commented on volcanoes and the internal heat of the Earth. The historical eruption of the Santorini Volcano may have initiated Plato's legend of the lost continent of Atlantis (Chap. 12). The Roman poet Virgil is thought to have described the 122 BC eruption of the Etna Volcano in the *Aeneid*. The 79 AD eruption of Vesuvius is a pivotal event in European history, in part due to the descriptions that Pliny the Younger (61–113 AD) provided to the Roman historian Tacitus (56–120 AD). William Hamilton, the British diplomat based in Naples for long periods during the latter part of the 18thC documented the historical eruptions of Vesuvius based on samples from the 1690 eruption of Vesuvius exhibited by the Catholic Church.

The prediction of eruptions, despite being a key component of volcanology, remains full of uncertainties. The Vesuvius Volcanological Observatory was established in 1841 with the primary aim of studying the Neapolitan volcanoes. The observatory has also contributed to global volcanological research (Di Vito et al. 2011). The absence of eruptions over long periods (including modern times) has encouraged the extensive settlement of the volcanic slopes surrounding the Gulf of Naples. The fertile volcanic soils are assisted by a relatively wet, yet hot and sunny climate, with well-defined seasons that promote almost perfect agricultural conditions. As many as four crops may be produced annually. The persistent erosion of the volcanic rocks almost negates the necessity of fertilization.

The last significant eruption of Campi Flegrei was in 1538. Vesuvius is quiescent after 300 years of almost continuous activity. Campi Flegrei and Somma-Vesuvius are possibly the most hazardous volcanoes in Europe as they periodically experience catastrophic, Plinian-style eruptions and are situated close to large urban centres.

9.3 Regional Geology

The regional geology is summarized on a simplified map (Fig. 9.3). The Mesozoic basement is exposed in the Southern Apennine Mountains and coastal locations such as the Amalfi Coast and the Sorrento Peninsula (Fig. 9.4a). The dominant lithologies are limestone, dolomite, and marble. The island of Capri includes high cliffs comprised of Mesozoic limestone (Fig. 9.4b). The Campanian Plain is underlain by marine, fluvial, and glacial sediments with a combined thickness of up to 3,000 m. The sediments were deposited during the Neogene and Quaternary. The Campanian Plain is part of a graben situated between the Southern Apennines and the Tyrrhenian Sea. Faulting commenced in the Neogene during extensional tectonism ascribed to release of stress associated with the active subduction (Panza and Suhadolc 1990) (Chap. 8). Tectonism persisted into the Quaternary. Some faults remain active. This is supported by historical and archaeological evidence which covers the past 5,000 years (Pescatore and Cinque 2004). The elevation of the Campanian Plain has been affected by both localized tectonic activity and regional variations in sea-level. There is evidence of higher sea levels during warmer periods of the Pliocene, Pleistocene, and Holocene (e.g., raised beaches and palaeo-surfaces). Sea-level changes were particularly pronounced during the Late Pleistocene Ice Ages.

9.4 Neapolitan Volcanoes

The Neapolitan volcanoes are part of the Campanian Volcanic Field which includes volcanic seamounts in the vicinity of the Gulf of Naples. The volcanic deposits of the Somma-Vesuvius and Phlegraean Districts are locally interbedded with Quaternary sediments on the Campanian Plain. The volcanism is associated with subduction of the Adriatic Microplate beneath the Eurasian Plate. Melting of the subducted slab occurs beneath the Tyrrhenian Sea, i.e., west of the Apennine-Maghreb Thrust (Panza and Suhadolc 1990; Cavazza and Wezel 2003; Orsi et al. 2004; Di Vito et al. 2011). This correlates approximately with the depth of the Benioff zone (Chap. 8). Extensional tectonics associated with the Campanian Graben promotes ascent of magmas. The oldest of the Neapolitan volcanoes is probably Ischia, the earliest eruption of which may date to 150,000 BP (Fig. 9.5). The maximum ages of the Campi Flegrei and Somma-Vesuvius Volcanoes is estimated at 60,000 BP and > 55,000 BP, respectively.

The Neapolitan volcanoes are characterized by eruption of alkali-rich basaltic and foidolite magmas (Chap. 8). A broad range of compositions have been identified, including nephelinite, potassium-rich phonotephrite and tephriphonolite (Somma-Vesuvius) and leucitite and sodium-rich basanite (Phlegraean District) (Orsi et al. 2004; Di Vito et al. 2011). The petrogenesis is complicated by eruption of magmas from shallow staging chambers. Batches of magma derived from depth may have mixed with the resident magma in a staging chamber (or a group of nested chambers). The injection of magma into the staging chambers either triggered an instantaneous eruption, or resulted in subsequent eruptions after cooling and differentiating.

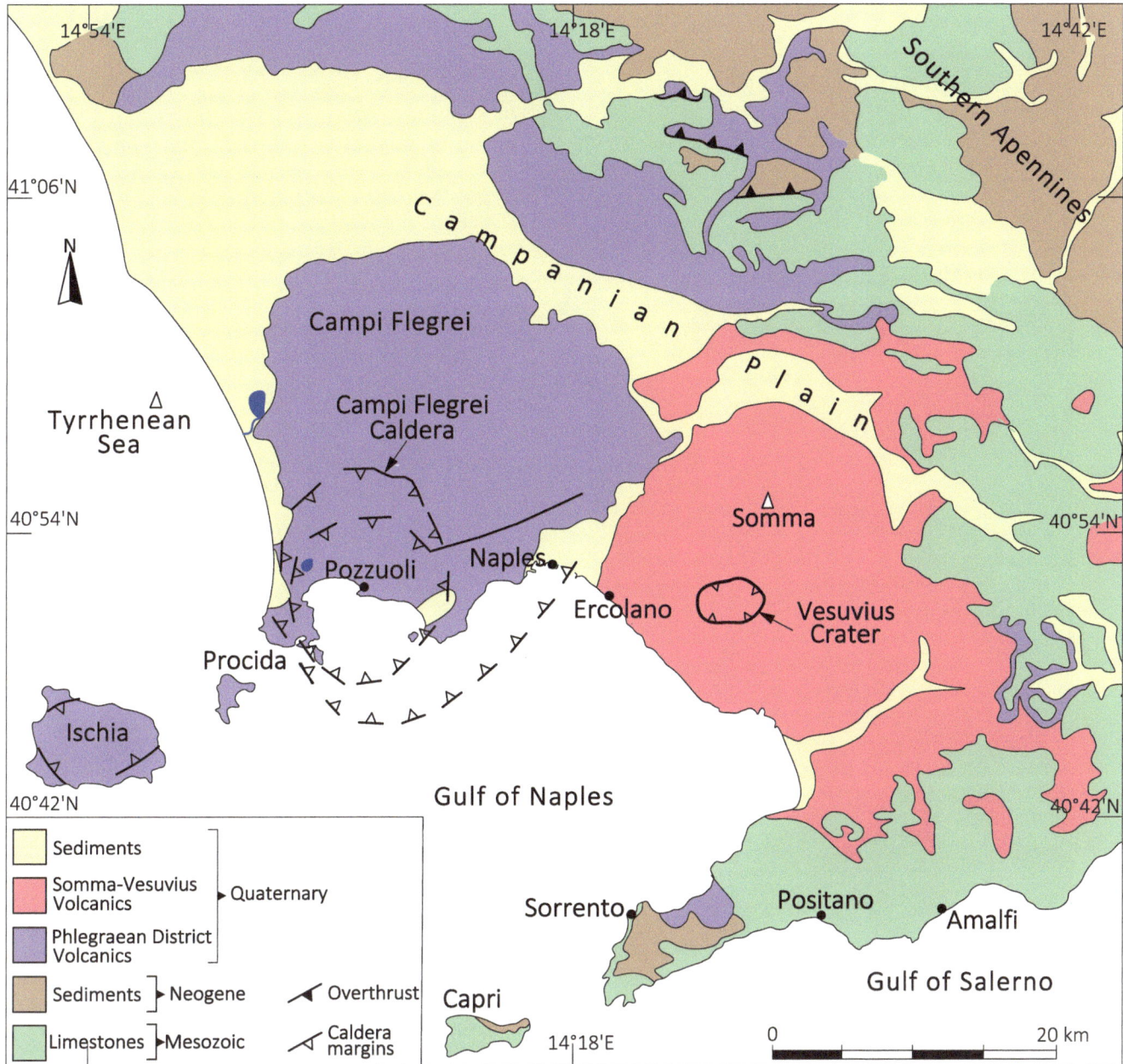

Fig. 9.3 Geological map of the area around the Gulf of Naples. *Source* Simplified after Orsi et al. (2004)

Derivation of the potassium-rich alkali magmas associated with the Somma-Vesuvius massif is influenced by three principal processes (Orsi et al. 2004; Di Vito et al. 2011): (i) partial melting of the deep lithosphere; (ii) interaction with fluids derived from subducted oceanic crust; and (iii) interaction with fluids derived from partial melting of continental crust. The variations in composition and differing styles of eruption are controlled by the relative dominance of the three processes. The introduction of fluids into the mantle wedge (process (ii)) was significant in some of the older activity, with process (iii) having influenced the more recent activity (typically younger than 39,000 BP), including

eruptions of Vesuvius. The introduction of fluids from melting of continental crust has given rise to many of the most catastrophic eruptions.

9.5 Campi Flegrei Volcano

The dominant feature of the Campi Flegrei Volcano is a large, nested caldera structure (Fig. 9.2b). The extent of the complex of calderas can be viewed from Camaldoli Hill (Fig. 9.6). Campi Flegrei has been linked to two Plinian events which included major episodes of magma withdrawal

a

b

Fig. 9.4 **a** The scenic town of Amalfi is overshadowed by steep cliffs of Mesozoic limestone; **b** The western side of the island of Capri includes cliffs and sea stacks consisting of Mesozoic limestone

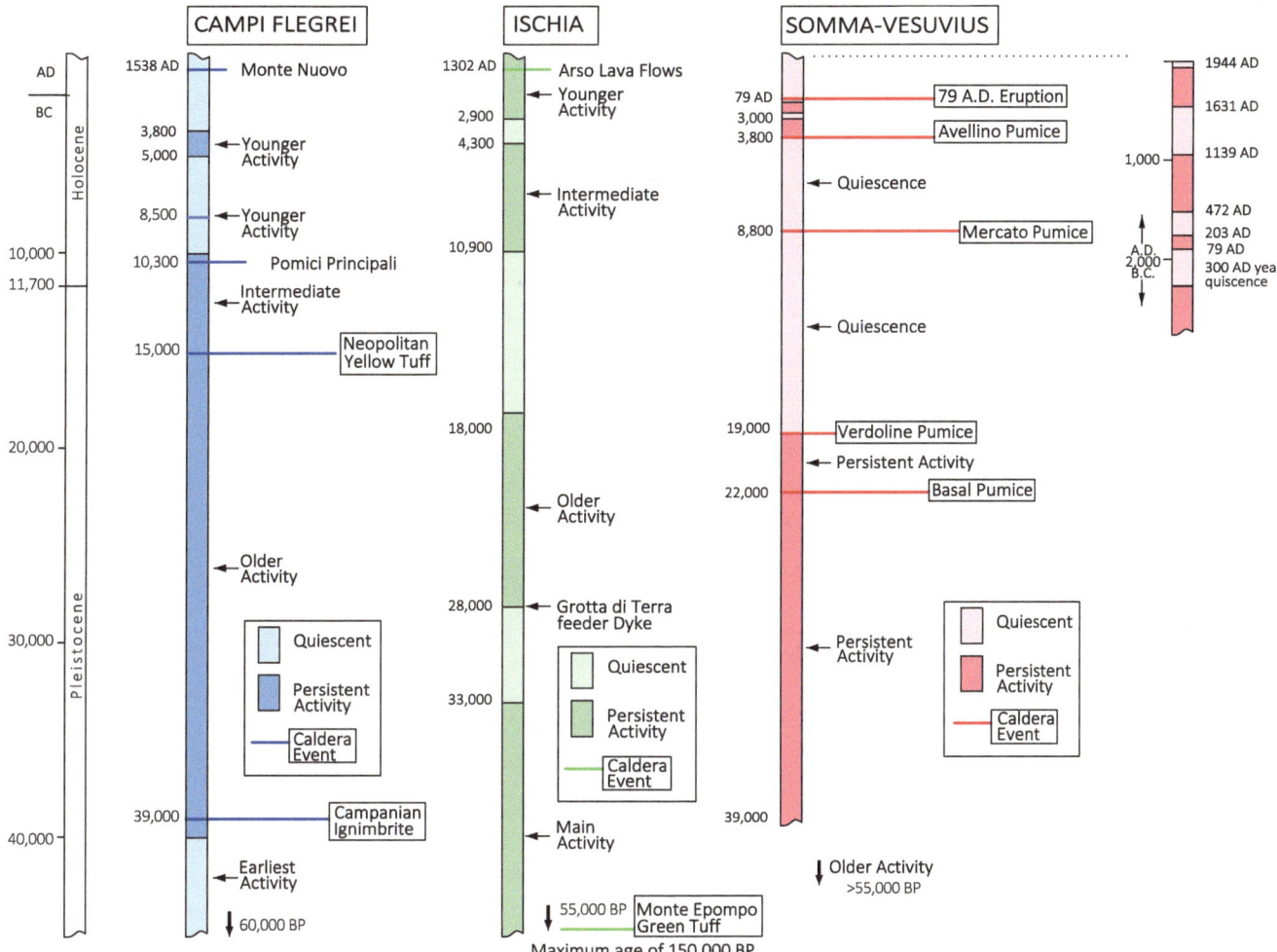

Fig. 9.5 Chronology of eruptions of the principal Neapolitan Volcanoes. *Source* Modified after Orsi et al. (2004) and Di Vito et al. (2011)

and caldera collapse (Fig. 9.5) (Orsi et al. 1992). The Plinian events are named after their principal deposits, the Campanian Ignimbrite (39,000 BP) and the Neapolitan Yellow Tuff (15,000 BP). These deposits can be examined in the Verdolino Quarry, located near the centre of Pozzuoli (Fig. 9.7). The coloration of the Neapolitan Yellow Tuff is ascribed to zeolitization. Both Plinian events were sufficiently catastrophic as to have triggered regional climatic changes that affected European civilizations. Ash fallout covers large parts of the central and eastern Mediterranean and extends as far as Russia (Polacci et al. 2003). Archaeologists correlate the Campanian Ignimbrite with a "cultural layer" used to divide the Early and Middle Palaeolithic.

Campi Flegrei experienced persistent activity in the period between the two Plinian events, i.e. the "older activity" depicted in Fig. 9.5. The period following the Neapolitan event until approximately 10,300 BP included several sub-Plinian events, i.e. the "intermediate activity". Deposits from these events occur on the outer slopes of the caldera complex. Deposits associated with the Principali Pumice (10,300 BP), a significant sub-Plinian event, extent onto the slopes of the Somma-Vesuvius massif. Sub-Plinian events also occurred at 8,500 BP and 5,000-3,800 BP, i.e., the "younger activity". This activity was followed by approximately 3,000 years of quiescence. The Monte Nuovo eruption in 1538 AD, the youngest of the sub-Plinian events, was the last significant activity of the Campi Flegrei Volcano.

9.5.1 Historical Tectonism

Geological information on tectonic activity associated with the Campi Flegrei calderas is supplemented by historical accounts. The 3rdC AD through to the 11thC AD was a period of subsidence. This has been displaced by episodic periods of uplift. Columns of the Serapeo Marketplace, a popular tourist attraction located near the harbour in Pozzuoli, provide detailed evidence in this regard (Fig. 9.8a).

Fig. 9.6 View from Camaldoli Hill shows the city of Pozzuoli nestling between the rim of the Campi Flegrei caldera

The columns were constructed by the Romans in the 1st-2nd C AD. During the period of subsidence noted above the columns sunk beneath the Gulf of Naples, prior to "emerging" in the 11th C AD. Uplift of between 5 m and 8 m occurred in 1538 AD, immediately prior to the Monte Nuovo eruption. Two periods of rapid uplift, each of approximately 2 m, were monitored between 1969-1972 and 1982-1984. The harbour wall at Pozzuoli, where the British fleet under Admiral Nelson was temporarily based during the Napoleonic Wars, has risen several metres in less than 200 years (Fig. 9.8b). Nelson first visited Naples in September 1793 when he viewed the glow of Vesuvius at night (presumably from a lava lake in the crater).

9.5.2 Magma Chamber

Geophysical surveys have shown that Campi Flegrei is underlain by a large magma chamber at an estimated depth of 10–20 km. Several shallower staging chambers have also been identified. The presence of geothermal features in the caldera complex is indicative of the high heat flow. The

calderas contain numerous craters, including the large Solfatara which contains sulphur deposits and active fumaroles (Fig. 9.9a). Lake Averno occurs in a scenic crater popular with locals and tourists (Fig. 9.9b). Campi Flegrei is considered as an extremely hazardous volcano, possibly even more so than Vesuvius, in part as the active or quiescent craters and vents occur in a densely urbanized area.

9.6 Ischia Volcano

The island of Ischia is part of a large volcanic ring complex, or caldera, which rises some 1,500 m above an unusually deep section of the Tyrrhenian Sea. The main caldera event occurred at approximately 55,000 BP (Orsi et al. 2004) (Fig. 9.5). The caldera event generated the Mount Epomeo Green Tuff, the most widespread of the volcanic deposits on the island. Mount Epomeo (height of 787 m) is a relict of the original cone prior to caldera collapse. The Ischia Volcano is quiescent, and future eruptions cannot be discounted. The potentially active component of the volcano has an area of 46 km². The most recent eruption – the Arso Lava - occurred in 1302 AD.

Fig. 9.7 The Verdolino Quarry at Pozzuoli exposes the two principal products of the Campi Flegrei Volcano, the Campanian Ignimbrite (39,000 BP) and the Neapolitan Yellow Tuff (15,000 BP)

9.6.1 Cliff Sections

A circumnavigation of the island of Ischia provides a unique opportunity to examine cliff sections of a relatively young volcano (Orsi et al. 2004; Carlino 2018). The section at Punta Imperatore, which may be broadly representative of the volcanic complex, consists of lavas, breccia deposits, a lower unit of pyroclastic rocks, the Mount Epomeo Green Tuff, deposits of the Scarrupo Di Panza Volcano, and an upper unit of pyroclastic rocks (Fig. 9.10). The section at Grotta di Terra consists of a sequence of lavas and pyroclastic rocks crosscut by a younger trachybasaltic dyke (Fig. 9.11). The trachybasaltic dyke is interpreted as a magma conduit, or feeder dyke, which has spread laterally as a sill in the upper part of the cliff. Sills probably form when dykes intersect contacts in the host rock (i.e., between the lavas and pyroclastics in this section). Once a sill is initiated it may evolve into a magma chamber through successive magma injections (Barnett and Gudmundsson 2014). For the sill complex to remain partially molten, and thus inflate to form a magma chamber, it must receive a constant replenishment of magma, implying a high dyke-injection rate.

9.6.2 Recent Activity

Ischia is seismically active and recent events may be related to movement of magma at depth. The earthquake of 21 August 2017 (which measured 4.7 on the Richter scale) caused some buildings to collapse. The two fatalities that unfortunately occurred provoked a debate on the poor state of buildings, many of which were erected with few safety concerns since the main tourist boom commenced in the 1950s.

9.7 Somma-Vesuvius Massif

The volcanology of the Somma-Vesuvius massif has been described in detail (e.g., Sigurdsson et al. 1985; Orsi et al. 2004; Di Vito 2011). The older Somma Volcano includes an

Fig. 9.8 **a** Roman columns at the Serapeo Market, Pozzuoli, buried in the Gulf of Naples for many years due to ground subsidence, abruptly reappeared in the 11ᵗʰC AD due to localized uplift; **b** The harbour at Pozzuoli includes an old wall used by the British Navy in 1799 (left of the car). The location relative to the current sea level is indicative of several metres of uplift during the past 200 years

Fig. 9.9 **a** The Solfatara, Pozzuoli, is a crater located in the Campi Flegrei caldera complex. The crater contains sulphur deposits and active fumaroles; **b** Lake Averno occurs in a scenic crater in the Campi Flegrei calderas

Fig. 9.10 Sequence of lavas and pyroclastics at Punta Imperatore, Ischia. 1: Lavas (< 75,000 BP); 2: Breccia deposits (75,000–55,000 BP); 3: Pyroclastic rocks (75,000–55,000 BP); 4: Mount Epomeo Green Tuff (55,000 BP); 5: Deposits of Scarrupo Di Panza Volcano; 6: Pyroclastic rocks (28,000–18,000 BP)

ancient caldera (diameter of >10 km) in which two younger cones, Somma and Vesuvius, are situated (Fig. 9.2b). The earliest activity commenced prior to 55,000 BP, although there are no accurate dates in this regard. The period between 39,000 BP and 22,000 BP was dominated by persistent effusive, low-energy eruptions (Fig. 9.5). This period was followed by multiple catastrophic, Plinian and sub-Plinian events. Details of the Plinian and sub-Plinian events are preserved in the thick sequences of pumice and ash that mantle the lower slopes of the massif.

9.7.1 Prehistoric Eruptions of Somma-Vesuvius

The Traianello Quarry on the lower slopes of the Somma-Vesuvius massif contains deposits associated with four Plinian events and one significant sub-Plinian event (Orsi et al. 2004) (Fig. 9.12). The ages quoted here are from Di Vito et al. (2011). The Basal Pumice is a thick pyroclastic deposit found at the base of the quarry (age of 22,000 BP). This deposit is overlain by the Verdoline Pumice a relatively thin unit of green pumice and ash derived from a sub-Plinian event (age of 19,000 BP), The Verdoline Pumice is in turn overlain by the products of two major Plinian events, the pink-grey coloured Mercato Pumice and the dark grey Avellino Pumice. These deposits have been radiometrically dated to yield ages of 8,800 BP and 3,800 BP, respectively. The Avellino Pumice discordantly cuts across the older deposits and partially fills palaeo-valleys. The ash deposits are generally very fine-grained. This is an indication of high explosive indices, a worldwide feature of volcanic ashes associated with Plinian or sub-Plinian eruptions (e.g., Barberi et al. 1989) (Chap. 8).

The Avellino Pumice was followed by eight smaller events, mostly Strombolian-style eruptions and/or sub-Plinian activity. This activity was followed by a 300 year-long period of quiescence. The period of quiescence is recognized from palaeosols, some of which have

Fig. 9.11 Trachybasaltic scoria cone and feeder dyke at Grotta di Terra, Ischia. 1: Lavas (<75,000 BP); 2: Pyroclastic rocks (75,000–55,000 BP); 3: Trachybasaltic dyke and sill (28,000 BP); 4: Pyroclastic rocks (<10,000 BP). The feeder dyke has spread laterally to form a sill part way up the cliff, exploiting the primary layering of the older volcanic deposits

been found to contain Early Bronze Age artifacts. This period of quiescence preceded the last catastrophic eruption of Vesuvius, i.e., the 79 AD event.

9.7.2 Periodicity

The prehistoric eruptions of Somma-Vesuvius can be summarized as involving either long periods of quiescence interrupted by catastrophic Plinian or sub-Plinian eruptions, or by periods of persistent activity with lava effusions and smaller-scale Strombolian style eruptions (Orsi et al. 2004). The alternation of Plinian and Strombolian events is attributed to open and closed conduit conditions, respectively. No systematic periodicity in the Plinian events, which could enable future eruptions to be predicted, has been identified.

9.7.3 79 AD Event

Historical data have yielded remarkably accurate information on the geology and timing of the 79 AD eruption of Vesuvius (Orsi et al. 2004; Pescatore and Cinque 2004). The initial activity on 24th August caused some Romans to evacuate villages near the volcano. Pliny the Younger, together with his uncle, Pliny the Elder, a Roman senator, travelled to Misenum (Fig. 9.2a). Pliny the Younger observed many details of the main eruption, which occurred on the 25th August, from Misenum. His letters and sketches were incorporated into historical documents by the Roman historian Tacitus, including descriptions of a huge ash plume with an estimated height of 32 km, the shape of which was compared to an umbrella pine tree (Box 8.2). The letters of Pliny the Younger also described details of the movements

Fig. 9.12 Deposits associated with Plinian and sub-Plinian deposits of Somma-Vesuvius are exposed in the Traianello Quarry. The Basal Pumice (22,000 BP) and the Verdoline Pumice (19,000 BP) are capped by the Principali Pumice, the latter associated with the Campi Flegrei Volcano. The Mercato Pumice (8,800 BP) and Avellino Pumice (3,800 BP) are discordantly overlain by ash-flows associated with the 79 AD event

and death of his uncle. Pliny the Elder travelled with the Roman Navy to Stabaie during the peak of the activity on 25th August. This was prompted by the need to provide assistance to the residents of Pompeii, who had been affected by ash falls during the night. Pliny the Elder died at Stabaie on the 25th August, together with probably all of his accomplices, from asphyxiation and inhalation of poisonous gases associated with eruption of one of the destructive pyroclastic flows.

9.7.4 Roman Cities

The 79 AD event had catastrophic consequences for the Roman cities and settlements to the south and west of Naples (Giacomelli et al. 2003; Pescatore and Cinque 2004) (Fig. 9.2a). The most severely affected city was Hercula-neum, located in the modern suburb of Ercolano at the foot of Somma-Vesuvius (Fig. 9.13). Despite being buried in

volcanic ash and pumice to a depth of more than 20 m, the archaeological site of Herculaneum has been excavated with great attention to detail. Many of the stone or brick walls are relatively intact and mosaic floors, frescoes, and even tim-bered structures survive (Fig. 9.14a). Frescoes reveal scenes of Vesuvius prior to the eruption when the single peak was covered with trees and vineyards. The Roman building of Oplontis Villa provides important details of the 79 AD eruption as it occupies an intermediate position relative to the volcano (Fig. 9.14b).

The archaic and Roman city of Pompeii is located on a basement high 10 km to the south of Somma-Vesuvius. The city suffered considerably less damage than Herculaneum. Pompeii was partially rebuilt as a tourist attraction during the 19[th]C, unfortunately with little input from archaeologists, and many historical details have been lost (Fig. 9.15). Pompeii had already been devastated by a major earthquake in 63 AD and restoration was still underway when the 79 AD eruption occurred. The city walls reveal evidence of

Fig. 9.13 **a** The Roman city of Herculaneum is located in the modern suburb of Ercolano, at the base of the Somma (left) and Vesuvius (right) cones; **b** Roman house at Herculaneum partially filled with volcanic ash (mostly of the A4 and A5 deposits) from the 79 AD eruption

a

b

Fig. 9.14 **a** House at Herculaneum with original wood beams which survived the 79 AD eruption; **b** Part of the reconstructed Oplontis Villa

Fig. 9.15 Part of the reconstructed city of Pompeii with views of Somma-Vesuvius in the background

several building phases (the oldest dates from the 6thC BC), as well as the effects of more recent earthquakes. Some walls are constructed from the Campanian Ignimbrite. Understanding of the geologic events that affected Pompeii has been assisted by drilling of more than 400 boreholes since the 1800s. The geological record of this area is very complex and includes volcanic eruptions, earthquakes, sea-level changes, and catastrophic floods.

9.7.5 Roman Beach Resorts

Archaeological evidence from Roman beach resorts on the Sorrento Peninsula, located close to the shoreline, were severely affected by ash and pumice deposits associated with the 79 AD eruption (Pescatore and Cinque 2004). Excavations at Marina di Equa have shown that the current sea-level is some 5 m higher than during Roman times (Fig. 9.16). Different phases of villa construction are recognized, each triggered by sea level changes and the 79 AD eruption. Sea

level changes are ascribed to localized tectonism (subsidence), rather than global variations.

9.7.6 Details of the 79 AD Volcanic Deposits

The distribution and thickness of the volcanic deposits associated with the 79 AD eruption is summarized after Sigurdsson et al. (1985) and Orsi et al. (2004). The volcanic activity commenced with intense phreatomagmatic eruptions. This activity resulted in formation of the Pompeii Pumice, a deposit that blankets large areas on the slopes of the Somma-Vesuvius massif (Fig. 9.17a). The Pompeii Pumice is overlain by fine-grained ash (deposit A1), which in turn is followed by successive layers of pumice several metres in thickness (deposits A2 and A3) (Fig. 9.17b). All of this activity occurred on the 24th August. The deposits and events that followed on 25th August include volcanic ash (deposits A4-A10), basal surges (surges S1-S6), and pyroclastic flows (flows F1 to F6). The eruption continued with

Recent Soils,
Colluvium

1st–2ndC AD Ash,
Alluvium

79AD Pumice
and Ash

Fig. 9.16 Ruins of Roman beach villas at the Marina di Equa, Sorrento Peninsula. The beach is comprised of volcanic sand. Deposits from the 79 AD eruption occur at the base of the cliff and partially infill the buildings

persistent phreatomagmatic activity for several weeks after 25th August.

Basal surges and their associated pyroclastic flows are characteristic of Plinian eruptions induced by highly explosive volcanoes. Basal surges typically precede eruption of pyroclastics flows and they can travel far greater distances. The basal surges that travelled down the slopes of the Somma-Vesuvius massif towards the Gulf of Naples attained speeds > 100 kph. Temperatures reached 350–400 °C. The basal surges are preserved in the geologic record as either planar surfaces or thin deposits of rock debris. The surfaces include evidence of blasts of hot air that can both burn and asphyxiate victims. Each basal surge and pyroclastic flow is generally followed by falls of fine-grained ash.

Herculaneum was almost instantaneously overrun by the first of the catastrophic events, i.e., the S1 basal surge and the F1 pyroclastic flow. These events occurred at 01.00 h on 25th August. The S1 surge and F1 pyroclastic flow were rapidly followed by a thick ashfall, deposit A4 (the numbering is rather confusing as the initial ashfall, the A1-A3 deposits, occurred prior to the S1 event). Five more basal

surges (surges S2-S6) and their associated pyroclastic flows (flows F2-F6) and ashfall (deposits A5-A9) affected Herculaneum. The overall thickness of the basal surges, pyroclastic flows, and ashfall deposits at Herculaneum exceeds 25 m. Evidence of the six basal surges, the six pyroclastic flows, and the A4 to A9 ashfall deposits are exposed in the 23 m-high wall adjacent to the entrance tunnel. Many of the buildings at Herculaneum are infilled by the relatively thick A4 and A5 deposits. The S1 layer at Herculaneum includes wood remains and human skeletons. This represents the height of the beach during Roman times. The poorly consolidated S2 layer includes a high proportion of building material, including bricks, roof tiles, and mosaics. The relatively thick deposit associated with the S5 surge includes cross lamination. The F1 through F6 pyroclastic flows constitute massive deposits, each of which is several metres in thickness at Herculaneum.

The S1 through S5 surges and the accompanying pyroclastic flows occurred during the night of the 25th August 79 AD. Pliny the Younger is thought to have documented the S6 surge and F6 pyroclastic flow, the last of the catastrophic

Fig. 9.17 **a** The Pompeii Pumice, the earliest deposit associated with the 79 AD eruption, is exposed at the Traianello Quarry (Photograph: Amelia Scoon); **b** Coarse-grained ash (A3 layer) associated with the 79 AD eruption, Herculaneum

Fig. 9.18 Section at the Oplontis Villa shows the air fall lithic-rich ash (A7) which is sharply overlain by the S5 surge. The S6 surge and F6 pyroclastic flow completely buried the villa

events. They occurred at approximately 0800 h on the morning of the 25th August. They were the most intense of the catastrophic events and probably led to the death of Pliny the Elder. The sketch by Pliny the Younger showing a pyroclastic flow as travelling across the surface of the Gulf of Naples, i.e., from Herculaneum towards Stabaie, is probably of the F6 event. The sketch was not always accepted as accurate but has been substantiated from the behaviour of pyroclastic flows associated with recent eruptions of Montserrat.

The sequence of pyroclastics and ash at the semi-proximal site of Oplontis Villa is considerably thinner (total thickness of 5 m) than at Herculaneum. Roof collapse at Oplontis is ascribed to deposition of the A3 pumice and ashfall on 24th August. The S1 surge destroyed vegetation next to the villa, but not the buildings. The S2 surge did not reach Oplontis. The villa was almost entirely buried by the thick deposit of A7 ash. The A7 deposit is sharply overlain by the S5 surge, a relatively thick deposit, which in turn is overlain by the S6 surge and F6 pyroclastic flow (Fig. 9.18).

The A7 ash deposit contains accretionary lapilli, indicative of phreatomagmatic activity (Fig. 9.19).

The S1 and S2 surges did not reach the more distal environment of Pompeii. Activity associated with these eruptions is restricted to the occurrence of reverse-graded deposits of grey ash (deposits A4 and A5). In reverse-graded deposits the grain-size coarsens upwards, a characteristic of sediments deposited by granular or debris flows. The first basal surge to reach Pompeii was the S3 event. Historical evidence suggests it was the S4 surge which triggered the evacuation. The S5 and S6 surges also affected Pompeii. The S3 through S6 surges were accompanied by extensive ashfall (deposits A6-A10). None of the pyroclastic flows reached Pompeii.

9.8 Vesuvius National Park

A section of the Somma-Vesuvius massif is protected in the Vesuvius National Park (area of 135 km²). The park was founded in 1995 to cope with the large number of tourists.

Fig. 9.19 Graded beds of air-fall ash (A7) at the Oplontis Villa

There have been a number of significant eruptions of Vesuvius since the catastrophic 79 AD eruption. Historical eruptions in 1631, 1760, 1794, 1858, 1861, and 1872 were recorded by means of paintings, engravings, and drawings. Many of the eruptions during the 20thC, specifically those in 1902, 1906, 1914, 1929, and 1933, were photographed. The book by Elio Abatino "Vesuvio: A volcano and its history" includes a comprehensive collection of historical paintings and photographs. The principal feature of the national park is a rounded and relatively deep summit crater (Fig. 9.20). The crater formed during the 1906 eruption. The near-vertical internal walls reveal details of the eruption of 1944 (Fig. 9.21) (Marianelli et al. 1999). This event, which was observed and described by British comedian Spike Milligan, partially filled the crater with ash and lava.

Fig. 9.20 The Vesuvius crater with part of the Somma caldera visible in the background. *Source* https://images.app.goo.gl/SLuoKJRayTHaKjHU9

Fig. 9.21 The northern wall of the Vesuvius crater is capped by deposits of the 1944 eruption. Lava overflowed the crater during this eruption, partially destroying the town of San Sebastiano to the east of Naples. Scale provided by a group of people (just visible on the horizon)

References

Barberi, F., Cioni, R., Rosi, M., Santacroce, R., Sbrana, A., & Vecci, R. (1989). Magmatic and phreatomag-matic phases in explosive eruptions of Vesuvius as deduced by grain-size and compositional analysis of pyroclastic deposits. *Journal of Volcanological and Geothermal Research, 38,* 287–307.

Barnett, Z. A., & Gudmundsson, A. (2014). Numerical modelling of dykes deflected into sills to form a magma chamber. *Journal of Volcanology and Geothermal Research, 281,* 1–11.

Carlino, S. (2018). Neapolitan Volcanoes: A Trip around Vesuvius, Campi Flegrei and Ischia. Geoguide, Springer, 312 p. ISBN 978-3-319-92877-7.

Cavazza, W., & Wezel, F. C. (2003). The Mediterranean Region – a Geological Primer. *Episodes, 26,* 160–168.

Di Vito, M. A., Piochi, M., Mormone, A., & Tramelli, A. (2011). Somma-Vesuvius: the Volcano and Observatory. Field Trip Guidebook – REAKT, 39 p.

Giacomelli, L., Perrotta, A., Scandone, R., & Scarpati, C. (2003). The eruption of Vesuvius 79 AD and its impact on human environment of Pompeii. *Episodes, 26,* 235–238.

Marianelli, P., Metrich, N., & Sbrana, A. (1999). Shallow and deep reservoirs involved in magma supply of the 1944 eruption of Vesuvius. *Bulletin of Volcanological Research, 61,* 48–63.

Orsi, G., D'Antonio, M., De Vita, S., & Gallo, G. (1992). The Neapolitan Yellow Tuff, a large-magnitude trachytic Phreatoplinian eruption: eruptive dynamics, magma withdrawal and caldera collapse. *Journal of Volcanological and Geothermal Research, 53,* 275–287.

Orsi, G., De Vita, S., Di Vito, M. A., & Isaia, R. (2004). The Neapolitan active volcanoes (Vesuvio, Campi Flegrei, Ischia): Science and impact on Human Life. Field Trip Guidebook B28, 32nd International Geological Congress, 44 p.

Panza, G. F., & Suhadolc, P. (1990). Properties of the lithosphere in collisional belts in the Mediterranean – a review. *Tectonophysics, 182,* 39–46.

Peccerillo, A. (2003). Plio-quaternary magmatism in Italy. *Episodes, 26,* 222–226.

Pescatore, T.S., & Cinque, A. (2004). Historical-geological events and their impact on Man. Field Trip Guidebook P14, 32nd International Geological Congress, 44 p.

Polacci, M., Pioli, L., & Rosi, M. (2003). The Plinian phase of the Campanian Ignimbrite eruption (Phlegraean Fields, Italy): evidence from density measurements and textural characterization of pumice. *Bulletin of Volcanological Research, 65,* 418–432.

Santacroce, R., Christofolini, R., La Volpe, L., Orsi, G., & Rosi, M. (2003). Italian active volcanoes. *Episodes, 26,* 227–234.

Sigurdsson, H., Carey, S., Cornell, W., & Pescatore, T. (1985). The eruption of Vesuvius in AD 79. *National Geographic Research, I,* 332–387.

Acropolis, Delphi, Meteora, Pindus Mountains, and Thermopylae

Abstract

Many antiquities in southeast Greece, including the Acropolis of Athens and Delphi, are among the most well known archaeological sites in the world. Antiquities are generally situated in localities where earthquake damage is minimized by building on outcrops of resistant limestone and marble. Historical sites in southeast Greece, such as the Plains of Marathon and the Pass of Thermopylae, are associated with events which have helped shaped European history. Antiquities in the remote Pindus Mountains of northwest Greece, including monasteries constructed in medieval times when defensive positions were crucial, may occur in seemingly inaccessible localities. The monasteries of the Meteora are well known as they have featured in a number of films. The regional geology of southeast and northeast Greece is dominated by NW-SE trending tectonic zones associated with the long drawn-out Alpine Orogeny. Alpine zones are associated with resistant metamorphic rocks (Palaeozoic-Mesozoic) and include thick sequences of limestone and marble. Crustal extension during the Cenozoic resulted in formation of basins and grabens that are partially infilled by flysch and molasse sediments. The eastern Mediterranean region is one of the most tectonically active regions on Earth and geological processes are continuously reshaping landforms. The most famous of the antiquities in southeast Greece is the Acropolis of Athens. The "hill city" is situated on a resistant block of limestone (Upper Cretaceous) which has been thrust over a sequence of younger and softer marls within the Athens Basin. Monuments such as the Parthenon and the Temple of Rome and Augustus are in part built from the high-quality Pentelic Marble derived from mountains to the north of Athens. The antiquity of Delphi occurs in a karstic (or "Delphic") landscape associated with the rugged Parnassus Mountains. Delphi is widely known for oracles offered by priestesses who inhaled hallucinogenic vapours derived from geothermal springs. Descriptions by the Greek historian Herodotus of narrow defiles (which favoured the "Three Hundred") and natural hot springs at the site of the Battle of Thermopylae (480 BC) have been corroborated by geomorphological reconstructions. Rapid rates of sedimentation in the Euboean graben have caused the shoreline to retreat northward by several kilometres. Antiquities and monasteries in the Pindus Mountains occur in some of the most remote regions of Europe. Glaciated landforms and spectacular limestone gorges are protected in several national and geoparks. The monasteries of the Meteora are perched on hilltops comprised of resistant beds of conglomerate (Miocene). The conglomerate is part of a sequence of molasse sediments derived from erosion of the older metamorphic rocks that border the Central Hellenic Basin.

Keywords

Acropolis • Delphi • Limestone • Marble • Meteora • Pindus Mountains • Thermopylae

10.1 Introduction

The mainland of southeast and northwest Greece contains a rich heritage of antiquities and archaeological sites, many of which are known for events that helped shaped European history. The Acropolis of Athens and the archaeological complex of Delphi were awarded UNESCO World Heritage status in 1987. The Plains of Marathon and the Battle at the Pass of Thermopylae are associated with the Greco-Persian Wars (499 BC-449 BC). This involved a series of conflicts between most of the Greece city-states and the Persian

---

Photographs not otherwise referenced are by the author.

The original version of this chapter was revised: Belated corrections have been incoporated. The corrections to this chapter are available at https://doi.org/10.1007/978-3-030-54693-9_18

Empire. The remoteness of the Pindus Mountains has resulted in some small towns and villages in northwest Greece having a unique history. There are several national parks and geoparks in the region where glacial landforms are well represented. Many of the monasteries in northwest Greece date from medieval times when it was essential to build in locations that could be easily defended. The monasteries at Meteora are located on hills and pinnacles that rise precipitously above the Thesalian Plain (Fig. 10.1). The Kipina Monastery near Ioannina occurs in an equally dramatic setting.

The antiquities and archaeological sites described here occur in the districts of Attica, Sterea Ellas, Thessaly, Epirus, and Macedonia (Fig. 10.2). The regional geology is dominated by the occurrence of NW-SE trending tectonic zones associated with the long drawn-out Alpine Orogeny. Large parts of southeast and northwest Greece include mountain ranges associated with resistant Palaeozoic-Mesozoic metamorphic rocks. Younger sediments occur in broad valleys associated with Cenozoic basins and grabens. The Pindus Mountains were severely glaciated during the Late Pleistocene Ice Ages. Active geological processes are reshaping the landforms in many of the areas described here.

10.2 Regional Geology

The eastern Mediterranean is one of the most tectonically active areas on Earth (Robertson and Mountrakis 2006). The complexity of the Alpine convergence has been widely discussed (Chap. 8). The Alpine Orogeny peaked in the Oligocene-Miocene (34–5.3 Ma). The orogeny resulted in formation of microplates with complex boundaries, including the Hellenic Microplate (Okay and Tüysüz 1999; Robertson and Mountrakis 2006; Papanikolaou et al. 2019). The Mediterranean microplates may have started to form in the Early Cretaceous, but the main development is thought to have been in the Oligocene-Miocene (i.e., concurrent with

Fig. 10.1 The monastery of St. Nicholas Anapausas in the Meteora, near Kalambaka, northwest Greece, is perched precariously on the crest of a near-vertical pinnacle of molasse sediments (Miocene)

Fig. 10.2 Map showing location of antiquities, archaeological sites and geosites in southeast and northwest Greece

the peak of the Alpine Orogeny). The Hellenic Microplate and the Eurasian Plate are separated by the W-E trending Central Hellenic Shear Zone (Robertson and Mountrakis, 2006). The area to the south of the shear zone includes the Peloponnese, a discrete physiographic component of southwest Greece (Chap. 11). The Hellenic Volcanic Arc is a significant component of the Hellenic Microplate (Chap. 12). The eastern Mediterranean is part of an accretionary convergent margin, rather than a typical subduction-related collision (Box 8.1).

10.2.1 Alpine Zones

The NW-SE trending tectonic zones of southeast and northest Greece are linked to the Alpine massifs of central Europe (Fig. 10.3). Alpine zones are aligned subparallel to the coastline and the near-shore archipelagos. Some zones are contiguous between the Eurasian Plate and the Anatolian Microplate, e.g., the Pelagonian and Vadar Zones (Greece) may connect with the Menderes Massif and Sakarya Zones (Turkey), respectively. Each of the Alpine zones is separated

Fig. 10.3 Simplified geological map of Greece showing the dominance of the NW-SE trending Alpine zones. *Source* Simplified from regional geological maps and references in the text

by complex boundaries which may include regional faults and thrusts. Some boundaries are complicated by the presence of large nappes. The Alpine zones are comprised of a multitude of primary geological terrains. They are dominated by ancient continental fragments, but may also contain sections of islands, ocean ridges, and ocean floors. The Ionian-Paxon and Gavrovo Zones consist of metamorphic rocks, mostly limestone and marble (Mesozoic). A characteristic feature is the occurrence of thinly-bedded limestones and marbles with near-vertical dips (Fig. 10.4a). The Pindus Zone includes thick sequences of limestone, marble, and chert (Mesozoic). The Parnassus Zone, which has a relatively restricted occurrence, is distinguished by the presence of Cenozoic limestone in addition to the Mesozoic metamorphic rocks. The Pelagonian Zone is similar to the Pindus Zone in that Mesozoic metamorphic rocks are prominent, but fragments of the Palaeozoic basement are a diagnostic feature. Both the Pindus and Pelagonian Zones contain

Fig. 10.4 **a** Thinly-bedded limestones and marbles with near-vertical dips are characteristic of the Alpine zones of northwest Greece (locality near Igoumenitsa); **b** The flysch of northwest Greece includes steeply-inclined deposits of schist (locality near Kalarrytes)

near-linear arrays of ophiolite intrusions. The ophiolite intrusions are obducted slabs of the Tethys Ocean floor with an average age of 170 Ma (Rassios and Smith 2000).

10.2.2 Basins and Grabens

The development of fore-arc and back-arc basins is an important component of accretionary convergent margins (Box 8.1 and Fig. 8.3). The Aegean Sea is associated with a back-arc basin which is expanding at estimated rates of 50 mm/year (Jolivet et al. 2013; Papanikolaou et al. 2019). Crustal extension, which impacted the Eurasian Plate in southeast and northwest Greece during the Cenozoic, is manifested by normal faulting and formation of basins and grabens. The indented coastline that is such a characteristic feature of Greece includes the Euboean Graben (the Euboean Gulf connects to the Aegean Sea). The Gulf of Corinth, which connects to the Ionian Sea, is related to an elongate basin (Chap. 11). Some grabens may have started to form in the Cretaceous (Papanikolaou et al. 2019). The main period of crustal stretching, however, probably occurred in the Neogene and Quaternary. Many graben faults are active and are periodically prone to seismic events.

The basins and grabens in the interior of southeast and northwest Greece constitute discrete physiographic features between the mountain ranges. They typically contain sediments which accumulated either during or after the peak of the Alpine Orogeny. Thick deposits of flysch (Late Palaeocene or Early Oligocene) formed in foreland basins at the base of the rising Pelagonian nappes. The flysch is comprised of marine sediments, including schist, marl and sandstone. The flysch exhibits moderate levels of deformation and bedding may be steeply-inclined (Fig. 10.4b). Molasse sediments (Late Oligocene and Miocene) accumulated in basins and grabens after the peak of the orogeny. The molasse is comprised of terrestrial sediment eroded from the older mountain ranges and deposited in extensive fluvial systems. The molasse is dominated by conglomerate and sandstone with low levels of deformation and flat-lying beds. The molasse is notably developed in the Meso-Hellenic Basin. Substantial deposits of gravel and alluvium (Quaternary) occur in most grabens.

10.3 The Acropolis of Athens

The Acropolis of Athens is situated in the heart of a major European metropolis. Athens occupies a topographic basin between interior mountain ranges and the Aegean Sea. The mountain ranges include Mount Penteli (1,109 m), part of a 20-km long range situated northeast of the city, Mount Parnes (1,412 m) located north of the city, and Mount

Hymettus (1,037 m) to the southeast. The mountains are comprised of metamorphic rocks (Palaeozoic-Mesozoic), primarily marble and schist, associated with the Pelagonian Zone (Fig. 10.3). The Athens Basin includes outcrops of the Tourkovounia Limestone and Athens Schist (Upper Cretaceous), an impure limestone (Pliocene), and unconsolidated deposits of alluvium and clay (Quaternary), as described by Gaïtanakis (1982) and Gaïtanakis and Dietrich (1992) (Fig. 10.5). The basin formed due to crustal extension on the margins of the Aegean Sea. The Athens Schist (estimated

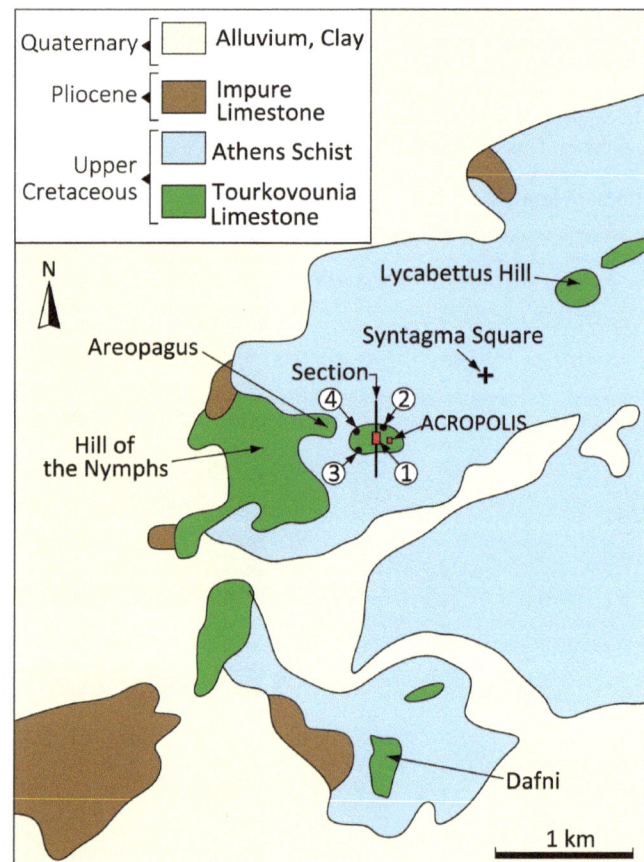

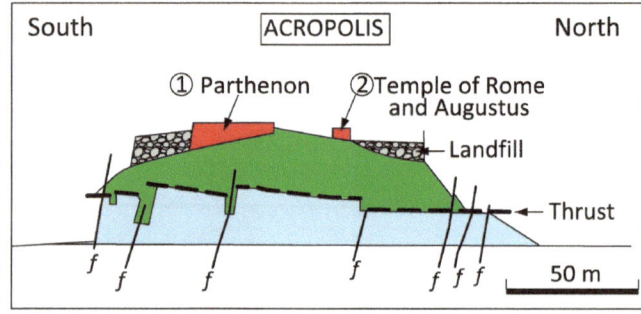

Fig. 10.5 Geological map of part of the city of Athens (1-Acropolis; 2-Temple of Rome and Augustus; 3-Theatre of Dionysus; 4-Klepsydra Spring) and section of the Acropolis. *Source* Map and section simplified from Gaïtanakis (1982), Andronopoulos and Koukis (1990) and Higgins and Higgins (1996)

age of 72 Ma) is a rather poorly named succession of relatively soft marls which is deeply eroded. The clay deposits were significant in historical times for the manufacture of bricks and pottery (Higgins and Higgins 1996).

Antiquities are mostly situated on the Tourkovounia Limestone (estimated age of 100 Ma), in part as the limestone forms hills which offer commanding positions, but also as these rocks are less affected by the seismic events to which the Athens Basin is frequently subjected. Areas of the city situated on the Athens Schist, the Pliocene limestone, and the Quaternary alluvium and clay are prone to suffer major earthquake damage.

10.3.1   Tourkovounia Limestone

The Tourkovounia Limestone is associated with the prominent hills which are such a feature of the city of Athens (Fig. 10.6). The hills include the thickly wooded, conical-shaped Lycabettus Hill (278 m), the bare, flat-topped Acropolis (150 m) or "Hill City", the Hill of the Nymphs (with the 147 m-high Philopappos Hill), and Dafni. The Areopagus is a prominent rocky outcrop west of the Acropolis (Fig. 10.7). The Tourkovounia Limestone is extremely resistant to erosion and is well exposed on a path that circles the Acropolis on the northern side. The limestone outcrops here as a grey, blocky rock, in part crystalline, with pockets of marble (Fig. 10.8). The prominent joints and fissures may be filled with marl or crystalline calcite. Some fissures have opened into small caves.

Natural springs are located at the base of many of the limestone hills, i.e., at or close to the contact between the limestone and the underlying, and less permeable, marls (Higgins and Higgins, 1996; Regueiro et al. 2014). Springs were probably more productive in classical times as infiltration has been negatively impacted by land use changes. The historically-important springs, e.g., the Klepsydra Spring on the northwest slopes of the Acropolis, have been described in the archaeological literature.

The unusual spatial relationship between the Tourkovounia Limestone and the underlying Athens Schist – the latter is estimated to be approximately 30 Ma younger - is related to thrusting associated with the Alpine Orogeny (Fig. 10.5) (Gaïtanakis and Dietrich, 1992; Regueiro et al. 2014). The Tourkovounia Limestone is associated with a series of nappes in the Athens Basin. The nappes probably

formed in the Oligocene and Miocene, i.e., during the peak of the orogeny. Erosion of the nappes has resulted in detached blocks that form isolated hills. The flat-topped nature of some hills, including the Acropolis, is a function of the near-horizontal bedding.

10.3.2   Antiquities

The most famous antiquity of the Acropolis of Athens is the Parthenon, the view of which dominates the skyline of Athens (Fig. 10.9a). The Parthenon is considered one of the most enduring symbols of the ancient Athenian democracy and is possibly the most important monumental building remaining from the Classical Greece period (500 BC–323 BC). The Parthenon is described as one of the principal celebrations of the Pan-Hellenic victory over the Second Persian Invasion (480 BC). Construction began in 447 BC and was completed by 438 BC. Recent phases of restoration began in 1982 and are ongoing. Material as similar to the original building stones as possible is used. One of the problems faced in the restoration is replacing marble frescos "looted" in the 19[th]C by Lord Elgin and housed in the British Museum (sentence based on a documentary video shown in the New Acropolis Museum, Athens).

The Parthenon is constructed of both limestone (Tourkovounia Limestone) and marble (Pentelic Marble). The limestone was derived from sites which may include the Hill of the Nymphs and the harbour city of Piraeus; the Acropolis was not defaced by quarries (Higgins and Higgins 1996). The marble sourced from Mount Penteli includes the high quality, white "Pentelic Marble" (which is relatively free of impurities) and a rather poor quality grey marble (which contains minute flakes of graphite). Ancient quarries on Mount Penteli have been incorporated into more recent operations, but some historical sites have been protected. High quality Pentelic Marble was used in the colonnade of Doric columns, the most prominent feature of the Parthenon (Fig. 10.9b). There are sixty nine columns in total (seventeen outer columns on the sides and eight on the ends; together with inner columns). The columns have fluted shafts which can be observed in the restoration work. The colonnade is built on a three-step platform (*stylobate*) with an exterior size of 69.5 m by 30.0 m emplaced directly into the bedrock limestone (Fig. 10.10a). The Temple of Rome and Augustus is situated on the bare rock pavement that

Fig. 10.6 a The conical-shaped
Lycabettus Hill (with the Church
of St. George), Athens, is
comprised of resistant
Tourkovounia Limestone;
b Monumental buildings of the
Acropolis of Athens are built on a
distinctive flat-topped hill of
Tourkovounia Limestone

a

b

characterizes the flat-top of the Hill City (Fig. 10.10b). The
Acropolis was, however, partially levelled by an artificial fill
which has a maximum thickness of approximately 14 m
(Higgins and Higgins, 1996).

The auditorium of the Theatre of Dionysus (5[th]C BC-5[th]C
AD) is the most important of the antiquities on the southern
slopes of the Acropolis (Fig. 10.11). The theatre is built into
the side of the limestone hill. A retaining wall has been partially rebuilt using a combination of limestone derived from
Piraeus and artificial blocks of conglomerate (or breccia). An
additional site of interest is the National Garden Shaft located
in Syntagma Square, part of the ventilation for the underground rail system. The shaft has exposed antiquities including a channel of the 6[th]C BC Peisistratean Aqueduct.

Fig. 10.7 The Areopagus is a prominent outcrop of the Tourkovounia Limestone, northwest of the Acropolis

Fig. 10.8 The Tourkovounia Limestone is a grey, blocky rock, in part crystalline, as observed on the Areopagus

10.4 Plains of Marathon

The Plains of Marathon are part of a broad coastal plain located approximately 42 km northeast of Athens (Fig. 10.2). The plain is situated between the northern side of the Penteli Range and the Aegean Sea. The coastal plain is associated with the active Euboean Graben and is underlain by Quaternary alluvium and gravel. The alluvium and gravel were derived from erosion of the Mesozoic metamorphic rocks in the enclosing hills. The area is synonymous with the Battle of Marathon (490 BC) which occurred during the First Persian Invasion. Many of the Persian soldiers died in the "Great Marsh", a swampy area which has been drained and is now the site of a large lake. The coastline in this area was probably located as much as 500 m further inland in 490 BC, as compared with the current position (Higgins and Higgins, 1996). Some of the Greek participants in the battle are buried in a large limestone tumulus.

A famous legend attached to the Battle of Marathon is associated with the runner *Pheidippides*, allegedly despatched to bring news of the victory to the city of Athens. According to the Greek historian Herodotus (484-425 BC), however, *Pheidippides* was sent not to Athens, but to Sparti in the Peloponnese. This misconception is recognized in Greece and commemorated by an annual event, the ultra-marathon of approximately 225 km between Athens and Sparti. The reason the Greek army returned to Athens from the Plains of Marathon was in reaction to information

Fig. 10.9 a The flat-topped
Acropolis Hill is dominated by
the partially-reconstructed
Parthenon; **b** The colonnade of
Doric columns in the Parthenon
consists of white Pentelic Marble
emplaced on a three-step platform
constructed of large blocks of
grey marble

a

b

that the Persian fleet was sailing for Piraeus. The Athenian
Navy, together with the Spartan Navy, subsequently defeated
the Persians during the Second Persian Invasion, approximately 20 years later at the Battle of Salamis (480 BC).

10.5 Delphi

Delphi is located approximately 120 km northwest of
Athens in the district of Sterea Ellas (Fig. 10.2). This is the
second most visited antiquity in Greece. Delphi was known
to the ancient Greeks as the centre of Earth, which may
explain why they named the relatively small River Pleistos
(which has a length of less than 30 km) as the "greatest".
(Charles Lyell named the Pleistocene epoch for the great
abundance of fossils.) The antiquities cling precipitously to
the southwestern slopes of the Parnassus Mountains; the
modern town of Delphi occurs at the base of the archaeological site (Fig. 10.12). The town is constructed on a narrow platform situated more than 600 m above the Gulf of
Corinth. The continental climate at Delphi is sufficiently
extreme as to result in hot, dry summers and snowy winters.
Mount Parnassus (2,457 m) is one of the primary skiing
centres in Greece.

Fig. 10.10 a The monumental buildings on the Acropolis are emplaced into the bedrock limestone; **b** The Temple of Rome and Augustus is situated on a limestone pavement on the crest of the Acropolis

10.5.1 Regional Geology

The Parnassus Mountains are associated with the Parnassus Zone, a relatively minor feature located between the more extensive Pindos and Pelagonian Zones (Fig. 10.3). Three ages of carbonate are recognized in the Parnassus Zone. The oldest group of limestone and marble (Triassic, Jurassic and Lower Cretaceous) is the most resistant and forms the highest peaks. The intermediate group is restricted to limestone (Middle Cretaceous) that is somewhat less resistant. The youngest group of limestone with subordinate chert (Upper Cretaceous and Palaeocene) is relatively soft and is restricted to the lower slopes. The carbonates were exposed to three erosional cycles in the Late Mesozoic and Early Cenozoic. Erosion occurred in tropical climates and is recognized by distinctive palaeosols. Thin layers of bauxite accumulated in some regions. This area of Greece is a significant producer of aluminium.

The platform on which the modern town of Delphi is situated developed due to erosion on the Pleistos Fault, a subordinate structure associated with the Corinth Basin. The dramatic topography is related to the ruggedness of the

Fig. 10.11 The Theatre of Dionysus is excavated into the Tourkovounia Limestone on the southern side of the Acropolis

Parnassus Mountains. The rapid rates of uplift on the northern shoulder of the Corinth Basin are a contributory factor (Chap. 11).

10.5.2   Delphic Landscape

The rugged topography of the Parnassus Mountains is known to archaeologists and classical scholars as a *Delphic Landscape*. Geologists prefer the description *karstic*, named after the limestone plateau in the region of Karst (southwest Slovenia and northeast Italy). The permeable nature of the limestone bedrock in karstic landscapes results in deep gorges and caves carved by subterranean rivers.

10.5.3   Antiquities

The antiquities of Delphi were most active between the 8thC and 4thC BC, but may have been used as long ago as 1600 BC (and as recently as 400 AD). The original name of Delphi is Pythia, or "House of Snakes", derived from Pytho referring to the decomposition of the body of the monstrous

python purported to have been slain here by the Greek God Apollo. The Sacred Way extends for several hundred metres on the lower slopes and connects the Temple of Apollo to the Athena Pronaia Sanctuary (Fig. 10.13a). Located behind the Temple of Apollo is the 6thC BC Polygonal Wall, the main face of which is some 90 m long and built of limestone blocks carved with ancient inscriptions. The Theatre of Delphi is situated on the upper slopes (Fig. 10.13b). The 5thC BC "Stadion" (stadium) is located even higher on the mountain side. The Stadion was extended by the Romans in the 2ndC AD to host the Pythian Games. Most buildings at Delphi are constructed from locally-derived limestone and marble.

10.5.4   Delphic Oracles

The priestesses at Delphi delivered the "Oracles of Antiquity", or *Delphic Oracles*, from sites adjacent to the Temple of Apollo (Fig. 10.14). The priestesses were also known as Sibyls. The Sibyl Rock is a pulpit-like outcrop of limestone bedrock, located on the Sacred Way. In a rocky cleft east of the Temple of Apollo is the Castalian Spring, used by the

Fig. 10.12 The modern town of Delphi is overlooked by the antiquities perched on the lower slopes of the Parnassus Mountains

prophetesses for washing, which feeds into the River Pleistos. The limestone cleft includes recesses carved for offerings. The Temple of Apollo is situated on two intersecting faults, the N-S Kerna Fault and the E-W Delphic Fault. Both faults are associated with natural springs that have precipitated travertine within the cleft and juxtaposed caves (Piccardi 2001; De Boer et al. 2001).

The effectiveness of the Delphic Oracles may be underpinned by toxic vapours associated with the natural springs (De Boer et al. 2000). The vapours contain hydrocarbon gases, including methane and ethylene. Ethylene is a hallucinogenic gas and even trace amounts are sufficient to cause psychedelic experiences. A maximum of 0.3 ppm ethylene has been found to occur in the waters of the Castalian Spring (the spring is drier than in Classical Times). The gases are derived from bituminous deposits, located at depth beneath the caves (De Boer et al. 2001; Spiller et al. 2008). The

release of hydrocarbon vapours increases during seismic events. Seismic events were relatively prevalent in the Classical Greece and Hellenistic periods and the decrease in the frequency of earthquakes is consistent with the waning in significance of prophesies during the Roman period.

10.6 The Pass of Thermopylae

The Pass of Thermopylae is located approximately 17 km south of the regional town of Lamia, in the district of Thessaly (Fig. 10.2). The translation of Thermopylae is "Hot Gates", which in this context refers to narrow defiles (passes) with natural hot springs. The defiles occur near the head of the Gulf of Malia, an extension of the North Euboean Gulf. The Battle of the Pass of Thermopylae (480 BC) occurred on a narrow strip of land at the base of Mount Kallidromom

Fig. 10.13 **a** The Sacred Way and Temple of Apollo, two of the more important antiquities at Delphi, are enclosed by steep, limestone cliffs of the Parnassus Mountains; **b** The rugged nature of the Delphic Landscape is apparent from the Theatre of Delphi which is located in the upper part of the archaeological site

a

b

(Fig. 10.15). The battle was part of the First Greco-Persian conflict. The Greek historian Herodotus described the battle as having occurred in a mountain pass perched precariously above the Aegean Sea. The pass was reputed to include three narrow "gates". Each gate was associated with a hot spring. Historians have reconstructed the locations where the Greek and Persian armies were encamped, together with the position of the three gates (Fig. 10.16). The narrowness of the terrain, in particular at the 20 m-wide Middle Gate, was described as having favoured the small number of Greek defenders ("The "Three Hundred"), who were arrayed against the much larger invading army of Persians.

Fig. 10.14 The Temple of Apollo, Delphi is constructed near several springs which contained vapours with hallucinogenic gases during the Classical Greece and Hellenistic periods

10.6.1 Modern Landscape

The mountains of Thessaly are part of the Pelagonian Zone and are generally comprised of Triassic-Jurassic limestone and marble, together with ophiolite complexes (Fig. 10.3). Mount Kallidromom, however, consists of Cretaceous limestone and conglomerate. There is no evidence of the narrow defiles described by Herodotus in the modern land-scape, and Mount Kallidromom is separated from the Gulf of Malia by a broad zone of reclaimed marshland. The marsh is part of a delta associated with the Spercheios River. Hot springs can be observed at the Middle and West Gates (the latter has a temperature of 40 °C), but are despoiled by modern piping and a bathing pool. Inconsistencies between historical descriptions and the modern landforms have caused some conflict among historians.

10.6.2 Geomorphological Reconstruction

Detailed geomorphological investigations including the drilling of seven boreholes into sediments on the shores of the Gulf of Malia have exposed many of the differences between the ancient and modern landforms (Kraft et al. 1987; Vouvalidis et al. 2010). The Gulf of Malia is part of a graben associated with the Euboean Gulf. This structure was initiated in the Miocene (at approximately 10 Ma), possibly as part of an early phase of extension of the Aegean Basin. Rapidly rising sea levels in the Early-Middle Holocene, following melting of the Late Pleistocene ice, exceeded the rate of sedimentation and large parts of the graben were submerged. It is estimated that at approximately 4,500 BC, the Gulf of Malia extended to the base of Mount Kallidromom (Fig. 10.16). The geomorphological evidence suggests

Fig. 10.15 The commemoration to the Battle at the Pass of Thermopylae (480 BC) is located on a broad floodplain at the head of the Gulf of Malia

that at 480 BC, the passes at Thermopylae may well have been as narrow as described by Herodotus, i.e., the sea extended inland to the base of Mount Kallidromom.

In the years since the Battle at the Pass of Thermopylae, sedimentation exceeded the rate of subsidence in the Euboean Graben. (Sea levels have been relatively constant during this period.) Progradation of the Spercheios Delta caused the shoreline to migrate northward. The thickness of sediments associated with the active subsidence may exceed 800 m. The extensiveness of the delta is, in part, related to deposition of travertine from the thermal springs which were probably more active in historical times. During specific periods of history (e.g., 1700-1300 BC; 300 BC-1100 AD), when sedimentation was either slowed or arrested, the passes may have been entirely inaccessible (Kraft et al. 1987). During the 19[th]C AD, the course of the Spercheios River was diverted towards the southern shore and sedimentation was enhanced by land use changes in the delta and interior plains. This broadened the coastal plain to as much as 5 km. The active geological setting, including rapid changes in accessibility, suggests Thermopylae was unlikely to have

been as significant a route between northern and southern Greece as historical accounts suggest.

10.7   Pindus Mountains

The Pindus Mountains in northwest Greece are part of the extensive Alpine fold mountains that dominate the Balkan Peninsula. The 160 km-long, NW-SE striking mountains in northwest Greece include parallel ranges with multiple peaks over 2,000 m (Fig. 10.17). This region constitutes one of the more remote parts of Europe. The regional centres are Ioannina (district of Epirus) in the South Pindus Mountains, and Grevena (district of Macedonia) in the North Pindus Mountains (Fig. 10.2).

10.7.1   Regional Geology

The Pindus Mountains transgress several tectonic zones, including the Ionian, Pindus, and Pelagonian (Fig. 10.18).

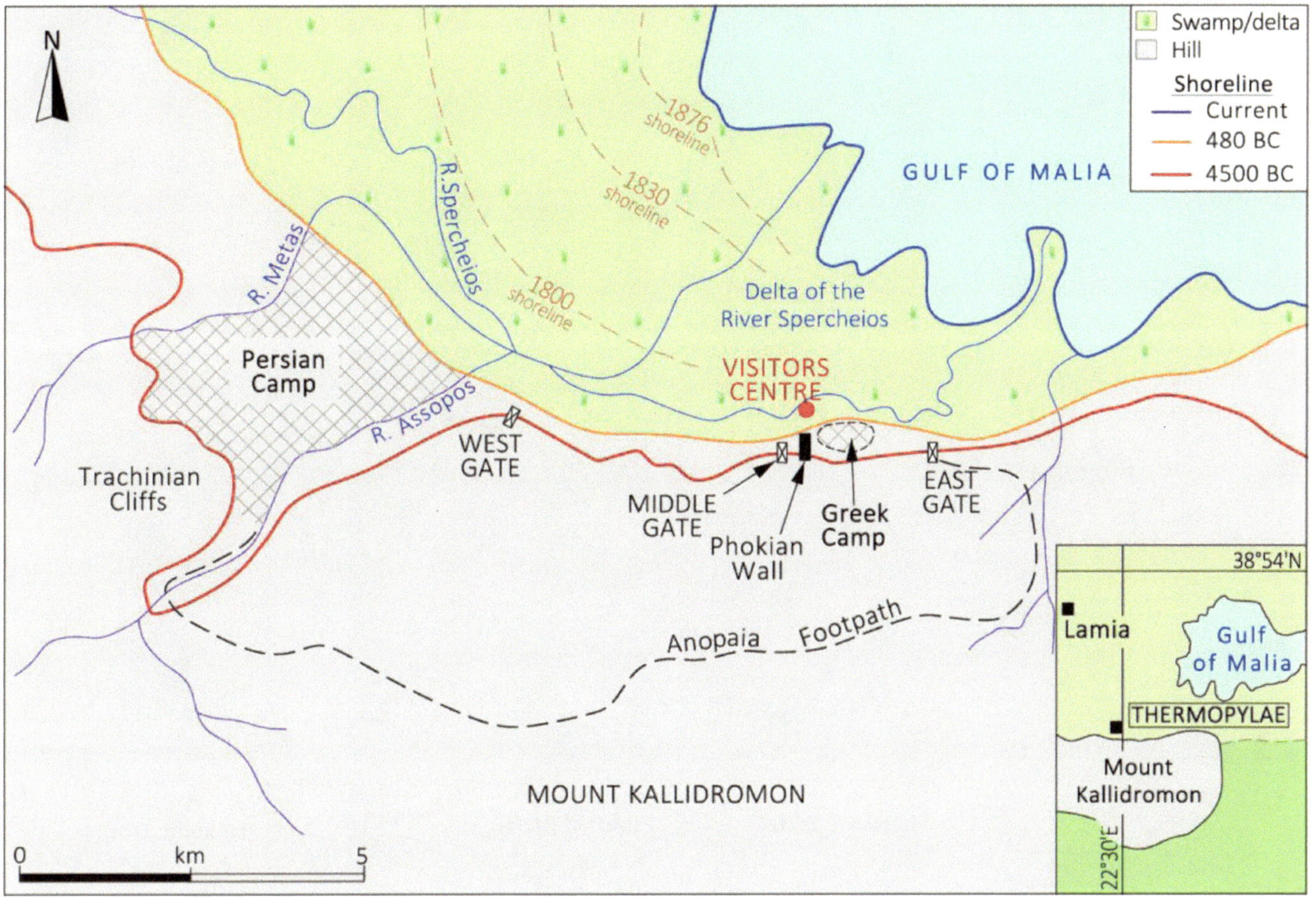

Fig. 10.16 Map of Thermopylae showing sea-level changes and location of the army camps relative to the three gates. *Source* After Kraft et al. (1987) and information boards at the site

Fig. 10.17 The Pindus Mountains, northwest Greece, reveal multiple ranges aligned parallel to the Ionian Sea coast and the island of Corfu (Kerkira)

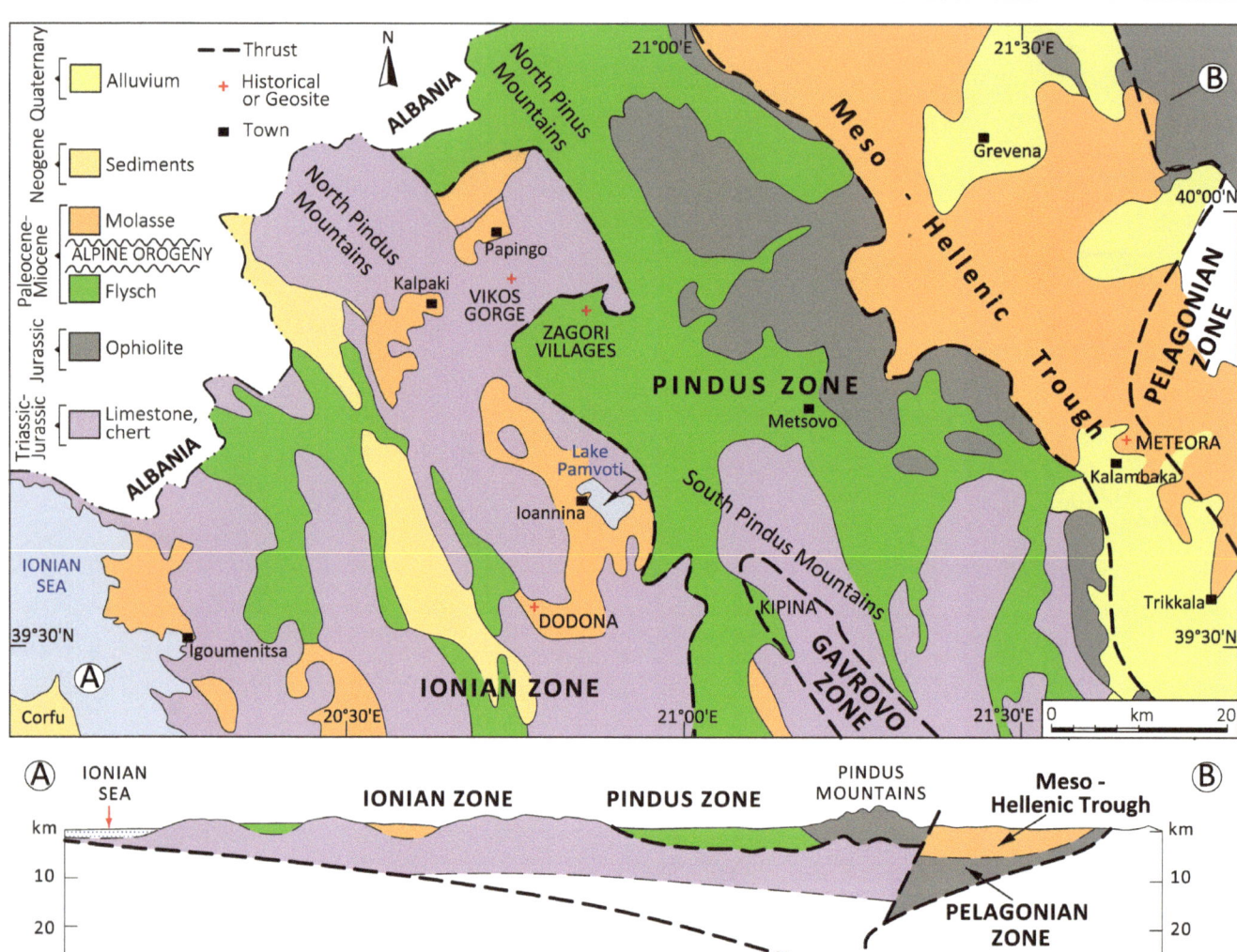

Fig. 10.18 Geological map and section of the Pindus Mountains, northwest Greece. *Source* Simplified from regional maps and Higgins and Higgins (1996)

The Palaeozoic-Mesozoic strata were severely deformed during the Alpine Orogeny. Tectonic zones reveal complex boundaries which include regional thrusts. The higher peaks are comprised of either Triassic-Jurassic limestone or Jurassic ophiolite complexes. The broad valleys and plains contain flysch (Palaeocene-Miocene) and molasse (Miocene-Pliocene) sediments. The molasse deposits occur extensively in the Meso-Hellenic Basin (Piper, 2006). The grabens in this area also contain substantial alluvium (Quaternary). Most grabens are Neogene-Quaternary features, although some structures may have started to form considerably earlier.

10.7.2   South Pindus Mountains

Ioannina is located on the western shores of Lake Pamvotis, at the foot of the South Pindus Mountains. Ioannina was

formerly the centre of a large mining industry that sustained ancient cultures renowned for production of silver jewellery. The silver was derived as a by-product of lead mining from Mississippian-style deposits in the Mesozoic carbonates. Ioannina is inextricably linked with Ali Pasha (1788–1822), a tyrant (in part known for his attempts to create an autonomous colony in East Africa), who used the mining wealth of the city for a fiefdom that stretched across much of northern Greece during the Ottoman Empire. The Perama Cave, located 4 km from Ioannina, is one of the largest in Greece with three levels and 1,100 m of pathways. The dazzling white stalagmites and stalactites alone are worth a visit. Surprisingly for such a large system, it was only discovered during World War II.

The most significant of the archaeological sites in the region is Dodona, situated 22 km west of Ioannina. Dodona occurs in a dry valley enclosed by high ramparts and scree-

Fig. 10.19 The Dodona antiquity includes a large theatre carved into the hill slopes

covered slopes of Jurassic limestone. Streams are trapped by sinkholes at the base of the hills (Higgins and Higgins, 1996). Dodona includes evidence of an old Bronze Age culture (2,500 BC), but the well-preserved theatre is part of the Classical Greece period. The theatre is constructed on a platform consisting of blocks of Palaeocene-Eocene limestone with distinctive veins of chert (Fig. 10.19).

Several 13–14th C AD monasteries are cut into the Triassic limestone in the mountains east of the village Kalarrytes. This area is approached via a narrow road that winds up the steep slopes of the South Pindus Mountains (Fig. 10.20). The road traverses thick deposits of flysch, steeply-dipping shales and marls. The Kipina Monastery is built across the entrance to a limestone cave, and appears to hang on a near-vertical cliff face (Fig. 10.21). The entrance to the monastery is via a narrow, rather exposed pathway and drawbridge.

10.7.3   North Pindus Mountains

The small town of Metsovo occurs at an altitude of 1,160 m in a remote part of the North Pindus Mountains (Fig. 10.22).

The town is accessed by the Via Egnatia highway which links the port of Igoumenitsa with Greece's second largest city, Thessalonica. The road includes a series of high bridges and tunnels from which the Palaeozoic-Mesozoic limestone, marble and ophiolite complexes of the Alpine zones can be observed. Metsovo has an unusual history, driven by the remote setting. During the Ottoman Empire, residents were granted self-autonomy and excused from taxes (in part as they provided a service by guarding the mountain passes to the north). This resulted in development of a socio-economic and cultural system known as "*Beneficence*", in which the richest residents supported the entire economy of the area. This system extended to the Zagoria, a group of 46 mountain villages that were entirely self-governing prior to establishment of the modern state of Greece. This region achieved considerable prosperity in comparison to surrounding areas. Metsovo and the Zagoria Villages have a unique architectural style using locally-derived limestone and slate for building. The twin villages of Megali Papingo and Mikro Papingo are reached by precipitous mountain tracks. The villages include large houses with fortified courtyards. The historic arched ("packhorse") bridges in the region are built of Mesozoic limestone.

Fig. 10.20 The village of Kalarrytes is approached via a road that snakes up the flanks of the South Pindus Mountains

Fig. 10.21 The Kipina Monastery, Kalarrytes, is cut into a cliff face comprised of resistant Triassic limestone

Fig. 10.22 Metsovo is dwarfed by the high peaks of the North Pindus Mountains

Fig. 10.23 Limestone pavement, Vikos-Aoös Geopark

A section of the spectacular Vikos Gorge in the North Pindus Mountains is protected in the Vikos-Aoös Geopark. Access is from the village of Megali Papingo. A strenuous trail through the limestone gorge reveals a karstic landscape with steep vertical walls (which rise to a height of almost 1,000 m), near-vertical pinnacles, and barren rock pavements (Fig. 10.23). Evidence of human settlement in limestone caves dating from 4,000 BC has been uncovered by archaeologists. The Provatina Cave has a depth of 407 m, possibly the second deepest cave in the world. The Dragon Lakes ("Drakolimni") are two high-altitude Alpine lakes related to the Late Pleistocene glaciation. One lake occurs in the Vikos-Aoös Geopark (at an altitude of 2,050 m) and the other lake is located on the western slopes of Mount Smolikas (2,637 m), the highest peak in the region.

The North Pindus National Park (area of 69 km^2) protects an area known as *Valia Calda* ("Warm Valley" in the local Aromanian language), northwest of the Aoös Gorge. Several peaks of over 2,000 m occur and the glaciated valley contains small Alpine lakes. The national park contains mixed forests of black pine and beech that characterize large parts of the Pindus Mountains. A conservation priority is protection of the Eurasian brown bear (*Ursus arctos arctos*) and the Balkan chamois (*Rupicapra rupicapra olympica*).

Many of the high peaks in the North Pindus Mountains are comprised of ophiolite complexes, rather than the ubiquitous limestone and marble. The principle rock type of the ophiolite complexes is grey- and green-coloured peridotite. The peridotite layers are altered to black or dark green serpentine. The peridotite or serpentine includes accessory Cr-spinel. The ophiolites contain small, podiform deposits of chromite.

The North Pindus Mountains are sufficiently high as to have been severely glaciated during the Late Pleistocene Ice Ages (Box 3.1). Features such as serrated ridges, cirques, moraines, and hanging valleys with Alpine lakes are characteristic. Three glacial stages are recognized (Hughes et al. 2006a). The oldest and most prominent stage produced extensive ice fields (approximate age of 350,000 BP). Glacial erosion was more effective on limestone, where pronounced glacial incision and sub-glacial and glacio-karst processes produced rugged topography (Hughes et al. 2006b). The glaciers formed on a range of slope orientations. Intriguingly the largest icefields developed on southward-facing slopes comprised of limestone. Glaciers which developed on the ophiolite complexes were mostly controlled by a northward aspect.

10.8 Meteora

The monasteries of the Meteora are located near the village of Kalambaka in the district of Thessaly (Fig. 10.2). Monasteries sit atop seemingly inaccessible hills and pinnacles and the translation of Meteora as "suspended in air" is descriptive (Fig. 10.1). Kalambaka is situated near the regional town of Trikkala. The settlements are located on the broad Thesalian Plain, a fertile and intensely farmed region, fringed by the South Pindus Mountains. The meandering Peneois River occupies a wide valley associated with a regional graben. The monasteries of the Meteora occur on a group of hills that rise to heights of 450 m on the northern flanks of the Peneois Graben (Fig. 10.24a). Vertical rock faces and pinnacles tower above the village (Fig. 10.24b).

Most monasteries date from the 14thC AD when it was essential to build in locations that could be easily defended. Access was traditionally made from baskets slung on rope windlasses, sights made famous as they have featured in several action films. Tourists now use pathways with staircases cut into the rock face, although some precarious-looking ropeways and skips are used by the local inhabitants.

10.8.1 Regional Geology

The Peneois Graben is a relatively old feature that was reactivated during the Neogene. The graben is part of the Meso-Hellenic Basin, a regional feature bordered to the east by the Pelagonian Tectonic Zone and to the west by the Pindus Tectonic Zone (Ferriere et al. 2011) (Fig. 10.18). The enclosing mountains are dominated by limestone, marble, and ophiolite complexes. The Peneois Graben is in part infilled by molasse, i.e., terrestrial sediments derived from erosion of the older metamorphic and igneous rocks. Extensive Quaternary sand and gravel deposits also occur.

10.8.2 Meteora Group

The hills and pinnacles of the Meteora consist of molasse sediments designated within the Meteora Group (Higgins and Higgins 1996; Ferriere et al. 2011). Two subdivisions are recognized, the Lower Meteora Formation and the Upper Meteora Formation. Both formations probably have a Lower Miocene age. The molasse was derived from erosion of the mountains east of the Peneois Graben. The Meteora Group correlates with a regional change in the sedimentation patterns of the Meso-Hellenic Basin, from accumulation of marine flysch to terrestrial sedimentation (molasse).

The steepest cliff sections consist of a thick unit of conglomerate with subordinate sandstone (Fig. 10.25a). The conglomerate is associated with the Lower Meteora Formation and was deposited in reworked deltas. Pebbles were sourced from the older limestone, marble, and serpentine. The conglomerate locally exhibits cross-bedding aligned at 15–20° southwest. The near even-spacing of the steep cliffs and columns is controlled by vertical jointing (Fig. 10.25b).

The Upper Meteora Formation consists of somewhat less-resistant conglomerate together with more abundant sandstone and marl. Bedding in the Upper Meteora Formation is generally near-horizontal. Erosion of the softer layers of sandstone and marl creates block-like cliffs. A distinctive

Fig. 10.24 **a** A pinnacle consisting of Lower Meteora Formation (Miocene) with view westward of the Peneois Graben and South Pindus Mountains; **b** Cliffs of resistant molasse sediments tower above the village of Kalambaka

Fig. 10.25 **a** Monasteries of the Meteora are perched on the crests of near-vertical, smooth cliff faces comprised of the Lower Meteora conglomerate; **b** Pinnacles and crenulated cliff faces at Meteora are associated with closely-spaced vertical joints in the Lower Meteora conglomerate

a

b

Fig. 10.26 a Upper Meteora conglomerate may include small hollows; **b** The conglomerate includes both rounded and angular boulders and pebbles

feature of the conglomerate of the Upper Meteora Formation is the presence of small hollows (Fig. 10.26a). Rounded boulders of poorly-sorted conglomerate may be perched precariously on the crest of some pinnacles (Fig. 10.26b). Differential erosion of the Meteora Group commenced soon after deposition, due to reactivation of the Peneois Graben.

References

Andronopoulos, B., & Koukis, G. (1990). Engineering problems in the Acropolis of Athens. In P. G. Marinos & G. C. Koukis (Eds.), Engineering Geology of Ancient Works, Monuments and Historical Sites (pp. 1819–1831). Rotterdam: Balkema.

De Boer, J. Z., Zeilinga, J., Rigby, J., Hale, J. R., & Spiller, H. A. (2000). "The Delphic Oracle: A Multidisciplinary Defense of the Gaseous Vent Theory. *Clinical Toxicology, 40,* 196.

De Boer, J. Z., Zeilinga, J., Chandon, J. P., & Hale, J. R. (2001). New evidence for the geological origins of the Ancient Delphic Oracle. *Geology, 29,* 707–711.

Ferriere, J., Chanier, F., Reynaud, J.-Y., Pavlopoulos, A., Ditbanjong, P., Migiros, G., et al. (2011). Tectonic control of the Meteora conglomeratic formations (Mesohellenic basin, Greece). *Bulletin Society Geologiques France, 182,* 437–450.

Gaïtanakis, P. (1982). Geological Map of Greece 1:50.000. Athinai-Pireus sheet. Institute of Geology and Mineral Exploration.

Gaïtanakis, P., & Dietrich, V. J. (1992). The Athenian Acropolis klippes: relics of early Tertiary large scale nappe. *Bulletin of the Geological Society of Greece, 28,* 41–42.

Higgins, M. D., & Higgins, R. (1996). A Geological Companion to Greece and the Aegean (p. 240). New York: Cornell University Press.

Hughes, P. D., Woodward, J. C., Gibbard, P. L., Macklin, M. G., Gilmour, M. A., & Smith, G. R. (2006a). The glacial history of the Pindus Mountains, Greece. *Journal of Geology, 114,* 413–434.

Hughes, P. D., Gibbard, P. L., & Woodward, J. C. (2006b). Geological controls on Pleistocene glaciation and cirque form in Greece. *Geomorphology, 88,* 242–253.

Jolivet, L., et al. (2013). Aegean tectonics: Strain localisation, slab tearing and trench retreat. *Tectonophysics, 597–598,* 1–33.

King, G., Sturdy, D., & Whitney, J. (1993). The landscape geometry and active tectonics of northwest Greece. *Geological Society of America Bulletin, 105,* 137–161.

Kraft, J. C., Rapp, G., Szemler, G. J., Tsiavos, C., & Kase, E. (1987). The Pass at Thermopylae. *Journal of Field Archaeology, 14,* 181–198.

Okay, A. I., & Tüysüz, O. (1999). Tethyan sutures of northern Turkey. *Geological Society London Special Publications, 156,* 475–515.

Papanikolaou, D., Nomikou, P., Papanikolaou, I., Lampridou, D., Rousakis, G., & Alexandri, M. (2019). Active tectonics and seismic hazard in Skyros Basin, North Aegean Sea, Greece. *Marine Geology, 407,* 94–110.

Piccardi, L. (2000). Active faulting at Delphi: seismotectonic remarks and a hypothesis for the geological environment of a myth. *Geology, 28,* 651–654.

Piper, D. J. W. (2006). Sedimentology and tectonic setting of the Pindus Flysch of the Peloponnese, Greece. *Geological Society London Special Publication, 260,* 493–505.

Rassios, A., & Smith, A.G. (2000). Constraints on the formation and emplacement age of western Greek ophiolites (Vourinos, Pindus, and Othris) inferred from deformation structures in peridotite. In: Ophiolites and Oceanic Crust: new insights from field studies and the ocean drilling program (Eds.) Dilek, Y., Moores, E.M., Elthon, D., Nicolas, A. *Geological Society of America Special Paper, 349,* 473–484.

Regueiro, M., Stamatakis, M., & Laskaridis, K. (2014). The geology of the Acropolis (Athens, Greece). *European Geologist, 38,* 45–52.

Robertson, A. H. F., & Mountrakis, D. (2006). Tectonic development of the Eastern Mediterranean region: an introduction. *Geological Society of London Special Publication, 260,* 1–9.

Spiller, H. A., De Boer, J. Z., Hale, J. R., & Chanton, J. P. (2008). Gaseous emissions at the site of the Delphic Oracle: assessing the ancient evidence. *Clinical Toxicology, 46,* 487–8.

Vouvalidis, K., Syrides, G., Pavlopoulos, K., Pechlivanidou, S., Tsourlos, P., & Papakonstantinou, M.-F. (2010). Palaeogeographical reconstruction of the battle terrain in Ancient Thermopylae, Greece. *Geodinamica Acta, 23,* 241–253.

Abstract

The Peloponnese is a mountainous region of southwest Greece restricted to a deeply indented peninsula surrounded by segments of the Mediterranean Sea. The Isthmus of Corinth is the only connection with southeast Greece. The Peloponnese is famous for antiquities and historical sites which are visited by large numbers of tourists. The archaeological record encompasses multiple civilizations. The Mycenaean civilization of the Upper Bronze Age incorporates the semi-mythical leaders and kings of Homer's *Iliad*. During the Archaic, Classical Greece and Hellenistic periods, the Peloponnese was subdivided into seven districts or city-states. Many of the names used for the inhabitants of the city-states, such as Corinthians and Spartans, are found in modern usage. The early civilizations occupied marine terraces and valleys, where fertile alluvium and abundant supplies of subterranean water from the interior mountains supported agricultural techniques introduced from Asia Minor. Many of the modern cities and towns in the Peloponnese, including Patras, Corinth, Kalamata, and Sparti, are situated proximal to the archaeological sites. The island of Zakynthos is the southernmost of the Ionian archipelago and has a similar history. The Peloponnese and the island of Zakynthos constitute one of the most tectonically-active regions of Greece. Tectonism is influenced by continuation of the continental collision, specifically between the African Plate and the Hellenic Microplate, and proximity of the active subduction zone. The interior mountains consist of resistant metamorphic rocks (Palaeozoic-Mesozoic) associated with the NW-SE trending Alpine zones. Many of the mountains form rugged, karstic landscapes due to the dominance of thick sequences of limestone and marble. The mountains are transacted by basins and grabens that formed in response to crustal extension during the Neogene and Quaternary. Formation of basins and grabens has dictated the irregular nature of the Peloponnese coastline. Grabens typically extend into the sea, where they form deep bays and secure natural harbours. Marine terraces and valleys associated with the grabens are underlain by poorly-consolidated sediments (Neogene) and alluvium (Quaternary). The discovery of gold artefacts and burial chambers at Mycenae in the Argolid authenticated descriptions in the *Iliad*, confirming the archaeological site as the Mycenaean capital. The coast of the Argolid is immortalised in mythology by the six labours that Theseus faced on his epic journey from Troezen to Athens. The antiquity of Epidaurus contains one of the best preserved theatres of the Hellenistic period. Historical sites in the district of Corinthia include Ancient Corinth, which is located on a marine terrace, the Acrocorinth, and the *Diolkos*, an ancient stone carriageway along which boats could be towed across the isthmus. The Corinth Basin is tectonically-active and the region is prone to earthquakes and tsunamis. The vertical sidewalls of the Corinth Canal expose a thick section of sediments which may include tsunami deposits. The antiquity of Olympia in the district of Elis is located on alluvium in the Alpheios Graben. Antiquities in the districts of Arcadia, Laconia, and Messinia, which include Ancient Sparta and Nestor's Palace, are located on either marine terraces or in major grabens. The island of Zakynthos experienced a catastrophic earthquake in 1953 and the main town, which is situated on a marine terrace associated with poorly consolidated sediments, was totally destroyed.

Keywords

Alpine orogeny • Antiquities • Corinth • Graben • Limestone • Mycenae

Photographs not otherwise referenced are by the author.

© Springer Nature Switzerland AG 2021
R. N. Scoon, *The Geotraveller*,
https://doi.org/10.1007/978-3-030-54693-9_11

11.1 Introduction

The Peloponnese is a mountainous region of southwest Greece restricted to a deeply indented peninsula surrounded by segments of the Mediterranean Sea. The peninsula is isolated from northwest Greece by the Gulf of Patras and the Gulf of Corinth, narrow, elongate channels linked to the Ionian Sea. The only connection with southeast Greece is via the Isthmus of Corinth. This narrow strip of land separates the Ionian Sea from the Saronic Gulf. The indented coastline includes deep bays, many of which constitute secure harbours, e.g., Navarinou Bay (Fig. 11.1). The Peloponnese is well known for antiquities and historical sites that form an important part of the cultural heritage of Greece. The antiquities of Epidaurus, Mycenae, and Troezen occur in the northeast part of the peninsula. This region is unusually rich in antiquities and also contains the archaeological sites of Ancient Corinth, the Acrocorinth, and Isthmia. The *Diolkos* is an ancient stone carriageway, located parallel with the modern Corinth Canal, along which boats could be towed to avoid the long passage

around the perimeter of the Peloponnese. The principal antiquity in the western Peloponnese is Olympia. Antiquities and historical sites in the southern Peloponnese include Ancient Sparta, Mystras and the Palace of Nestor. The Peloponnese has a vibrant tourist industry with large numbers of visitors attracted to the antiquities and archaeological sites.

The physiography of the Peloponnese is consistent with historical descriptions of the peninsular as the "Islands of Pelops" ("Pelop" was a mythical king). The Peloponnese is constrained on four sides by different segments of the Mediterranean Sea (Fig. 11.2). To the west is the Ionian Sea. To the north are the Gulfs of Patras and Corinth. The Saronic Gulf is located to the northeast and the Aegean Sea to the south. The island of Zakynthos is the southernmost of the Ionian archipelago, and is located off the northwest coast of the Peloponnese. Most settlements in the Peloponnese and on the island of Zakynthos are located on narrow coastal terraces. These include the major city of Patras and regional centres such as Corinth, Kalamata, Nafplio, and Zakynthos Town. Some regional towns in the interior of the

Fig. 11.1 The narrow entrance to Navarinou Bay, southwestern Peloponnese, is the site of two famous battles in ancient and modern Greek history. The bay is associated with a sub-graben of the Kalamata Graben, which is partially infilled by poorly-consolidated sediments (Neogene). The islands of Pylos (upper left) and Sfaktira (upper right) occur on the western flanks of the sub-graben and consist of resistant limestone (Upper Cretaceous)

Fig. 11.2 Location map showing antiquities and historical sites of the Peloponnese. The Corinth Canal transects the Isthmus of Corinth. *Source* Satellite Image of Europe based on NASA MODIS data, processed by Philip Eales, Planetary Visions/DLR

Peloponnese, including Sparta and Tripoli, are located in broad valleys.

The mountainous interior of the Peloponnese consists of resistant metamorphic rocks (Palaeozoic-Mesozoic) associated with the NW-SE trending Alpine zones. The valleys correlate with basins and grabens that formed in response to crustal extension during the Neogene and Quaternary. Elongate basins are associated with the offshore gulfs. The regions proximal to the offshore gulfs are separated from the interior mountains by extensive marine terraces. Many of the interior grabens extend into the sea, where they form deep bays and secure natural harbours. The grabens and marine terraces are underlain by poorly-consolidated sediments (Neogene) and alluvium (Quaternary). The Peloponnese and the island of Zakynthos constitute one of the most tectonically-active areas of Greece.

11.2 Early Civilizations

The coastal areas of the Peloponnese, together with the Ionian Islands, were settled by some of the earliest European civilizations. Archaeological evidence incorporates the Neolithic Age (7000 BC–3000 BC) and the Early/Middle Bronze Age (3000 BC–1600 BC). Archaeological sites relating to the Mycenaean civilization, which occurred in the Late Bronze Age, are of particular significance (Box 11.1). The Mycenaean civilization was followed by the Archaic, Classical Greece, and Hellenistic periods. The seven districts or city-states of the Peloponnese occurred during these periods (Fig. 11.3). The Argolid (Argolida) and Corinthia (Korinthia) occur in the northeast of the peninsula. Achaea (Achaia) is situated in the north. Elis (Ilia) occurs in the

Fig. 11.3 Ancient districts and capitols of the Peloponnese during the Archaic and Classical Greece periods

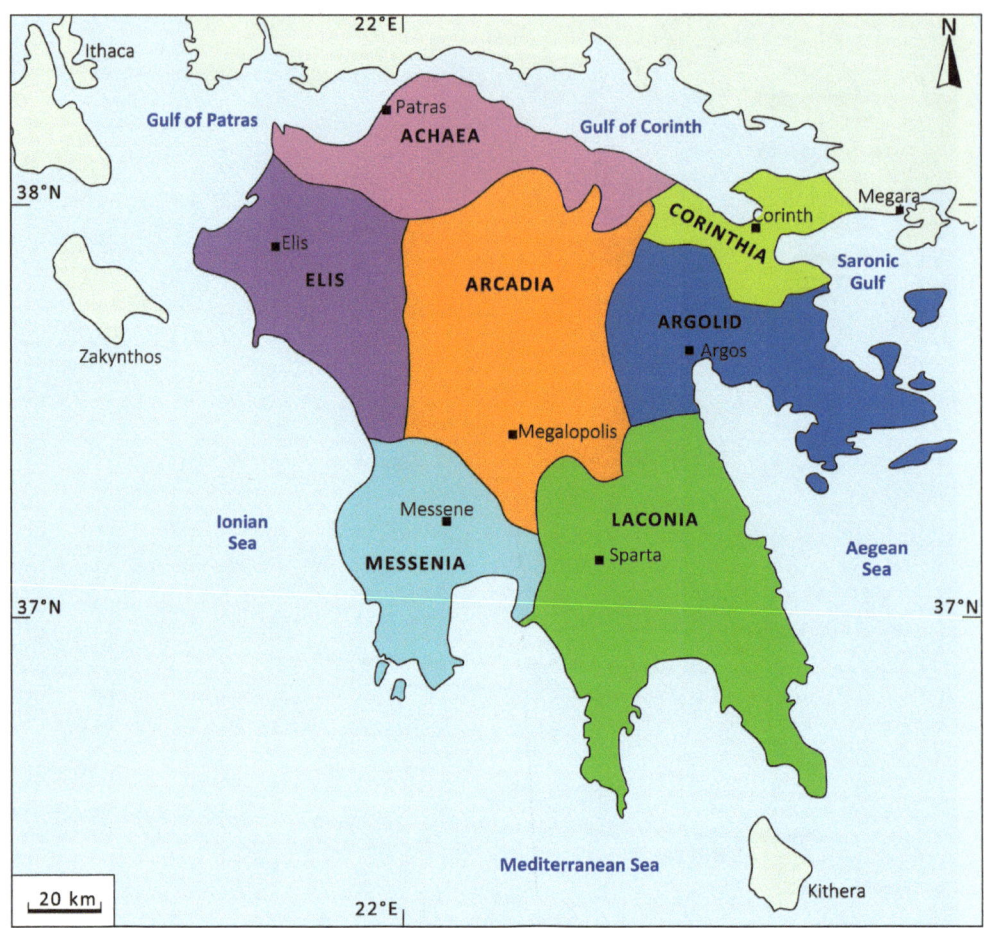

northwest and Arcadia is located in the mountainous interior. Laconia (Lakonia) and Messenia (Messinia) are situated in the southeast and southwest, respectively.

These subdivisions correlate reasonably well with modern districts, although there is little consistency in spelling of geographic localities (arising from problems translating the Greek alphabet). The approach taken here is based on various sources, and includes Higgins and Higgins (1996), Mythical Peloponnese (2018), Heikell and Heikell (2014).

Antiquities and archaeological sites are described from each of the seven districts following an approximately anticlockwise pattern commencing with the Argolid. The frequency and intensity of seismic events, together with tsunamis, has meant many of the antiquities have suffered major earthquake damage.

Box 11.1: Historical Overview

The oldest of the Ancient Greece civilizations is Mycenaean (1600 BC-1100 BC), named after the antiquity of Mycenae (Argolid), and equated with the Late Bronze Age. The Mycenaeans were colonisers who probably migrated from Asia Minor, displacing,

and integrating with the Early/Middle Bronze Age inhabitants throughout the Balkan Peninsula. It is the Mycenaeans that the Greek poet Homer described in the famous poems, the *Iliad*, and the *Odyssey*. Homer probably lived in the 8[th]C BC and wrote the poems based on oral accounts handed down over the generations. Homer used the description "Achaean" in a collective, Panhellenic sense, although strictly this should describe inhabitants of the district of Archaea. Geologists use the name "Archaean" as the oldest of the chronological grouping of rocks on Earth. The Mycenaean civilization is well known from multiple localities in the Peloponnese. The semi-mythical kings and leaders based in the Argolid include *Agamemnon*, *Menelaus* and his wife *Helen* ("Helen of Troy), and *Heracles*. *King Nestor* lived in the western Peloponnese, near Pylos. *Odysseus* was described as the king of the Ionian island of Ithaca.

The collapse of the Mycenaean civilization ushered in the "Greece Dark Ages" (1100 BC–800 BC). This period corresponds in part to the transition from the Late Bronze Age to the Iron Age. The Mycenaean alphabet (Linear B script) was lost at approximately 1100 BC and a new form of literacy, the Phoenician alphabet, was invented at approximately 1050 BC. The Phoenician alphabet was probably incorporated into Greece in the 9th C BC and was subsequently modified to create the Modern Greek alphabet (the ancestor of Latin and Cyrillic scripts). The Greek Dark Ages were followed by the Archaic (800 BC-500 BC) and the Classical Greece (500 BC-323 BC) periods with the commencement of city-states in the Peloponnese (and mainland Greece). The Hellenistic period (323 BC-31 BC), derived from "Hellas", the original name for Greece, started with the death of Alexander the Great and lasted until the emergence of the Roman Empire. The latter is signified by the Battle of Actium, a naval engagement in the Ionian Sea, northwest Greece. The Roman and Byzantine Empires continued for more than a thousand years, during which time many of the Hellenistic cities were occupied and extended. Parts of the Peloponnese were controlled for short periods by the Normans and Venetians.

Reference to the Peloponnese city-states has entered modern culture. The unique social system and constitution of the Spartans have been widely incorporated into philosophy and literature. In modern usage *Corinthians* infers ethical sporting prowess. Possibly the most famous of the antiquities in the Peloponnese is Olympia, where four-yearly sporting events were held over a period of more than a thousand years (776 BC-393 AD). Olympians were held in high esteem during historical times (as they still are).

11.3 Regional Geology

The tectonic framework of the Peloponnese is closely linked to that of southeast and northwest Greece, hence the considerable overlap between the description of the regional geology in this and Chap. 10. The eastern Mediterranean is one of the most tectonically active areas on Earth (Robertson and Mountrakis 2006). The complexity of the Alpine convergence was discussed in detail in Chap. 8. The Alpine Orogeny peaked in the Oligocene-Miocene (34–5.3 Ma). The orogeny resulted in formation of microplates that include the Hellenic Microplate (Okay and Tüysüz 1999; Robertson and Mountrakis 2006; Papanikolaou et al. 2019). The Mediterranean microplates may have started to form in the Early Cretaceous, but the main development is thought to

have been in the Oligocene-Miocene (i.e., concurrent with the peak of the Alpine Orogeny). The Hellenic Microplate and the Eurasian Plate are separated by the W-E trending Central Hellenic Shear Zone (Fig. 8.3) (Robertson and Mountrakis 2006). The Peloponnese is a discrete physiographic component of southwest Greece located to the south of the Shear Zone.

The compressional regime of the Alpine Orogeny was displaced in the Peloponnese during the Cenozoic by extensional tectonics. Crustal stretching triggered normal faulting and formation of elongate basins and grabens. The Aegean Sea is associated with a major back-arc basin within the Hellenic Microplate, which is expanding at estimated rates of 50 mm/year (Jolivet et al. 2013; Papanikolaou et al. 2019). The basins and grabens are largely responsible for the highly irregular nature of the Peloponnese coastline. Offshore bays and channels are a significant component of the rifting.

Subduction associated with the Alpine Orogeny is ongoing and proximity of the active component of the Hellenic Trench to the west coast of the Peloponnese and the southern islands of the Ionian archipelago is a notable feature (Fig. 11.4). The Kefalonia Fault is a major transform fault located northwest of Zakynthos. The Ionian Sea in the region to the west of Zakynthos drops rapidly into the deep water characteristic of an ocean trench, but the near-shore channels are relatively shallow (typically < 70 m). The Ionian Islands were connected to the mainland of Greece during the Late Pleistocene Ice Ages. Movement on the Hellenic Trench and on the Kefalonia Fault is responsible for major seismic events which impact the Ionian Islands and the Peloponnese (e.g., Hasiotis et al. 2002; Lagios et al. 2007). The Gulf of Corinth and the Gulf of Patras are highly susceptible to tsunamis (Papadopoulos 2003; Ambrasey and Synolakis 2010).

11.3.1 Alpine Zones

The Alpine zones in the Peloponnese are a continuation of the NW-SE tectonic zones found in northwest Greece (Fig. 11.4). Four zones are represented, the Ionian and Gavrovo Zones (west and south), the Pindus Zone (Arcadian Mountains), and the Parnassus Zone (northeast). The boundaries are constrained by thrusts and nappes. The Ionian Zone is dominated by Upper Cretaceous limestone (with minor Jurassic limestone). The Gavrovo Zone includes Triassic-Eocene limestone and marble. The Pindus and Parnassus Zones contain Upper Jurassic-Cretaceous limestone, marble, and dolomite. The latter may include Jurassic-age ophiolite complexes (e.g., in the Isthmus of Corinth). Some sections of the Peloponnese are underlain by an older

Fig. 11.4 Simplified geological map of the Peloponnese and Zakynthos showing the Alpine trend of the four tectonic zones and the younger basins and grabens. *Source* Geological Map of Europe (https://geoviewer.bgr.de) and articles referenced in the main text

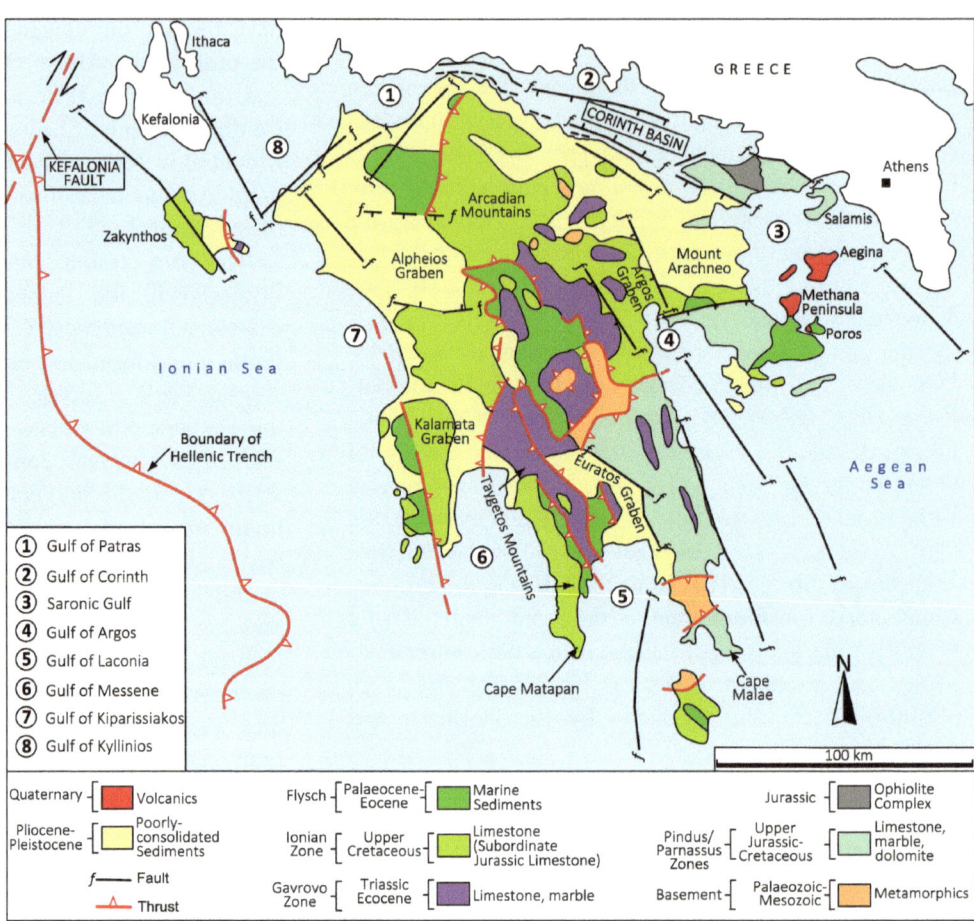

11.3.2 Basins and Grabens

Prior to formation of the basins and grabens, parts of the Peloponnese saw development of extensive deposits of flysch (Fig. 11.4). The marine sediments, which typically reveal modest levels of metamorphism, accumulated either during or after the peak of the Alpine Orogeny. Several basins in the northern part of the Peloponnese are aligned with the W-E trending Central Hellenic Shear Zone. The Patras Basin is aligned W-E and connects with the Gulf of Kyllinios in the Ionian Sea. The Corinth Basin is aligned WNW-ESE and is part of a rift associated with the Saronic Gulf. Many of the grabens in the central and southern parts of the Peloponnese are aligned NNW-SSE. These grabens may reflect reactivation of older Alpine structures (Papanikolaou et al. 2019). They include the Alpheios Graben, the Argos Graben, the Eurotas Graben, and the Kalamata Graben. Most basins and grabens in the Peloponnese are partially infilled by poorly-consolidated sediments (Pliocene and Pleistocene),

together with alluvium (Quaternary). Extensive marine terraces developed adjacent to some of the offshore basins. All of the interior valleys are probably associated with grabens. Basins and grabens developed in the Mid Cenozoic and continued to form during the Quaternary. Many of the graben faults remain active and are associated with localized seismic events. The Methana Peninsula and some islands in the Saronic Gulf, e.g., Aegina, are part of the Hellenic Volcanic Arc (Chap. 12).

11.4 The Argolid

The Argolid is subdivided by Mount Arachneo (1,199 m) into two regions, the coastal plain and the broad Argolid Peninsula are located to the east and south, and the Argos Graben is situated to the west (Fig. 11.4). Mount Arachneo and the coastal plain consist of limestone, marble, and dolomite (Upper Jurassic-Cretaceous) of the Parnassus Zone. The coastal landforms are exposed on the road that links this region with Corinth. Sections include a corniche that hugs the mountainous shoreline. The Argolid Peninsula is dominated by flysch deposits (Palaeocene-Eocene). The Argos Graben

basement terrain that includes Palaeozoic and Mesozoic metamorphic rocks.

persists into the Gulf of Argos. The Argos Graben contains the Archive Plain, one of the most fertile parts of the Peloponnese. Large sections of the graben are infilled by alluvium (Quaternary). The shoreline of the Gulf of Argos has changed considerably since historical times, e.g., the position of Lake Lerna, a historical site featuring in Greek mythology, was part of a lagoon prior to 5,000 BP (Higgins and Higgins 1996).

11.4.1   Ancient Humans

Franchti Cave located south of Nafplio on the Argolid Peninsula has yielded evidence of the earliest inhabitants of the Peloponnese. The cave was used by *Homo neanderthalensis* (40,000 BP) and early *Homo sapiens* (30,000 BP). Neolithic and Early Bronze Age cultures used the cave until approximately 3,600 BP. The cave is developed in Triassic-Jurassic limestone with a depth of 150 m. The entrance has a width of approximately 30 m. The "Dolines of Didyma" consists of two circular sink holes (locally known as "craters") which occur in the limestone a few kilometres north of the Franchti Cave. One sink hole was occupied by people during the Neolithic Age and the other is sufficiently large as to contain two Byzantine churches.

11.4.2   Antiquities

The most important antiquity in the Argolid is the archaeological site of the ancient city of Mycenae (1600 BC–1100 BC). Mycenae is located on the flanks of the Argos Graben, between Nafplio and Corinth (Fig. 11.2). In 1876, the German archaeologist Heinrich Schliemann discovered numerous gold artefacts (masks, vessels, and ornaments) in shafts and ornamental tombs. These finds indicate the city may be the capitol of the Mycenaean civilization (Homer described Mycenae as "Rich in Gold"). There is, however, ongoing debate as to the ages of the artefacts at Mycenae (and even the provenance), as well as to the historical accuracy of the accounts in the *Iliad*. Some of the most well known treasures, e.g., the "Mask of Agamemnon" (*Agamemnon* led the Greek forces which laid siege to Troy) and "Nestor's Cup" are displayed in the National Museum of Athens. The source of the gold at Mycenae remains a mystery (Egypt or Macedonia are possibilities), but the organized nature of the society, which included paved roads to access harbours on the Gulf of Argos, a rich agricultural sector, and a strong bureaucracy (evidenced by stone tablets with the Linear B script), could have supported a significant trading empire. Trading routes which distributed obsidian (for stone tools) from the island of Milos during the Neolithic Ages further support the sophisticated nature of some of the oldest communities in the Peloponnese.

Fig. 11.5 Lion gate, Mycenae. *Source* Andreas Trepte (Own work, CC BY-SA 2.5, https://commons.wikimedia.org)

The citadel at Mycenae is a complex of buildings located on low hills comprised of limestone (Triassic-Jurassic). The limestone bedrock provided a secure location and minimized the potential of earthquake damage (Higgins and Higgins 1996). The limestone also provided a local source of building material and was used in constructions such as the Lion Gate (Fig. 11.5). The three-level museum in the citadel accesses some of the shafts and burial tombs excavated by Schliemann. The main part of the city occurs on a lower-lying site associated with marls and conglomerate within the graben.

Other historical sites in the Argive Plain include Tiryns (1600–1100 BC), where a citadel is built on a limestone knoll (Lower Cretaceous) and where the mythical hero *Heracles* is purported to have lived. Both Mycenae and Tiryns had secret water supplies in case of sieges; water was channelled from the limestone bedrock. Argos is possibly the oldest city in Greece and became the dominant city of the Argolid in 1,000 BC. Historical sites at the regional town of Nafplio (or Nauplia) include the Hill of Palamidi, named after the Homeric hero of *Palamedes*. Nafplio is located on a limestone horst (Upper Cretaceous) within the centre of the Argos Graben.

The route which follows the coastal region of the Argolid is immortalised in mythology by the six labours that *Theseus* faced on his epic journey from his birthplace at Troezen to Athens. Troezen is an extensive archaeological site located near the modern village of Trizina spread over several square kilometres in a hilly section of the Argolid Peninsula. The limestone rock that *Theseus* was required to lift as the first of his tasks can be observed here (Fig. 11.6a). Marble slabs showing multiple stages of calcite veins, some of which reveal ptygmatic or sinuous folding, an indication of plastic deformation in which the vein has a greater competency than

Fig. 11.6 **a** The limestone rock that *Theseus* was required to lift prior to commencing his epic journey to Athens can be examined at Troezen; **b** Marble slab at Troezen showing ptygmatic veins of calcite

the host rock, were used for some constructions (Fig. 11.6b). The hills at Troezen are associated with flysch deposits, with some of the higher mountains, such as at the "Devils' Bridge", a natural spur of limestone spanning a narrow gorge, associated with older and more competent limestone and marble. The gorge has developed on the contact between limestone-marble and a thin schist belt. Large boudins of marble can be observed in the less competent schist.

The archaeological site of Epidaurus is situated in the foothills of Mount Arachneo (Triassic-Jurassic limestone) and includes one of the best preserved theatres, the Asklepieion (340 BC–330 BC), of the Hellenistic period in Greece (Fig. 11.7). The site offers both a commanding position and a source of building material. The theatre is constructed of limestone breccia (probably cave deposits), which contain clasts of red and grey limestone. Extensive deposits of Triassic volcanic ash occur in the region; this distinctive green, water-lain tuff can be observed in the foundations of some of the subordinate buildings at Epidaurus (Higgins and Higgins 1996).

11.5 Corinthia and Achaea

Some of the geological features of the Corinthia and Achaea districts can be examined from the modern highway that connects Athens and Igoumenitsa (northwest Greece). The section between Corinth and Patras is located close to the southern shores of the Gulf of Corinth. Marine and lacustrine terraces located between the interior mountains and the Corinth Basin constitute extensive areas of badlands, (Fig. 11.8). The badlands have formed from erosion of the poorly-lithified Neogene-Quaternary sediments. Erosion rates are enhanced due to rapid uplift on the flanks of the Corinth Basin. The frequency of seismic events in the region has necessitated building numerous rock avalanche shelters and tunnels to protect the highway.

The Straits of Rion and Andírrion is a narrow channel known as the "Little Dardanelles" that separates the Gulf of Corinth from the Gulf of Patras. The straits are crossed by a 2.9 km-long suspension bridge, the Rion-Andírrion Bridge, which was opened in 2004 (Fig. 11.9). The bridge required remarkable engineering solutions to overcome problems such as pillar support for the suspended cableways (pillars are located in the poorly consolidated sediments on the marine terraces) and risks attached to seismic activity and tsunamis. Expansion of the distance between pillars has required retrospective engineering: the Corinth Basin is estimated to be widening at a rate of 30 mm/year.

The narrow gauge Diakofto cog railway, located approximately 50 km east of Patras and built in the latter part of the 1800s, ascends the steep slopes of the Vouraikos Gorge prior to terminating in the historical village of Kalavryta (the site of a World War II massacre). The 20 km-long gorge is part of the Mount Chelmos National Park, which includes several hiking trails where the Upper Cretaceous limestone can be observed (Fig. 11.10). The coastal ranges, which include Mount Panachaiko (south of Patras), Mount Chelmos (south of Diakofto), and Mounts Aroania and Killini (southwest of Corinth), merge southward into the Arcadian Mountains (Fig. 11.2).

11.5.1 Corinth Basin

The Corinth Basin is situated within the Central Hellenic Shear Zone, a broad suture which developed during the Miocene at the boundary between the Hellenic Microplate and the Eurasian Plate. Rifting associated with the Corinth Basin may have commenced in the Early Miocene (Bull 2009), but the majority of faulting occurred during the Pliocene and Pleistocene (Roberts et al. 2011). The first

Fig. 11.7 The antiquity of Epidaurus is situated on the eastern slopes of Mount Arachneo (Triassic-Jurassic limestone) and includes the Asklepieion, one of the best preserved theatres of the Hellenistic period

generation of older faults have been progressively abandoned, and the structure is currently restricted to a relatively narrow, 30 km-wide rift (Jackson 1999) . Subsidence in the Gulf of Corinth, which correlates with the centre of the basin, is estimated to be as much as 3 km. Uplift of the rift shoulders has resulted in the Neogene and younger sediments occurring at elevations as high as 1,200 m. The rift is an asymmetric half-graben and most of the active faults are concentrated on the southern shores (Collier and Dart 1991; Ford et al. 2017). The asymmetry has resulted in a relatively linear coastline on the southern side, with a markedly indented coastline on the northern side. During the low stands of the Late Pleistocene Ice Ages, the Rion-Andírrion narrows were exposed and the Gulf of Corinth formed a shallow lake.

11.5.2 Antiquities

Corinthia and Achaea have a remarkable history with settlement dating from at least 1600 BC. Many of the well known antiquities occur close to the modern town of Corinth, which is located on the shores of the Gulf of Corinth

(Fig. 11.11). Ancient Corinth occurs several kilometres inland on a Pleistocene marine terrace aligned approximately parallel with the Gulf of Corinth (Fig. 11.12a). The terrace consists of sandy limestone with an estimated age of 250,000 BP (Higgins and Higgins 1996). During the Archaic and Classical Greece periods, Ancient Corinth was a strategic centre of great power, as the position close to the Isthmus of Corinth placed it with control over important trade routes. Ancient Corinth was destroyed by the Romans in 146 BC and rebuilt a century later. The antiquities are constructed primarily of locally-derived Pleistocene oolitic and shelly limestone. The limestone blocks used for the 7 m-high columns in the Temple of Apollo, the most significant of the monumental buildings (Roman), were derived from a local quarry. Ancient Corinth was the first region to re-develop (in the Archaic period) the stone architecture originally used by the Mycenaeans (Hayward 2003). This skill had been lost during the Greek Dark Ages. The Corinthians exported dressed blocks of oolitic limestone for construction of many antiquities in the Peloponnese, e.g., Epidaurus, as well as to other parts of Greece, e.g., Delphi (Chap. 10).

Fig. 11.8 a Marine and lacustrine terraces on the southern shores of the Gulf of Corinth include extensive badlands (Neogene-Quaternary sediments) located beneath interior mountains (Upper Cretaceous limestone); **b** Terraces observed next to the highway between Corinth and Patras form deeply eroded, pale buff-coloured cliffs

The ancient harbour of Lechaion was the main base for the Macedonian fleet of Philip V (Hellenistic period) and was also used during the Roman and Byzantine Empires. Lechaion has been subjected to repeated seismic activity which caused landslides and tsunamis (Hadler et al. 2011). Excavations have yielded evidence of historical tectonism. An island monument was destroyed by a major earthquake at approximately 70 AD that affected large parts of the Roman Empire. By the 6[th]C AD, the Byzantine Empire had constructed a new harbour, measuring approximately 40,000 m² (the older basins had been filled by sediment). Subsequent earthquakes may have raised the Byzantine harbour of Lechaion by more than a metre.

Marine terraces at Lechaion and surrounding areas are correlated with specific historical periods, e.g., the Ancient Corinth Terrace reveals an elevation change from 60 m near Acrocorinth to 360 m near Xylokastro (Roberts et al. 2011). Uplift on this terrace is estimated at 1.5 mm/year. The resort of Loutraki to the north of Corinth is well known for hot springs. The brines have been funnelled by a fault scarp associated with the steep cliffs on the landward side of the narrow marine terrace.

Fig. 11.9 View looking northeast of a section of the Rion-Andírrion Bridge which crosses the narrows between the Gulfs of Corinth and Patras

Fig. 11.10 The Vouraikos Gorge on the northern slopes of Mount Chelmos occurs in Upper Cretaceous limestone

The prominent block of Jurassic limestone that towers above Ancient Corinth is the site of the Acrocorinth, an old fortress with extensive stone walls on the crest of the 575 m-high hill (Fig. 11.12b). The archaeological relics at Acrocorinth cover the Hellenistic period through to the Byzantine and Venetian occupations.

11.5.3 Corinth Canal

The Corinth Canal transects the narrowest part of the Isthmus of Corinth (Fig. 11.11). The canal provides a link between the Ionian Sea (the Gulf of Corinth) and the Aegean Sea (Saronic Gulf), saving a 700 km circumnavigation of the Peloponnese. The possibility of a canal through the isthmus was first conceptualised by Periander, the Athenian leader in the 7th C BC. In the 6th C BC, this concept was replaced by construction of the *Diolkos*, a stone carriageway along which boats could be towed (Fowler and Stillwell 1935). Details of the *Diolkos* are described below. In 67 AD, Emperor Nero initiated excavation of a canal that resulted in a 50 m-wide trench being cut over a length of 700 m. This project was subsequently abandoned. The Hexamillion ("six-mile wall") is a defensive structure built by the Romans in 408–450 AD to the west of the Corinth Canal. The wall protected the only land route between Attica and the Peloponnese. Remnants of the wall are considerably older, and may date to the Mycenaean period. The archaeological site at Isthmia, at the southeastern end of the canal, includes parts of a ruined stadium used for the Isthmian Games, an important component of the Panhellenic events held every year before and after the Olympic Games.

Construction of the 6 km-long modern Corinth Canal started in 1881 and was completed in 1893. The canal has a width of only 25 m and is accessed by small cruise ships, tour boats, and yachts. The near-vertical side walls in the central part of the canal have a height of 70 m. The canal includes several high-level bridges, which provide popular view sites (Fig. 11.13). Floating bridges occur at the southeast (Isthmia) and northwest (Posidhonia) extremities. Problems encountered during building of the canal included partial collapse of the side walls and the different state of the tides in the two gulfs (there are no locks). A current of up to 5.5 km/h can flow either way in the canal (Heikell and Heikell 2014).

The sidewalls of the Corinth Canal expose a thick sequence of Pliocene and Pleistocene lacustrine sediments that accumulated in the Corinth Basin (Mariolakis and Stiros 1987; Higgins and Higgins 1996). Pale buff-coloured marls, sandstones, and conglomerates (Upper Pliocene) are exposed in the central part of the canal (Fig. 11.13a). Dark

Fig. 11.11 Location map showing historical sites in the vicinity of the modern town of Corinth. The Corinth Canal cuts through the narrowest point of the Isthmus

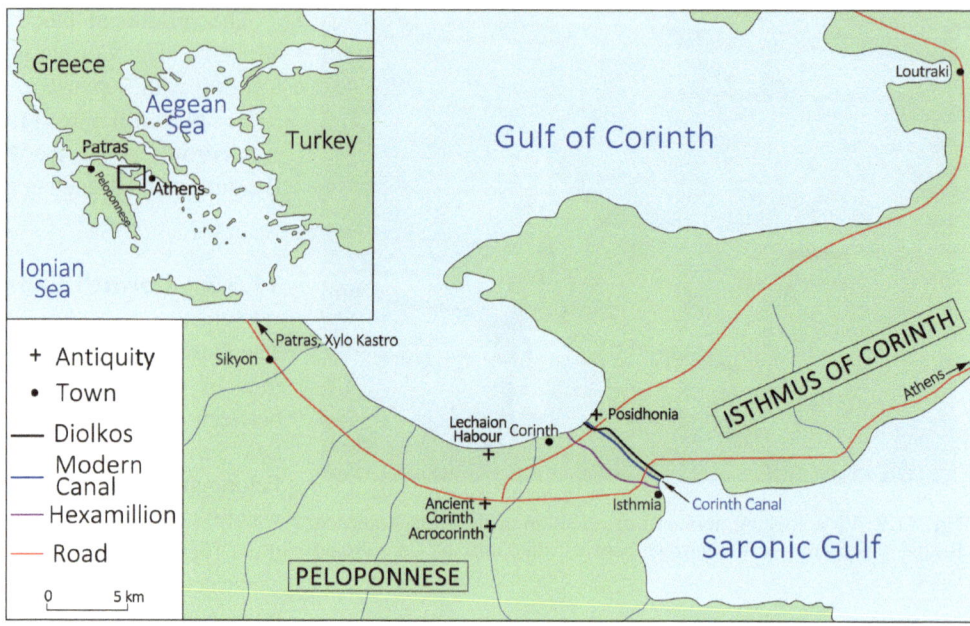

brown sandstones and limestones (Pleistocene) can be observed to unconformably overlie the older rocks at the extremities of the canal (Fig. 11.13b). Steeply inclined, normal faults, consistent with persistent seismicity, can also be observed. The sediments were deposited in a palaeo-lake when the Gulf of Corinth was isolated from the Ionian Sea by the Zakynthos-Kefalonia barrier. This barrier lasted until the Late Pleistocene (250,000 BP). Uplift of the sediments on the margins of the basin is a recent phenomenon.

Part of the historical stone carriageway, the *Diolkos*, has been excavated at Posidhonia (Fig. 11.14). Rectangular blocks of limestone (measuring 1.6 m by 1.1. m) occur within a calcified sequence of Holocene sands and gravels, or "beachrock". The stone blocks are covered by 30 cm of sediment, an indication that some of the beachrock accumulated in the previous 2,500 years, i.e., after the carriageway had been submerged. The sequence exposed here, i.e., the beachrock plus the stone blocks, has been subjected to several stages of uplift and submersion (Higgins and Higgins (1996). The current position of the carriageway suggests a recent phase of uplift greater than a metre. The beachrock extends approximately 300 m from the shoreline (as compared with 15–20 m at Xylokastro), an indication that the sequence at Posidhonia may have accumulated from tsunamis.

11.5.4 Earthquakes and Tsunamis

The active nature of the graben faults associated with the Corinth Basin triggers frequent earthquakes in the districts of Corinthia and Achaea. The problems include coastal instability and an unusually high potential for tsunamis (Hasiotis et al. 2002; Papadopoulos 2003; Ambrasey and Synolakis 2010). Numerous historical earthquakes have been documented. Catastrophic events in the vicinity of Corinth occur on average approximately every 300 years. The historical site of Ancient Corinth was severely damaged by two large earthquakes in the 6[th]C AD. A tsunami associated with this event may have destroyed the harbour at Lechaion. After a major earthquake in 1858, the town of Corinth was moved to its current position. Eleven seismic events with a magnitude > 6 have occurred in the vicinity of Corinth during the past 100 years. Three earthquakes affected Corinth in 1981, including a magnitude 6.6 tremor which caused severe damage and loss of life within the town. The city of Patras is also prone to earthquakes, with the most recent, a magnitude 6.6 event, having occurred in 2008.

11.6 Elis

The most widely visited antiquity in the district of Elis is Olympia. The popularity of the site has resulted in development of the modern town of Olympia, accessed by either road or rail from the regional town of Pyrgos. The interior mountains of Elis are dominated by limestone of the Ionian Zone (Upper Cretaceous), with substantial areas underlain by flysch deposits (Palaeocene-Eocene) (Fig. 11.4). Neogene-Quaternary grabens in the region are aligned either NNW-SSE (parallel with the narrow strait between the Peloponnese and Zakynthos) or W-E (parallel with the Central Hellenic Shear Zone) (Papanikolaou et al. 2007).

Fig. 11.12 a Ancient Corinth is situated on a marine terrace (Pleistocene) at the foot of a prominent limestone hill (Jurassic); **b** View of the archaeological site of Acrocorinth located on the hill of Jurassic limestone. View overlooks the marine terrace with the Gulf of Corinth visible in the background

Olympia is situated in a broad, fertile valley associated with the approximately W-E trending Alpheios Graben. The graben contains thick deposits of shelly limestone (Lower Pliocene) which formed in shallow marine conditions. When the rate of sedimentation in the graben exceeded subsidence, thick sequences of terrestrial sediment, i.e., sand, silt, clay and pebbles (Upper Pliocene-Pleistocene) were deposited by the westward-flowing River Alpheios. The terrestrial sediments are overlain by thick sequences of alluvium (Quaternary).

The broad coastal plain in the vicinity of Pyrgos is underlain by marine terraces (Neogene). Pyrgos was subjected to a major earthquake in 1993, ascribed to movement on the graben fault (Koukouvelas et al. 1996). Pyrgos is located close to the current exit of the Alpheios River, but in historical times the river flowed into the Ionian Sea north of Katakolon. Katakolon is a harbour used by cruise ships which include a visit to Olympia. The low cliffs behind the harbour are constructed of poorly-consolidated sands and gravels (Quaternary).

11.6.1 Olympia

The antiquity of Olympia is situated on alluvial deposits within the Alpheios Graben (Fig. 11.15a). This is a rather

Fig. 11.13 a Pale buff-coloured marls, sandstones, and conglomerates (Upper Pliocene) are exposed in the central part of the Corinth Canal; **b** Dark brown sandstone and limestone (Pleistocene) unconformably overlie the Upper Pliocene strata in the extremities of the canal, as seen here under the railway bridge. Both views looking southeast towards the Saronic Gulf

unusual setting, as most historical sites in Greece are constructed on resistant metamorphic rocks. There is some debate as to the thickness of the alluvium at Olympia as it is part of a fan associated with the confluence of two rivers, the Alpheios and the Kladeos (Higgins and Higgins 1969). The low, rounded hills on the floodplain, including the historically-important Hill of Kronos, are built of poorly consolidated terrestrial deposits (Upper Pliocene-Pleistocene). These hills erode to form unstable slopes. Small landslides are a common occurrence within and adjacent to,

the Olympia site. The older shelly limestone (Lower Pliocene) can be observed in the hills and ridges which surround the Olympia site (Fig. 11.15b).

Olympia is primarily a Hellenic site that may have been first occupied in 3,000 BC. Between 776 BC and 393 AD a four-yearly sporting event that initially included all of the Greece city-states, and then the Roman Empire, took place. The Olympic events were terminated in 393 AD by Theodosius I, the first of the Christian Roman Emperors. The site was largely destroyed in 425 AD by Theodosius II. Despite

Fig. 11.14 **a** Excavations next to the Corinth Canal at Posidhonia have uncovered parts of the 6ᵗʰC BC stone carriageway, the *Diolkos*; **b** Recent deposits of beachrock covering the stone carriageway accumulated during the previous 2,500 years when the carriageway was submerged. The carriageway has subsequently been uplifted by more than 1 m. The Posidhonia Floating Bridge is visible in the background

Olympia being located in a region where earthquakes are considerably less frequent than in the districts of Corinthia and Achaea, major seismic events in 522 AD and 551 AD caused catastrophic damage. The antiquity was subsequently affected by floods and landslides (Higgins and Higgins 1996).

Many of the monumental buildings at Olympia are built of shelly limestone quarried from hills south of the Alpheios River (Fig. 11.16a). The most famous component is the 5ᵗʰC BC, oval-shaped stadium (interior length of 150 m) which is surrounded by grassy banks and includes a monumental entrance at the western end (Fig. 11.16b). The site of the stadium may originally have been a natural channel of the River Alpheios. Other features of interest are the Temple of Hera (constructed in 650 BC) and the Temple of Zeus (constructed in 465 BC). The latter originally contained the colossal gold and ivory Statue of Zeus, one of the Seven

Wonders of the Ancient World. Columns of the Temple of Zeus probably lie where they collapsed in the 6ᵗʰC AD earthquakes. Some of the tiles and sculptures at Olympia consist of imported marble, either Parian or Pentelic in origin.

11.7 Arcadia

The mountainous western part of Arcadia is associated with two tectonic zones (Fig. 11.4). The Gavrovo Zone is dominated by limestone and marble (Triassic-Eocene). The Pindus Zone includes limestone, marble, and dolomite (Upper Jurassic-Cretaceous). Alpine thrusting has exposed sections of the Palaeozoic basement in the Arcadian Mountains. This region is drained by the River Alpheios, which flows westward into the district of Elis. The primary

Fig. 11.15 a The antiquity of
Olympia is situated in the
Alpheios Graben, which is
enclosed by hills and ridges of
shelly limestone (Lower
Pliocene); **b** The stadium is
situated in an old, alluvium-filled
river channel (Quaternary) and
overlooked by the Hill of Kronos
(Upper Pleistocene terrestrial
sediments)

source of the river is springs located at the base of the
Arcadian Mountains. The wetlands and swamps of Lake
Ladon are associated with some of the springs. The Lousios
River, a tributary of the Alpheios River, contains a
15 km-long gorge which is protected as an archaeological
site as it contains numerous churches.

The eastern part of Arcadia includes broad valleys and
plains associated with Neogene-Quaternary grabens. The
abundance of subterranean water in the limestone of the
Arcadian Mountains, together with the fertile nature of the
valleys, may explain why the Mycenaeans and Arcadians
flourished in this part of the Peloponnese. The regional
town of Tripoli is located in one such structure, as are the
antiquities of Megalopolis ("Big City"), which contains the
ruins of the largest theatre in Greece, and Tegea, the latter
being the most important centre in the Arcadian city-state
during the Archaic and Classical Greece periods. Tegea is
famous for a high-quality, sparkling grey-white marble,
derived from the slopes of Mount Parnon (Higgins and
Higgins 1996).

Fig. 11.16 a Slab of shelly limestone (Lower Pliocene) at Olympia; **b** The monumental western entrance to the stadium at Olympia

11.8 Laconia and Messenia

The southern Peloponnese is divided by the north-south trending Taygetos Mountains into two districts (Fig. 11.3). Laconia is located to the east and Messenia to the west. The Taygetos Mountains rise to a height of 2,300 m and are the highest range in the peninsula. They consist of limestone and marble (Triassic-Eocene), part of the Gavrovo Zone. Steep, dry gorges, typical of karstic landforms, can be viewed from a winding mountain road with tunnels and overhangs, which links the regional towns of Sparti (Laconia) and Kalamata (Messenia) (Fig. 11.17a).

Modern Sparti occupies the fertile valley of the Eurotas River, part of a large NNW-SSE trending graben (Fig. 11.4). The eastern side of the Eurotas Graben extends to the Parnonas Range, which consists of similar rocks to those within the Taygetos Mountains. The graben is partially infilled by alluvium (Quaternary) and includes inliers of somewhat more resistant marls and clays (Pliocene and Pleistocene).

The Kalamata Graben is the principal geological feature of Messenia. This structure is aligned parallel to the Eurotas

Graben. The regional town of Kalamata (famous for Kalamata olives) is situated close to the eastern side of the graben and is constructed on marls, sandstones, and conglomerates (Pliocene). The town of Kalamata was subjected to a major earthquake in 1986 (magnitude of 5.8) associated with the graben fault at a depth of 4 km.

11.8.1 Antiquities in Laconia

The Mycenaean palace of *Menelaus and Helen* (the Menalaion) is situated south of Sparti in the broad valley associated with the Eurotas Graben. The antiquity is constructed on a bluff of marls and clays (Pliocene-Pleistocene) that rises 50 m above the valley (Higgins and Higgins 1996). The stepped topography (or cuesta landform) of the valley is characteristic of an area underlain by gently-dipping strata. Ancient Sparta is located in the northern part of the modern town of Sparti. Sparta was possibly the most iconic of the Peloponnese city-states. Similar building styles to other localities of this period were used, with building stones, primarily limestone and grey marble, derived from quarries in the Taygetos Mountains. The antiquity occupies a low mound comprised of marls and clays (Pliocene-Pleistocene), similar to the Menalaion, and which would have provided a defensive position (Fig. 11.17b). Ancient Sparta was largely destroyed by an earthquake in 466 BC. This earthquake created a 10-12 m-high scarp on the western graben fault at the base of the Taygetos Mountains. Sparta was resettled and extended by the Romans (information boards at the antiquity provide details of the different phases of occupation).

The historical site of Mystras is located 5 km east of Sparti on the steep slopes of Kastro Hill, a resistant block of dolomite (Triassic-Jurassic). Mystras is a popular tourist destination as extensive remains of Byzantine (13–15[th]C AD) and Venetian (16–17[th]C AD) buildings can be observed on the hillside (Fig. 11.18).

The southern Peloponnese includes two prominent peninsulas (Fig. 11.4). The Matapan Peninsula is an extension of the Taygetos Mountains. The limestone at Cape Tanairo (the southernmost part of the Balkan Peninsula) contains a sea cave, the mythological home of *Poseidon*, the Greek God of the Seas. The extremity of the Malae Peninsula includes scenic landforms protected in the Kavomalias (Cape Malias or Cape Malae) Geopark. The Malae Peninsula was the source of red, green, and grey marbles (Upper Jurassic-Cretaceous) used for antiquities in Laconia and also exported for monumental buildings in Rome (Cooper 1981). Some of the ancient quarries, including the Rosso Antico Quarries (where a red to purple marble, the coloration linked to specks of hematite, was obtained), are protected sites. The headlands associated with both peninsulas are potentially

Fig. 11.17 a The road linking
Sparti and Kalamata crosses the
Taygetos Mountains, which
consist of resistant limestone and
marble (Triassic-Eocene);
b Ancient Sparta is situated on a
low knoll (right) consisting of
marl and clay (Plio-Pleistocene).
Part of modern Sparti and the
western edge of the Eurotas
Graben are visible in the
background

dangerous localities for small boats and yachts. Both regions
are subjected to frequent, albeit localized storms and Cape
Malae is affected by a turbulent wave pattern derived from
interaction of the Aegean Sea and Ionian Sea (Chap. 8). This
explains the mythology underpinning *Poseidon* and the
reason for cutting the Corinth Canal.

The Gulf of Laconia which separates the Matapan and
Malae Peninsulas is an extension of the Eurotas Graben. The
town of Gythion, located at the head of the gulf, was the
main harbour for Ancient Sparta. The small island of Kranae
is linked to Gythion by a causeway, which is where Homer
describes *Paris* and *Helen* as spending their first night on
their way to Troy. The island is constructed of resistant, pale
grey marble (a Triassic horst), whereas the town is built on
softer sediments (Pliocene) within the graben. The island
of Kranae was possibly a more dominant feature in historical times; geomorphological changes have affected the
coastline.

The historical town of Monemvasia is located on the
steep southern slopes of a rocky outcrop to the east of the
Malae Peninsula (Fig. 11.19). The outcrop consists of
Triassic-age dolomite which is notably resistant to erosion.

Fig. 11.18 a The historical site of Mystras is constructed on Kastro Hill, a resistant block of dolomite (Triassic-Jurassic); **b** The view from Mystras (with the dolomite visible in the foreground) overlooks the Eurotas Graben with Sparti visible in the background

Monemvasia is accessed by a narrow peninsula and was an important and relatively easily defended locality in historical times. The town was established in the 6[th]C AD by Laconians fleeing Slavic invasions, and remained an important trading centre throughout medieval times and beyond. The exposure to the prevailing Meltemi winds, which blow almost all summer in this part of Aegean Sea, yield a healthy climate (disease was a major problem in the Mediterranean during medieval times).

11.8.2 Antiquities in Messenia

Historical sites associated with early civilizations in Messenia can be observed at Kalamata and Korone. The recently discovered archaeological site of Thouria, near Kalamata, includes a well preserved theatre (4[th]C BC) which overlooks the fertile plain of Messenia. The coastal town of Methoni contains a Venetian castle which forms a prominent landmark at the entrance to the Gulf of Messene (or

Fig. 11.19 The historical town of Monemvasia is constructed on the steep southern slopes of a rocky outcrop consisting of resistant dolomite (Triassic)

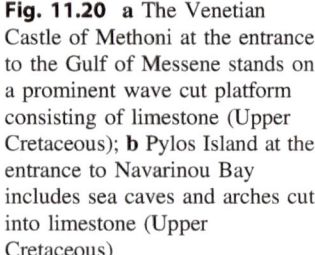

Fig. 11.20 **a** The Venetian Castle of Methoni at the entrance to the Gulf of Messene stands on a prominent wave cut platform consisting of limestone (Upper Cretaceous); **b** Pylos Island at the entrance to Navarinou Bay includes sea caves and arches cut into limestone (Upper Cretaceous)

Messiniakos) (Fig. 11.20a). The scenic road which links Kalamata and Pylos accesses Neda Gorge, known for remarkable variations in flow rate and depth (thus the female form of the name), probably arising from the underground source of much of the water in the limestone bedrock (Upper Cretaceous).

Pylos is located in the southern part of Navarinou Bay, the largest and safest natural harbour in Greece (Fig. 11.1). The harbour is protected on the outer (western) side by two islands, the elongate Sfaktira Island (Spacteria in historical references) and Pylos Island. The islands consist of resistant limestone (Upper Cretaceous). Sea caves and natural arches are prominent features (Fig. 11.20b). Navarinou Bay occupies part of a subsidiary graben (measuring 10 km by 4 km and with a depth of up to 60 m) on the western side of the Kalamata Graben. The western graben fault is aligned with

a

b

Fig. 11.21 The Mycenaean walls (13th C BC) of the Palace of Nestor are protected by a roof in an archaeological site which overlooks a broad plateau situated above Navarinou Bay

the outer coast of the two islands. The northern part of the bay includes a large lagoon. The western part of the lagoon abuts against limestone cliffs which contain two historical sites, Nestor's Cave and a historical castle ("Palaiokastro"). A crescent-shaped indentation on the western side of the lagoon, Voidokoila Bay (or Bouphrus Bay), includes a remote, scenic beach connected to the Ionian Sea. The lagoon is fringed by an extensive dune field.

Sea level changes have had a pronounced effect on the extent of Navarinou Bay and the shoreline is estimated to have been several kilometres farther inland in the Early-Middle Holocene (Kraft et al. 1980). In historical times, the geomorphology was impacted by sedimentation patterns and the shoreline in the southern and northern parts of the bay differed substantially to that currently observed. The dune field associated with the lagoon is thought to have been far more extensive and the description by Homer of "Sandy Pylos" was probably accurate (Kraft et al. 1980). Historical evidence suggests Pylos was originally located at the northern end of the bay.

Navarinou Bay is famous for two historical battles. During the Peloponnesian Wars, the Athenians attacked a Spartan army based on Sfaktira (Battle of Spacteria, 425 BC). In 1827, a large, combined British, French and Russian fleet, under the control of Admiral Codrington, defeated an even larger (but less modern) fleet assembled by the Ottoman Empire and Egypt. The battle was unusual as this was the last sea battle fought entirely between sailing ships, although most remained at anchor (the bay has only one navigable entrance). Some of the sunken ships can be observed beneath the placid waters of the bay. Memorials have been constructed in Pyrgos and on the two islands. The battle of 1827 is commemorated by a local public holiday as it subsequently led to the independence of the modern state of Greece.

The plateau to the northeast of Navarinou Bay is an erosion surface (the "Kampos") underlain by Neogene sediments. The surface was uplifted by approximately 400 m during the Pleistocene (Higgins and Higgins 1996). The Palace of Nestor is a Mycenaean archaeological site with an estimated age of 1300 BC, located on the plateau 17 km north of Pylos. The

Fig. 11.22 Shipwreck Beach is enclosed by steep cliffs of white limestone (Cretaceous), typical of the north coast of Zakynthos

ancient walls are protected by a roofed enclosure and accessed by walkways (Fig. 11.21). The view west over the rolling hills towards Navarinou Bay fits historical descriptions of this famous antiquity, as does the location which catches the cooling sea-breezes during the summer heat.

11.9   Zakynthos

The island of Zakynthos (area of 406 km^2) is one of the most popular tourist destinations in Greece and is famous for the wooded and hilly interior, sandy beaches, and rocky coastline. Zakynthos was known historically for extensive cultivation, in part due to the relatively high annual rainfall. The town of Zakynthos (or Zante) is the only large settlement on the island. More than a million visitors arrive annually, the majority attracted by the beaches of Lagana Bay in the southern part of the island. Lagana Bay includes a sanctuary for preserving loggerhead sea turtles (*Caretta caretta*). The northern part of the island includes sea arches and caves renowned for patches

of translucent turquoise water, e.g. Shipwreck Beach, Peristero Cave and the Blue Caves at Agios Nikolaos (Fig. 11.22).

11.9.1   Geological Framework

The geology of Zakynthos was first remarked upon by Strickland (1840), who identified three principal features: a mountainous spine of older limestones, younger sediments on the lower-lying coastal areas, and mineral springs (Fig. 11.23). The northern and western parts of the island are underlain by Cretaceous limestone (Avramidis et al. 2017a). These rocks form the Vrachionas Mountains (756 m) and can also be observed in many of the sea cliffs. The spine of the mountain range correlates with a regional anticline. The broad coastal plain on the eastern flanks of the anticline reveals successive outcrops of Eocene-Oligocene limestone, Miocene limestone, and poorly-consolidated Pliocene and Pleistocene sandstones, conglomerates, and marls.

Fig. 11.23 Simplified geological map of the island of Zakynthos. *Source* Avramidis et al. (2017a)

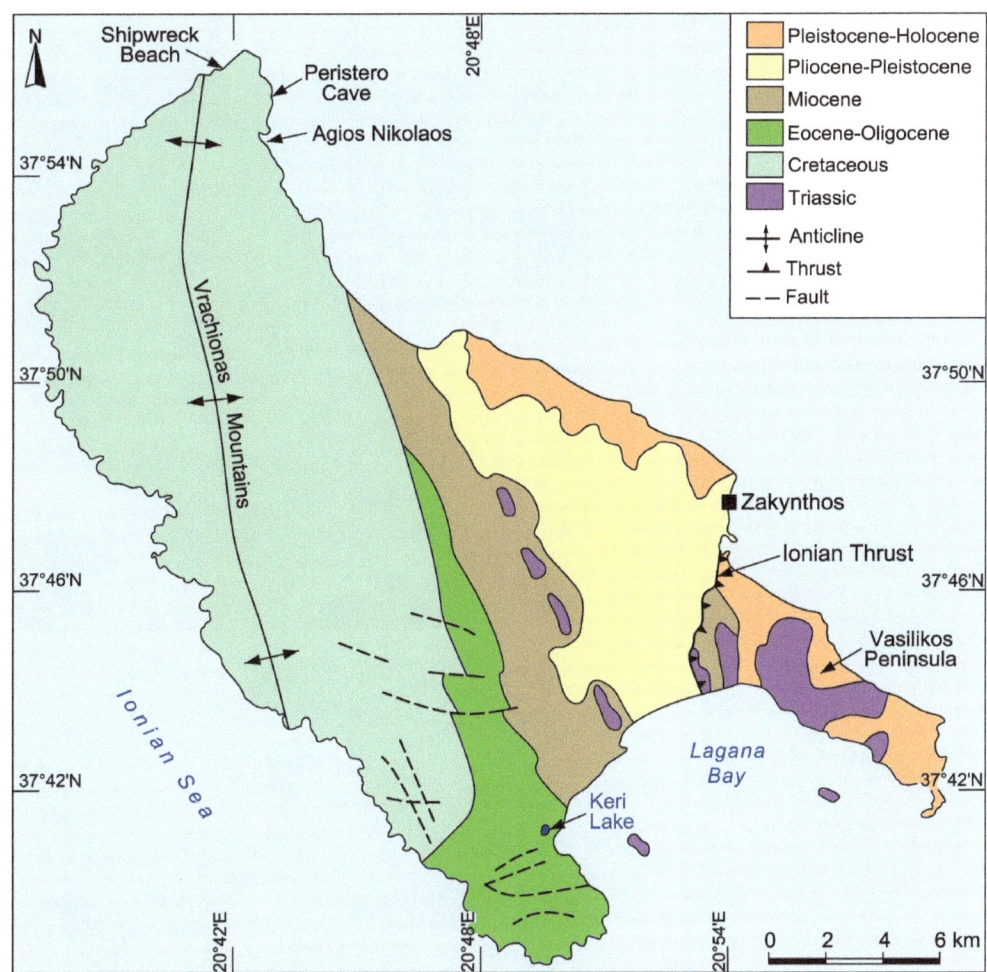

The town of Zakynthos is built around a large harbour located at the base of pale coloured cliffs comprised of Pliocene and Pleistocene sandstone and marl (Fig. 11.24). The sandstone and marl erode to form extensive areas of badlands. The cliffs are capped to the north of the town by a thin layer of relatively hard conglomerate on which a Venetian castle is located. Deposits of poorly consolidated sands and gravels (Quaternary) occur in some coastal locations.

The Ionian Thrust is an important structural feature in the southern part of Zakynthos (Fig. 11.23). The thrust extends northward towards the town of Zakynthos. Thrusting is related to compressional tectonism associated with the waning stages of the Alpine Orogeny. The thrust has an approximate age of 75,000 BP (Avramidis et al. 2017a). The Vasilikos Peninsula is part of a thrust sheet of older rocks, including Triassic limestones, which overlies the younger strata. Prior to this event, the Vasilikos Peninsula was separated from Zakynthos by a sea channel. This division has

consequences, not only for geological studies, but also for archaeological findings.

11.9.2 Earthquake of 1953

The location of Zakynthos proximal to the Hellenic Trench and the Kefalonia Fault means the island is unusually susceptible to earthquakes. The Ionian Earthquake of August 1953 included four major seismic events that caused massive destruction on the island, as well as on the islands of Ithaca and Kephalonia. The epicentre of the third event (which measured 7.3 on the Richter scale) was located on the southern tip of Kefalonia. This event destroyed the town of Zakynthos. Only three or four buildings survived. The location of the town on poorly consolidated sediments is problematic. Historic settlements located in the mountainous parts of the island were notably less impacted. Zakynthos has experienced additional events since 1953. A 5.9 event

Fig. 11.24 **a** View of the town of Zakynthos showing the large harbour with limestone hills (Triassic) of the Vasilikos Peninsula in the background. The Ionian Thrust is aligned parallel with the shoreline (foreground) and extends southward to Lagana Bay (background, right); **b** The pale grey cliffs of poorly consolidated sandstone and marl (Pliocene-Pleistocene) located above the town of Zakynthos are severely eroded and form extensive areas of badlands

occurred in 2006 and a 6.4 event in 2008. Damage to the new town has been minimal as strict building guidelines, introduced after the 1953 event, are strictly enforced.

11.9.3 Springs

Zakynthos includes a number of natural springs. Sulphurous hot springs occur at Xigia Beach in the northern part of the island. The Spring of Herodotus is an antiquity named after the Greek historian, located near the resort of Keri, Lagana Bay. Keri is situated near the high limestone cliffs of Megali Myzithra. The spring is part of Keri Lake (area of 3 km^2), a low-lying area (elevation of 1 m) separated from the sea by a sand barrier (Avramidis et al. 2017b). The lake occupies a Neogene-age depression. The primary limnic environment was impacted by recent tectonism. Accumulation of peat during the previous 4,000 years caused the ecosystem to change from marsh to fen. Prior to this, the depression was inundated by the sea. The Spring of Herodotus is famous for seepage of bitumen or "pitch" (Fig. 11.25). Bitumen was highly sought after in historical times (primarily for caulking

Fig. 11.25 The seepage of bitumen into Keri Lake, Zakynthos was described by the Greek historian Herodotus

boats). The bitumen is derived from deeper-lying petroleum deposits in the Cretaceous limestone. The fresh water in Keri Lake originates from the limestone catchment.

References

Ambrasey, N., & Synolakis, C. (2010). Tsunami catalogs for the eastern Mediterranean, revisited. *Journal of Earthquake Engineering, 14*, 309–330.

Avramidis, P., Iliopoulos, G., Konstantinos, N., Kontopoulos, N., Koutsodendris, A., & Van Wijngaarden, G. J. (2017a). Holocene sedimentology and coastal geomorphology of Zakynthos Island, Ionian Sea: A history of a divided Mediterranean island. *Palaeogeography, Palaeoclimatology, Palaeoecology, 487*, 340–354.

Avramidis, P., Kalaitzidis, S., Iliopoulos, G., Papadopoulou, P., Nikolaou, K., Papazisimou, S., Christanis, K., & Van Wijngaarden, G. J. (2017b). The so called "Herodotus Springs" at "Keri Lake" in Zakynthos Island western Greece: A palaeoenvironmental and palaeoecological approach. *Quaternary International, 439*, 37–51.

Bull, W. (2009). Tectonically active landscapes (p. 326). Wiley-Blackwell.

Collier, R. E., & Dart, C. J. (1991). Neogene to Quaternary rifting, sedimentation, and uplift in the Corinth basin, Greece. *Journal of the Geological Society, 148*, 1049–1065.

Cooper, F. A. (1981). A source of ancient marble in the southern Peloponnese. *American Journal of Archaeology, 85*, 190–191.

Ford, M., Hemelsdael, R., Mancini, M., & Palyvos, N. (2017). Rift migration and lateral propagation: evolution of normal faults and sediment-routing systems of the western Corinth rift (Greece). In C Childs, R. E. Holdsworth, C. A. L. Jackson, T. Manzocchi, J. J. Walsh & G. Yielding (Eds.), The geometry and growth of normal faults. *Geological Society London Special Publications, 439*, 131–168.

Fowler, H. N., & Stillwell, R. (1935). Corinth: Introduction, topography (p. 239). Architecture: Harvard University Press, Cambridge.

Hadler, H., Vött, A., Koster, B., Mathes-Schmidt, M., Mattern, T., Ntageretzis, K., Reicherter, K., Sakellariou, D., & Willershäuser, T. (2011). Lechaion, the Ancient Harbour of Corinth (Peloponnese, Greece) destroyed by tsunamigenic impact. In 2nd INQUA-IGCP-567 international workshop on active tectonics, earthquake geology, archaeology and engineering, Corinth, Greece.

Hasiotis, T., Papatheodorou, G., Bouckovalas, G., Corbau, C., & Ferentinos, G. (2002). Earthquake-induced coastal sediment instabilities in the western Gulf of Corinth, Greece. *Marine Geology, 186*, 319–335.

Hayward C. L. (2003). The geology of Corinth: Study of a basic resource. In C. K. Williams & N. Bookidis (Eds.), Corinth XX: The centenary, 1896–1996. American school of classical studies (pp. 15–42).

Heikell, R., & Heikell, L. (2014). Greek waters pilot (12th ed.) (p. 568). Imray, Laurie: Norie and Wilson Ltd.

Higgins, M. D., & Higgins, R. (1996). A geological companion to Greece and the Aegean (p. 240). New York: Cornell University Press.

Jackson, J. (1999). Fault death: a perspective from actively deforming areas. *Journal of Structural Geology, 21*, 1003–1010.

Jolivet, L., et al. (2013). Aegean tectonics: Strain localisation, slab tearing and trench retreat. *Tectonophysics, 597–598*, 1–33.

Koukouvelas, I., Mpresiakas, A., Sokos, E., & Doutsos, T. (1996). The tectonic setting and earthquake ground hazards of the 1993 Pyrgos earthquake, Peloponnese, Greece. *Journal of Geological Society, 153*, 39–49.

Kraft, J. C., Rapp, G., & Aschenbrenner, S. E. (1980). Late Holocene palaeo-geographical reconstructions in the area of the Bay of Navarinou: Sandy Pylos. *Journal of Archaeological Sciences, 7*, 187–210.

Lagios, E., Sakkas, V., Papadimitriou, P., Parcharidis, I., Damiata, B. N., & Chousianitis, K. (2007). Crustal deformation in the central Ionian Islands (Greece): results from DGPS and DInSAR analyses. *Tectonophysics, 444*, 119–145.

Mariolakis, L., & Stiros, S. C. (1987). Quaternary deformation of the Isthmus and Gulf of Corinth. *Geology, 15*, 225–228.

Mythical Peloponnese (2018). Road—tourist map and guide (scale 1:550,000). www.mythicalpeloponnese.

Okay, A. I., & Tüysüz, O. (1999). Tethyan sutures of northern Turkey. *Geological Society London Special Publication, 156*, 475–515.

Papadopoulos, G. A. (2003). Tsunami hazard in the eastern Mediterranean: Strong earthquakes and tsunamis in the Corinth Gulf, Central Greece. *Natural Hazard, 29*, 437–464.

Papanikolaou, D., Fountoulis, I., & Metaxas, Ch. (2007). Active faults, deformation rates and Quaternary paleogeography at Kyparissiakos Gulf (SW Greece) deduced from onshore and offshore data. *Quaternary International, 171–172*, 14–30.

Papanikolaou, D., Nomikou, P., Papanikolaou, I., Lampridou, D., Rousakis, G., & Alexandri, M. (2019). Active tectonics and seismic hazard in Skyros Basin, North Aegean Sea, Greece. *Marine Geology, 407*, 94–110.

Roberts, G., Papanikolaou, I., Vött, A., Pantosti, D. & Hadler, H. (2011). Active Tectonics and Earthquake Geology of the Perachora Peninsula and the area of the Isthmus, Corinth Gulf, Greece. Field Trip Guide. In 2nd INQUA-IGCP 567 International Workshop on Active Tectonics, Earthquake Geology, Archaeology and Engineering, Corinth (p. 70).

Robertson, A. H. F., & Mountrakis, D. (2006). Tectonic development of the Eastern Mediterranean region: An introduction. *Geological Society of London Special Publication, 260*, 1–9.

Strickland, H. E. (1840). On the geology of the island of Zante. *Transactions of the Geological Society of London, S2(5)*, 403–409.

Volcanoes of the Hellenic Volcanic Arc, Greece

Methana, Milos and Santorini

Abstract

The active and quiescent volcanoes associated with the Hellenic Volcanic Arc are located in the Methana Peninsula (Peloponnese) and islands of the Cycladic archipelago. The historical eruption of the Mavri Petra Volcano, the youngest component of the volcanism in the Methana Peninsula, was described by the Greek historian Strabo (260 BC). The volcanic islands of Milos and Santorini, which occur in the southern part of the Aegean Sea, consist of multiple, accreted volcanic centres. The volcanism on the island of Milos includes two recent eruptive centres, of which the youngest is the Fyriplaka Volcano (0.14–0.09 Ma). During Palaeolithic times, Milos was the principal source of obsidian (volcanic glass) for production of stone tools in the eastern Mediterranean. The group of islands known as Santorini is dominated by a giant volcanic edifice with a central, partially submerged caldera. The Santorini Caldera formed during the Minoan Event, one of the most explosive volcanic eruptions recorded in historical times (1600–1627 BC). This eruption may have initiated Plato's legend of the lost continent of Atlantis. The volcanic eruptions and formation of the Kameni Islands within the Santorini Caldera was documented during historical times. The Hellenic Volcanic Arc is a 500 km long curvilinear feature located approximately 200 km north of the Hellenic Trench. The Hellenic Trench, which occurs to the south of Crete, demarcates the position where the African Plate is currently being subducted beneath the Hellenic Microplate. The subducted slab dips at an angle of 10–20° N with melting occurring at a depth of 160–180 km. The majority of the volcanism is Pliocene and younger. Active centres in the southern Aegean Sea include, in addition to Milos and Santorini, several of the Dodecanese Islands. Most of the volcanic centres report evidence of highly explosive, bimodal volcanism which is characteristic of arc-related activity. The petrogenesis is, however, complicated by the recognition of shallow staging chambers. The Methana Volcanic Complex consists of clusters of andesite and dacite lava domes and lava flows that attain maximum dimensions of a few kilometres. The intensely hydrothermally-altered rhyolitic ashes and pyroclastics on the island of Milos contain deposits of industrial minerals, including bentonite, kaolin, and perlite. Milos offers an unparalleled opportunity for geotourism and mining heritage, with geotrails accessing sites including deposits of pumice and agglomerate in sea cliffs. The Santorini volcanic edifice, which is dominated by potassium-rich rhyodacite magmas, has experienced twelve major Plinian, or ultra-Plinian events, each of which involved near-instantaneous emptying of a shallow magma chamber (which subsequently reformed) and formation of a caldera. The periodicity is estimated at approximately 20,000–40,000 years. The Minoan city of Akrotiri was entirely buried by the most recent of the Plinian events which included formation of the youngest of the multiple calderas. The island of Nea Kameni, an active volcano associated with effusive activity, and the Columbo Seamount (located to the NE of Santorini) are monitored as potentially hazardous. The continuous reshaping of Santorini by catastrophic eruptions provides a vivid demonstration of both the creative and destructive forces of volcanism.

Keywords

Akrotiri • Island arc volcanism • Mavri Petra • Methana Peninsula • Milos • Nea Kameni • Santorini

Photographs not otherwise referenced are by the author.

The original version of this chapter was revised: Belated corrections have been incoporated. The corrections to this chapter are available at https://doi.org/10.1007/978-3-030-54693-9_18

12.1   Introduction

The active and quiescent volcanoes associated with the Hellenic Volcanic Arc are located in the Methana Peninsula and on islands in the southern Aegean Sea. The Methana Volcanic Complex is the dominant feature of a peninsula that projects into the Saronic Gulf from the Peloponnese, southwest Greece. The Mavri Petra Volcano, the most recent of the separate volcanic centres in the peninsula, is accessed by a hiking trail, and there is increasing awareness of the potential for geotourism in this region. The most recent eruption of Mavri Petra was witnessed in Athens and described by the Greek historian Strabo, at approximately 260 BC. Two islands of the Cycladic archipelago in the Aegean Sea contain active and quiescent volcanoes. Milos consists of multiple, accreted volcanic centres, the youngest of which is the Fyriplaka Volcano. Milos has a long history of mining and was a major source of obsidian, or volcanic glass, during Palaeolithic times. Many of the volcanic landforms on Milos can be accessed by geotrails which include the sea cliffs and beaches (Fig. 12.1). The group of islands popularly known as Santorini (or Thira) is dominated by a giant volcanic edifice.

The principal feature of Santorini is a centrally situated, partially submerged caldera. The caldera is fringed by near-vertical cliffs which attain a height of 300 m. The eruption and formation of the Kameni Islands, which are protected in a national park, within the Santorini Caldera was documented by classical scholars during historical times. The Santorini Caldera formed during the catastrophic Minoan Event, a Plinian-style eruption that destroyed the ancient city of Akrotiri. This was one of the most explosive volcanic eruptions recorded in historical times and may have initiated Plato's legend of the lost continent of Atlantis.

The Hellenic Volcanic Arc incorporates volcanic centres on the mainland of Greece, several islands in the Saronic Gulf, two islands of the Cycladic Archipelago, and some of the Dodecanese Islands near the coast of Turkey (Fig. 12.2). The volcanoes are aligned on a curvilinear feature which has a length of 500 km. The majority of the volcanism is Pliocene and younger. The active and quiescent volcanoes described here are restricted to the Methana Peninsula and the islands of Milos and Santorini. The Cyclades is an archipelago of approximately thirty islands restricted to the southern part of the Aegean Sea. The islands appeared to the

Fig. 12.1 Pedestals on a wave-cut platform at Cape Pelekouda, Milos, consist of boulders of resistant andesite lava (dark grey-green) deposited by a volcanic debris avalanche on a base of soft pyroclastics (pale grey-white)

Fig. 12.2 Satellite image showing the extent of the Hellenic Volcanic Arc. Active or quiescent volcanoes are located in the Methana Peninsula, the islands of Milos and Santorini in the Cycladic archipelago, and the island of Nisryos in the Dodecanese. *Source* Satellite Image of Europe based on NASA MODIS data, processed by Philip Eales of Planetary Visions/DLR

ancient Greeks as a wheel, or cycle, with the sacred island of Delos located at the centre. The majority of the Cycladic archipelago is comprised of metamorphic rocks; Milos and Santorini are the only islands dominated by volcanic centres. The extinct Sousaki Volcano (Isthmus of Corinth) and the extinct volcanic centres on the islands of Aegina and Poros (Saronic Gulf) are not described here. The active or quiescent volcanoes of the Dodecanese Islands are also not included, although for completeness it is noted that the

island of Nisyros is dominated by a large volcanic cone which includes an active summit crater.

12.2   Regional Geology

The Pliocene and Quaternary volcanoes associated with the Hellenic Volcanic Arc generally overlie metamorphic terrains which represent the southeast continuation of the

Alpine zones of mainland Greece (Robertson and Moun-trakis 2006). The tectonic zones were metamorphosed and folded during the Alpine Orogeny. The Alpine Orogeny probably peaked in this region in the Oligocene-Miocene (34–5.3 Ma). The Methana Peninsula is part of the Parnas-sus Zone and the Cyclades are associated with either the Pelagonian Zone or the Atticacycladic Zone (Chap. 10). The islands in the central and southern part of the Cycladic archipelago, including Ios, Naxos and Paros, contain Neo-proterozoic rocks as well as the Palaeozoic-Mesozoic rocks that characterize the Pelagonian Zone on the mainland of Greece. Islands in the northern Cyclades, e.g., Mykonos and Delos consist almost entirely of Mesozoic limestone. Crete has a broadly similar geological framework to the islands of the central and southern Cyclades, although a number of additional Alpine zones are recognized. This has important implications for the tectonic setting. The volcanic deposits associated with the Hellenic Volcanic Arc are in part inter-calated with sedimentary rocks which formed after the peak of the Alpine Orogeny.

12.2.1 Aegean Basin

The continental collision in the Mediterranean constitutes an accretionary convergent margin (Box 8.1). The Aegean Sea is associated with a back-arc basin that developed between the colliding plates, i.e., north of the active Hellenic Trench (Fig. 12.3). The Hellenic Trench is active, and is migrating southward (Flower and Dilek 2003; Jolivet et al. 2013). The Mediterranean Ridge is a relatively large accretionary prism located south of Crete, part of a horst situated within the

fore-arc basin. The continental crust associated with the Aegean Basin has been drastically thinned (this is a char-acteristic feature of accretionary margins), and has a thick-ness of approximately 20–30 km. This may be compared with a thickness of 40–50 km for the crust underlying mainland Greece and Turkey.

The tectonics of the region has been controlled since the Late Cretaceous by a combination of the convergence of the continental plates and subduction of the oceanic lithosphere to form relatively small, linear basins (Jolivet et al. 2013). The oceanic lithosphere has, however, been almost com-pletely subducted. The continuation, or rejuvenation, of the orogeny (which includes closure of the basins) has resulted in a number of different orogenic belts developing. The Aegean Basin is part of the Hellenic Microplate which is being strained by its position relative to the African Plate (south), the Eurasian Plate (north), and the Anatolian Microplate (east). The Aegean Basin is actively extending southward and northward at estimated rates of 50 mm/year (Jolivet et al. 2013; Papanikolaou et al. 2019).

12.2.2 Subduction Zone

The extensional tectonics of the Aegean Basin (which appears to be at odds with the general nature of the conver-gent margin) is in part driven by roll back of the subducted slab, i.e. the ancient oceanic crust which is being compressed between the African and Eurasian Plates (Jolivet et al. 2013). Rollback may have started as long ago as 30 Ma. Despite the southward roll back of the trench, however, the subducted slab is moving northward at an estimated rate of 35 mm/year.

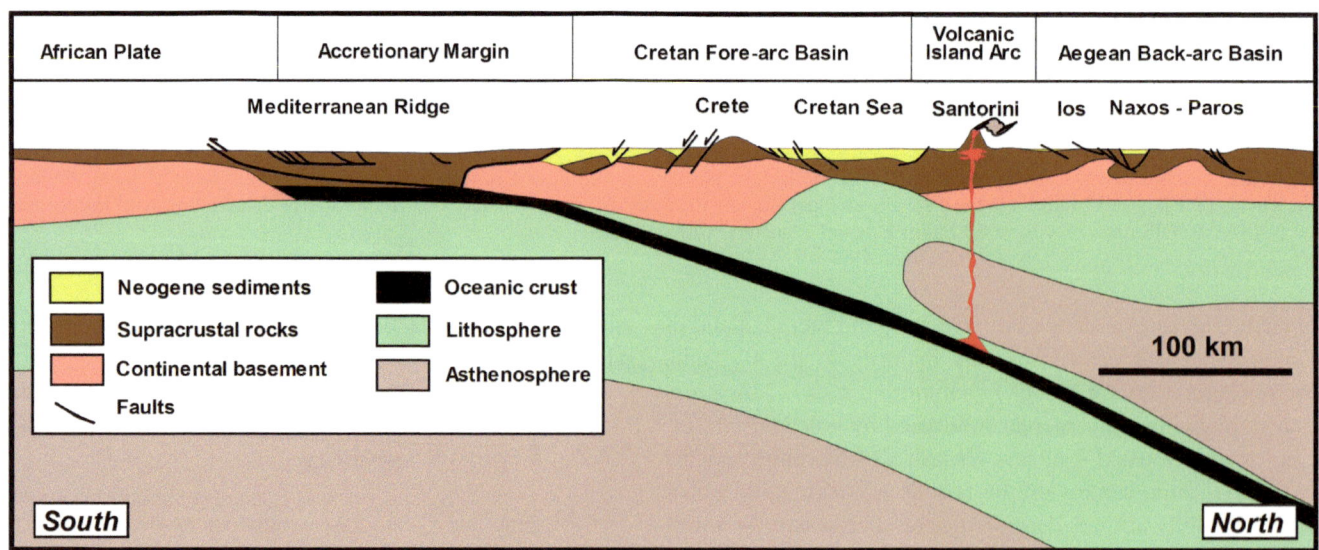

Fig. 12.3 Cross-section centred on Santorini showing the subducted slab of oceanic crust and thinned nature of the continental crust (dominated by supracrustal rocks) beneath the Hellenic Volcanic Arc. Plumbing system of the volcanoes associated with the Hellenic Volcanic Arc (e.g., Santorini) may include shallow crustal chambers. *Source* Simplified after various sources including Meier et al. (2007)

The majority of the oceanic lithosphere was destroyed by the longevity of the collision, as noted above, and an unusual feature of the current collision is that thick sequences of evaporites, rather than basaltic ocean crust, are being subducted (Flower and Dilek 2003). The evaporites developed as part of the Messinian salinity crisis (Cita 2006). The unusually shallow taper of the Mediterranean Ridge may be related to the dominance of evaporites in this area.

The current position of the subduction zone to the south of Crete is demarcated by the Hellenic Trench (Chap. 8). In the Ionian Sea, the collision is represented by a more-or-less continuous trench, but to the east of Crete the collision is associated with disjointed structures that include the Pliny Trench and the Strabo Trench (Le Pichon and Angelier 1979; Flower and Dilek 2003; Jolivet et al. 2013). These features are depicted in Fig. 8.3. The Hellenic Trench is related to thrusting (the oceanic slab is being subducted beneath the continental crust), but the Pliny and Strabo Trenches are dominantly strike-slip in nature. Subduction of the oceanic slab in the Aegean commenced at approximately 20-15 Ma; the active margin has since retreated by approximately 35–500 km south-westward (Thomson et al. 1998).

12.2.3 Island Arc Volcanism

The active volcanoes of the Hellenic Volcanic Arc (which has a width typically less than 20 km) are located approximately 200 km from the Hellenic Trench (Fig. 8.3). The distance of the volcanoes from the subduction zone is dictated by the angle of the subducted slab, as described by Druitt et al. (1999). The subducted slab dips at approximately 10–20° north and partial melting occurs at a depth of 160–180 km (Fig. 12.3). The location of the volcanic centres is controlled by lines of weakness (mostly aligned NE-SW) in the tectonic structure of the Aegean Basin (Le Pichon and Angelier 1979; Fytikas et al. 1976). Volcanism in central and northern Greece may have started in the Oligocene (these volcanic centres are extinct) and migrated southward, reaching the Aegean region in the Pliocene (at approximately 3–4 Ma). Volcanism has persisted into the Quaternary and there are several active centres.

The Oligocene and Early Miocene volcanism in central and northern Greece was dominated by calc-alkaline magmas, although by the Mid Miocene there was a gradational change to alkali basaltic magmas (Fytikas et al. 1984). The

Pliocene and Quaternary volcanoes of the Aegean region are dominated by bimodal volcanism, i.e., alternating eruption of dacitic and andesitic magmas. This style of bimodal volcanism is characteristic of arc-related activity. The recent volcanism of Milos and Santorini includes potassium-rich alkali magmas, including rhyodacite. The volcanism is associated with magma derived from partial melting of the subducted slab and of the thinned continental crust underlying the Aegean Basin. Most magma associated with the Santorini volcanic edifice, and possibly the centres on Milos, appears to have been erupted from shallow crustal chambers (Druitt et al. 1999).

12.3 Methana Peninsula

The Methana Peninsula projects northward from the Peloponnese into the Saronic Gulf. The connection to the Peloponnese constitutes a 300 m-wide gooseneck. The peninsula is dominated by the Methana Volcanic Complex which outcrops in rugged hills and sea cliffs. The region includes an unusual botanical diversity, including endemic species, in part a reflection of the volcanic soils. The principal town, Methana, is located in the southeast of the peninsula. The town is constructed around a spa driven by geothermal water. A strong odour permeates the seafront and hot waters discolour parts of the harbour (Fig. 12.4). A tourist centre with open-air pools has been developed on the largest of the springs. Geothermal springs also occur in the northern part of the peninsula (D'Alessandro et al. 2008). The mountains, beaches and isolated villages of Methana attract tourists and there is increasing awareness of the potential for geotourism.

12.3.1 Geological Framework

The Methana Volcanic Complex unconformably overlies metamorphic rocks of the Parnassus Zone. The latter is dominated by Mesozoic limestone with subordinate conglomerate. These rocks outcrop in the narrow gooseneck and in the northwest part of the peninsula; the contact with the volcanic complex can be observed near Palaeo Kastro (Fig. 12.5). The Methana Volcanic Complex encompasses a subordinate group of Pliocene volcanics, but is primarily a Pleistocene feature with an estimated age of 1–2 Ma. Volcaniclastics and alluvium occur in some of the sea cliffs.

Fig. 12.4 The town of Methana nestles on the lower slopes of rugged volcanic hills, typical of the Methana Peninsula. Discoloration in the harbour is caused by geothermal water discharged from hot springs

12.3.2   Methana Volcanic Complex

The Methana Volcanic Complex is characterized by explosive bimodal activity (Pe 1974; Fytikas et al. 1988; Gaitanakis and Dietrich 1995; D'Alessandro et al. 2008; Pe-Piper and Piper 2013). The dominant feature is a cluster of andesite and dacite lava domes and lava flows. Individual features reveal maximum dimensions of a few kilometres. Thirty or more discrete volcanic centres have been identified. The domes and lava flows of andesite generally form relatively smooth landforms, including the highest point on the peninsula, Mount Chelona (742 m). The siliceous dacite is associated with more rugged topography, including hills near the village of Paleo Loutra (Fig. 12.6a). The lava domes tend to form the highest hills. The domes and lava flows are enveloped by the Mantling Ash, a thick deposit of volcanic ash, pyroclastic flows, and volcaniclastics (Fig. 12.6b). The Mantling Ash is associated with lower ridges which may be thickly vegetated.

The volcanic rocks are used for building stones, and can be observed in harbour walls and dry stone walls in the village of Agios Nikolaos (Fig. 12.7). The dark green-grey andesite and brick red dacite are readily identified. Both lithologies contain phenocrysts of white or pale grey plagioclase. Andesite agglomerate with small volcanic bombs can also be observed. The volcanic rocks are also found in archaeological sites, such as the acropolis of Palaeo Kastro which is constructed from blocks of andesitic lava.

The Methana volcanism is associated with magmas which exploited fault systems related to extension of the Aegean Basin (Pe 1974). During the Pliocene, magma was emplaced along N-S striking listric faults associated with an early

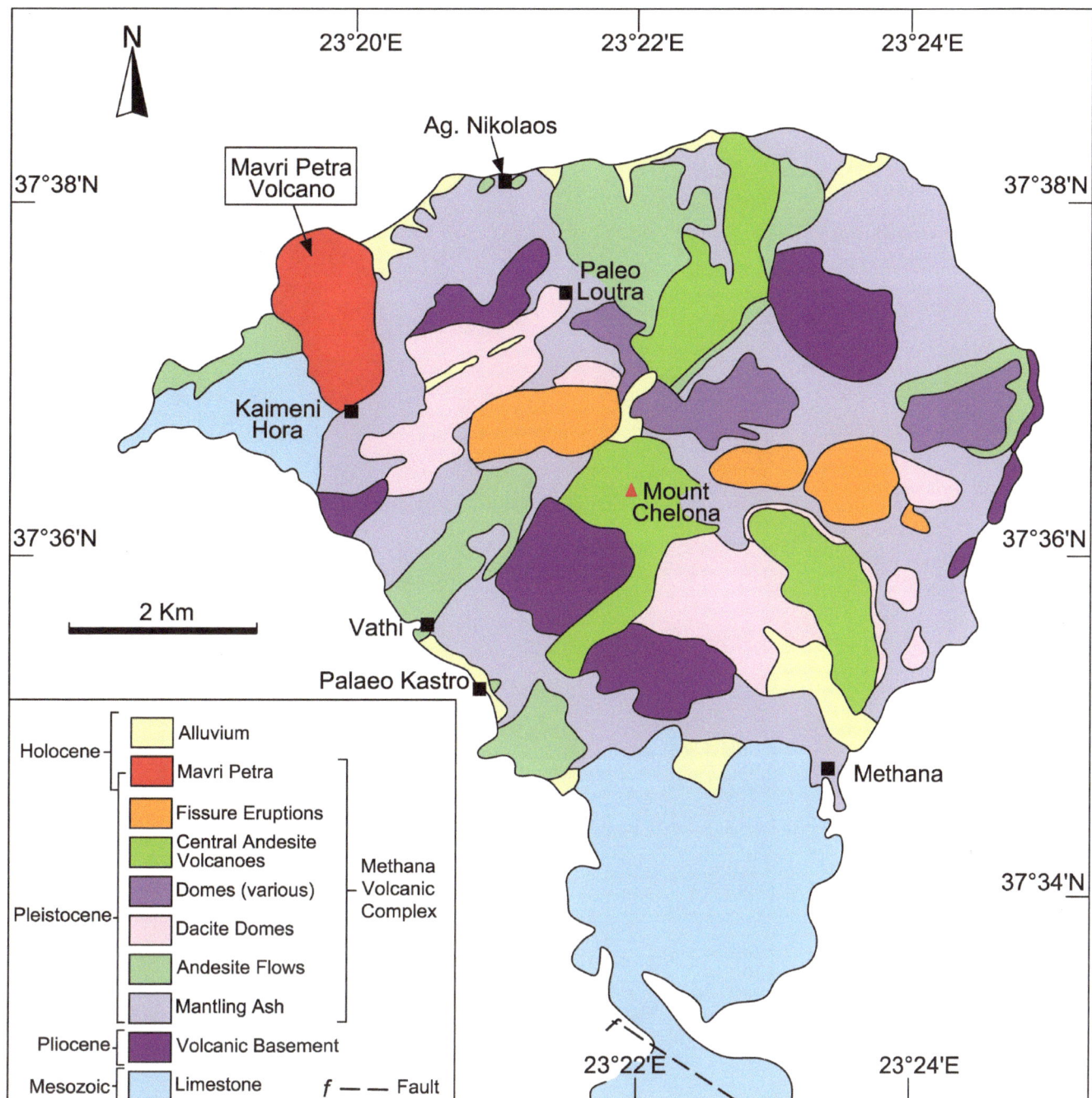

Fig. 12.5 Geological map of the Methana Peninsula. *Source* Simplified from Gaitanakis and Dietrich (1995), Pe-Piper and Piper (2013)

phase of W-E extension (Pe-Piper and Piper 2013). In the Early Pleistocene, NE-SW crustal-scale strike-slip faulting, triggered by resurgence of the continental collision, initiated the main phase of volcanism. The ascent of magma in the Late Pleistocene and Holocene was controlled by W-E striking faults.

12.3.3 Mavri Petra Volcano

The Mavri Petra Volcano is situated near Kaimeni Hora ("Burnt Village"), in the northwest part of the peninsula (Fig. 12.5). The volcano consists of an andesite dome that was erupted at approximately 0.2 Ma together with a

Fig. 12.6 a The village of Agios
Nikolaos is located at the base of
thickly vegetated hills associated
with the Mantling Ash
(foreground) with lava domes
forming the higher hills
(background) in the northern part
of the Methana Peninsula;
b Sub-horizontal beds of red
dacitic ashes and pyroclastics
(Mantling Ash) are exposed near
Palaeo Kastro, Methana Peninsula

1 km-long flow andesite lava flow associated with a historical eruption (Pe 1974). The eruption was triggered by collapse of the lava dome to form a crater which subsequently
overflowed flow (Pe-Piper and Piper 2013; Hurni et al.
1993). The lava flow has a width of 2 km and thickness of
up to 150 m and constitutes a dark, unvegetated swathe on

the slopes above the village (Fig. 12.8). The eruption was
described by the Greek historian Strabo. The date of the
report has been estimated to be between 276 BC and 239 BC
(Strothers and Rampino 1983), but evidence based on ice
cores from Greenland has refined the age to 260 BC. The ice
cores contain anomalous amounts of sulphuric acid. The lava

Fig. 12.7 **a** Boulders of andesite (dark grey-green) and dacite (red) in the seawall at Agios Nikolaos; **b** Dry stone walls are constructed of andesite and dacite lava; **c** Blocks of andesitic agglomerate include plagioclase laths (white) and small volcanic bombs (dark)

dome is accessed by the "Volcano Trail". The dome consists of lava and agglomerate in the lower parts, with large blocks of cinder and agglomerate near the summit. The latter are sufficiently large and loosely assembled as to enclose caves.

12.3.4 Pausanias Submarine Volcanism

The Pausanias Submarine Volcanic Field is located west of the Methana Peninsula in an anomalously deep section of the Saronic Gulf. High-resolution imagery from an underwater survey has revealed the unusual nature of the bathymorphological setting (Foutrakis and Anastasakis 2018). The seabed is extremely rugged with cone-shapes which may correlate with lava domes. The rapid increase in depth close to the shoreline near Vathi, a scenic village with a small harbour, can be detected by the change in the water colour. There may have been a small, offshore eruption in this region at approximately 1700 AD, but this remains unconfirmed.

12.4 Milos

The island of Milos in the southern Aegean has an unusual, amoeboidal shape that is the result of accretion of multiple volcanic centres (Fig. 12.9). The central parts of the island include volcanic plateaus and rugged hills associated with the more resistant of the volcanic rocks. The rather barren landscape is in part related to the semi-arid climate and in part due to the relative youthfulness of the volcanism. The most recent of the volcanic centres, the Trachilas Volcano and the Fyriplaka Volcano, are almost unvegetated. The principal town of Adamas is constructed around a harbour located in a sheltered bay, but many of the older centres, including the second town of Plaka, are situated on the volcanic plateaus (Fig. 12.10). The location of the older settlements may have been influenced by the potential impact of tsunamis (e.g., as experienced from the eruption of Santorini during the Minoan period). The scenic beaches on Milos support a vibrant tourist industry and the island has considerable potential for both geotourism and for those interested in historical mining. The ancient obsidian workings at Nychia may rank as the longest period a single mine site has ever operated.

The volcanism on Milos is interpreted to be quiescent, rather than extinct, and there are several active geothermal systems. The unusual nature of the indented coastline suggests that the island is gradually subsiding. This interpretation is supported by historical and archaeological evidence. Estimates suggest as much as 7 m of subsidence occurred during the past 3,000 years. Subsidence may be related to the cessation of the main peak of the volcanism followed by cooling and contraction of the crust (Higgins and Higgins 1996).

Fig. 12.8 The historical lava flow of the Mavri Petra Volcano forms a dark swathe on the hill slopes above the village of Kaimeni Hora, Methana Peninsula

12.4.1   Geological Framework

The Pliocene and Quaternary volcanic centres on Milos overlie older metamorphic and sedimentary rocks in two localities (Fig. 12.11). Green schist and marble (Late Proterozoic) occur east of the Fyriplaka Volcano. Conglomerate with subordinate limestone (Upper Miocene-Lower Pliocene) occur west of the Fyriplaka Volcano. Two grabens are recognized, the large Adamas Graben (aligned NW-SE) which is in part occupied by Adamas Bay, an indentation of the Aegean Sea, and the small Zephyria Graben (aligned N-S) which is partially infilled by recent alluvium and gravel. Both grabens include geothermal activity (fumaroles and hot springs). Most settlements on the islands are located on the stable horst blocks, but an exception is the village of Zephyria. Zephyria has been subjected to a number of earthquakes during part of its history (1350 AD–1750 AD). The relatively small earthquake of 1992 may have been associated with the Adamas Graben. Most seismic activity reported on Milos can, however, be related to the Columbo Seamount, an active submarine volcano located near Santorini (Sakellariou et al. 2010).

12.4.2   Volcanism

The volcanic history of Milos has been described in detail by Fytikas et al. (1986a, b; 1990). The older of the volcanic centres outcrop in the southwest of the island (Fig. 12.11).

Fig. 12.9 The unusual shape of Milos is related to accretion of multiple volcanic centres. The two youngest volcanoes, Fyriplaka and Trachilas, reveal relatively barren landforms. The irregular patches of white dotted over large parts of the island are quarries associated with production of industrial minerals. *Source* Google Earth

The Basal Pyroclastics (Pliocene) consist of thick beds of submarine pyroclastics capped by andesitic pillow lavas (age of 3.5–3.1 Ma). The highest peak on the island (Profitis Ilias: 748 m) is situated here. The subaerial volcanics (Upper Pleistocene) in the northwest of the island consist of andesite and dacite lava domes and lava flows (age of 2.5 Ma), similar to the Methana Volcanic Complex. During the Middle Pleistocene, the volcanic activity switched to the eastern side of the island. The eruption of thick deposits of pyroclastics (age of 1.7–1.0 Ma) was followed by the formation of rhyolite domes, submarine andesitic flows, and additional pyroclastic activity. Phreatic activity resulted in clusters of explosive craters. The Green Lahar is an extensive debris avalanche deposit (or DAD). The Trachilas Volcano (0.38 Ma) and the Fyriplaka Volcano (0.14–0.09 Ma) consist of rhyolitic ashes and pyroclastics that have experienced intense hydrothermal alteration.

12.4.3 Geotrails

Geotrails have been established on Milos that access geosites and historical mining locations. Brochures with large scale maps and detailed descriptions can be obtained from the Milos Mining Museum at Adamas. The museum includes information on the geology, the mining history and the archaeology of the island, together with an extensive collection of locally-derived minerals and obsidian flakes.

12.4.4 Historical Mining

The earliest settlements on Milos were established at approximately 5,000 BC to exploit the obsidian deposits at two localities, Fyriplaka ("red stone") and Nychia ("nails"). Melian obsidian has, however, been found in archaeological

Fig. 12.10 View looking east over the Bay of Adamas (an active graben) towards the town of Plaka on the island of Milos. Plaka is situated above the potential impact of tsunamis on a volcanic plateau. The cone-shaped hills on the plateau are lava domes. The harbour town of Adamas (centre right) and the north-east coastline of the island (background) are also visible

sites on mainland Greece dated at 7,000 BC and the quarrying of obsidian on Milos may pre-date the settlement by several thousand years. The obsidian is used for manufacture of stone tools and, as the flakes can give a sharper edge than bronze, continued to be an important commodity during the Early Bronze Age. The obsidian workings at Nychia are located on the western edge of the Plaka Plateau (Miloterranean Geo Walks 6–7 2012). The pits, trenches and spoil heaps cover an area of 90 hectares (Fig. 12.12a). Spoil heaps are several metres thick, consistent with the presence of tens of thousands of tonnes of obsidian waste which accumulated over thousands of years. The obsidian is associated with flow-banded rhyolite (Lower Pleistocene). The banding is an indication of the lower viscosity of the final melt (glass) as compared with the primary rhyolite (magma). The obsidian is readily identified by the black, glassy appearance and presence of conchoidal fractures (Fig. 12.12b). Spoil heaps include nodules of obsidian up to 15 cm in length. Smaller flakes and shards located on the spoil heaps are relicts of partially-shaped tools. The archaeologists have reported that

only the initial stage of beneficiation took place on Milos, as the sharpened tools were easily damaged during transportation.

Deposits of potassium-rich alunite (also known as alum stone) were quarried on Milos in historical times. Alunite is generally found associated with potassium-rich volcanic rocks. The potassium-rich variety of alum was prized for medical usage in historical times. Milos was a major source of sulphur in the 19–20[th]C AD. During the ancient Greek and Roman times, Milos may have been the largest producer of sulphur in the Mediterranean. The Paliorema sulphur mine occurs in a deep ravine on the south coast (Fig. 12.13a), accessed by a geotrail that requires driving on rough tracks east of Zephyria (Miloterranean Geo Walks 3 2012). The mine last operated between 1928 and 1958, during which time 125,000 tonnes of sulphur was exported. The yellow crystals of sulphur occur in subvertical veins and fractures located in the hydrothermally-altered rhyolitic lavas and ashes (Lower-Middle Pleistocene) (Fig. 12.13b).

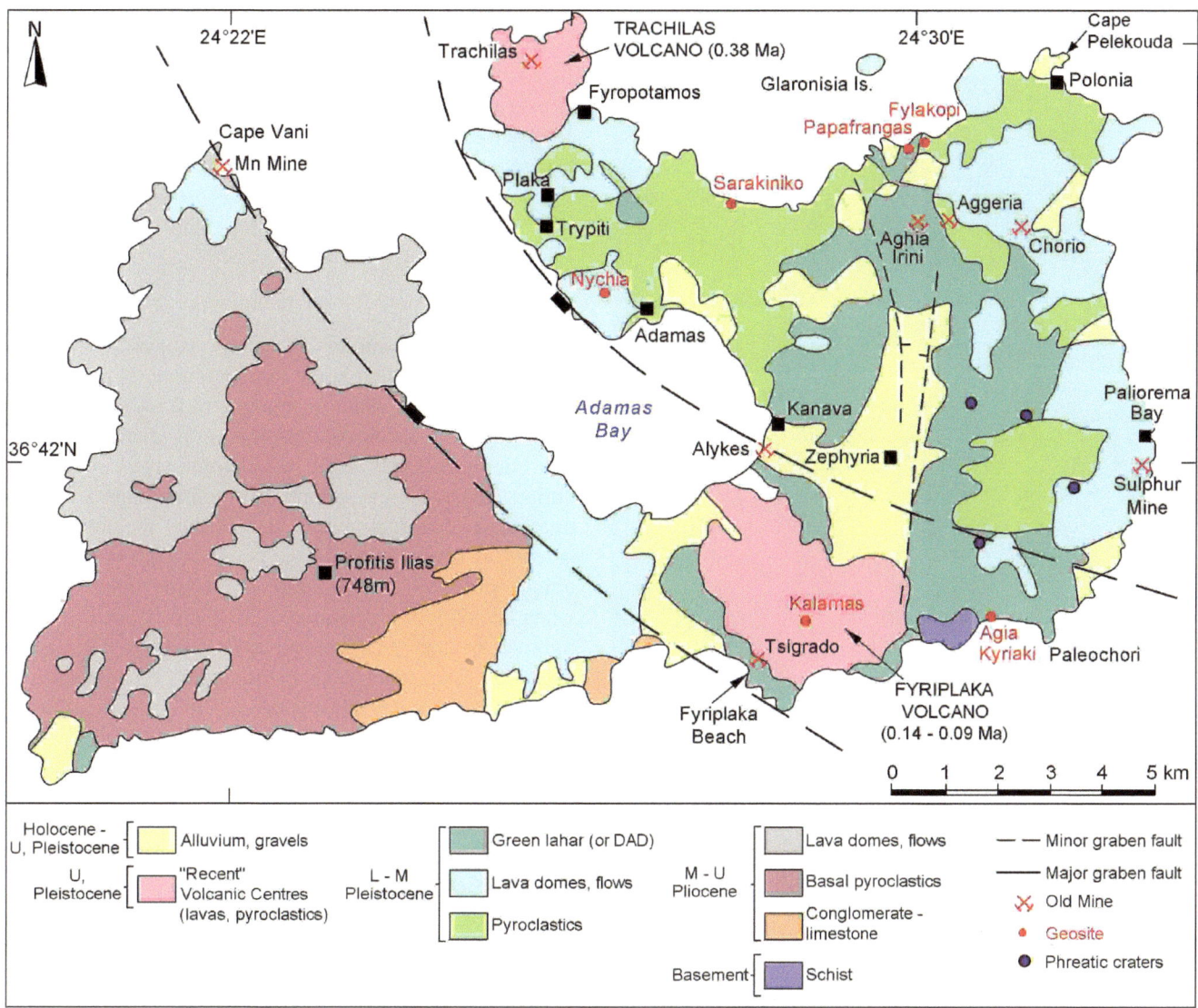

Fig. 12.11 Geological map of Milos including geosites and mines referred to in the text. *Source* Simplified from Fytikas et al. (1986b)

The Kanakas salt workings to the south of Adamas operated from medieval times until 1981 (Miloterranean Geo Walks 2 2012). The salt was collected from seawater which circulated in shallow basins. The enhanced geothermal heat flow in the Adamas Graben may have increased the precipitation rates of the salt. The defunct Cape Vani manganese mine in the northwest of Milos exploited a deposit located in hydrothermally-altered tuffs (Miloterranean Geo Walks 1 2012). The orange and red rocks exposed at the mine site contain manganese minerals such as pyrolusite and manganite. "Melian Earth" (probably kaolin) was produced

from the island in historical times. Iron ore and barytes were exploited in recent times, although these workings are all currently closed.

12.4.5 Archaeological Sites

The Bronze-age settlements at Fylakopi in the northeast of Milos cover the period between approximately 3000 BC and 1100 BC (Miloterranean Geo Walks 4 2012). The site has been partially destroyed by earthquakes. The youngest of the

Fig. 12.12 a Extensive spoil heaps at Nychia, Milos illustrate the remarkable scale of obsidian mining during Palaeolithic times; **b** Obsidian occurs in banded rhyolite (Lower Pleistocene); rubble and flakes also visible

three settlements, Fylakopi III, was influenced by the Late Bronze Age Minoan civilization of Santorini and Crete. The building material is locally-derived, dark grey-green, andesite lava, sufficiently fissile as to form tabular stones. The settlement of Plaka was established during the Minoan times. The Aghia Kyriaki archaeological site on the south coast of Milos dates from the Late Hellenistic and Roman periods (200 BC–200 AD). The sea cliffs here consist of highly-altered pyroclastics. Alunite and kaolin were exploited here during historical times. The Aghia Kyriaki site may have been partially destroyed by tsunamis. A thin deposit of gravel caps some of the cliffs and ancient texts refer to events in which the seabed was abruptly exposed. The Milos

catacombs at Trypiti ("full of holes"), near Plaka, are a subterranean complex of Christian burial chambers (12[th]C AD) with a total length of over 200 m. The chambers occur in pale grey tuffs (Pleistocene) which are friable and poorly compacted. Trypiti includes an ancient theatre (3[rd]C BC) carved out of tuff (although lined with imported marble).

12.4.6 Current Mining

Mining of industrial minerals, specifically perlite, kaolin (china clay), and bentonite, is an important contributor to the local economy. Milos is the largest source of perlite in the world. Perlite is a white or transparent form of natural volcanic glass, similar in composition to obsidian, but found in rhyolitic pyroclastics rather than rhyolite lavas. Perlite forms during subaqueous eruptions of rhyolitic magma. The distinctive saccharoidal texture is described in Greek as "zaharopetra" (sugar rock). In the granular form perlite has the unique property of expanding in volume up to 20 times when heated to temperatures of 800–950 °C, thus its use in lightweight aggregates and insulation boards. Deposits are extracted from quarrying Pleistocene-age rhyolitic pyroclastics, at Trachilas and Fyriplaka.

Large deposits of kaolin are quarried at the Tsigrado mine. The kaolin is associated with rhyolitic ashes and pyroclastics (Fyriplaka Volcano) that have been hydrothermally-altered. The formation of kaolin is related to vapour, enriched in hydrogen sulphide, fractionated from the cooling rhyolite magma. The white clay is the principal resource of kaolin and the orange and red clays, which are associated with sulphur-rich and iron-rich solutions, respectively, cause some loss of resources.

Deposits of bentonite clay are derived from hydrothermal alteration of feldspar and volcanic glass in deposits of pyroclastics that formed in shallow marine environments. The bentonite can only be mined during the dry summer months on Milos (due to slipperiness of the clay when wet) and no explosives are used in the quarries. Spoil heaps of some of the worked out bentonite mines have recently been reclaimed. Large deposits of bentonite occur in the northeast of the island (Miloterranean Geo Walks 4 2012). The Aggeria mine is exploiting a deposit with surface dimensions of 2.5 km by 0.7 km. The bentonite occurs in altered

Fig. 12.13 a The sulphur mines at Paliorema Bay occur in a narrow ravine on the south coast of Milos. Multi-coloured cliffs reveal hydrothermal alteration of lavas and ashes (Lower-Middle Pleistocene); **b** Veins and fractures infilled by yellow crystals of sulphur

pyroclastics intercalated with thin beds of limestone (supportive of the submarine origin of the pyroclastics). The quarries require large areas as the clay is spread out for drying prior to transportation. The abundance of both kaolin and bentonite on Milos may be explained by the relatively long duration of hydrothermal activity, i.e., several hundred thousand years.

12.4.7   Geosites

The transition from submarine pyroclastics to subaerial volcanic ashes can be observed in the Lower-Middle Pleistocene volcanics between Adamas and Kanakas (Miloterranean Geo Walks 1 2012). The outcrop reveals marine organisms and fossilized burrows, together with rounded pumice fragments.

This geotrail continues to the sparsely populated northwest part of Milos where rugged landforms associated with the Pliocene submarine tuffs and andesitic lavas occur. At Cape Vani, the bimodal nature of the Lower-Middle Pleistocene volcanics is revealed by the presence of dark grey-green andesite lavas and white dacitic domes.

The geotrail to the Fyriplaka Volcano includes views of the scarp face associated with the Zephyria Graben (Miloterranean Geo Walks 2 2012). The primary feature of

the volcano is the Kalamos crater, which has a diameter of 1.7 km and depth of 200 m. The crater is rimmed by gently dipping sequences of pyroclastics that include white rhyolitic ash with tiny fragments of perlite. Hydrothermal vents on the crater rim are associated with multi-coloured volcanic ashes and agglomerates (Fig. 12.14a). The fumaroles here emit vapour at temperatures as high as 1,000°C. The smell of hydrogen sulphide gas is pronounced. Sulphur crystals can be observed in veinlets and small pockets. The rugged cliffs

Fig. 12.14 **a** Active fumarole at Kalamos, Fyriplaka Volcano; **b** Fyriplaka Beach is rimmed by multi-coloured cliffs of hydrothermally altered rhyolitic pyroclastics (Upper Pleistocene)

Fig. 12.15 **a** Cape Pelekouda is comprised of lavas and pyroclastics capped by the Green Lahar (a debris avalanche deposit with large boulders) (Lower-Middle Pleistocene); **b** The cleft and caves at Papafrangas are comprised of agglomerates with clusters of welded fragments, or volcanic bombs (Lower Pleistocene)

Fig. 12.16 **a** Dark grey-green andesite lavas (Lower-Middle Pleistocene) forms the cliffs near the Fylakopi archaeological site. The andesite is overlain by a thin debris avalanche deposit; **b** Columnar jointed andesite lavas, Glaronisia Islet

at Fyriplaka Beach reveal a sequence of multi-coloured rocks, i.e., the hydrothermally-altered rhyolitic pyroclastics (Fig. 12.14b). The cliffs are capped by a DAD which contains large angular boulders, including andesitic lava. The DAD is prone to landslips and sections of the cliffs are unstable.

The Aggeria Geotrail starts at Polonia, near Cape Pelekouda, northeast Milos, where a thick sequence of Pleistocene-age bedded pumice tuffs and tephra can be observed (Miloterranean Geo Walks 4 2012). The tuffs at the base of the cliffs erode to form undercuts where small boats are stored. The cliffs are capped by the Green Lahar (Fig. 12.15a). Small pedestals on the wave cut platform contain resistant boulders of andesite (Fig. 12.1). The Papafrangas locality is a narrow cleft and cave carved out of pale grey-white agglomerates near the Fylakopi archaeological site (Fig. 12.15b). The agglomerate contains large welded fragments, or "bombs". To the east of Fylakopi, the pumice or agglomerate gives way to thick beds of grey-green andesite lava (Fig. 12.16a). Columnar jointing in the andesite is a prominent feature of the Glaronisia Islet (Fig. 12.16b).

The Trachilas Volcano forms a prominent headland north of Plaka (Miloterranean Geo Walks 6–7 2012). The crater contains extensive deposits of perlite. To the east of the Upper Pleistocene volcano, the coastal cliffs give way to interbedded deposits of Lower-Middle Pleistocene lavas, ashes, and pyroclastics. Distinctive beds of white, subaqueous pumice are a prominent feature in the vicinity of Fyropotamos (Fig. 12.17a). The pumice beds are intercalated with layers of white diatomite (aqueous algae). The diatomite may reveal a reddish-brown colour due to leaching of iron and manganese oxides (Fig. 12.17b). The white rock pavements, caves, sea arches, and crystal clear waters at Sarakiniko attract large numbers of tourists. Deposits of Lower-Middle Pleistocene pumice reveal graded beds, indicative of multiple, subaqueous eruptions (Fig. 12.18). To the east of Sarakiniko, the pumice is interbedded with a diatomite deposit (as much as 100 m in thickness), which in turn is overlain by a DAD with large irregular boulders of volcanic and metamorphic rocks.

Fig. 12.17 **a** The cliffs at Fyropotamos reveal a section of lavas, ashes, and pyroclastics (Lower-Middle Pleistocene) in which the distinctive white rocks are subaqueous pumice beds; **b** Pumice beds are intercalated with layers of altered reddish-brown diatomite, road cutting near Fyropotamos

Fig. 12.18 **a** At Sarakiniko, the white pumice deposits (Lower-Middle Pleistocene) form a barren, unvegetated coastline; **b** Graded beds in the pumice are indicative of multiple, subaqueous eruptions

12.5 Santorini

The dominant geological feature of Santorini is a large volcanic edifice with a centrally-located, partially submerged caldera. The Santorini Caldera has a depth of 370 m and is partially enclosed by the large, crescent-shaped island of Thera and the subsidiary islands of Therasia and Aspronisi (Fig. 12.19). The two uninhabited Kameni Islands occur within the centre of the caldera. Santorini is one of the most popular tourist destinations in Greece, although access to the main island of Thera is inhibited by the steepness and height of the caldera walls. Clusters of white house's cling like snowfields to the 300 m-high cliffs on the western side of the island (Fig. 12.20a). Most tourists arrive by either ferry or cruise ship. The advantage of accessing via the new harbour of Port Athinias is the opportunity to examine the rock exposures in the access road (Fig. 12.20b). The new harbour is too small for the cruise ships and, as the caldera is too deep for

anchoring, tourists are transferred by small boats to the old port. Access to the cliff tops is provided by a funicular, although some tourists prefer to use the original donkey trail (Santorini Guidebook 2012). The sight of cruise ships alternating between steaming slowly against the wind and drifting in the caldera is an unusual feature of Santorini. The NE Meltemi wind regularly blows gale force in the summer months and this results in both the old port and the new harbour being regularly closed for periods of three or four days.

Many visitors to Santorini are fascinated by the Atlantis legend and the historical connection with the Minoan civilization. This is generally recognized as the first advanced culture in Europe having extended through large parts of the Bronze Age. The Minoan period commenced at approximately 2700 BC and flourished until at least 1450 BC. The civilization was rejuvenated between approximately 1400 BC and 1100 BC. The Minoans were centred in Crete with important settlements on some of the Cycladic Islands, including Thera.

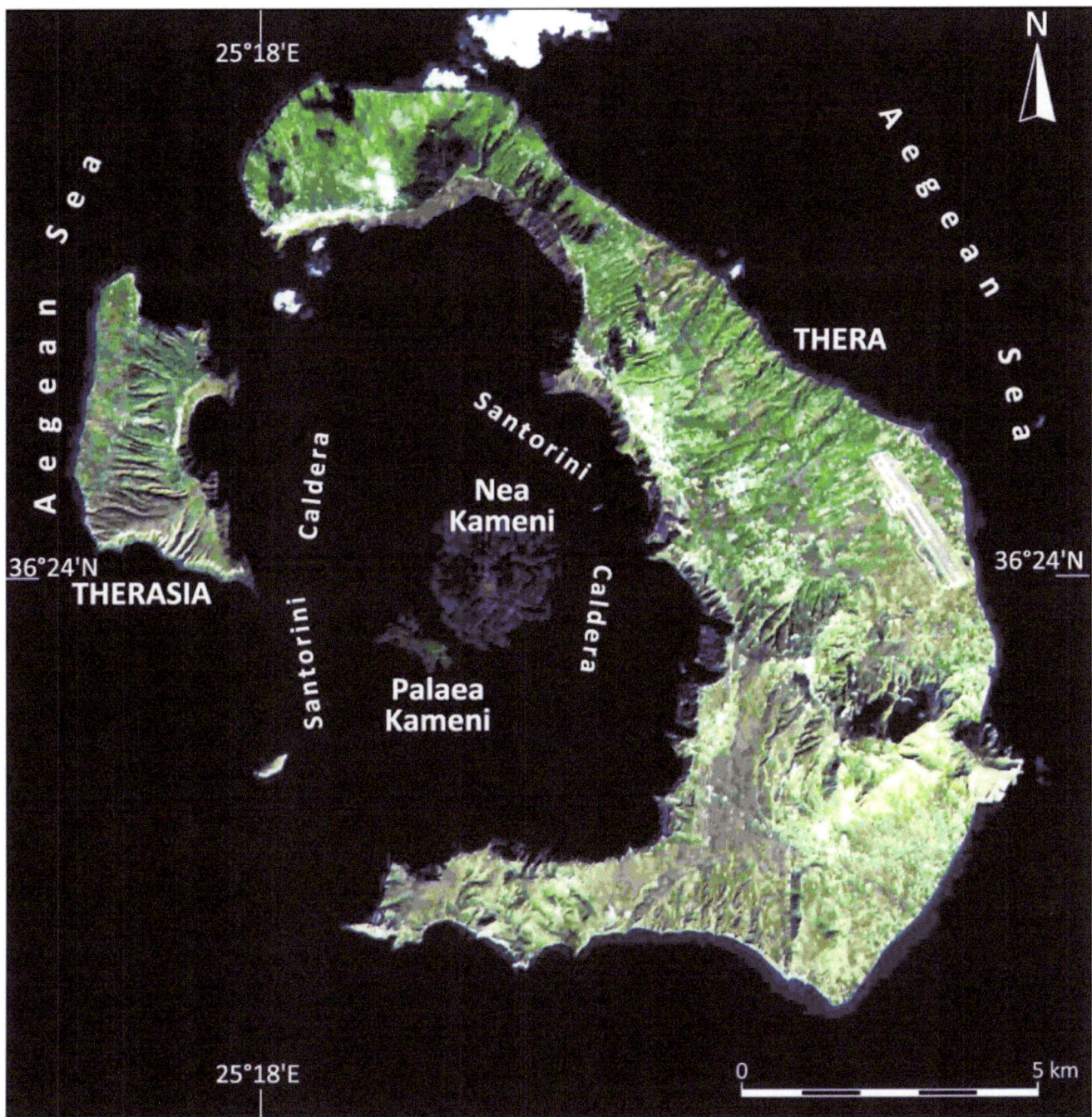

Fig. 12.19 The partially submerged Santorini Caldera is located between the crescent-shaped island of Thera (east) and the subsidiary island of Therasia (west). The Kameni Islands are situated in the centre of the caldera. The steep walls of the caldera are clearly visible on the western side of Thera. *Source* Image from an ASTER instrument on NASA's Terra Satellite (21st November 2000) of the Santorini Islands (18 km in width) processed by Philip Eales of Planetary Visions/DLR and first published in "Map: satellite" by Dorling Kindersley (2007)

Fig. 12.20 **a** The caldera wall
on Thera (view looking south)
reveals multiple layers of volcanic
ashes, lavas, and pumice deposits;
b The road from the new harbour
exposes some of the older caldera
events. Deposits associated with
the Minoan Event are represented
by the distinctive pale-grey or
white layer capping the plateau in
both (**a**) and (**b**)

12.5.1   Geological Framework

The Santorini edifice is probably the largest of the Quaternary volcanoes in the Aegean region (Druitt et al. 1999). The
group of five islands in the archipelago is situated on a
NE-SW lineament that links the volcanism of the Christiania
Islands (20 km southwest) and the Columbo Seamount
(7 km northeast) (Fig. 12.2). Basement rocks associated
with the Pelagonian Tectonic Zone are exposed on Thera in
several sections of the caldera wall, e.g., in the vicinity of
Port Athinias, as well as at Mount Profitis, which with an
altitude of 552 m is the highest point on the islands

(Fig. 12.21). The basement consists of Mesozoic limestone
and mica schist, together with Lower Cenozoic limestone
and metapelites.

The volcanism on Santorini commenced with submarine
activity at approximately 0.65 Ma. Six broad groups of
volcanic deposits are identified (Druitt et al. 1999). (i) The
oldest activity is associated with the submarine volcanic
centres of the Akrotiri Peninsula, southwest Thera (0.645–
0.584 Ma). (ii) Lavas associated with the Peristeria Volcano
dominate the caldera wall in the northern part of Thera and
on the eastern part of Therasia (551–308 Ma). (iii) A group
of cinder cones occur in the Akrotiri Peninsula (0.522–

Fig. 12.21 Geological map of the Santorini archipelago. The majority of the deposits depicted as the Thera Pyroclastics are associated with the Second Explosive Event. The plateaus that cap the islands of Thera and Therasia are primarily underlain by tephra associated with the Minoan Event. *Source* Simplified after Druitt et al. (1999)

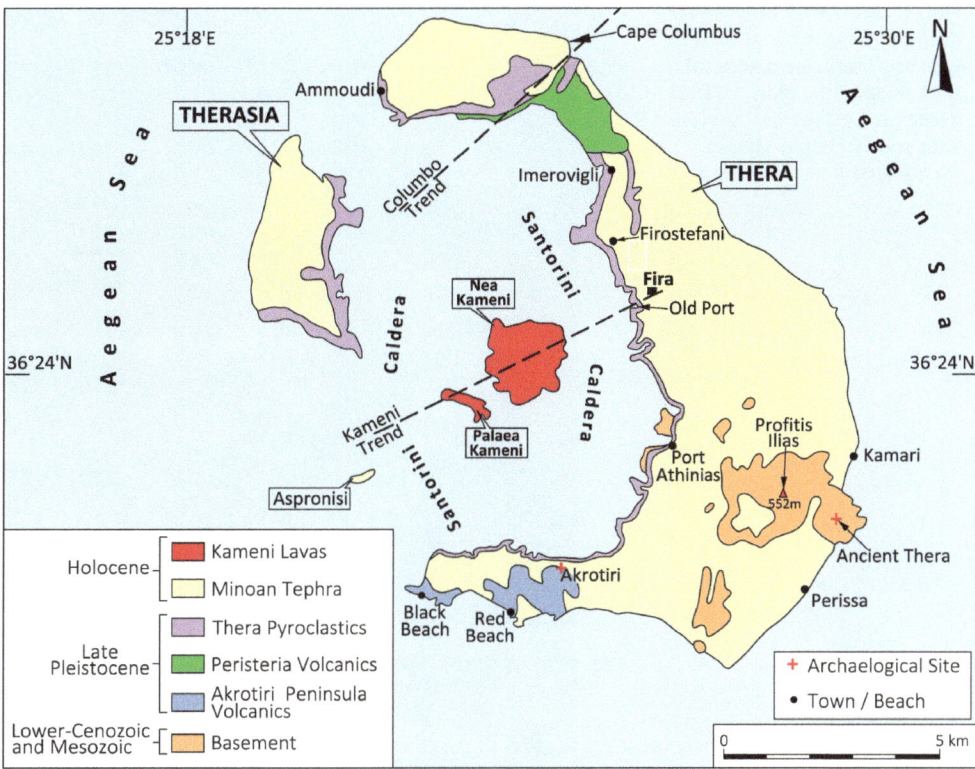

0.34 Ma). (iv) Volcanic deposits of the "First Explosive Cycle" drape over many of the cliffs in southern Thera (0.36–0.18 Ma). These deposits are dominated by pyroclastic deposits and are associated with five separate Plinian events. An important observation in recognizing the individual explosive events (which generally last a maximum of a few days) is the use of pyroclastic deposits as stratigraphic markers. (v) Volcanic deposits of the "Second Explosive Cycle" occur on the cliffs in southern Thera (0.18 Ma–3,600 BP). These deposits include lavas and tephra, in addition to pyroclastics, and are associated with seven discrete Plinian events. The products associated with both of the explosive cycles are informally described as the "Thera pyroclastics" (although they also occur on the eastern cliffs of Therasia). The youngest of the explosive events, which severely affected the Minoan civilization, is known as the Minoan Event. The Minoan Tephra covers large sections of the plateaus on the two large islands. The Santorini Caldera is a composite structure resulting from at least four of the explosive eruptions (four discrete, flat-floored components are recognized), including the Minoan Event. (vi) Lavas of the Kameni volcanoes (Holocene).

12.5.2 Explosive Cycles

The volcanism of Santorini is characterized by periodic Plinian or ultra-Plinian eruptions (Druitt et al. 1999). Twelve

catastrophic events are recognized (Events 1–5 are grouped into the First Explosive cycle; Events 6–11, together with the Minoan Event, are grouped into the Second Explosive Cycle). In addition, more than 100 minor explosive cycles and thousands of smaller eruptions are recognized. Each of the twelve Plinian or ultra-Plinian events generated huge volumes of ashes, lavas, and pyroclastics. The volume of material in each event varied between 1 km^3 and 30 km^3. Each event records cycles of growth of the volcanic cone followed by partial and complete discharge of a relatively shallow magma chamber. The rapid discharge of the magma chamber results in tectonic collapse of the volcanic cone and formation of a new caldera. The rejuvenation and growth of the volcanic cones corresponds to periods of slow cooling and partial crystallization of a shallow crustal magma chamber. The magma chamber periodically reforms. The periodicity of the Plinian events (unusually frequent) is estimated at 20,000–40,000 years. Each would most likely have generated multiple tsunamis in the eastern Mediterranean. The complex sequences of pyroclastics and lava flows on the walls of the Santorini Caldera is related to the multiplicity of the Plinian events. The chaotic nature of the layering in some areas is a marked feature (Fig. 12.22a).

The Plinian events repeatedly reshaped the geography of the Santorini archipelago. During the Upper Pleistocene, Santorini probably consisted of several small islands. At approximately 0.20 Ma, however, one large island (Strongili) with a central caldera developed. Santorini was comprised of

Fig. 12.22 **a** Chaotic volcanic deposits in the walls of Santorini Caldera, Thera, are associated with some of the older caldera events; **b** Agglomerate associated with one of the pre-Minoan events, Red Beach near Akrotiri

a single island and a centrally-submerged caldera prior to the Minoan Event (Druitt et al. 1999). The Santorini volcanism is linked to melting of continental material overlying the subducted slab. The magmas range in composition from alkali basalt and andesite to highly siliceous rhyolite and dacite (Druitt et al. 2015). They are invariably potassium-rich. The older centres of the Akrotiri Peninsula include domes of rhyodacite; they are diagnosed by the presence of abundant amphibole. The Peristeria Volcanics are mostly andesite lava flows. The Minoan Event is associated with rhyodacite, as is the active Kameni volcanism. The petrogenesis of the parental magmas is, however, relatively complex. The recognition of compositionally-zoned olivine xenocrysts is evidence of mixing of parental magmas (derived from depth) and evolved magmas in the shallow crustal magma chambers (Higgins 1996).

The beaches on Santorini reflect the diversity of volcanic deposits. Beaches on the eastern side of Thera, e.g., in the vicinity of the resorts of Kamari and Perissa consist of black sand. The black sand is comprised of fine-grained volcanic

Fig. 12.23 The white or pale grey tephra correlated with one of the pre-Minoan events contains numerous lithic fragments and small volcanic bombs, indicative of a highly explosive, Plinian-style eruption. Exposure in road cutting in cliffs above Port Athinias

ash and lava derived from multiple centres. Red-coloured agglomerate and scoria are spectacularly exposed at the "Red Beach" in the Akrotiri Peninsula (Fig. 12.22b). Some of the volcanic deposits associated with both the pre-Minoan explosive events and the Minoan Event are exposed in the cliffs located above Port Athinias (Fig. 12.23).

12.5.3 Minoan Event

The age of the Minoan Event remains a contentious issue and, in part as it represents an important marker in Bronze Age chronology, continues to be debated by archaeologists and historians. The archaeological evidence suggests the event occurred at approximately 1500 BC, but radiocarbon dates of 1600–1627 BC are now preferred (e.g., Ramsey et al. 2004; Sigl, et al. 2015). Evidence includes dating of an olive branch preserved in volcanic ash (Manning et al. 2006), although data from Greenland ice cores that contain ash associated with a volcanic eruption dated at 1642 BC have been questioned. Dendrochronology based on analysis of tree rings has been used to correlate the eruption with a climatic event in North America (Pearson et al. 2018). The eruption is constrained by the tree-ring data within the period 1600-1525 BC, with the approximate date of 1560 BC preferred by Pearson et al. (2020).

Evidence based on underwater surveys suggests the Minoan Event may have been far larger than previously thought. A magma volume of 55 km^3 is estimated (Vougioukalakis 2005). If correct, this makes the Minoan Event one of the largest historical eruptions and of a similar scale

to the ultra-Plinian Tambora Event, Indonesia (1815). The largest of the ash columns may have attained a height of approximately 40 km. Four distinct cycles are recognized in the Minoan Event (Druitt et al. 1999). The deposits associated with the first cycle include a layer of white-grey tephra that caps the caldera walls and mantles large parts of the pre-Minoan land surface (Fig. 12.20a). This layer attains a maximum thickness of 60 m. The tephra is underlain by a very thin layer of ash, which is interpreted as the first evidence of the eruption, and would have served as a warning to the inhabitants of the island. The latter overlies an old soil horizon and the volcanic ash near the base of the deposit may contain charred relics of trees. A 7 m thick layer of pumice and ash is associated with the most intense phase of activity during the first cycle. The ash contains lithic fragments, locally both sufficiently large and numerous as to constitute block-and-ash deposits. The pumice and ash is exposed in several quarries on Thera. The fine-grained component of the ash was widely dispersed by westerly winds throughout the eastern Mediterranean. The second and third cycles were dominated by pyroclastic surges and lava fountains. The third cycle initiated the start of the collapse of the caldera. The fourth cycle included a range of volcanic activity including basal surges, lava flows, and combined ignimbrite flows and extensive ashfall. The final phase of the caldera collapse occurred in the fourth cycle.

The possibility of tsunami waves having affected the entire eastern Mediterranean adds credence to the catastrophic nature of the Minoan Event. Sheet-like tsunami deposits of suitable age occur on the shores of Crete and Kos, as well as Israel (Goodman-Tchernov et al. 2009).

Fig. 12.24 a Part of the Minoan city of Akrotiri which was entirely buried by the Minoan Event, Thera; **b** The fractured stair case at Akrotiri is attributed to a powerful earthquake several months before the main eruption

Tsunamis may have attained heights of 35–150 m. This hypothesis has, however, been questioned by Dominey-Howes (2004) who suggested the tsunamis would have been relatively small.

12.5.4   Historical and Archaeological Sites

The most well known and widely visited of the historical and archaeological sites on Santorini is the ancient city of Akrotiri. Archaeologists have been excavating the city since 1969 and the site is a world heritage site protected by a roofed enclosure (Fig. 12.24a). Akrotiri may have been first settled in Neolithic times (approximately 5,000 BC), but is principally known as one of the most important of the Minoan cities. The city was entirely buried by volcanic ash during the Minoan Event. The city has an organized street pattern with sophisticated houses, many of which contain paintings and frescoes protected by the ashfall. Similarities with the Roman city of Herculaneum are evident, but Akrotiri is far older.

The destruction of Akrotiri and surrounding sites by a catastrophic volcanic eruption is thought to have influenced Greek scholars' views of the Ancient World, and possibly

Fig. 12.25 Ruins at Mesa Vouno or Ancient Thera showing the use of decorative slabs of red agglomerate in some of the limestone walls and columns

resulted in Plato's legend of the lost continent of Atlantis. The novel by Georges Koukoulas "Atlantis never lost" may be of interest in this regard. Some historians suggest the eruption sparked the collapse of the Minoan culture, but geological evidence indicates the citizens would have been pre-warned by repeated seismic activity. No human remains have been discovered and the thin ash deposit described above, together with evidence of seismic activity, are significant in this regard. The fractured staircase described by Vougioukalakis (2005) is particularly convincing (Fig. 12.24b).

There is a suggestion that huge rafts of pumice may have blocked harbours throughout the eastern Mediterranean. This would have inhibited the sea-trade, fundamental to the Minoan economy, for several years, and the psychological or spiritual damage may have been more significant than the physical effects. The possibility that the eruption caused several years of acid rainfall on the island of Crete, which could have devastated the Minoan economy, is described by Pearson et al. (2020).

The archaeological site of Ancient Thera, or Mesa Vouno ("city on the hill"), was founded by post-Minoan colonists in the 7thC BC. The city is located on a hill of metamorphic rocks in the eastern part of Thera and overlooks a broad, volcanic ash-covered plain. Selection of the site may have been influenced by knowledge of the tsunamis associated with the Minoan Event. Mesa Vouno was inhabited throughout the Classical Greece, Hellenistic, and Roman periods. Archaeological excavations have unearthed paved streets and buildings constructed of Mesozoic limestone (Fig. 12.25). Large slabs of dark-coloured andesite lava and red agglomerate were used for decorative purposes.

12.5.5 Kameni Islands

The uninhabited Kameni Islands are among the youngest landforms on Earth (Fig. 12.26a). The eruption and formation of Palaea Kameni in the Santorini Caldera was documented by classical scholars during historical times. Pliny the Elder reported a new island as having emerged from the sea in 19 AD, although the Roman historian Cassius Dio suggested the new islet near Thera formed in 47 AD. Palaeo Kameni is linked to two main periods of activity, 46 AD–47 AD and 726 AD (Fig. 12.26b). The paucity of vegetation on the younger of the two islands, Nea Kameni, (with the exception of a thin covering of succulents in some parts) is indicative of the youthfulness of this volcano. Moreover, Nea Kameni has a rounded shape (diameter of approximately 2 km) which is consist with eruptions fed from a central conduit. Eruptions on Nea Kameni occurred in well-defined periods, 1570–1573, 1707–1711, 1866–1870, 1925–1928, 1939–1941, and 1950. They typically involved small-scale Strombolian activity preceded by substantial phreatic explosions. The 1925–1928 activity began with jets of water shooting up from the surrounding sea. This was followed by lava fountains and explosive activity which included clouds of steam reaching heights of 3 km. Small pyroclastic flows were also generated during this activity. A long repose period was followed by extrusive activity which terminated this cycle. The 1939–1941 eruption began with a submarine explosion on the western side of the island. Activity then shifted to the centre of the island, where lava built up multiple flows and domes. The 1950 activity was preceded by phreatic volcanism. A small lava flow was extruded as the final burst of activity.

The Kameni Islands have been incorporated into a national geological park. The geotrail on Nea Kameni accesses the Mikra Kameni Crater which formed in 1950 (Fig. 12.27). The blocky nature of lava flows that radiate outward can be observed from the crater rim (height of 70 m). Volcanic bombs from recent eruptions occur on the flanks of the cone. Subsidiary craters contain active hydrothermal vents. Iron oxide-rich hydrothermal fluids discolour the sea in some of the inlets adjacent to Palaeo Kameni. Recent bathymetric imagery of the Santorini Caldera has revealed previously unknown lava flows and submarine extensions of historical flows associated with the Kameni Islands (Nomikou et al. 2019). A large volcaniclastic apron protrudes from the base of the islands in the Santorini Caldera. The total volume of erupted material is estimated at 4.85 km^3. There appears to be a linear correlation between erupted volume and the length of the repose period proceeding each eruptive cycle.

Fig. 12.26 **a** Satellite image of the Kameni islands, Santorini. *Source* Image processed by Philip Eales of Planetary Visions/DLR; **b** Simplified geological map showing ages of eruptions on the Kameni Islands. *Source* Druitt et al. (2015)

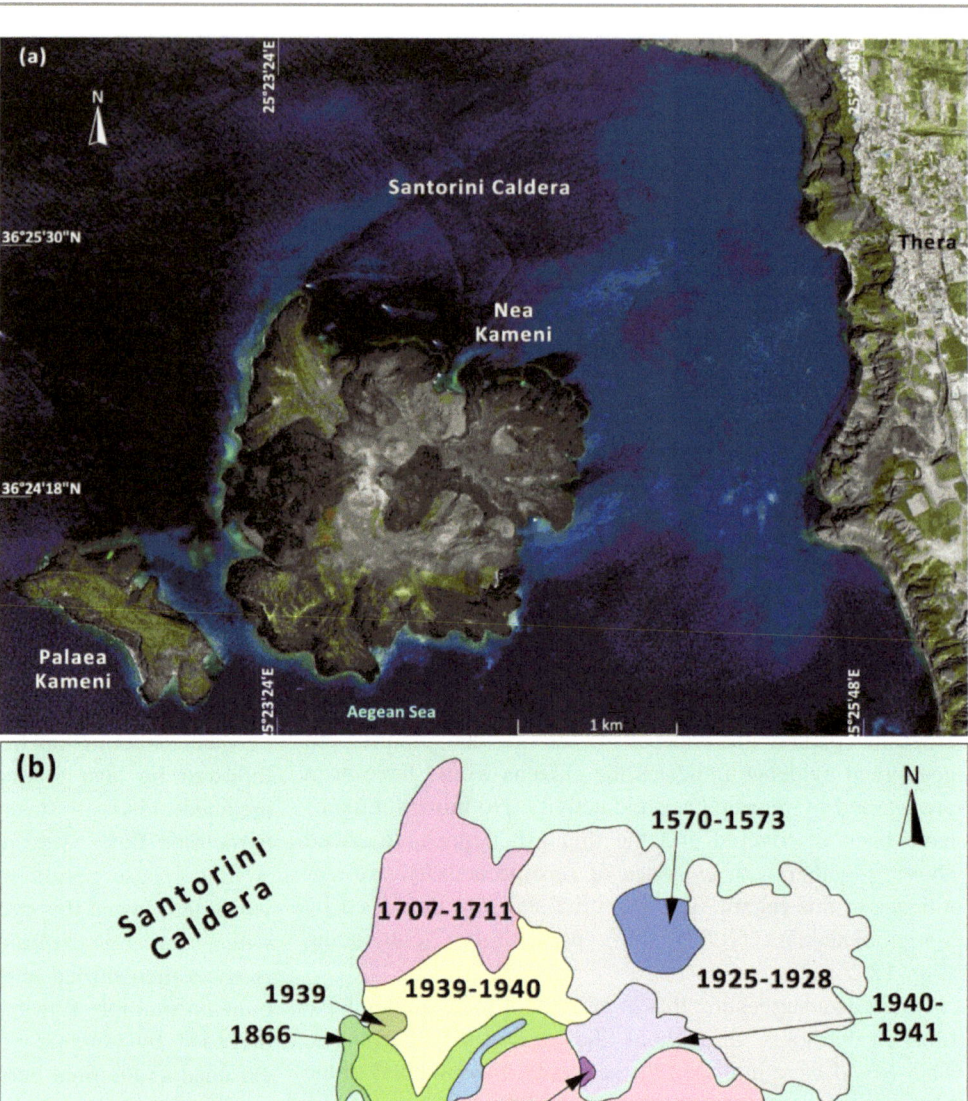

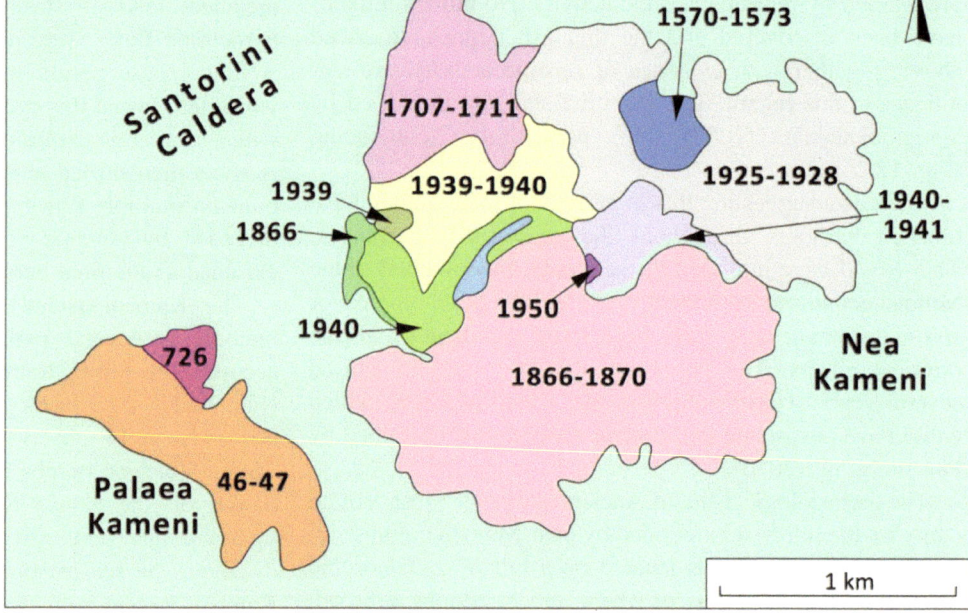

12.5.6 Columbo Seamount

The Columbo (or Kolombo) Seamount is one of a group of approximately twenty seamounts that extend to the northeast of Santorini (Fig. 12.2) The seamount includes a caldera with a diameter of 3 km and height of 500 m. In 1685, this previously unknown volcano suddenly and violently erupted; the event was witnessed from a ship bound from Western Europe to the Middle East. An estimated volume of 2 km³ of ash was ejected, some of which reached Turkey. The eruption of the seamount breached the surface, and generated pyroclastic flows that swept across the surface of

Fig. 12.27 The Mikra Kameni Crater on the island of Nea Kameni

the sea as far as Thera. Gas associated with the eruption caused multiple fatalities on Thera and tsunami waves of over 100 m were reported along some coastlines. In the final stages of the eruption, the caldera collapsed beneath the surface of the sea. The shallowest point of the caldera occurs at a depth of 10 m, but the average depth is approximately 500 m. The seamount includes an active hydrothermal vent field (temperatures attain 220 °C). Some of the hydrothermal vents include chimneys with polymetallic sulphides. These and other features have been documented using submersible robots by the well known underwater explorer Robert Ballard.

12.5.7 Recent Activity

In 2011, after sixty years of quiescence, the island of Thera experienced significant seismic activity. Data from terrestrial GPS measurements revealed the eastern and western edges of the Santorini Caldera had spread by 14 cm. Magma or gas was shown to have inflated to within 4 km of the sea floor. Despite the paucity of earthquakes in 2012, researchers are

currently investigating whether this activity deformed the seabed (as reported in Nature, 20th July 2012). Instrumentation includes a tiltmeter and underwater pressure sensors. The temperature and composition of hydrothermal vents on Nea Kameni are monitored as these features can assist with predicting new eruptions. The Columbo Seamount is also monitored as this feature is potentially hazardous (Hensch et al. 2009). The continuous reshaping of the Santorini Islands from the repetition of catastrophic caldera events provides a vivid demonstration of both the creative and destructive forces of volcanism.

References

Cita, M. B. (2006). Exhumation of Messinian evaporites in the deep-sea and creation of deep anoxic brine-filled collapsed basins. *Sedimentary Geology, 188–189,* 357–378.

D'Alessandro, W., Brusca, L., Kyriakopoulos, K., Michas, G., & Papadakis, G. (2008). Methana, the westernmost active volcanic system of the south Aegean Arc (Greece): Insight from fluids geochemistry. *Journal of Volcanology and Geothermal Research, 178,* 818–828.

Dominey-Howes, D. (2004). A re-analysis of the Late Bronze Age eruption and tsunami of Santorini, Greece, and the implications for the volcano–tsunami hazard. *Journal of Volcanology and Geothermal Research, 130*, 107–132.

Druitt, T.H., Edward, L., Mellors, R.M., Pyle, D.M., Sparks, R.S.J., Lanphere, M., Davies, M., & Barreiro, B. (1999). Santorini Volcano. *Geological Society Memoir, 19*, 169.

Druitt, T. H., Francalanci, L., & Fabbro, G. (2015). *Field guide to Santorini Volcano* (p. 57). Santorini: MeMoVolc short course.

Flower, M.F.J., & Dilek, Y. (2003). Arc–trench rollback and forearc accretion: 1. A collision–induced mantle flow model for Tethyan Ophiolites. In: Dilek, Y. and Robinson, P.T. (Eds.) Ophiolites in Earth History. *Geological Society London Special Publications, 218*, 21–41.

Foutrakis, P. M., & Anastasakis, G. (2018). Bathy-morphological setting of the Submarine Pausanias Volcanic Field, South Aegean Active Volcanic Arc. *Journal of Maps, 14*(2), 341–347.

Fytikas, M., Innocenti, F., Marinelli, G., & Mazzuoli, R. (1976). Geochronological data on recent magmatism in the Aegean Sea. *Tectonophysics, 31*, T29–T34.

Fytikas, M., Innocenti, F., Manetti, P., Peccerillo, A., Mazzuoli, R., & Villari, L. (1984). Tertiary to Quaternary evolution of volcanism in the Aegean region. *Geological Society London Special Publications, 17*, 687–699.

Fytikas, M., Giuliani, O., Innocenti, F., Kolios, N., Manetti, P., & Mazzuoli, R. (1986a). The Plio-Quaternary volcanism of Saronikos area (western part of the active Aegean volcanic Arc). *Annales Geologiques des Pays Helleniques, 33*, 23–45.

Fytikas, M., Innocenti, F., Kolios, N., Manetti, P., Mazzuoli, R., & Poli, G., et al. (1986b). Volcanology and petrology of volcanic products from the island of Milos and neighbouring islets. *Journal of Volcanology and Geothermal Research, 28*, 297–317.

Fytikas, M., Kolios, N., & Vougioukalakis, G. (1990). Post-Minoan volcanic activity of the Santorini volcano: Volcanic hazard and risk, forecasting possibilities. In D. A. Hardy (Ed.), Thera and the Aegean World III, vol 2, Earth Sciences (pp. 183–198). Thera Foundation: Essex, UK.

Gaitanakis, P., & Dietrich, V. (1995). Geological map of Methana Peninsula, 1:25 000. Zurich: Swiss Federal Institute of Technology.

Goodman-Tchernov, B. N., Dey, H. W., Reinhardt, E. G., McCoy, F., & Mart, Y. (2009). Tsunami waves generated by the Santorini eruption reached Eastern Mediterranean shores. *Geology, 37*, 943–946.

Hensch, M., Dahm, T., Hort, M., Huebscher, C., & Dehghani, A. (2009). On the interrelation of fluid-induced seismicity and crustal deformation at Columbo Seamount (Aegean Sea, Greece). The Smithsonian/NASA Astro-physics Data (abstract).

Higgins, M. D. (1996). Magma dynamics beneath Kameni Volcano, Thera, Greece, as revealed by crystal size and shape measurements. *Journal of Volcanology and Geothermal Research, 70*, 37–48.

Higgins, M. D., & Higgins, R. (1996). A geological companion to Greece and the Aegean (p. 240). New York: Cornell University Press.

Hurni, L., Dietrich, V. J., & Gaitanakis, P. (1993). Computer-aided geological cartography: 3-dimenisonal modelling of the Methana Volcanoes. *Bulletin of the Geological Society of Greece, 2893*, 515–518.

Jolivet, L., et al. (2013). Aegean tectonics: Strain localisation, slab tearing and trench retreat. *Tectonophysics, 597–598*, 1–33.

Le Pichon, X., & Angelier, J. (1979). The Hellenic arc and trench system: a key to the neotectonic evolution of the Eastern Mediterranean area. *Tectonophysics, 60*, 1–42.

Manning, S. W., Ramsey, C. B., Kutschera, W., Higham, T., Kromer, B., Steier, & P., et al. (2006). Chronology for the Aegean late bronze age 1700–1400 B.C. *American Association for the Advancement of Science, 312*, 565–569.

Meier, T., Becker, D., Endrun, B., Rische, M., Bohnhoff, M., Stockhert, B., & Harpjes, H-P. (2007). A model for the Hellenic subduction zone in the area of Crete based on seismological investigations. In: T. Taymaz, Y. Yilmaz & Y. Dilek (Eds.), The geodynamics of the Aegean and Anatolia. *Geological Society London Special Publications, 291*, 183–199.

Miloterranean Geo Walks 1 (2012). Vani. ISBN 978-960945685-2.

Miloterranean Geo Walks 2 (2012). Volcano. ISBN 978-960945686-9.

Miloterranean Geo Walks 3 (2012). Sulfur Mines. ISBN 978-960945687-6.

Miloterranean Geo Walks 4 (2012). Aggeria. ISBN 978-960945688-3.

Miloterranean Geo Walks 6–7 (2012). Sarakiniko and Nychia. ISNB 978-9609455690-8.

Nomikou, P., Hübscher, C., & Carey, S. (2019). The Christiana-Santorini-Kolumbo volcanic field. *Elements, 15*(3), 171–176. https://doi.org/10.2138/gselements.15.3.171.

Pearson, C. L., Brewer, P. W., Brown, D., Heaton, T. J., Hodgins, G. W. L., et al. (2018). Annual radiocarbon record indicates 16th century BCE date for the Thera eruption. *Science Advances, 4* (eaar8241), 7.

Pearson, C.L., Salzer, M., Wacker, L., Brewer, P., Sookdeo, M., & Kuniholm, P. (2020). Securing timelines in the ancient Mediterranean using multiproxy annual tree-ring data. www.pnas.org/cgi/doi/10.1073/pnas.1917445117.

Pe, G. G. (1974). Volcanic rocks of Methana, South Aegean Arc, Greece. *Bulletin Volcanologique, 38*, 270–290.

Pe-Piper, G., & Piper, D. J. W. (2013). The effect of changing regional tectonics on an arc volcano: Methana, Greece. *Journal of Volcanology and Geothermal Research, 260*, 146–163.

Ramsey, C. B., Manning, S. W., & Galimberti, M. (2004). Dating the volcanic eruption at Thera. *Radiocarbon, 46*, 325–344.

Robertson, A. H. F., & Mountrakis, D. (2006). Tectonic development of the Eastern Mediterranean region: an introduction. *Geological Society London Special Publications, 260*, 1–9.

Santorini Guidebook (2012). Editor: M. Synodinou. Print and Web Guidebooks Ltd., Santorini, p. 98.

Sakellariou, D., Sigurdsson, H., Alexandri, M., Carey, S., Rousakis, G., & Nomikou, P., et al. (2010). Active tectonics in the Hellenic Volcanic Arc: the Kolumbo submarine volcanic zone. *Bulletin of the Geological Society of Greece, 63*, 1056–1063.

Sigl, M., et al. (2015). Timing and climate forcing of volcanic eruptions for the past 2,500 years. *Nature, 523*, 543–549.

Strothers, R. B., & Rampino, M. R. (1983). Volcanic eruptions in the Mediterranean before AD 630 from written and archaeological sources. *Journal of Geophysical Research, 88*, 6357–6371.

Thomson, S. N., Stöckhert, N., & Brix, M. R. (1998). Thermochronology of the high-pressure metamorphic rocks of Crete, Greece: implications for the speed of tectonic processes. *Geology, 26*, 259–262.

Vougioukalakis, G. (2005). Santorini Geological Guide. Thira Municipal Corporation and the Institute for the study and monitoring of the Santorini volcano, p. 79

Antiquities and Archaeological Sites of Western Turkey

Aphrodisias, Assos, Ephesus, Gallipoli Peninsula, Pergamum and Troy

Abstract

Western Turkey is rich in antiquities and archaeological sites that encompass a long history of civilization. Possibly the most evocative of the antiquities is the ancient city of Troy, situated close to the southern shores of the Dardanelles. Descriptions of the last ten years of the Trojan Wars in Homer's *Iliad* have assisted archaeologists in identifying historical sites that include the landing place of the Greek army and the defensive positions of the Trojans. The nine levels excavated in the city of Troy correspond to multiple civilizations, including the Late Bronze Age (or Homeric) city. The Greco-Roman antiquities of Assos, Ephesus, and Pergamum contain well-preserved monumental buildings and theatres that attract large numbers of visitors. The antiquities of Aphrodisias, Colossae, and Laodicea have suffered historical earthquake damage, but are nonetheless of considerable interest. Many of the antiquities became Christian centres of historical significance during the Roman times. The Gallipoli Peninsula Historical National Park on the northern shores of the Dardanelles was established to commemorate the Dardanelles Campaign of World War I. Western Turkey is located in a region of the eastern Mediterranean subjected to active deformation and has a long history of devastating earthquakes. The tectonic evolution can be summarized as including the effects of active subduction along the Hellenic Trench, continental collision in eastern Turkey (Anatolian Microplate and Arabian Microplate) and in the Caucasus (Anatolian Microplate and Eurasian Plate). The Anatolian Microplate forms a buffer between the active orogenic belts and the subduction zone. The strain is taken up in regional sutures such as the North Anatolian Fault Zone and the Izmir-Ankara Fault Zone. The antiquities and archaeological sites described here occur in discrete geological terrains. The Thrace Basin, the Sakarya Tectonic Zone, and the Menderes Massif are each subdivided by the regional sutures. Some of the antiquities occur in basins and grabens which developed as post-collisional crustal extension displaced the compressive tectonism. Extension was triggered by westward movement of the Anatolian Microplate and development of the back-arc basin associated with the Aegean Sea. Landforms at some of the archaeological sites differ markedly from historical descriptions. Geomorphological reconstructions, including modelling of active deltas and river estuaries in the basins and grabens, suggest active faulting and recent sedimentation patterns are significant. The coastline at Troy has migrated several kilometres northward due to accumulation of estuarine sediments in the Dardanelles Graben. Erosion of the poorly-consolidated Neogene sediments in the Gallipoli Peninsula, notably at Anzac Cove, has created extensive areas of badlands. The antiquities of Assos and Pergamum are situated on hills consisting of resistant volcanic lavas (Miocene). Active sedimentation in the Küçük Menderes Graben has effectively destroyed the harbour of Ephesus, as well as the Artemision, one of the seven wonders of the ancient world. The theatre and other monumental buildings at Ephesus have survived relatively intact as they are located on the margins of the graben. Quarries in the vicinity of Aphrodisias, Colossae, and Laodicea exploited high-quality *Carian* marble (Palaeozoic) from the Menderes Massif. The marble was used for the black and white statues associated with the "School of Aphrodisias".

Keywords

Deltas • Earthquakes • Geomorphology • Grabens • Greco-Roman antiquities • Iliad • Sedimentation

The original version of this chapter was revised: Belated corrections have been incoporated. The corrections to this chapter are available at https://doi.org/10.1007/978-3-030-54693-9_18

13.1   Introduction

Western Turkey is rich in antiquities and archaeological sites which encompass a long history of civilization. Some antiquities are situated near the extensive coastline of the Aegean Sea, whilst others occur in the mountainous hinterland. Possibly the most evocative of the Greco-Roman sites is the ancient city of Troy. Troy is widely known from historical descriptions of the Trojan Wars (13thC BC) in the *Iliad*, probably written by the Greek poet Homer in the 8thC BC. Both Troy and the Gallipoli Peninsula Historical National Park are located on the shores of the Dardanelles. Ephesus and Pergamum are two of the best preserved Greco-Roman antiquities in western Turkey. They include monumental buildings and theatres which attract large numbers of visitors (Fig. 13.1). The antiquity of Assos includes a well-preserved theatre located on a rocky outcrop which overlooks the Aegean Sea and the island of Lesbos. The antiquities of Aphrodisias, Colossae, and Laodicea include archaeological sites that cover large areas.

The antiquities and archaeological sites described here can be reached from the regional centres of Çanakkale, Izmir, Selçuk, and Denizli (Fig. 13.2). Thrace and Anatolia are historical names for the European and Asian sections of Turkey. The division is associated with the Bosphorus, the Sea of Marmara, and the Dardanelles. Anatolia was divided into historical districts, of which the largest was Lydia. Thrace and Lydia were incorporated into the Roman Empire and many of the Ancient Greek antiquities were extended and rebuilt by the Romans. The time lines associated with the different civilizations, including the Mycenaean period which predated the Archaic and Classical Greece times, is summarized in Box 11.1. Many of the Greco-Roman antiquities in western Turkey became Christian centres of historical significance during the Roman times.

Western Turkey is located in a region subjected to active deformation, and has a long history of devastating earthquakes. Probably all of the antiquities have been subjected to major earthquake damage during historical times (Ambraseys 2009). Seismic events are centred on regional sutures,

Fig. 13.1 The Great Theatre at the antiquity of Ephesus is partially built into the side of Mount Pion, a horst block located within the Küçük Menderes Graben. In the background is a ridge of resistant metamorphic rocks, part of Mount Preon, the horst block located on the southern side of the graben

Fig. 13.2 Map showing location of antiquities and archaeological sites in western Turkey described here. Thrace is a historical name for the European section of Turkey. The historical name for the Asian section of Turkey is Anatolia. The approximate eastern extent of the Roman Empire is depicted

including in the tectonically-active Aegean Sea, as well as with active faults associated with basins and grabens. The tectonic framework is related to the ongoing continental collision (convergence) and subduction. The antiquities and archaeological sites described here occur in discrete geological terrains. Terrains are defined by regional sutures. Some of the antiquities occur in basins and grabens which developed as post-collisional crustal extension replaced the compressive tectonism. Landforms differ markedly from historical descriptions, and geomorphological reconstructions, including modelling of active deltas and river estuaries in the basins and grabens, suggest active faulting and recent sedimentation patterns are significant. Many of the coastal sites have been subjected to extensive sedimentation (Kraft et al. 2003; Marriner et al. 2010).

13.2 First Century Christianity in Asia Minor

Many of the Greco-Roman antiquities in western Turkey became Christian centres of historical significance during the Roman times. In the 1stC AD, the region of Anatolia, which

was known as Asia Minor, was divided into well-defined districts based on historical regions (Fig. 13.3). Asia Minor features prominently in the New Testament section of the Christian Bible, particularly due to the missionary journeys of the apostle Paul in the years following the death of Jesus of Nazareth. Paul's letters to the Christian communities in the cities of Ephesus and Colossae, as well as in the region of Galatia, are included in the New Testament. Colossae was originally the capital of the ancient region of Phrygia, one of the most celebrated cities of southern Anatolia prior to being incorporated into Lydia (in approximately 7thC BC). The Epistle to the Colossians, an early Christian text which identifies its author as Paul the Apostle, is addressed to the church in the city. Revelation, the final book of the New Testament, also refers to Christian communities in Asia Minor, and was written by the apostle John while he was in exile on the island of Patmos. As a prelude to the description of his apocalyptic vision in Revelation, John (who was also the author of one of the four Christian gospels) issued a series of prophetic warnings to communities in the western part of Asia Minor. John specifically wrote to the "Seven Churches of Revelation", also known as the "Seven

Fig. 13.3 The regions and towns of Asia Minor in the 1stC AD. The letters that the apostle Paul (a native of Tarsus in the region of Cilicia) wrote to Christian communities in the region are included in the New Testament. The seven churches of Ephesus, Smyrna, Pergamum, Thyatira, Sardis, Philadelphia and Laodicea (indicated in red) were addressed by the apostle John in the book of Revelation. John was in exile on the island of Patmos. The majority of the region formerly known as Asia Minor is incorporated into the modern country of Turkey. The city of Smyrna is renamed Izmir. Patmos and the other islands in the Aegean Sea and eastern Mediterranean are part of Greece. *Source* Map draughted by Andrew Mitchell

Churches of Asia" and the "Seven Churches of the Apocalypse". Early Christian and medieval legends, based on geological events, are attached to several of the antiquities in western Turkey, including the apparition of Archangel Michael at Colossae (Piccardi 2007).

13.3 Regional Geology

The active deformation and tectonism which is such a feature of western Turkey is in part related to the occurrence of microplates which formed in response to the convergence of the African and Eurasian Plates (Taymaz and Yilmaz 2007). Three microplates are recognized. Western Turkey is part of the Anatolian Microplate, which is wedged between the Hellenic Microplate (west) and the Arabian Microplate (southeast) (Fig. 8.3). The long drawn out nature of the regional collision associated with the northward-pushing African Plate has resulted in the Alpine Orogeny having peaked at different times in different regions of the Mediterranean. The orogeny commenced relatively late in the eastern Mediterranean, probably in the Early Cenozoic

(Okay and Tüysüz 1999; Bozkurt et al. 2000; Taymaz and Yilmaz 2007). The peak occurred in the Miocene and Pliocene with formation of a series of disconnected fold mountains that include the Taurus Mountains (southeast Turkey) and the Caucasian Mountains.

The tectonic evolution can be summarized as including the effects of active subduction along the Hellenic Trench together with continental collision in eastern Turkey (Anatolian Microplate and Arabian Microplate) and in the Caucasus (Anatolian Microplate and Eurasian Plate) (Taymaz and Yilmaz 2007). The active component of the Hellenic Trench, which includes several subsidiary features, e.g., the Pliny Trench and the Strabo Trench, is located south of Crete (Fig. 13.4). The Anatolian Microplate forms a buffer between the orogenic belts and the active subduction zone.

13.3.1 Regional Sutures

The Anatolian Microplate is subdivided by regional sutures which were probably most active during the Neogene, but may have started to form in the Early Cenozoic (Okay and

Fig. 13.4 The tectonic framework of western Turkey is dominated by major curvilinear sutures ascribed to the differential movement of the Anatolian Microplate and the surrounding terrains. For details of the plate boundaries and trenches the reader is referred to Fig. 8.3. *Source* Satellite Image of Europe based on NASA MODIS data, processed by Philip Eales of Planetary Visions/DLR with tectonics in part after Bozkurt (2001)

Tüysüz 1999; Bozkurt et al. 2000; Taymaz and Yilmaz 2007; Yolsal-Çevikbilen et al. 2012). Sutures continued to develop in the Quaternary. Sutures take up a significant amount of the current stress associated with the ongoing collision and the south-westward movement of the Anatolian Microplate. Many of the most destructive earthquakes in Turkey are associated with the regional sutures. The discrete geological terrains of interest here, include the Thrace Basin, Sakarya Zone, and Menderes Massif (Fig. 13.4).

The North Anatolian Fault Zone (NAFZ) is a regional suture associated with major strike-slip faults that demarcate the boundary between the Anatolian Microplate and the Eurasian Plate (Fig. 13.4). The NAFZ encloses the Sea of Marmara. The Dardanelles is associated with a subsidiary graben within the NAFZ. The NAFZ formed at approximately 5 Ma and reveals lateral displacement of up to 85 km (Yolsal-Çevikbilen et al. 2012). The Izmir-Ankara Fault Zone (IAFZ) is an equally significant (and laterally continuous) suture located to the south of the NAFZ. Both the NAFZ and the IAFZ may terminate westward against the

Vadar Suture, a prominent NW-SE lineament in the Balkan Peninsula. The Lycian Nappes are regional thrusts that affect the southern part of western Turkey, including the Anatolid Block in the central part of the microplate. The West Anatolian Shear Zone is a discontinuous series of faults associated with the poorly-defined boundary between the Anatolian Microplate and the Hellenic Microplate (the latter incorporating the Aegean Sea).

13.3.2 Southwest Migration of the Anatolian Microplate

The subduction of ancient oceanic crust and occurrence of relatively recent evaporites in the Hellenic Trench, as described in Chap. 12, is driving extension of the continental crust in western Turkey and volcanism in the Aegean back-arc basin. Eastern Turkey, however, has experienced crustal shortening and thickening as a result of northward motion of the Arabian Microplate. The resulting combination of forces (the "pull" from the rollback of the subduction zone to the southwest and the "push" from the convergent zone to the southeast) is causing the Anatolian Microplate to move south-westward (Taymaz and Yilmaz 2007). The strain is taken up in regional sutures. Large scale crustal deformation and the associated seismicity and volcanism in western Turkey is ascribed to the relative motions of the adjoining plates and microplates.

13.3.3 Basins and Grabens

Basins and grabens developed in western Turkey as post-collisional crustal extension displaced the earlier compressive tectonism (Altunel and Hancock 1993; Bozkurt 2000; Bozkurt et al. 2000; Yilmaz et al. 2000; Bozkurt and Mittwede 2005). Extension was in part triggered by westward movement of the Anatolian Microplate and development of the back-arc basin associated with the Aegean Sea. Many of the basins and grabens are aligned with older (Alpine) sutures. The post-collisional extension and formation of grabens has exhumed the core-parts (Proterozoic-Palaeozoic inliers) of the Menderes Massif (Çemen et al. 2006). Most of the lower-lying regions in the terrains described here are associated with basins and grabens.

Graben faults trend either W-E or NW-SE. Many of the basins and grabens continue to subside and the rift shoulders, or horsts, are subjected to considerable uplift. Some faulting may have commenced in the Oligocene, but most faults are Neogene and Quaternary features. The basins and grabens are partially infilled by either shallow marine or terrestrial sediments, including conglomerate, limestone and sandstone. Most of the sediment was deposited by rivers flowing from the hinterland. Some grabens contain up to 30 m of sediment located below present sea level. Seismicity in western Turkey is in part associated with the active graben faults, although the major catastrophic earthquakes are related to the regional sutures.

13.3.4 Geological Terrains

The geological terrains in western Turkey contain relicts of the pre-Alpine basement together with younger deposits. Typically, the older metamorphic rocks are exposed in the hilly areas between the regional sutures and younger grabens. The valleys and coastal terraces are associated with Neogene-Quaternary sediments.

The tectonic zones in western Turkey can be correlated with the NW-SE trending Alpine zones of the Balkan Peninsula, although in Turkey the alignment is generally W-E (Bozkurt et al. 2000) (Fig. 13.4). The southern boundary of the Thrace Basin is poorly defined, but the

approach taken here is to position the division with the southern side of the Dardanelles Graben (Fig. 13.5). The Dardanelles Graben is aligned with the NAFZ. The graben started to form in the Pliocene due to release of stress associated with the regional suture (Yaltirak et al. 2000).

The region between the Dardanelles Graben and the southernmost expression of the NAFZ is equated with the Rhodope-Serbomacedonian Massif of northeast Greece. The terrain located between the NAFZ and the IAFZ constitutes the Sakarya Zone. The Sakarya Zone extends into central Turkey and may be correlated with the Vadar Zone in Greece. The Bakırçay Graben is a prominent Neogene-age feature of the Sakarya Zone. Both of these regions, i.e., the area between the NAFZ and the Thrace Basin, as well as the Sakarya Zone, contain a broad range of geological terrains (Fig. 13.5). Segments of the Proterozoic-Palaeozoic basement (relatively high grade metamorphic rocks) are represented together with Mesozoic metamorphic rocks and Jurassic ophiolite complexes. Cenozoic features include intrusive bodies of granite and the West Anatolian Volcanic Province.

The Menderes Massif covers a large area of southwest Turkey ($>500,000$ km^2) and can be correlated with the Pelagonian and Atticacycladic Zones of Greece (Fig. 13.4). The southern boundary of the massif is defined by regional thrusts associated with the Lycian Nappes. The massif contains inliers of Proterozoic-Palaeozoic basement dominated by granite-gneiss and amphibolite (Bozkurt and Oberhänsli 2001) (Fig. 13.6). Some of the Palaeozoic-Mesozoic metamorphic terrains contain high quality marbles, together with quartzite and mica schist. The marble, quartzite, and mica schist may occur in complex thrust belts. The thrust belts also contain Mesozoic rocks, including limestone and low quality marble, together with Jurassic ophiolite complexes. The Menderes Massif is subdivided by prominent W-E trending grabens into northern, central, and southern sections. The Küçük Menderes Graben is enclosed by horst blocks consisting of resistant limestone and marble (Seyitoglu and Işik 2009). The city of Denizli occurs in an intracratonic basin associated with multiple grabens, including the Büyük Menderes Graben which is aligned with the older W-E Alpine trend (Bozkurt 2000).

13.3.5 Sea Level Changes

Sedimentation patterns in the basins and grabens of western Turkey were affected by sea-level changes during the Late Pleistocene glaciations (Box 3.1). The sea level was as much as 120 m lower than currently observed. During this time, the Sea of Marmara and the Black Sea were freshwater lakes and the Dardanelles and Bosphorus were part of a south-westward flowing, palaeo-river system (Yaltirak et al. 2000). The palaeo-river system included deeply-incised

Fig. 13.5 Simplified geological map of part of western Turkey showing alignment of major sutures and grabens relative to the antiquities and historical sites of Troy, Gallipoli Peninsula National Park, Assos, Kazdağı National Park, and Pergamum. The Dardanelles Graben demarcates the southern limit of the Thrace Basin. The remainder of the area confined by the NAFZ is correlated with the Rhodope-Serbomacedonian Massif of northwest Greece. The area south of the NAFZ is part of the Sakarya Tectonic Zone. *Source* Geological Map of Europe (*https://geoviewer.bgr.de*) and articles referenced in the main text

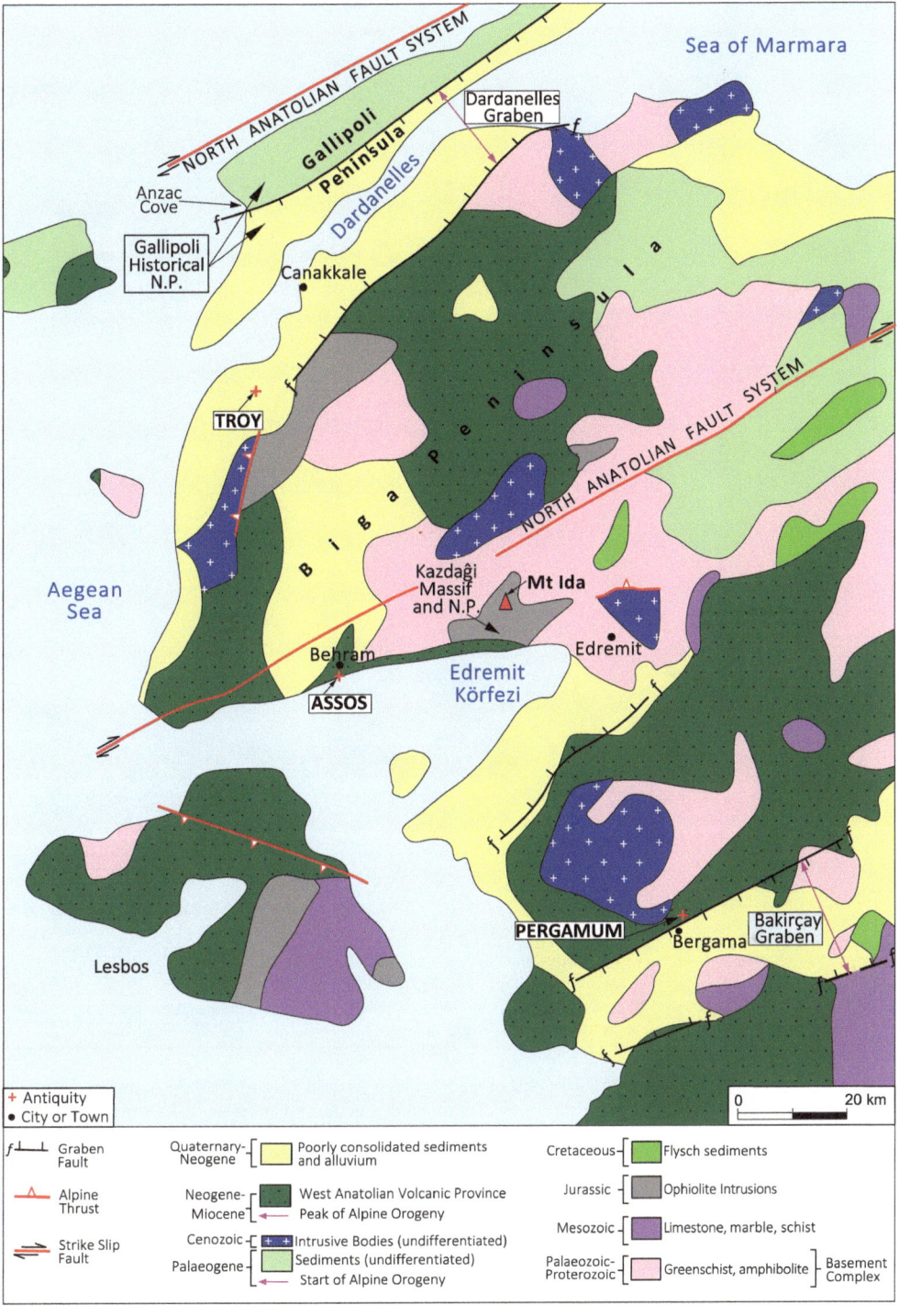

channels due to flooding associated with draining of the Sea of Marmara and of the Black Sea. Thick sequences of sediment accumulated on the coastal plains adjacent to the Dardanelles during this period. A rapid increase in sea-level occurred at the termination of the Pleistocene Ice Ages. During the Early Holocene, the Aegean Sea and the Dardanelles encroached on many of the coastal plains creating broad estuaries. The sea level attained the relatively stable stand currently observed at approximately 5,000 BP. Subsequent variations in sea level, e.g., associated with climatic swings such as Medieval Warming and the Little Ice Age, have been relatively minor.

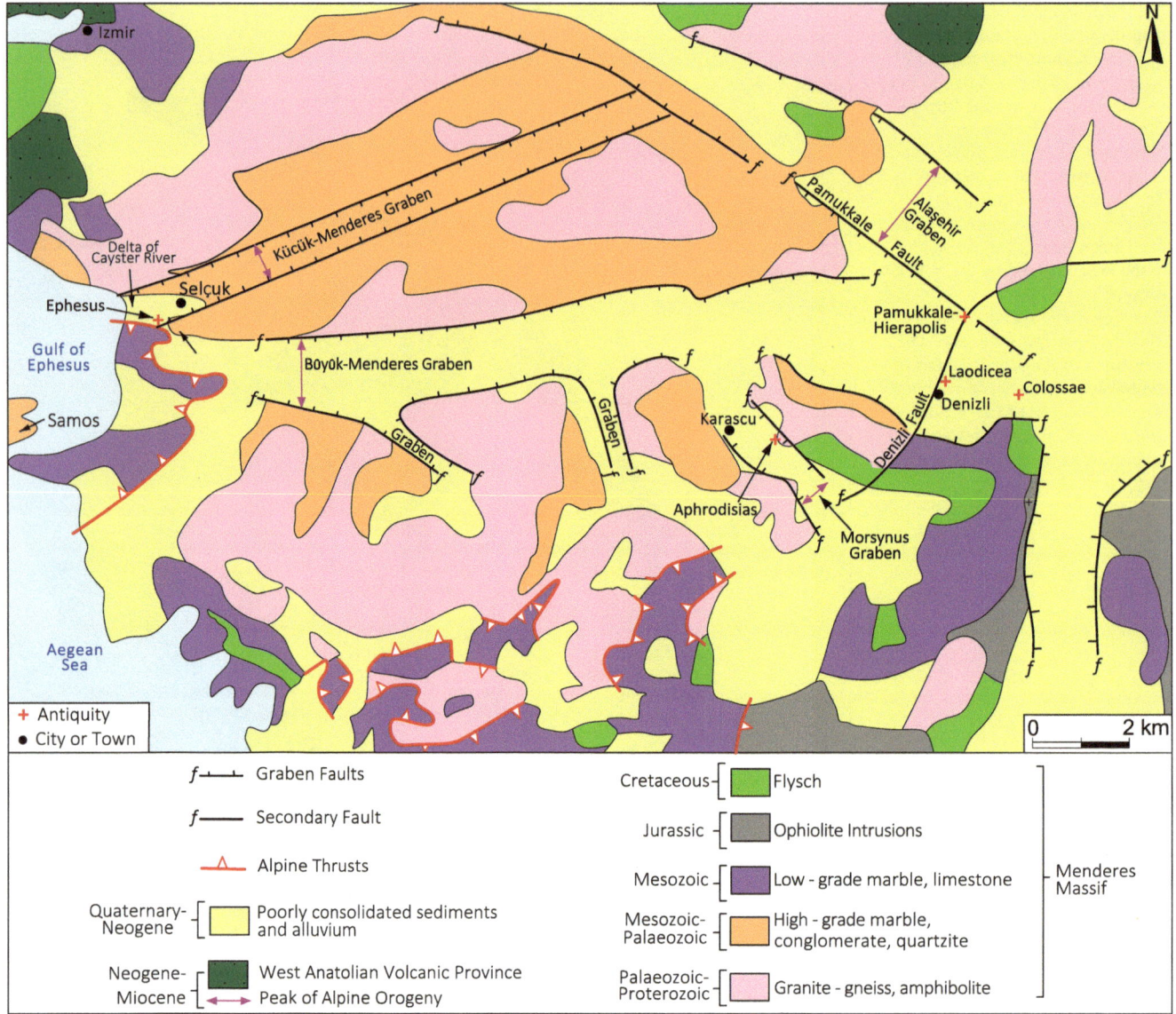

Fig. 13.6 Simplified geological map of part of the Menderes Massif, western Turkey, showing alignment of major sutures and grabens relative to the antiquities and historical sites of Ephesus, Aphrodisias, Laodicea, Colossae, and Pamukkale-Hierapolis. *Source* Geological Map of Europe (*https://geoviewer.bgr.de*) and articles referenced in the main text

13.3.6 Geomorphological Changes

Differences between the geomorphology of landforms currently observed and historical accounts of many of the antiquities and archaeological sites suggest that localized patterns of sedimentation in basins and grabens are more significant than sea level changes (Kraft et al. 2001). The progradation and aggradation of deltas associated with the rivers that drain into the Aegean Sea and Dardanelles are particularly significant. These processes commenced at approximately 5,000 BP, i.e. the start of the Lower Bronze Age. The Lower Bronze Age initiated a period of relatively intensive agriculture and major deforestation in western Turkey which coincided with stabilization of the current sea level. The impact of cycles of coastal erosion and deposition, from the Late Bronze Age through to Byzantine times, is particularly well represented at Ephesus.

13.4 Troy

The archaeological site of Troy (or Troia) is situated in the Biga Peninsula, near the regional town of Çanakkale (Fig. 13.2). The historical name of the peninsula is the Troad. The archaeological site is situated on the 30-m high Hisarlik Hill ("castled place") located on the margins of a broad coastal plain. Troy was formerly a seaport of strategic importance but is now situated approximately 4.5 km from the shoreline (Fig. 13.7a). The Dardanelles is known in Turkey as the Straits of Çanakkale, but in classical times was described as the Hellespont ("Sea of Helle"). The straits have a length of 61 km and width of 1.2 km at the narrowest part. The Hellespont was one of the great crossroads of the ancient world. The out-flowing (westerly) current and prevailing northeast wind caused most ships to stop over in Besika Bay (or Beşik Bay), on the Aegean shoreline. Ships may have had to wait weeks here for favourable sailing conditions. The strong westerly current observed in the Dardanelles is driven by the continued draining of the Sea of Marmara and the Black Sea.

13.4.1 Geological Setting

Hisarlik and the other low hills in the vicinity of Troy are underlain by Neogene-Quaternary sediments deposited in the Dardanelles Graben (Kraft et al. 1980; Yaltirak et al. 2000) (Fig. 13.5). Basement rocks equated with the Rhodope-Serbomacedonian Massif can be observed on the low plateau to the south of Troy. The coastline in the vicinity of Troy has migrated northward during the previous 5,000 years (Fig. 13.8) (Kraft et al. 1980). The shape and size of Troia Bay has changed considerably. Troia Bay was most prominent in the Early Bronze Age (2500 BC), but by the Late Bronze Age, or Homeric times (1250 BC), was considerably smaller. By Roman times (1 AD), Troy was located 3 km inland and Troia Bay had almost disappeared.

The archaeological site is currently separated from the Dardanelles by river estuaries and floodplains. Estuarine sediments were deposited by the Kara Menderes River and Dümrek River (Kraft et al. 1980). They consist of distal and proximal prodelta muds covered by thick deposits of alluvium. The estuarine sediments have a thickness of approximately 20 m. Radiocarbon dating of fossils obtained from drill cores has provided reliable ages for the sediment. The estuaries and floodplains started to develop after stabilization of the sea level at approximately 5,000 BP. Historical sedimentation rates were relatively rapid, although current sedimentation rates of the rivers flowing into the Dardanelles are relatively low. The landscape in the vicinity of Troy has changed considerably during the previous 5,000 years (Kraft et al. 2003). The shallow marine environment of the Mid-Holocene changed initially into a prominent estuary, and then into a brackish lagoon, and finally into areas of seasonal swamps.

13.4.2 Ancient Troy

There is a broad acceptance among archaeologists and historians that part of the antiquity of Troy corresponds to *Ilias*, the city of the Homeric classic. The *Iliad* described historical events from a Greek perspective which occurred towards the end of the Mycenaean period. Some of these events were subsequently described by the Roman historian Strabo (in the 1^{st}C AD). The initial excavations at Troy were undertaken in the 1870s by German archaeologist Heinrich Schliemann. The selection of the site was based on historical descriptions pertaining to the last ten years of the Trojan Wars provided by Homer and Strabo. The "Schliemann Trench" revealed an ancient city with multiple layers (Fig. 13.7b). Despite damage to some relics, misappropriation of gold and valuable artefacts, and misidentification of the "*Homeric Layer*", recent excavations led by Turkish archaeologists have to a large extent substantiated the claims made by Schliemann. Troy was occupied over several millennia, and nine different levels have been identified, each of which corresponds with a unique historical city (Fig. 13.9). Not all of the information presented on the city map is in agreement with the international literature as there is a lack of consensus on the dates of some historical events.

The oldest settlement at Troy is estimated at 2920 BC and the most recent at 500 AD. Level I is associated with an Early Bronze-age culture (2920 BC–2450 BC) that had spread rapidly throughout large parts of the eastern Mediterranean. The Homeric city is associated with Level VI (1700–1250 BC). The Trojan Wars were a series of long drawn-out conflicts which may have lasted from 1260 BC until 1180 BC. The Greek polymath Eratosthenes (276–195 BC) suggested the ten years of events described in the *Iliad* occurred between 1194 BC and 1184 BC. Other researchers, however, suggest the Trojan Wars ended at approximately 1250 BC (e.g., Kraft et al. 1980). The walls of the Homeric city are similar to structures observed at Mycenae in the Peloponnese (Fig. 13.10a). Level VI was, however, almost entirely rebuilt in 1250 BC after damage caused by a catastrophic fire and a severe earthquake. Level VII overlaps with the Late Bronze Age and Early Iron Age (1250 BC–

a

b

Fig. 13.7 **a** View from the antiquity of Troy looking northwest over the coastal plain toward the Dardanelles with hills of the Gallipoli Peninsula visible in the background; **b** The Schliemann Trench cut down from the Roman period (red bricks, upper right) to some of the oldest parts of Troy (low walls in the base of the excavation)

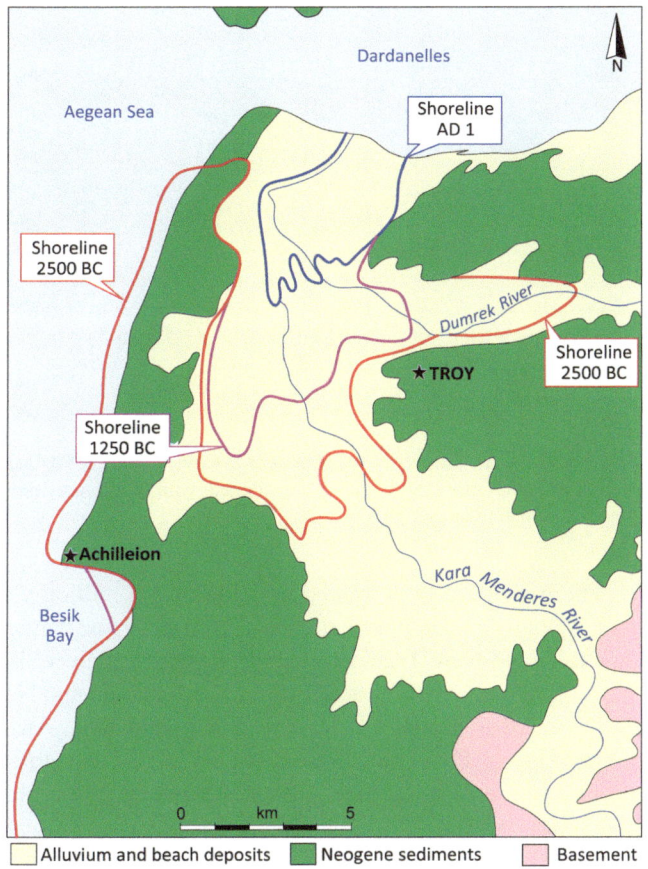

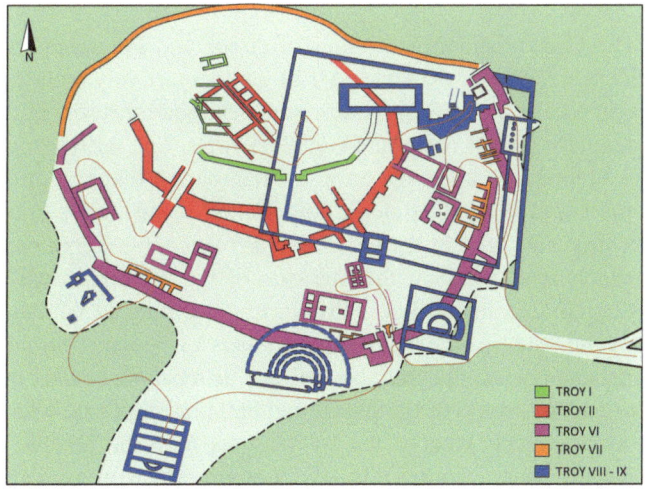

Fig. 13.8 Simplified geological map of the area surrounding the antiquity of Troy with reconstruction of the coastline in the Early Bronze Age (2500 BC), in the Late Bronze Age (1250 BC), and in Roman times (1 AD). *Source* Modified after Higgins and Higgins (1996) and Kraft et al. (1980)

Fig. 13.9 The antiquity of Troy contains nine cities, Level I through Level IX. The Homeric city is correlated with Level VI and the Greek and Roman cities with Level VIII and Level IX, respectively

Fig. 13.10 a One of the few walls remaining of the Homeric city (Level VI) at Troy; **b** Parts of the Greek Level VIII (lower left) and the Roman Level IX (centre and upper right) occur in the Sanctuary at Troy

1000 BC). Level VII had some continuity with the Homeric city. Level VIII (700 BC–85 BC) correlates with the Greek city of *Ilium* (Fig. 13.10b). Level IX (85 BC–500 AD) is associated with the Roman and Byzantine city of *New Ilium*.

13.4.3 Historical Reconstruction

Many of the historical sites described in the *Iliad* have been identified by Kraft et al. (2003). The attacking Greek or Mycenaean army camped west of the city on the shores of the Aegean Sea (Fig. 13.11). This location is consistent with historical reports suggesting the army camp was based approximately 4 km from the city. Moreover, the occurrence of marshland and saltwater lagoons at the landing site were noted by Strabo. The Kesik Cut is a defensive line next to the army camp that includes a wall and ditch, both of which have been unearthed. The attacking army of the Greeks forded the Scamander River (renamed the Kara Menderes River) on the floodplains south of Troy (the Scamandrian Plain). The classical site of Achilleion at Besika Bay was

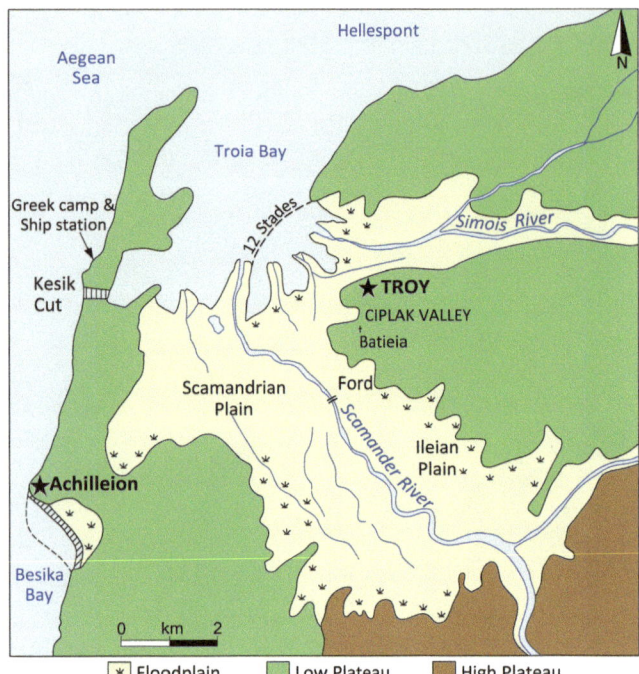

Fig. 13.11 During the Late Bronze Age or Homeric times (1250 BC) Troy was perched on a hill located adjacent to Troia Bay. The city was enclosed on the northern and southern sides by marshes associated with the Scamander and Simois Rivers. Historical sites include Achilleion (Besika Bay), the landing place of the Greek armies, the Kesik Cut, and the ford over the Scamander River. The Trojans mounted their defence at Batieia (Thorn Hill). *Source* Kraft et al. (2003)

famous in antiquity for its association with Achilles, after whom it was named, and also subsequently with Alexander the Great. The Batieia (Thorn Hill) is a high mound where the Trojans mounted their defence.

13.5 Gallipoli Peninsula National Park

The Gallipoli Peninsula is situated in the historical region of Thrace, on the northern shore of the Dardanelles (Fig. 13.2). The Gallipoli Peninsula contains archaeological sites which date back to 4,000 BP. In 1973, part of the region was incorporated into the Gallipoli Peninsula Historical National Park (area of 33,000 ha). The national park was established to commemorate the Dardanelles Campaign of 1915–1916 (World War I) and honours the approximately 500,000 lives lost. The campaign was an unsuccessful attempt by the Allied Powers, mostly, but not exclusively, Australia, France, New Zealand, and the United Kingdom, to wrestle control of the Dardanelles sea route away from Turkey. The national park contains statues, memorials, and cemeteries that are immaculately maintained, with equal status afforded to both the Turkish and Allied Powers.

Fig. 13.12 View looking north from Anzac Cove, Gallipoli Peninsula National Park, showing badlands associated with rapid erosion of Neogene-Quaternary sediments

The national park is visited by more than a million visitors annually and holds particular significance for Turkey, Australia and New Zealand. One of the most widely visited parts of the park is the beaches where the Australia and New Zealand Army Corps (ANZAC) landed, including Anzac Cove (Fig. 13.12). The campaign failed in part due to the rugged nature of the terrain, which is apparent in the cliffs and scrub-covered ridges inland from Anzac Cove. This area is underlain by poorly-consolidated sands and gravels (Neogene-Quaternary). Rapid erosion enhanced by uplift on the margins of the Dardanelles Graben has created badlands topography.

13.6 Assos and Pergamum

The Greco-Roman antiquities of Assos and Pergamum are located between 80 km and 120 km, respectively, south of Çanakkale (Fig. 13.2). Assos is a small coastal resort in the southern part of the broad Biga Peninsula. The nearest town is Behram. Assos is the only safe harbour on this part of the coast and was of strategic importance in historical times. Assos is located in the Çanakkale Province, but Pergamum, which is situated on the outskirts of the regional town of Bergama, is in the Izmir Province. According to Christian legend, the apostle Paul visited Assos on journey through Asia Minor. Assos probably fell into disrepair soon after this period and some of the pillars and building stones may have been sold off. Most of the art treasures from the Temple of Athena were acquired by the Louvre Musuem in Paris. Pergamum was the northernmost of the seven churches of Asia Minor. The Kazdağı National Park (area of 21,542 ha) occurs between the two antiquities. This is a mountainous area formerly known as Mount Ida. The park includes several peaks (maximum altitude 1,574 m), together with streams, waterfalls, and Alpine pastures.

13.6.1 Geological Setting

The antiquities of Assos and Pergamum occur in the Sakarya Zone, i.e., they are situated between the NAFZ and the IAFZ (Fig. 13.5). Both sites are unusual in that they are constructed on mesas or plateaus consisting of volcanic lavas. (Most of the Greco-Roman antiquities in western Turkey are built on limestone or poorly-consolidated graben sediments). The volcanic lavas are associated with the Dikili-Çandarli Volcanic Suite, part of the extensive Western Anatolian Volcanic Province (Karacik et al. 2007). The volcanic province is similar to the volcanism observed at active continental margins. The mantle-derived magmas probably experienced significant crustal contamination (Karacik et al. 2007). This part of western Turkey includes prominent W-E aligned grabens. Pergamum lies on the northern edge of a broad plain associated with the Bakırçay Graben. The magmas associated with the Dikili-Çandarli volcanics exploited faults concomitantly with activation of the Neogene grabens.

The two components of the volcanic suite, the Dikili Group and the Çandarli Group, formed in the Lower-Mid Miocene and Upper Miocene-Pliocene, respectively. The volcanic plateau at Pergamum is associated with an andesitic lava dome, part of the older Dikili Group. The mesa at Assos consists of trachytic lava, part of the younger Çandarli Group. The dark coloration of the trachytic lava, together with the occurrence of white-coloured crystals of plagioclase, are distinctive features (Fig. 13.13).

The Kazdağı National Park is mostly underlain by the Kazdağı Massif, the lower part of which is comprised of an ophiolite complex, or obducted oceanic crust (Erdoğan et al.

Fig. 13.13 The dark-coloured trachytic lava at Assos contains prominent phenocrysts of plagioclase (white)

2013). The ophiolite is unconformably overlain by metamorphic rocks, together with an intrusive body of granite-gneiss (dated at 230 Ma). The peak of the metamorphism associated with the Alpine Orogeny in this region occurred at approximately 25 Ma.

13.6.2 Assos

The antiquity of Assos is located on a steep sided hill above the modern village (Fig. 13.14a). The Aegean Sea and island of Lesbos can be observed from the hill (Fig. 13.14b). The natural cleavage of the trachytic lavas forms terraces in the cliff face which were used for the earliest constructions. Assos was founded at approximately 1000 BC. The monumental buildings include a well-preserved theatre and two large columns from the Hellenistic period. The Temple of Athena illustrates how the monumental buildings were constructed directly on the bedrock (Fig. 13.15). Establishment of the Academy of Assos in the 4[th]C BC ushered in a golden period when a number of famous philosophers, including Aristotle (384 BC–322 BC), moved to the city. Important advancements were made in philosophy and science. Aristotle undertook the first scientific study of biology after he had moved from Assos to the nearby island of Lesbos.

Aristotle is considered as one of the fathers of geology, and he reported that geological changes are too slow to be observed in a single lifetime. According to the founder of British geology Charles Lyell, Aristotle commented on examples such as lakes "that had dried up", "deserts that had become watered by rivers" (such as the Nike Delta), and "the up heaving of one of the Aeolian Islands previous to a volcanic eruption". It is likely that Aristotle was aware of the silting up of harbours such as Troy and Ephesus.

13.6.3 Pergamum

The acropolis complex at Pergamum is built on a flat-topped hill, or mesa, with a height of 450 m above sea level, length of approximately 5.5 km, and width of 1 km (Fig. 13.16). The mesa is part of a volcanic plateau situated on the northern edge of a broad plain associated with the Bakırçay Graben. (The Bakırçay River was known as the Caicus River in ancient Greek times). The mesa has steep slopes on the north, west, and east sides, characteristic of the rugged nature of volcanic plateaus (Fig. 13.17). Many of the monumental buildings at Pergamum are constructed on natural terraces on the sides of the mesa. Pergamum was one of the richest and most influential cities in the region of Lydia during the Classical Greece period. Pergamum was controlled by the Attalid dynasty during the Hellenistic

a

b

Fig. 13.14 a The coastal resort of Assos is located at the base of a steep hill consisting of dark-coloured trachytic lava. The antiquity is visible on the crest of the hill; **b** The antiquity of Assos overlooks the Aegean Sea with the island of Lesbos visible in the background

Fig. 13.15 The Temple of Athena at Assos is built on the trachytic lava bedrock

period, at which time the city continued to prosper, and became the capital of an extensive region in this part of Lydia. Pergamum was subsequently occupied by the Romans and the Ottomans. Many of the antiquities reflect the changes and rebuilding which occurred over these long periods of time.

A reconstruction of the acropolis complex at Pergamum shows some of the most important features (Fig. 13.18a). The theatre, which dates from the Hellenistic period, included a monumental entrance and could accommodate 10,000 people (Fig. 13.18b). The most famous of the antiquities is the "Pergamum Altar", a monumental construction located on the terrace to the southeast of the theatre. The Pergamum Altar was built in 190 BC during the reign of Eumenes II. The building measured 35.6 m in width and 33.4 m in

Fig. 13.16 The antiquity at Pergamum is located on a volcanic plateau on the northern side of the Bakırçay Graben. The acropolis occurs on the highest point of the mesa which, despite the relatively steep sides, includes natural terraces where many of the larger monumental buildings were constructed. *Source* Google Earth image

Fig. 13.17 View of part of the rugged topography characteristic of the volcanic plateau at Pergamum

depth. Only the foundations of the original building remain (Fig. 13.19a). Some of the pillars together with the large frescoes were taken to Germany in the latter part of the 18[th]C AD, where they form part of a major exhibit in the Berlin Musuem. The reconstruction has been used for various propaganda reasons and remains a sensitive issue. The Temple of Trajan is a major attraction of the site as many of the original columns and pillars remain intact (Fig. 13.19b). The Asklepion at Pergamum (sanctuary of the Greek God of healing) has undergone multiple stages of construction and restoration, including under the Roman emperor Hadrian (117–138 AD). The reconstruction includes an extensive portico (Fig. 13.20a). The Lower Agora (an open space for civic meetings, including for sale of goods) is situated at the base of the acropolis (Fig. 13.20b).

13.7 Ephesus

The Greco-Roman antiquity of Ephesus is located near the small town of Selçuk in the Central Aegean region of western Turkey (Fig. 13.2). This region includes broad, fertile valleys and rugged hills. Ephesus was the main commercial centre of the eastern Mediterranean during historical times. The antiquity is a popular tourist destination and attracts large numbers of visitors every year, in part due to the easy access from beach resorts and cruise ships. The site covers several hectares and excavations have revealed centuries of history, from the Attic colonists (10[th]C BC) through to the Byzantine period. Ephesus was partially destroyed by a major earthquake in 614 AD, but the main reason for its decline was the persistent silting up of the "Sacred Harbour". The Roman historian Strabo reported attempts to modify the entrance to the harbour by building of a long mole, but this increased the problem of sedimentation.

13.7.1 Geological Setting

Ephesus is located on the southern flanks of a broad valley associated with the Küçük Menderes River ("Little Meander") (Fig. 13.21). "Menderes" is the Turkish translation of the Greek noun meander and is applied to many snake-like (or meandering) rivers. The river was known to the Ancient Greeks as the Kaystros River (or Cayster River) (Higgins and Higgins 1996). The valley is associated with the

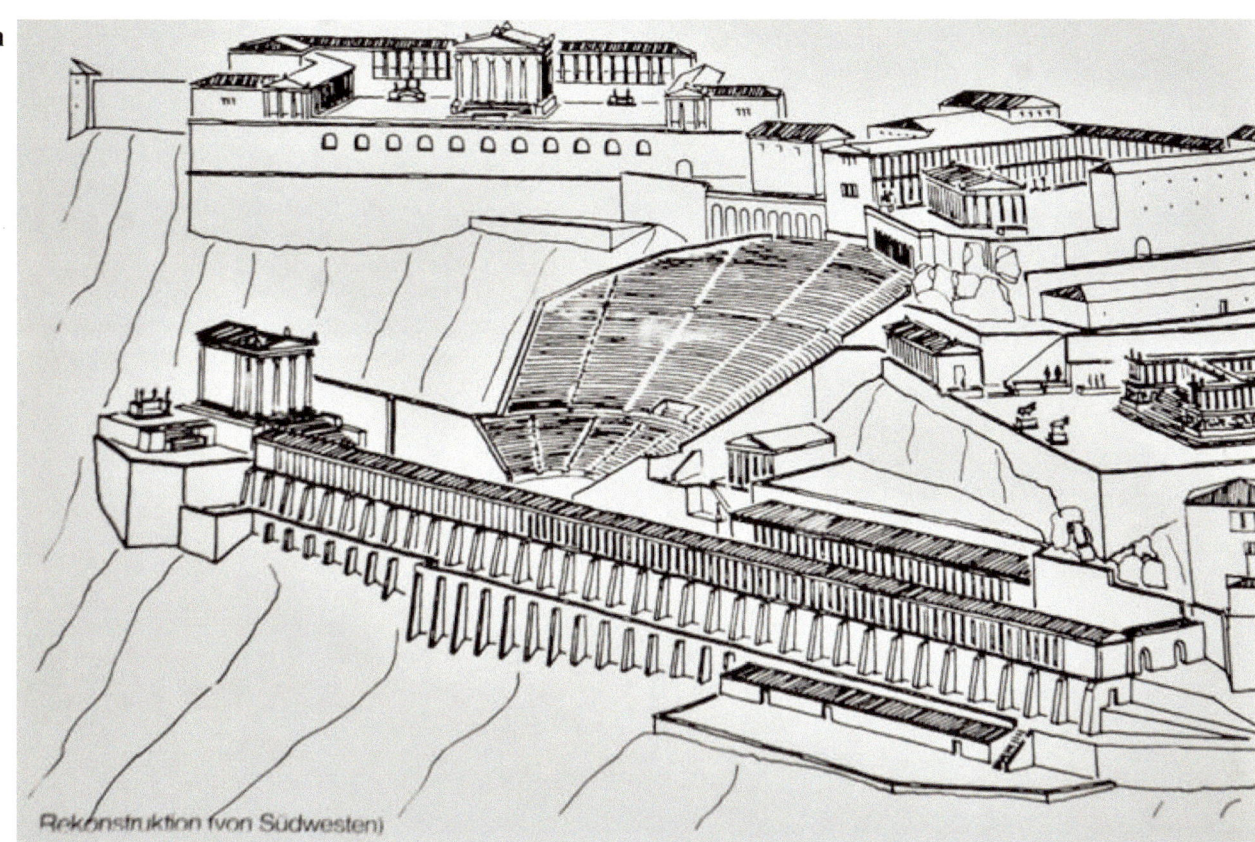

Fig. 13.18 **a** Reconstruction of part of the acropolis complex at Pergamum (view looking east) with the theatre situated on a lower terrace (foreground) and the Pergamum Altar on an upper terrace (centre right); **b** View of the theatre at Pergamum looking southwest over the modern city of Bergama (situated wholly within the Bakırçay Graben). Outcrops of grey-green andesite lava visible on the mesa

Fig. 13.19 a The foundations of the Pergamum Altar; **b** The foundations and bases of the columns of the Temple of Trajan at Pergamum are constructed from dark grey-green andesite lava; the columns are of white marble

Fig. 13.20 **a** The extensive portico is part of the reconstructed Asklepion at Pergamum; **b** The Lower Agora at Pergamum is situated at the base of the acropolis complex

Fig. 13.21 The archaeological site of Ephesus is located situated close to the southern margins of the Küçük Menderes Graben. The foundations of the Artemision occur in swampy part of the valley but many of the other monumental buildings occur on the flanks of the more resistant horst block. *Source* Google Earth image

approximately W-E trending Küçük Menderes Graben (Fig. 13.6). The graben dissects the northern part of the Menderes Massif (Seyitoglu and Isik 2009). The horst blocks adjacent to the graben consist of Proterozoic-Palaeozoic metamorphics, including gneiss, mica schist, quartzite and marble. The main part of the archaeological site at Ephesus is situated close to the southern margins of the graben (Fig. 13.22a). Extensive Hellenistic walls occur on the hillsides of Mount Bülbül (or Mount Preon), a prominent horst. Some of the monumental buildings, including the Great Theatre, are constructed on the western and southern flanks of Panayir Daği (or Mount Pion) (Fig. 13.22b). Mount Pion is a small horst block located within the graben.

13.7.2   Antiquities

Ephesus is particularly well known for the association with the Classical Greece (500 BC–323 BC), Hellenistic (323 BC–31 BC) and Roman periods. Historical descriptions position the city on the shores of a large estuary, but the archaeological

site is situated approximately 5 km from the shoreline (Fig. 13.22a). The most famous building at Ephesus during these periods was the Artemision, one of the Seven Wonders of the Ancient World. The earliest building of the Artemision was completely destroyed by floods and a second building, constructed between approximately 550 BC and 540 BC, was destroyed by arson in 356 BC. Construction of the most recent version commenced in 323 BC, is estimated to have taken more than a hundred years to complete, and may have survived for more than 600 years. The building is thought to have been located close to Selçuk, although only the foundations remain (Fig. 13.21). The swampy nature of this part of the valley—the Artemision was built entirely within the graben - has, however, resulted in debates concerning the exact location. The Artemision was formerly linked to the Great Theatre by the upper part of the Sacred Way.

The advantage of accessing Ephesus by the Koressos Gate is that the southwestern section of the archaeological site contains the main highlights (Fig. 13.21). The monumental buildings located on, or partially on, the resistant horst blocks are well preserved compared with those located wholly

Fig. 13.22 **a** View looking west along Harbour Street, Ephesus, towards a grassy area which corresponds with the historical location of the Sacred Harbour. The harbour was originally part of a large estuary, but the Aegean Sea (upper right) is now several kilometres to the west. This part of the archaeological site is located in the Küçük Menderes Graben, the southern margin of which is fringed by resistant horst block, such as Mount Preon (upper left); **b** The Great Theatre at Ephesus is partially cut into the slopes of Mount Pion, a horst block located within the graben. View includes Marble Street in the foreground

Fig. 13.23 **a** View of a section of Marble Street at Ephesus; **b** The Library of Celsius at Ephesus includes columns built of marble breccias

within the graben. The Great Theatre at Ephesus was constructed during the Hellenistic period and subsequently rebuilt by the Romans (Fig. 13.1). The theatre is linked to the Sacred Harbour by Harbour Street (or Arcadian Street), a 500 m long and 11 m wide road which has been partially excavated (Fig. 13.22a). The existence of the Sacred Harbour, however, remains unproven (although reports of a famous siege in 190 BC provide corroborating descriptions).

The antiquities south of the Great Theatre are reached by Marble Street, which is partly lined by columns and frescoes (Fig. 13.23a). Marble frescoes include the famous sculptures of *Nike* (the Greek god of Victory). This section of the complex includes the Lower Agora and the Library of Celsius (built during the Roman period at approximately 125 AD), the eastern facade of which has been carefully rebuilt (Fig. 13.23b). The archaeological complex at Ephesus covers a large area and many of the buildings, including Roman temples and baths, are located on the southern slopes of Mount Pion. These include sections of a large housing complex (which dates to the Archaic Greek times: 800 BC–500 BC), as well as the Upper Agora, the Odeon, and the East Gymnasium located near the Magnesian Gate. The housing complex is the subject of a major excavation and is protected under a roofed enclosure. Marble slabs were used for decorative purposes in the houses (Fig. 13.24a). The use of book-matched marble slabs illustrates how ancient this technique is (Fig. 13.24b).

Fig. 13.24 **a** Marble slabs were used for decorative purposes in the housing complex at Ephesus; **b** Book-matched slabs of marble breccia at Ephesus may date from the Archaic Greek period

13.7.3 Geomorphological Reconstruction

Over the two thousand years of inhabitation, Ephesus was continuously being adapted to suit landform changes related to sedimentation in the active graben. Mapping of sediments in the lower part of the valley and floodplains associated with the Küçük Menderes River enabled Kraft et al. (2001) to reconstruct the ancient shoreline (Fig. 13.25). The city was moved several times to accommodate the changing position of the Sacred Harbour. During the Hellenistic and early Roman times, the Aegean Sea extended to the foot of Mount Pion. The progradation of the delta has continuously pushed the coastline westward. The reconstruction raised questions about the Artemision, which is generally positioned in the northeastern part of the site, but this may be incorrect. The location may have been closer to the estimated position of the Sacred Harbour.

The archaeological site of Ephesus has been affected by numerous earthquakes (Ambraseys 2009). A number of significant earthquakes occurred between 17 AD and 614 AD, as well as in the 10th and 11thC AD. The largest event was the catastrophic Southwest Anatolia Earthquake, which probably occurred in 262 AD, and devastated most of the Greco-Roman cities proximal to the west coast of Anatolia.

The epicenter was in the seismically-active southern Aegean Sea (Chap. 12). Many sites in coastal Anatolia were flooded, presumably by tsunamis. Historical accounts report that buildings at Ephesus had to be reconstructed after the earthquake and the Library of Celsius may have suffered a major fire related to the event.

13.8 Antiquities in the Denizli Basin

The regional city of Denizli in the southwestern part of Turkey occurs in a fertile region with a temperate climate. The region is dominated by broad valleys which are in marked contrast to the mountainous and semi-arid areas located farther inland. The Denizli Basin contains a number of Greco-Roman and Christian antiquities, including Aphrodisias, Colossae, Laodicea, and Kibyra (or Cibyra) (Fig. 13.3). Aphrodisias, Colossae, and Laodicea are located approximately 30 km west, 20 km east, and 6 km north of Denizli, respectively, and Kibyra is situated 110 km to the southwest near the town of Gölhisar (Fig. 13.2). The antiquities include archaeological sites spread over large areas and include monumental buildings as spectacular as those at Ephesus and Pergamum, but there are far fewer visitors.

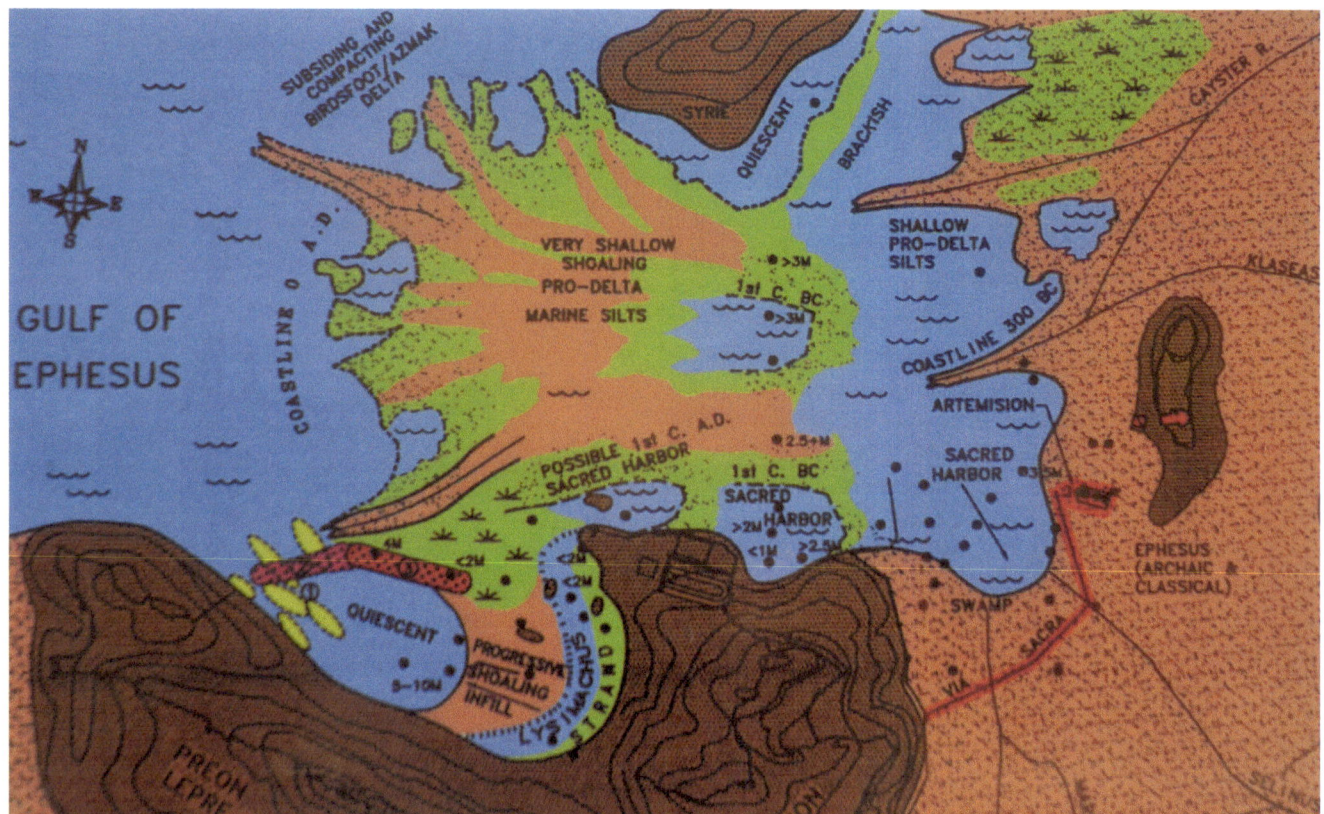

Fig. 13.25 Geological map mounted in a display board at Ephesus showing westward migration of the Sacred Harbour in the region to the north of the main archaeological site (width of view approximately 5 km). The Great Theatre is located near the southern end of the Sacred Way. *Source* Map after Kraft et al. (2003)

13.8.1   Geological Setting

Denizli is located in a large basin associated with several intersecting grabens (Fig. 13.6). The most prominent feature is the Buyuk-Menderes Graben, which in the vicinity of Denizli is bounded to the northeast by the Pamukkale Fault and to the southwest by the Babadağ-Honaz Fault (Seyitoglu and Işik 2009; Hançer 2013). The NW-SE aligned Pamuk-kale Fault is a relatively well-defined structure, but the Babadağ-Honaz structure is a complex zone of faulting comprised of several W-E aligned segments. Subsidiary grabens occur on the southern flanks of the Buyuk-Menderes Graben e.g., the Morsynus Graben. The grabens are defined by normal faults. Faulting was initiated during the Early-Late Miocene. Many faults in the region are currently active. The Buyuk-Menderes Graben has been subjected to severe earthquakes during historical times and the possibility of future seismic events is discussed by Hancer (2013).

The horst blocks adjacent to the Buyuk-Menderes Graben (and the subsidiary grabens) in this section of the Menderes Massif contain inliers of Proterozoic-Palaeozoic basement (Bozkurt and Oberhänsli 2001; Seyitoglu and Işik 2009).

The inliers are dominated by granite-gneiss and amphibolite (Fig. 13.6). Some of the Palaeozoic-Mesozoic metamorphic terrains (including the Ortaköy Formation) contain high quality marbles, together with quartzite and mica schist (Çemen et al. 2006).

13.8.2   Aphrodisias

The archaeological site of Aphrodisias is located in a broad valley associated with the Morsynus River west of the Buyuk-Menderes Graben (Fig. 13.6). The valley occurs in a hilly region where metamorphic rocks of the Menderes Massif outcrop over large areas. Aphrodisias was an important Greco-Roman city but fell into disuse after being severely affected by earthquakes in the 4thC AD and 7thC AD. The seismic activity affected the water table, and parts of the city were flooded. The main feature of the antiquity is a complex of partially-restored buildings, including the Bouleuterion (a well-preserved theatre), the Temple of Aphrodite, several Roman baths and large civic structures (Fig. 13.26a). Some of the seats reserved for dignitaries in

Fig. 13.26 a The archaeological site of Aphrodisias includes sections of several temples located in a broad valley (the Morsynus Graben) enclosed by rolling hills (horst blocks); **b** The Bouleuterion at Aphrodisias includes a colonnaded stage constructed of locally-quarried marble

Fig. 13.27 The unusual ovoid-shaped stadium at Aphrodisias includes seats built into raised earthworks

the Bouleuterion at Aphrodisias are built of white marble (Fig. 13.26b). The stadium at Aphrodisias is a large and impressive structure which had a capacity of 30,000 with thirty rows of seats constructed on grassy banks (Fig. 13.27). The ovoid shape (length of 270 m and width of 60 m) is unusual. The stadium was probably built around a natural depression.

13.8.3   Laodicea

The antiquity of Laodicea is located in a valley associated with tributaries of the Lycus River, close to the boundary faults of the Buyuk-Menderes Graben. The proximity to active faults has resulted in major earthquakes (e.g., Bozkurt et al. 2000). Laodicea was largely destroyed by a significant earthquake at approximately 60 AD, but was soon rebuilt and restored as a major centre. (Laodicea was referred to in historical texts as recently as 1119 AD). The extensiveness and number of antiquities is remarkable. Outer walls enclose paved streets, some of which are flanked by colonnades and pedestals, together with baths and temples, a gymnasium, relics of theatres, and a Senate House. The main antiquities

date to the period 261 BC to 253 BC, although there is evidence of older settlements. During Roman times, Laodicea was one of the most important centres in Asia Minor due to its position on a significant trade route.

13.8.4   Colossae and Kibyra

The archaeological sites of Colossae and Kibyra (which the author has not visited) are situated near active fault zones and have suffered extensive earthquake damage. Colossae is located on the main road leading from the Aegean region of Anatolia toward the central parts and was described by the historian Herodotus as a major town in 480 BC. Colossae is located at the southeastern extremity of the Denizli Basin where the main border faults converge (Piccardi 2007). The most significant structure is the active fault that transects the site of Honaz (ancient Chonae), at the base of Mount Honaz, approximately 3 km south of Colossae. Field observations suggest that much of the damage at Colossae can be ascribed to the earthquake of 60 AD (Michetti and Hancock 1997). The Early Christian and medieval legend attached to the destruction of the antiquity by Pagans attests that the

Fig. 13.28 The theatre at Kibyra. *Source* https://images.app.goo.gl/ykVDc99aXmwY8Wqx5

Archangel Gabriel appeared, and by striking a rock with his staff, opened a new fissure in the rock which diverted a water course (Piccardi 2007). The name of the new settlement, Chonae, was derived from the fissure (referring to the fissure as a funnel), and the sanctuary constructed here was visited by pilgrims for many centuries. The fissure may be related to the earthquake of 60 AD.

Kibyra occurs in a NE-SW trending fault zone which was ruptured during both historical and recent times. The ancient theatre at the archaeological site has undergone considerable restoration (Fig. 13.28). The majority of the antiquities reveal blocks and columns which have been broken and tilted. Sinistral faults have offset seat rows in the ancient stadium by up to 50 cm (Akyüz and Altunel 2001).

13.8.5 Marbles

The wealth of the ancient cities in the region of Denizli was in part derived from their proximity to extensive deposits of high-quality marble (Bruno et al. 2009). The occurrence of marble sculptures is a major feature of Aphrodisias. The most famous of the black and white statues associated with the "School of Aphrodisias" are of the Greek god Aphrodite. Some of the sculptures and statues are preserved on the archaeological site, with others displayed in local and international museums. The white and blue-grey Aphrodisian or Carian marble was extensively quarried during historical times from hill slopes near the cities. The name Carian is derived from the historical name of the region (modern day quarries export high-quality "*Caria marble*").

The Aphrodisian marble was used in many of the antiquities of Aphrodisias and Laodicea for constructions, including colonnades and facades. Historical quarries include the Göktepe locality, where a high-quality, fine-grained, white marble was obtained, and the Karahisar (or city) quarries which are included in the archaeological site of Aphrodisias. Large, stepped blocks of fine-or medium-grained white and grey marble can be observed here. The dimensions and shapes of the cut blocks are consistent with techniques used in Roman times (Bruno et al. 2009). The majority of the marble produced from the quarries in the region was probably exported to Rome. The Aphrodisian marble replaced usage of the well-known Parian and Pentelic marbles during the latter parts of the Roman Empire.

References

Akyüz, H. S., & Altunel, E. (2001). Geological and archaeological evidence for post-Roman earthquake surface faulting at Cibyra, SW Turkey. *Geodinamica Acta, 14,* 95–101.

Ambraseys, N. (2009). Earthquakes in the Mediterranean and middle east: A multidisciplinary study of seismicity up to 1900. Cambridge University Press ISBN 978-0521872928.

Bozkurt, E. (2000). Timing of extension on the Büyük Menderes Graben, Western Turkey, and its tectonic implications. In: Bozkurt, E., Winchester, J.A., & Piper, J.D.A. (Eds.). Tectonics and magmatism in Turkey and the surrounding area. *Geological Society of London Special Publication, 173,* 385–403.

Bozkurt, E. (2001). Neotectonics of Turkey—A synthesis. *Geodinamica Acta, 14,* 3–30.

Bozkurt, E., & Mittwede, S. K. (2005). Introduction: Evolution of Neogene extensional tectonics of western Turkey. *Geodinamica Acta, 18,* 153–165.

Bozkurt, E., & Oberhänsli, R. (2001). Menderes Massif (western Turkey): Structural, metamorphic and magmatic evolution—A synthesis. *International Journal of Earth Sciences, 89,* 679–708.

Bozkurt, E., Winchester, J. A. & Piper, J. D. A. (2000). Tectonics and magmatism in Turkey and surrounding areas. *Geological Society of London Special Publication, 173,* 521.

Bruno, M., Elçi, H., Yavuz, A. B., & Attanasio, D. (2009). Unknown ancient marble quarries of western Asia Minor. In: Gutierrez Garcia-M, A., Mercadal, P. L. and Roda de Llanza, I. (Eds.) Interdisciplinary Studies on Ancient Stone, Proceedings of the IX Association for the Study of Marbles and Other Stones in Antiquity (ASMOSIA Conference, Tarragona), pp. 562–572.

Çemen, I., Catlos, E. J., & Gögüs, O. (2006). Postcollisional extensional tectonics and exhumation of the Menderes Massif in the Western Anatolia extended Terrane, Turkey. *Geological Society of America Special Paper, 409,* 353–379.

Erdoğan, E., Akay, E., Hasözbek, A., Satir, M., & Siebel, W. (2013). Stratigraphy and tectonic evolution of the Kazdağı Massif (NW Anatolia) based on field studies and radiometric ages. *International Geology Review, 55*(16), 2060–2082.

Hancer, M. (2013). Study of the structural evolution of the Babadağ-Honaz and Pamukkale Fault Zones and the related earthquake risk potential of the Buldan region in SW Anatolia, East of the Mediterranean. *Journal of Earth Science, 24,* 397–409.

Higgins, M. D., & Higgins, R. (1996). A geological companion to Greece and the Aegean (p. 240). New York: Cornell University Press.

Karacik, Z., Yilmaz, Y., & Pearce, J. (2007). The Dikili-Çandarli volcanics, Western Turkey: Magmatic interactions as recorded by petrographic and geochemical features. *Turkish Journal of Earth Sciences, 16,* 493–522.

Kraft, J. C., Kayan, I., & Erol, O. (1980). Geomorphic reconstruction in the environs of ancient Troy. *Science, 209,* 776–782.

Kraft, J. C., Rapp, G., Kayan, I., & Luce, J. V. (2003). Harbor areas at ancient Troy: Sedimentology and geomorphology complement Homer's Iliad. *Geology, 31,* 163–166.

Kraft, J. C., Kayan, I., Buckner, G., & Rapp, G. (2001). A geologic analysis of ancient landscapes and the harbors of Ephesus and the Artemision in Anatolia. *Jahreshefte des Österreichischen Archaologischen Institues in Wien, 69,* 175–234 (in English).

Marriner, N., Morhange, C., & Goiran, J. P. (2010). Coastal and ancient harbour geoarchaeology. *Geology Today, 26,* 21–27.

Michetti, A. M., & Hancock, P. L. (1997). Paleoseismology: Understanding past earthquakes using Quaternary geology. *Journal of Geodynamics, 24,* 3–10.

Okay, A. I., & Tüysüz, O. (1999). Tethyan sutures of northern Turkey. *Geological Society London Special Publication, 156,* 475–515.

Piccardi, L. (2007). The AD 60 Denizli Basin earthquake and the apparition of Archangel Michael at Colossae (Aegean Turkey). In: Piccardi, L. & Masse, W. B. (Eds.) Myth and Geology. *Geological Society, London, Special Publications, 273,* 95–105.

Seyitoglu, G., & Işik, V. (2009). Meaning of the Küçük Menderes graben in the tectonic framework of the central Menderes metamorphic core complex (western Turkey). *Geologica Acta, 7,* 323–331.

Taymaz, T., Yilmaz, Y., & Dilek Y. (2007). Introduction to the geodynamics of the Aegean and Anatolia: Introduction. In: Taymaz, T., Yilmaz, Y. and Dilek Y., (Eds.) Geodynamics of the Aegean and Anatolia. *Geological Society London Special Publications, 291,* 1–16.

Yaltirak, C., Alpar, B., Sakinç, M., & Yüce, H. (2000). Origin of the Strait of Çanakkale (Dardanelles): Regional tectonics and the Mediterranean-Marmara incursion. *Marine Geology, 153,* 17–40.

Yilmaz, Y., Genç, S.C., Gürer, F., Bozcu, M., Yilmaz, K., Karacık, K., Altunkaynak, S., & Elmas, A. (2000). When did the western Anatolian grabens begin to develop? In: Bozkurt, E., Winchester, J. A., Piper, J. D., (Eds.) Tectonics and Magmatism in Turkey and the Surrounding Area. *Geological Society London Special Publications, 173,* 353–384.

Yolsal-Çevikbilen, S., Berk Biryol, C., Beck, S., Zandt, G., Taymaz, T., Adiyaman, H. E., et al. (2012). 3-D crustal structure along the North Anatolian Fault in north-central Anatolia revealed by local earthquake tomography. *Geophysical Journal International, 188,* 819–849.

Abstract

The Hierapolis-Pamukkale archaeological and geosite is located near Denizli in southwest Turkey. The site was awarded UNESCO World Heritage status in 1988 and receives more than 2.5 million visitors annually. The translation of Pamukkale is "Cotton Castle", an apt description of the white deposits of travertine that are so extensive they form an extensive ridge. Travertine is a hard, compact variety of limestone deposited from hot springs or geothermal waters. The ancient city of Hierapolis is situated on the crest of the travertine ridge and was originally built as a Greek colony in the 2ndC BC. The city was enlarged during the Roman Empire and became an important spa and burial ground. The archaeological and historical data suggest the city had to be entirely rebuilt after the major earthquake of 60 AD and was abandoned after an earthquake in 1354 AD. The Hierapolis-Pamukkale site is located in the Denizli Basin, a significant feature of the Menderes Massif. The basin has developed where two large grabens intersect, the W–E trending Büyük Menderes Graben and the NW–SE trending Alaşehir Graben. Four main centres of active travertine deposition are recognized, the distribution of which is influenced by the NW–SE trending graben faults. The travertine deposits at the Hierapolis-Pamukkale site are associated with hot springs located on the Hierapolis Fault, a subsidiary feature of the regional Pamukkale fault system. The grabens are primarily Neogene features related to crustal extension and many of the faults in the Denizli Basin remain active. The basin is thermally active and the occurrence of hot springs and groundwater circulation through limestone bedrock is a key component in formation of the travertine. The upwelling hot waters are supersaturated in calcium carbonate. Seventeen hot water springs with temperatures ranging from 35 °C to 100 °C are currently active at Pamukkale. The deposition of the travertine is contemporaneous with recent fault activity and the deposits have a maximum age of 400,000 BP. Most of the travertine is younger than 60,000 BP. Five principal varieties of travertine are recognized at Pamukkale, the most abundant types of which are fissure-ridge and terrace-mound travertine. The fissure-ridge travertine forms the largest of the deposits and has been extensively quarried. The majority of the buildings and tombs at the Hierapolis archaeological site are built of either locally-derived marble or fissure-ridge travertine. Large blocks of white travertine were used in the Roman theatre for decorative purposes. Most of the sacred sites at Hierapolis were constructed directly above the active Hierapolis Fault. These include the Temple of Apollo and Plutonium. The site was probably chosen for the occurrence of natural gases associated with the thermal waters. Priests are thought to have deceived visitors by appearing to be immune to the toxic vapours. Historical legends in the region are linked to catastrophic earthquakes which may describe the major event of 60 AD. Movement on the fault zone at Hierapolis occurred during the Denizli Earthquake of 1965 and toxic gases, similar to those identified in the ancient Plutonium, are emitted from the rupture.

Keywords

Active faults • Geothermal water • Grabens • Limestone • Plutonium • Roman Empire • Travertine

---

Photographs not otherwise referenced are by the author.

The original version of this chapter was revised: Belated corrections have been incoporated. The corrections to this chapter are available at https://doi.org/10.1007/978-3-030-54693-9_18

© Springer Nature Switzerland AG 2021, corrected publication 2021
R. N. Scoon, *The Geotraveller*,
https://doi.org/10.1007/978-3-030-54693-9_14

14.1 Introduction

The Hierapolis-Pamukkale archaeological and geosite in southwest Turkey was awarded UNESCO World Heritage status in 1988. The translation of Pamukkale is "Cotton Castle", an apt description of the white deposits of travertine which are so extensive they form an extensive ridge (Fig. 14.1). The ancient city of Hierapolis is situated on the crest of the travertine ridge. The city was originally built as a Greek colony in the 2nd C BC, during the latter part of the Hellenistic period, but may have been inhabited since 1900 BC. The antiquity is thought to have been named after *Hiera* or *Hiero*, the wife of *Telephus*, the mythical ancestor of the Pergamenes. Hierapolis was enlarged during the Roman Empire and became an important spa and burial ground. The ancient buildings, sanctuaries, and tombs have been excavated from the travertine deposited since the site was abandoned towards the end of the Byzantine Empire. An Early Christian and medieval legend, based on geological events, is attached to Hierapolis (Piccardi 2007). The apocryphal account of the arrival of the apostle Philip at Hierapolis was commemorated by building of a sanctuary which attracted pilgrims through the centuries. Hierapolis-Pamukkale is one

of the most popular tourist attractions in Turkey with more than 2.5 million visitors in 2018.

The combined Hierapolis-Pamukkale locality is situated approximately 16 km north of the regional city Denizli (Fig. 14.2). This region of Turkey is particularly fertile, with a temperate climate, and is dominated by broad valleys and plains. The modern town of Pamukkale is located at the eastern margin of a broad valley associated with the Büyük Menderes River. The valley is part of the Denizli Basin, a significant feature of the Menderes Massif, the dominant geological terrain of southwestern Turkey (Chap. 13). The travertine at Pamukkale is deposited from upwelling hot waters, supersaturated in calcium carbonate. The hot waters exploit the active graben faults.

14.2 Travertine

Travertine is a hard, compact variety of limestone deposited from hot springs or geothermal waters which are supersaturated in calcium carbonate. The name is a corruption of "*Tiburtinus*" after the occurrence of extensive deposits at Tibur, near Rome. Travertine is distinguished from a softer,

Fig. 14.1 The snow white deposits of travertine form an extensive ridge above the village of Pamukkale

Fig. 14.2 The Hierapolis-Pamukkale archaeological and geosite is situated near the regional city of Denizli, southwest Turkey. Most of the antiquities at Hierapolis are from the Roman period

porous variety of limestone known as tufa, which is also deposited from geothermal waters. The precipitation of the calcium carbonate is triggered by degassing of carbon dioxide when the hot waters reach surface. Travertine is invariably restricted to areas of limestone bedrock where the hot waters leach the calcium carbonate. Hot waters may be related to magmatic or volcanic heat (e.g., Yellowstone: Chap. 4), or more commonly to surficial waters circulating through active faults. The heat is derived from the depth to which the waters penetrate. Travertine deposits generally reveal long periods of formation. The deposits at Pamukkale have a maximum age of approximately 400,000 BP and sections of the deposits remain active. The active nature of the travertine formation is in turn related to the active faults.

14.3 Geological Setting

The collisional tectonism associated with the Alpine Orogeny during the Early and Mid Cenozoic was replaced in the Neogene by crustal extension, resulting in development of large basins and grabens in southwest Turkey (Okay and Tüysüz 1999; Bozkurt 2000; Bozkurt et al. 2000). Crustal extension was driven by westward movement of the Anatolian Microplate and development of the Aegean back-arc basin. The formation of basins and grabens exhumed the core-parts (Proterozoic-Palaeozoic inliers) of the Menderes Massif (Çemen et al. 2006). The majority of the basins and grabens are aligned with the older, W-E trending Alpine sutures, with subordinate structures trending NW–SE or NE-SW. Faulting may have commenced in the Oligocene, but the main phase of faulting occurred in the Neogene and persisted into the Quaternary (Yilmaz et al. 2000). Many of the graben faults remain active. The grabens or rift valleys continue to subside and the rift shoulders (horsts) are subjected to ongoing uplift. Grabens are partially infilled by either shallow marine or terrestrial sediments, including conglomerate, limestone and sandstone. The majority of sediment was deposited by rivers flowing from the hinterland. Some grabens contain up to 30 m of sediment located below present sea level.

14.3.1 Denizli Basin

The Denizli Basin developed where the Büyük Menderes Graben and the Alaşehir Graben intersect (Fig. 13.6). The W–E aligned Büyük Menderes Graben trends approximately parallel with the Alpine sutures (Bozkurt 2000; Piccardi 2007; Gürer et al. 2009; Hançer 2013). In the region of Denizli, the Büyük Menderes Graben is bounded to the southwest by the Babadağ Fault and to the southeast by the Honaz Fault. These are complex zones of faulting which delimit isolated W-E aligned segments of the basement. The NW–SE trending Alaşehir Graben is bounded on the southwestern side by the Pamukkale Fault.

Faulting associated with the Büyük Menderes Graben includes both oblique and normal fault systems which record two separate (and complex) tectonic events (Gürer et al. 2009). The first of the events is defined by N–S compression (and W–E extension). This resulted in development of NW–SE and NE–SW trending faults. These structures are mostly related to oblique faults which developed in the Early and Mid Miocene. Sediment deposited in the grabens between the Late Miocene and Quaternary has been subjected to folding, uplift, and erosion. These features are preserved in the geological record as unconformities. The second of the tectonic events reflects a change to W–E compression (and N–S extension). Three pulses of deformation accompanied the second tectonic event in the Büyük Menderes Graben: (i) exhumation of the Menderes Massif within the earlier-formed W–E trending graben during the Late Pleistocene (the minimum age of this process is controlled by the occurrence of Plio-Pleistocene fluvial sediments); (b) rapid deposition of alluvial deposits during the Holocene in association with formation of high-angle W-E trending normal faults (or secondary listic faults) on the northern margin of the graben; (iii) migration of the graben axis which is marked by diachronous activity on the steeper of the listric faults. The latter faults are the most seismically active structures in southwest Turkey.

The fault patterns described above are probably applicable to the subsidiary grabens associated with the Denizli Basin. This region has been subjected to severe earthquakes during historical times and the possibility of future seismic events is discussed by Hançer (2013). The Denizli Basin is thermally active, a key component in formation of the travertine deposits (Altunel and Hancock 1993).

14.3.2 Pamukkale Travertine Deposits

The location of the travertine deposits at Pamukkale is controlled by the NW–SE trending Pamukkale Fault (Altunel and Hancock 1993) (Fig. 14.3). The Pamukkale Fault demarcates the southwestern boundary of the Alaşehir Graben. The Gediz Graben is a subsidiary feature of the boundary. The Pamukkale travertine deposit is located where the NNE-SSW trending Denizli Fault cuts across the Pamukkale Fault. The active hot springs are aligned on the Hierapolis Fault, a subsidiary feature of the Pamukkale Fault (Piccardi 2007) (Fig. 14.4). Four broad geological terrains are identified in this section of the Denizli Basin (Altunel and Hancock 1993). First, the horst blocks on the margins of the grabens are comprised of Palaeozoic-Mesozoic metamorphic rocks. They are dominated by marble and limestone. Second, Neogene clastic sediments, primarily limestone, occur on the margins of the horst blocks and in

parts of the grabens. Third, the grabens are partially infilled by poorly consolidated Quaternary fluvial and colluvial deposits. Fourth, the Late Pleistocene and Holocene travertine deposits are located proximal to the NW–SE (or NNW-SSE) trending faults associated with the Alaşehir Graben. The distribution of the travertine is controlled by the faults, but the occurrence of limestone bedrock is equally significant.

Four main centres of active travertine deposition are recognized in the vicinity of Pamukkale (Altunel and Hancock 1993). They cover a combined area of approximately 10 km^2 (Fig. 14.3). Pamukkale is the largest and most well known of the four deposits. The deposits have a maximum age of 400,000 BP. Most are younger than 60,000 BP. The radiometric data are consistent with long periods of formation. Historical dates imply a maximum age of 14,000 years for some localities. Most of the deposits are currently active. Geophysical surveys have confirmed that the deposition of travertine is contemporaneous with recent fault activity (Brogi et al. 2015). Fissures associated with the extensional tectonism have created major, and irregularly-spaced, conduits, an interpretation which explains why the travertine does not occur along the entire strike of the active faults. An understanding of the travertine fissure-ridges may assist with reconstruction of palaeo-tectonic activity throughout the Denizli Basin.

14.3.3 Formation of the Travertine Deposits

The travertine at Pamukkale is formed by upwelling hot water supersaturated in calcium carbonate ($CaCO_3$). Seventeen hot water springs with temperatures ranging from 35 °C to 100 °C are currently active. The precipitation of calcium carbonate is triggered by degassing of carbon dioxide (CO_2) from the hot waters. This process continues until the carbon dioxide in the hot water is in equilibrium with the carbon dioxide in the air. The reaction can be quantified as $H_2O + CO_2 + CaCO_3 = Ca (HCO_3)_2$. The calcium carbonate is initially deposited as a soft jelly-like substance, but this rapidly hardens into a hard, compact variety of travertine. Hardening is dictated by temperature of the water and air, as well as by flow rate. The water sourced from the hot springs at Pamukkale typically reports atmospheric levels of 725 mg/l of carbon dioxide. The concentration of the carbon dioxide decreases to 145 mg/l as the water flows across the already formed travertine terraces. This is accompanied by a decrease in the concentration of calcium carbonate, from approximately 1,200 mg/l–400 mg/l, resulting in deposition of 500 mg of calcium carbonate for each litre of hot water. Assuming a constant flow rate of 465 l/s, this equates to deposition of 44 kg of

Fig. 14.3 Generalized map showing the structural setting of travertine deposits in the vicinity of Pamukkale. *Source* Simplified after Altunel and Hancock (1993) and Geological Map of Europe (https://geoviewer.bgr.de)

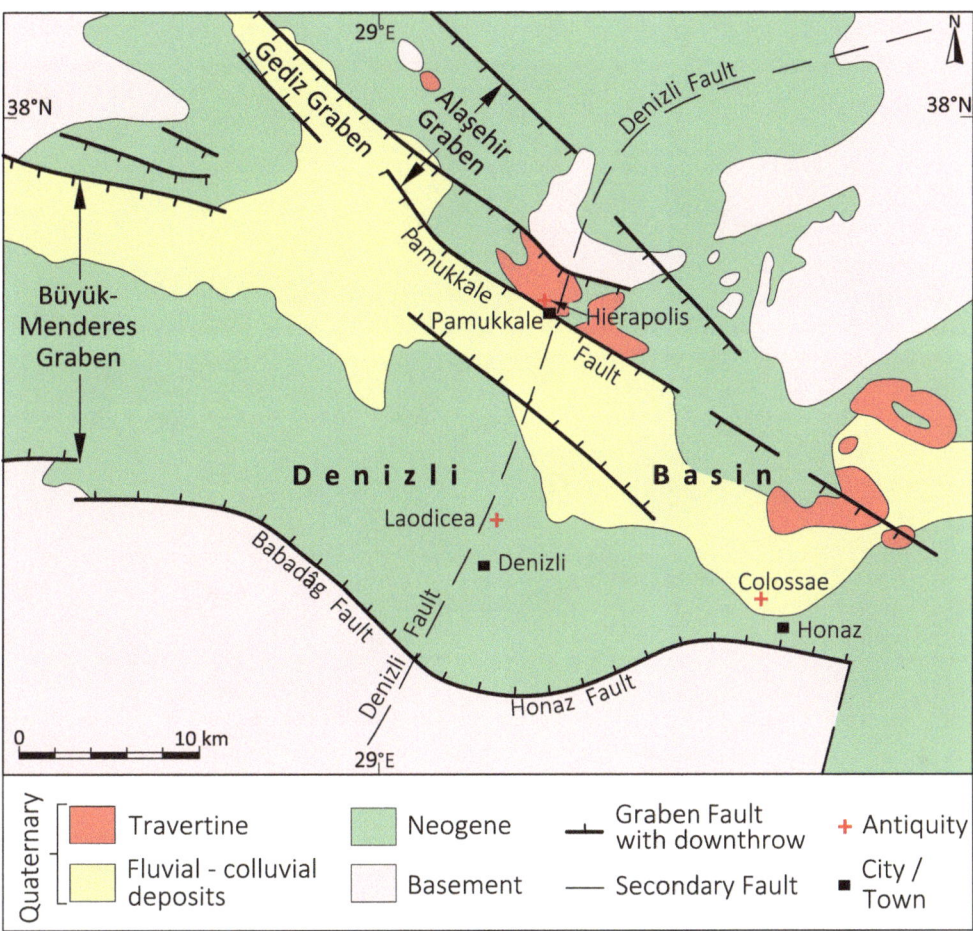

calcium carbonate daily. The calcium carbonate is deposited as a thin skin over the older travertine and may cover an area of as much as 13,500 m^2.

14.4 Pamukkale Geosite

The hillside at Pamukkale is covered almost entirely by the white and grey travertine deposits (Fig. 14.4). The newer white deposits can be seen to cover the older grey deposits on the flanks of the ridge (Fig. 14.5). The hot springs and travertine pools at Pamukkale have probably been used as a spa for thousands of years. By the mid-20thC, hotels and a road were built over part of the site, causing considerable damage. After Pamukkale was declared a World Heritage Status, some hotels were demolished and the road was replaced with artificial pools. Most visitors now stay overnight in hotels located in the adjacent towns and villages. Access to the Pamukkale site is controlled by a system of walkways and visitors are only allowed to walk and bathe in

selected pools. The wearing of shoes on the travertine terraces is prohibited. The hot water from several of the active springs associated with the graben fault on the crest of the ridge is fed into a 300 m-long channel to provide hot water for a manmade bathing pool. This pool, which is partially lined with ancient stone tablets and blocks derived from the Hierapolis archaeological site, is located partly on the site of the original "Sacred Pool".

Five principal varieties of travertine have been identified at Pamukkale (Altunel and Hancock 1993). The most abundant form is **fissure-ridge travertine** that forms the largest deposits which may have strike lengths of 1.5 km, widths of 400 m, and heights of 20 m. Some of the fissure-ridge deposits have been extensively quarried. The high point of the fissure-ridge deposits corresponds to localized areas of high water flow. Ridges typically contain a central fissure (the width varies between 2 cm to 5 m), from which the water was, or is, sourced. The travertine ridges are interpreted as the main linkage of the brittle structures which influence circulation of hot waters and location of thermal

Fig. 14.4 Image showing the travertine deposits at Pamukkale. The modern village of Pamukkale is located at the base of the travertine ridge near the small lake (lower left). The Hierapolis archaeological site occurs on the crest of the travertine ridge (centre) with the necropolis extending northward on Frontinus Street (a Roman road). The crest of the ridge corresponds to the alignment of hot springs where travertine is actively forming, including the modern pool which may correlate with the location of the Sacred Pool. The eastern part of the ridge is dominated by older sections of the travertine deposit. *Source* Google Earth

springs (Yalçiner 2013). The colour of the fissure-ridge travertine is typically white, pale orange, or buff (Fig. 14.6). Individual beds have thicknesses of 20–70 m and dip outward from the source at angles of 5–30°. A vermicular texture demonstrates how travertine layers build up incrementally cm by cm (Fig. 14.7a). Some travertine deposited from fissures occurs in a botryoidal form (Fig. 14.7b).

The second most abundant variety of travertine at Pamukkale is **terrace-mound travertine** (Altunel and Hancock 1993). The terrace mounds build up around thermal springs, either as point sources or laterally along faults. The majority of the deposit forms proximal to, and down slope from fault zones, on the flanks of ridges. Terrace-mound travertine may include ornate forms with steep, curving structures overhanging pools (Fig. 14.8). The primary colour of the terrace-mound travertine is white, but the deposits turn initially buff and then grey upon oxidation. Active terrace-mound deposits that mantle the steep slopes at Pamukkale attain a vertical thickness of 80 m. A pseudo layering may be observed (gently inclined beds are a few m in thickness), caused by channels that cover the terrace-mounds. The scale of these terraces is from a few mm to several m.

Fig. 14.5 a The ridge at Pamukkale consists almost entirely of white deposits of travertine; **b** The new deposits (white) drape downward from the plateau, covering the older deposits (grey) at the base of the ridge

Three additional morphologies of travertine have been identified at Pamukkale (Altunel and Hancock 1993). The **range-front travertine** and **eroded-sheet travertine** include massive beds, typically located near the base of ridges. Their relative age (the travertine has ceased forming) is attested to by the extensiveness of erosion. Some are ascribed to ancient Karstic landforms which have been incised by present day streams. The **self-built channel travertine** is associated with streams of calcium carbonate-rich hot water which flows away from the fissure-ridge and terrace-mound deposits.

14.5 Hierapolis Archaeological Site

The archaeological site of Hierapolis is located close to the active fault zone associated with the hot springs and pools at Pamukkale (Fig. 14.4) The site covers an extensive area on the crest of the ridge (Fig. 14.9). There are several access points. The central locality is near the theatre, in the heart of the complex (Fig. 14.10). The theatre was constructed in the 3^{rd}C AD, mostly of light grey marble derived from the metamorphic rocks exposed in the nearby horst block. Paths also access the antiquity from the base of the ridge and/or the

Fig. 14.6 The fissure-ridge deposits of travertine reveal pale orange or buff colours at Pamukkale

Fig. 14.8 The terrace-mound deposits of travertine at Pamukkale are continuously forming as the hot waters supersaturated in calcium carbonate drip down the flanks of pools

Fig. 14.7 a Vermicular-textures demonstrate how the travertine layers in the fissure-ridge deposits at Pamukkale build up incrementally cm by cm; **b** Botryoidal form of travertine at Pamukkale

southern gate near Pamukkale. The access from the northern gate which includes a longish walk along the crest of the ridge is recommended. The hot springs at Hierapolis were active in historical times and the antiquity was developed for the healing properties of the waters. Hierapolis became an important burial ground during Roman times and the Northern and Southern Necropolis are significant components of the archaeological site. The archaeological and historical data suggest the antiquity had to be entirely rebuilt after the earthquake of 60 AD and was finally abandoned after a large earthquake in 1354 AD (Bean 1971; Piccardi 2007).

The majority of the buildings at Hierapolis are built of fissure-ridge travertine. This includes the use of large blocks. White travertine was used in the theatre for decorative purposes, multi-coloured and banded fissure-ridge travertine was particularly sought after. Historical quarries can be observed in the area as deep, vertical-sided trenches. Trenches are linked to extinct or active fissures. Large parts of the antiquity are developed adjacent to Frontinus Street, a long road constructed in the 1stC AD which connects the Agora and Northern Necropolis (Fig. 14.11). Frontinus Street was covered with deposits of travertine some 2 m-thick prior to being excavated in the 19thC (the travertine had built up since the site was abandoned in the Byzantine period). Some of the tombs in the Northern and Southern Necropolis remain covered by the terrace-mound travertine, although others have been unearthed (Fig. 14.12a). The Southern (Byzantine) Gate reveals the selective use of travertine for decorative purposes (Fig. 14.12b).

Fig. 14.9 Historical map of the antiquity of Hierapolis. The travertine deposit is depicted by area "B". Other sites of interest are the Agora (7), Frontinus Street (8), Theatre (11), and the Southern Gate (14). *Source* Information Board at the site

Fig. 14.10 The Roman theatre at Hierapolis is built from blocks of light grey marble with white travertine used for decorative purposes (notably in the gallery and stage)

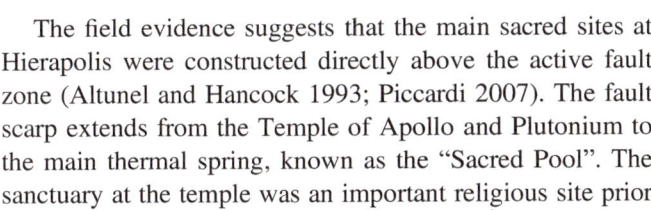

Fig. 14.11 Part of Frontinus Street which connects the Agora and Northern Necropolis at Hierapolis. The street was covered with a 2 m-thick deposit of travertine prior to recent excavations

The field evidence suggests that the main sacred sites at Hierapolis were constructed directly above the active fault zone (Altunel and Hancock 1993; Piccardi 2007). The fault scarp extends from the Temple of Apollo and Plutonium to the main thermal spring, known as the "Sacred Pool". The sanctuary at the temple was an important religious site prior to the Greek and Roman cities being established. The site was chosen for the occurrence of natural gases (mostly carbon dioxide and hydrogen sulphide) in the hot waters. In the Classical Greek times, the sanctuary was described as the "Plutonium" (place of the god *Pluto*). The Plutonium was excavated in 1965 and a dark chamber, approximately 3 m

Fig. 14.12 a Some burial chambers adjacent to Frontinus Street at Hierapolis remain partially covered by deposits of terrace-mound travertine; **b** The Southern (Byzantine) Gate at Hierapolis includes a monolithic arch supported by walls built from large blocks of marble (light grey) with selective use of travertine (white)

square, with a natural cleft at the back from which a fast stream of hot water with toxic vapours, was discovered (Bean 1971; Piccardi 2007). The chamber corresponds to classical descriptions, including those by Strabo, and it is probable that priests deceived visitors to the site, in the same way as at Delphi (Chap. 10), by holding their breath and appearing to be immune to the toxic gases.

A well known historical legend describes the slaying of Echidna, a snake goddess (similar to the cult of Pythia at Delphi), at Hierapolis (Piccardi 2007). The legend is linked to earthquakes and the martyrdom of the Apostle Philip who arrived in 70–80 AD. Apparitions describe a dragon associated with a strong earthquake.

14.6   Earthquakes

Hierapolis was affected by several earthquakes in Roman times, many of which necessitated major rebuilding programmes. The most severe earthquake probably occurred in 60 AD and is related to the apparations described above

(Michetti and Hancock 1997; Piccardi 2007). The earthquake opened a major rupture along the trace of the fault in the city, and also triggerred catstrophic lightning storms. The travertine deposits at Pamukkale have also been affected, and in some case partially destroyed, by both historical and recent seismic events (Hancock et al. 2000). The two large earthquakes of 60 AD and 1354 AD were particularly significant. Evidence of recent activity on the Hierapolis Fault is illustrated by a profile between the floor level of the Temple of Apollo and Plutonium and the travertine plateau to the west of the site (Piccardi 2007). The surface is offset by between 7.5 m and 9.5 m. Assuming the fault zone has a vertical displacement of approximately 1.5–2 m since Roman times (Altunel and Hancock 1993). The fault displays a slip rate of approximately 0.75–1 mm/year. The throw of 9.5 m may, therefore, have taken as long as 10,000 years to develop. Activity associated with the Denizli Earthquake of 1965 (5.7 on the Richter scale) opened a small chasm close to the Nymphaeum and the Sacred Pool (Piccardi 2007). Toxic gases, similar to those identified in the ancient Plutonium, are emitted from the rupture.

References

Altunel, E., & Hancock, P. L. (1993). Morphology and structural setting of Quaternary travertines at Pamukkale, Turkey. *Geological Journal, 28*, 335–346.

Bean, G. (1971). Turkey beyond the Meander (p. 236). London: E. Benn.

Bozkurt, E., Winchester, J. A. & Piper, J. D. A. (2000). Tectonics and Magmatism in Turkey and surrounding Areas. *Geological Society of London Special Publication, 173*, 521.

Brogi, A., Capezzuoli, E., Ruggieri, G., Gandin, A., Alicek, M.C., Liotta, D., Meccheri, M., Yalciner, C.C. & Büyüksarac, A. (2015). Travertine fissure-ridges as proxies of tectonically controlled hydrothermal fluids fed by a carbonate reservoir: special emphasis on Denizli basin of SW Anatolia, Turkey. Abstract 31st IAS Meeting of Sedimentology, Krakow.

Çemen, I., Catlos, E. J., & Gögüs, O. (2006). Postcollisional Extensional Tectonics and Exhumation of the Menderes Massif in the Western Anatolia Extended Terrane, Turkey. *Geological Society of America Special Paper, 409*, 353–379.

Gürer, O. F., Sarica-Filoreau, N., Özburan, M., & Sangu, E. (2009). Progressive development of the Büyük Menderes Graben based on new data, western Turkey. *Geological Magazine, 146*, 652–673.

Hançer, M. (2013). Study of the Structural Evolution of the Babadağ-Honaz and Pamukkale Fault Zones and the Related Earthquake Risk Potential of the Buldan Region in SW Anatolia, East of the Mediterranean. *Journal of Earth Science, 24*, 397–409.

Hancock, P. L., Chalmers, R. M. L., Altunel, E., Cakir, Z. & Becher-Hancock, A. (2000). Creation and destruction of travertine monumental stone by earthquake faulting at Hierapolis, Turkey. In: The Archaeology of Geological Catastrophes. *Geological Society of London Special Publication, 171*, 1–14.

Michetti, A. M., & Hancock, P. L. (1997). Paleoseismology: Understanding past earthquakes using Quaternary geology. *Journal of Geodynamics, 24*, 3–10.

Okay, A. I., & Tüysüz, O. (1999). Tethyan sutures of northern Turkey. *Geological Society London Special Publication, 156,* 475–515.

Piccardi, L. (2007). The AD 60 Denizli Basin earthquake and the apparition of Archangel Michael at Colossae (Aegean Turkey). In: L. Piccardi & W. B. Masse (Eds.), Myth and Geology. *Geological Society London Special Publications, 273,* 95–105.

Yalçiner, C. ç. (2013). Investigation of subsurface geometry of fissure-ridge travertine, Pamukkale, Turkey. *Journal of Geophysical and Engineering, 10*(3). https://doi.org/10.1088/1742-2132/10/3/035001.

Yilmaz, Y., Genç, S. C., Gürer, F., Bozcu, M., Yilmaz, K., Karacık, K., Altunkaynak, S., & Elmas, A. (2000). When did the western Anatolian grabens begin to develop? In: E. Bozkurt, J. A. Winchester & J. D. Piper (Eds.), Tectonics and Magmatism in Turkey and the Surrounding Area. *Geological Society London Special Publications, 173,* 353–384.

Cappadocia, Central Turkey

Volcanic Plateau, Natural Rock Monoliths and Pinnacles, and Underground Rock Dwellings

Abstract

The Cappadocia region of central Turkey contains innumerable natural rock monoliths and pinnacles, together with the largest concentration of underground rock dwellings reported in the world. Parts of Cappadocia have been declared UNESCO World Heritage sites and are visited by an estimated 2.5 million tourists annually. The principal physiographic feature of the region is the Cappadocia Plateau, a rugged landscape with extensive badlands. The geological setting is dominated by the Cappadocia Volcanic Province. Extrusion of huge volumes of lava, volcanic ash and pumice subsumed the older land surface to produce a plateau with an average height of between 1,400 m and 1,500 m. Nine giant stratovolcanoes are recognized in the region, including the snow- and ice-capped peak of Mount Erciyes (3,917 m). The Erciyes Volcano is quiescent, but poses a potential hazard to the regional city of Kayseri, in part as instantaneous melting of the ice sheets on the upper slopes could create catastrophic floods. An additional natural hazard to permanent residents of the underground rock dwellings in some of the smaller settlements in the Cappadocia region is the high incidence of mesothelioma. The volcanism at Cappadocia commenced in the Upper Miocene, peaked in the Pliocene, and persisted intermittently into the Quaternary. The volcanic outpourings were centred on a group of stratovolcanoes. Major caldera events generated extensive pyroclastic flows. Thick sheets of ignimbrite are a characteristic feature. Differential erosion of the flat-lying volcanic deposits, primarily during the Quaternary, has produced spectacular landforms. The volcanic lavas and ignimbrite sheets are resistant to erosion in the semi-arid continental climate, and cap the plateau and mesas. The volcanic ash and pumice is relatively soft and is found in the deeply eroded valleys. The regional town of Göreme includes a mix of ancient and modern structures carved out of natural rock monoliths and pinnacles. Some hotel rooms are excavated into the volcanic ash and pumice. The Göreme Historical National Park contains extensive areas of badlands, including natural pinnacles known as "fairy chimneys". Clusters of pinnacles at localities such as Paşabağ can attain heights of 40 m, although the average height is between 2 m and 5 m. The columns or pinnacles typically consist of volcanic ash and pumice, with basaltic lava or ignimbrite forming the distinctive caps. Üchisar Castle is a natural monolith of ignimbrite with a remarkable concentration of underground dwellings. The 14 km-long hike of the Ihlara Gorge is an additional highlight of a visit to the region, as the rock walls, which are eroded from thick sheets of ignimbrite and volcanic ash, contain innumerable rock dwellings. The underground cities of Derinkuyu and Kaymakli have multiple levels with passageways and stairs carved out of the volcanic ash and pumice to depths of as much as 60 m.

Keywords

Cappadocia Plateau • Ignimbrite • Monoliths • Pinnacles • Stratovolcanoes • Underground cities • Volcanism

15.1 Introduction

The Cappadocia region (pronounced *Kapadokia*) in central Turkey is characterized by a landscape of vast plains with rolling hills and high volcanic peaks. Parts of Cappadocia were declared a UNESCO World Heritage site in 1985 and are visited by an estimated 2.5 million tourists annually.

Photographs not otherwise referenced are by the author.

The original version of this chapter was revised: Belated corrections have been incoporated. The corrections to this chapter are available at https://doi.org/10.1007/978-3-030-54693-9_18

Cappadocia is famed for clusters of natural rock monoliths and pinnacles, together with ancient, underground rock dwellings. Most tourists arrive in the summer months, but the number of winter visitors has increased, in part as activities that include skiing on the volcanic peaks are increasingly popular. The snow-and ice capped peak of Mount Erciyes (3,917 m) is a quiescent stratovolcano. The principal physiographic feature is the Cappadocia Plateau, a rugged landscape with extensive badlands. The badlands contain natural monoliths and pinnacles which at first glance may appear to be man-made. The unusual shapes of the columns of volcanic ash, pumice, and lava are related to differential erosion of the near horizontal strata (Fig. 15.1). A distinctive feature is the occurrence of caps of relatively resistant volcanic lava.

The Cappadocia Plateau is situated in the Anatolian region of central Turkey (Fig. 15.2). The regional centre is Kayseri, and the tourist areas, which include several national parks and the world heritage centres, are served by networks of good quality roads. The main areas of interest are the Göreme Historical National Park, the Ihlara Gorge, and the Derinkuyu and Kaymakli underground cities. There are a number of small towns with accommodation and the opportunity to have a hotel room carved out of volcanic ash is unique. The plateau has an average elevation of between 1,400 m and 1,500 m. The semi-arid, continental-style climate is influenced by the altitude and central location. Hot and dry summers alternate with cold and wet winters that include snowfall. Late autumn is possibly the best time to visit for tourists who intend undertaking the hiking trails.

The geological setting is dominated by the Cappadocia Volcanic Province. Huge outpourings of volcanic ash, pumice, and lava subsumed the palaeo-landscape. The plateau was built up layer-by-layer from the near horizontal strata. The stepped surface is characteristic of volcanic plateaus that are deeply eroded, and the topography may be compared with the Columbia River basalt province in Oregon, USA. The Cappadocia Plateau is intersected by the northward-flowing Kizilirmak River, the longest river in Turkey. The name of the river is derived from erosion of the red volcanic soil.

Fig. 15.1 The "Fairy Chimneys" at Paşabağ in the Göreme Historical National Park, Cappadocia, are natural rock pinnacles that consist of relatively soft volcanic ash and pumice (pale grey). The distinctive caps consist of horizontal layers of resistant basaltic lava (dark)

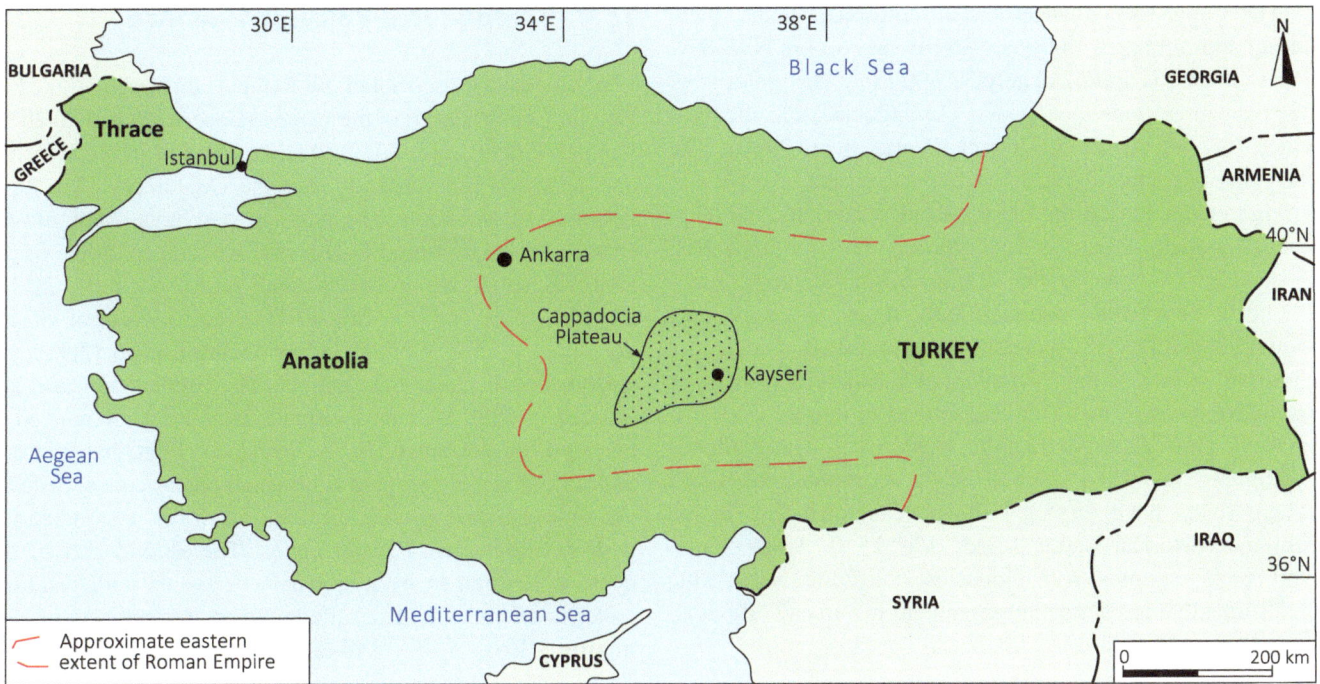

Fig. 15.2 Map showing location of the Cappadocia Plateau in central Turkey

15.2   Historical Activity

The Cappadocia region was ruled by the Hittites between 1800 BC and 1200 BC. A 13th C BC rock carving located on Mount Erciyes, which depicts a storm god situated above three mountain gods, is an important archaeological site related to this period. A man-made tunnel near the summit of Mount Erciyes has been linked to a place of Hittite worship. There is a suggestion that a link exists between the Greek legend of *Typhon* and *Zeus* and the volcanoes of Cappadocia. The region was successively invaded and colonized by the Persians, the Greeks under Alexander the Great, and the Romans. Historical volcanic eruptions in the Cappadocia region were witnessed by Roman leaders and poets, including Strabo (63 BC-21 AD) and Claudius Claudianus (370-410 AD).

Cappadocia is located east of the area controlled by the Roman Empire and the region became an important refuge for Christians in Asia Minor during the 1st–3rd C AD (Chap. 13). Many of the dwellings and underground cities located in the relatively soft volcanic ash and pumice were carved by the Christian communities. Archaeological and historical sites were also developed during the Byzantine Empire (4th–11th C AD). In the 12th C AD, the Muslim Turks invaded the entire region of Anatolia and reshaped the history, although Cappadocia remained a largely Greek-speaking community until the 15th C AD. The location of Cappadocia on the ancient Silk Road from China to Europe is reflected in local cultural activities, including making of world-class ceramics and the use of silk in manufacture of some carpets. Handmade carpets of a very high quality are a speciality of the region.

15.3   Regional Geology

Cappadocia is part of a complex tectonic zone affected by collision of the Anatolian Microplate and the northward-pushing Arabian Microplate (Bozkurt et al. 2000; Ocakoglu 2004). The NE-SW trend of the Cappadocia Plateau is a pronounced feature (Fig. 15.2). The pre-volcanic basement consists of Palaeozoic and Mesozoic metamorphic rocks, similar to the Menderes Massif of southwest Turkey. The W-E strike of the basement complexes is characteristic of the Alpine tectonic zones throughout Turkey. The basement includes limestone, marble, and schist, together with ophiolite complexes. A period of crustal extension occurred after the peak of the Alpine tectonism. Intracratonic basins and grabens commenced forming in the Cappadocia region during the Lower Miocene. Basins and grabens continued to form into the Quaternary (Ocakoglu 2004). Some graben faults remain active. The basins and grabens are partially infilled by shallow marine or terrestrial sediments, including conglomerate, limestone and sandstone.

The Cappadocia Plateau has an ellipsoidal shape, measuring approximately 300 km NE–SW and 60 km NW–SW. The total area is approximately 32,500 km². The majority of the plateau is underlain by the Cappadocia Volcanic Province (Şen et al. 2003; Viereck-Goette et al. 2010). The flat-lying deposits of volcanic ash, pumice, and lava have to a large extent buried the older basement terrains, although coeval graben sediments are exposed in some localities. A notable example is the Ürgüp Basin, a small, fault-bounded depression partially infilled by Neogene age lacustrine and fluvial deposits (Göz et al. 2014). The deeply dissected nature of the Cappadocia Plateau, which includes extensive areas of badlands, is ascribed to intense cycles of erosion, particularly during the Quaternary. The badlands topography is well developed on the extremities of mesas where rivers have dissected the plateau, including in the Göreme Historical National Park (Fig. 15.3). Areas of natural rock columns and pinnacles constitute remarkable landforms that are almost impossible to traverse (Zorlu et al. 2011) (Fig. 15.4).

15.4 Cappadocia Volcanic Province

The volcanism associated with the Cappadocia Volcanic Province commenced in the Upper Miocene (Şen et al. 2003; Viereck-Goette et al. 2010). Activity peaked in the Pliocene and persisted intermittently into the Quaternary. The huge volumes of volcanic ash, pumice, and lava were mostly extruded from a group of giant stratovolcanoes. Some of the stratovolcanoes are associated with prominent peaks which may be snow-and ice capped (Fig. 15.5). Several of the volcanoes in the region, including Mount Erciyes (3,917 m), are quiescent, rather than extinct. The Erciyes Volcano last erupted in 6880 BC. Two volcanic cycles are recognized in the style of volcanism in the province. First, basaltic and andesitic flows are erupted from small cinder cones. Second, volcanic ash and pyroclastic deposits form in conjunction with development of major cones. The second part of the cycle is terminated by major caldera events with resulting collapse of cones and accumulation of thick sheets of ignimbrite (Box 8.2). Ash-fall deposits associated with

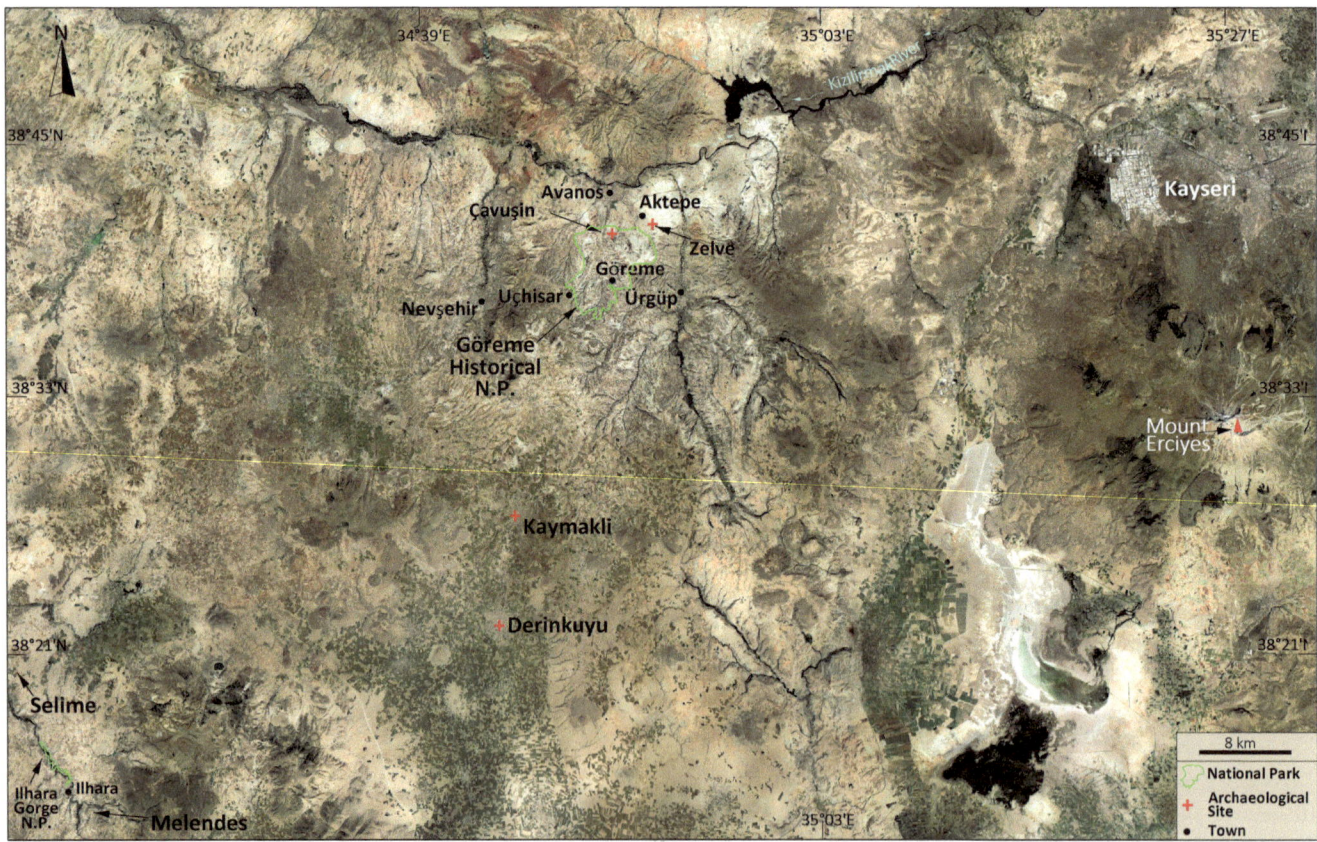

Fig. 15.3 The deeply dissected Cappadocia Plateau contains snow- and ice-capped volcanic peaks, including Mount Erciyes, and is cut on the northern side by the Kizilirmak River. *Source* Google Earth

Fig. 15.4 The Cappadocia Plateau contains extensive areas of badlands dominated by natural rock monoliths and pinnacles

15.4.1 Petrogenesis and Volcanic Centres

The Cappadocia Volcanic Province is associated with the calc-alkaline volcanic trend (Chap. 8). Magma was derived from melting of ocean floor subducted during collision of the Anatolian and Arabian Microplates (Şen et al. 2003). The principal magma-types are basalt and andesite (Viereck-Goette et al. 2010). Nine major stratovolcanoes have been recognized, together with 19 discrete eruptive centres. The latter have diameters of up to 40 km and are mostly located proximal to major faults (Şen et al. 2003; Ocakoglu 2004; Sorkhabi 2011). Magma exploited regional strike-slip and extensional faults associated with the post-collisional crustal extension. Two fault systems have been identified, one bounding the volcanic province, and the other occurs parallel to the axis.

Fig. 15.5 The snow- and ice-capped peak of Mount Erciyes with views of the deeply eroded Cappadocia Plateau in the foreground

15.4.2 Ignimbrite Sheets and Ash-Fall Deposits

Plinian events may occur more than 50 km from individual centres. The most recent of the Plinian events is dated at 2.8 Ma. Deposits of volcaniclastics are also extensively developed in the region.

The presence of extensive ignimbrite sheets is a characteristic feature of the Cappadocia Volcanic Province (Temel

Fig. 15.6 The tranquil setting of Ihlara Gorge contains a small river in a wooded setting. The gorge is carved into thick sheets of ignimbrite. The ignimbrite sheets reveal poorly developed columnar jointing

et al. 1998). The ignimbrite has a distinctive pale grey or red-pink coloration and occurs in stacked sheets, often with poorly-developed columnar jointing (Fig. 15.6). The ignimbrite contains welded fragments (Fig. 15.7a). The majority of the ignimbrite sheets have been dated at between 9 Ma and 1 Ma (Aydar et al. 2012). The ignimbrite sheets are intercalated with thick deposits of volcanic ash which may contain abundant lithic fragments, indicative of the explosive nature of the volcanic activity (Fig. 15.7b). The flat-lying layered deposits of volcanic ash typically reveal knife-sharp contacts (Fig. 15.8a). Cross bedding in some of the finely laminated deposits of volcanic ash may be enhanced by discoloration due to weathering (Fig. 15.8b). Some of the pumice beds intercalated with the ash-fall deposits contain distinctive lapilli, and constitute marker layers in areas distal to the eruptive centres where the ignimbrite sheets do not occur.

15.5 Göreme Historical National Park

The Göreme Historical National Park (area of approximately 100 km^2) is surrounded by the small towns of Çavuşin, Göreme, Üchisar and Ürgüp (Fig. 15.9). Nevsehir, the regional centre is located approximately 5 km to the west of the national park. Erosion of the mostly flat-lying strata has produced successions of mesas and ridges dissected by steep-sided valleys (Fig. 15.10). The orange and yellow colours of the ash-fall deposits are pronounced in the semi-arid environment. The valleys are characterized by extensive areas of badlands, where erosion of the volcanic ash and pumice volcanic deposits has resulted in clusters of monoliths and pinnacles. The mesas and ridges are capped by resistant basaltic lavas and sheets of ignimbrite. Erosion is aided by the semi-arid climate and paucity of vegetation

Fig. 15.7 **a** The pale grey ignimbrite at Ihlara Gorge includes welded fragments; **b** The volcanic ash from Göreme contains abundant lithic fragments, indicative of Plinian-style eruptions

(Zorlu et al. 2011). Most of the larger valleys in the national park, and by inference areas where the eroded landforms and monoliths are concentrated, are probably related to catastrophic flood events. Floods may have been triggered by seismic activity associated with the active faults (Göz et al. 2014).

The geological processes in the region may be compared with differential erosion of the Colorado Plateau in the southwestern part of the USA (Chap. 1), although the stratigraphy at Göreme is dominated by volcanic sequences rather than sandstones and shales. The occurrence of intercalated sequences of lava, ash and pumice deposits, with the

Fig. 15.8 a Some of the flat-lying layered deposits of volcanic ash exposed in underground rock dwellings at the Ihlara Gorge reveal knife-sharp contacts; **b** Cross bedding in finely laminated deposits of volcanic ash in the Ihlara Gorge is enhanced by orange-brown discoloration due to weathering

more resistant lava forming the resistant capping, can also be compared with the role of dolerite sills in the Karoo region of South Africa.

15.6   Differential Erosion

Differential erosion of the intercalated and flat-lying sequences of relatively soft volcanic ash and pumice together with resistant lava has created spectacular monoliths and pinnacles in the Göreme Historical National Park. The monoliths and pinnacles cluster in valleys. The town of Göreme is located in a valley where the buildings are constructed around, and partially into, the monoliths and pinnacles (Fig. 15.11). Monoliths are particularly well developed at the Paşabağ locality, where the ash and pumice deposits are interbedded with thin layers of basaltic lava. Monoliths include the "fairy chimneys", i.e., pinnacles or columns with heights of up to 40 m. The fairy chimneys are comprised of columns of relatively soft volcanic ash and

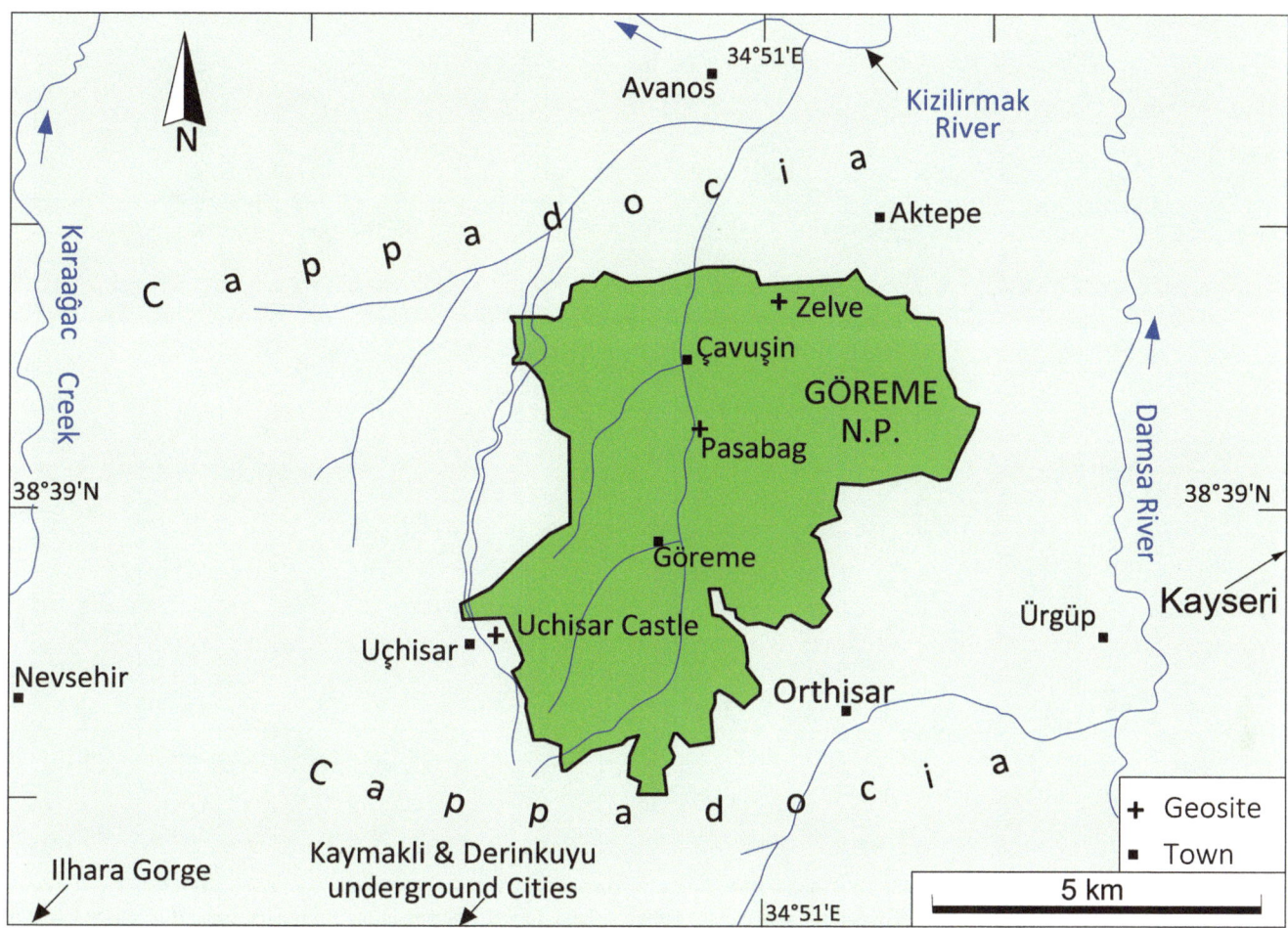

Fig. 15.9 Map of the Göreme Historical National Park and surrounding areas of the Cappadocia Plateau

pumice (Fig. 15.1). Resistant layers of basaltic lava (or in rare cases ignimbrite) form the distinctive cap rock. Differential erosion has also resulted in formation of cone and mushroom-shaped pinnacles. The Camel Rock is an unusual shaped natural monolith located in the Dervent Valley, near Zelve (Fig. 15.12). This locality is also known as Imagination Valley or Pink Valley due to the abundance of pink-coloured monoliths which resemble the shapes of animals and birds.

The occurrence of extensive deposits of obsidian (numerous obsidian tools have been found) is significant. The obsidian is exploited today from underground workings where small pockets of metre-sized or even larger lumps are derived. Onyx is also extensively quarried, and is used to produce dimension stone, floor tiles, and other objects. Many of the dwellings are associated with the early Christians, as noted above, and some underground structures were used as churches.

15.7 Human Settlement

Ancient communities took advantage of the relatively soft nature of the volcanic deposits at Cappadocia to excavate the greatest concentration of underground rock dwellings known in the world. They include private dwellings, monasteries and churches, as well as entire underground towns. Human settlement dates back to Neolithic times (possibly 8000 BC).

15.7.1 Üchisar Castle

Üchisar Castle is a rock monolith with a myriad of excavations carved out of a resistant outcrop of ignimbrite (Fig. 15.13). The 60 m-high monolith occurs on the outskirts of the small town of Üchisar, situated on the southwest flanks of the Göreme Historical National Park. Some of the natural structures in the "Castle" may have been used by the

Fig. 15.10 Landforms at the Göreme Historical National Park are dominated by flat-lying strata which outcrop in elongate ridges. Extensive areas of badlands erosion occur in the valleys. The yellow and pinkish-brown colours reflect changes in the iron oxide content of the volcanic ash, pumice, and lava

Hittites, but the majority of the excavations were developed in the 7thC AD during the Byzantine Empire. There are underground passageways and rooms in addition to the shallow excavations. Some of the rooms served as residential areas and cloisters. It is estimated that approximately 1,000 people lived here.

15.7.2　Ihlara Gorge

The hiking trail (length of 14 km) in the Ihlara Gorge National Park, located southwest of the regional town of Nevsehir, follows the Melendez River. The trail commences near the small town of Selime and exits close to the village of Ihlara. The tranquil setting of a small stream and wooded valley, together with the opportunity to enjoy traditional teashops located on pontoons or rafts, is recommended. Whereas many gorges are eroded into limestone, Ihlara is unusual in that the cliffs are dominated by thick sheets of ignimbrite (Fig. 15.6). The ignimbrite is interbedded with extensive deposits of volcanic ash (Fig. 15.8). The walls of the gorge contain churches, excavations and rock dwellings (Fig. 15.14). Some of the more than one hundred churches are open to visitors. The lower section of the gorge is known as the "Valley of the Dovecotes" (or Pigeon Valley) due to the occurrence of excavations with interior niches that are used by birds for roosting. The churches and excavations date from the Byzantine Period. Settlement was assisted by the plentiful supply of water and the relatively soft nature of the ignimbrite and volcanic ash.

15.7.3　Underground Cities

More than 200 underground cities of at least two levels have been discovered in the area between Kayseri and Nevsehir. Many of the locations remain relatively unexplored. The underground cities were excavated as hiding places during

Fig. 15.11 **a** Göreme is situated on the edge of an extensive area of badlands erosion protected in a national park; **b** The town of Göreme is a mix of ancient and modern structures carved out of natural volcanic monoliths and pinnacles, with some hotel rooms excavated into the soft ash and pumice

times of raids and wars. Two of the most well known locations are Derinkuyu and Kaymakli. Derinkuyu is located 35 km south of Göreme and has multiple levels with passageways and stairs carved to a depth of 60 m that could shelter as many as 20,000 people, together with livestock and provisions, during sieges. The city is entered by a winding staircase cut into the volcanic ash and pumice. The lowest levels are accessed by vertical stairs and include a ventilation shaft and water well. Archaeological evidence suggests the oldest caves may have been excavated by the

Fig. 15.12 The Camel Rock is a distinctive rock monolith located in the Dervent Valley. Lava (grey) forms the resistant capping to the pinkish volcanic ash which dominates the feature

Phrygians (Indo-European people) in the 7–8thC BC. These inhabitants were replaced by Greek-speaking people of the Byzantine Empire, probably by the 5thC AD. Derinkuyu was probably used most extensively in the period of the Arab and Byzantine wars (780–1180 AD). The city was connected to other underground cities by extensive tunnels over many kilometres. Derinkuyu was used again by Christians during incursions by the Mongolians in the 14thC AD. During the early days of the Ottoman Empire, Derinkuyu and the other underground cities were used as refuges by some of the local inhabitants. They may have been used as recently as 1909 in response to the massacre of Cappadocian Greeks at Adana.

The underground cities were finally abandoned in 1923 after the forced removal of the Cappadocian Greeks as part of the agreement reached between the modern states of Greece and Turkey. This followed the Greek genocide which lasted between 1913 and 1922, in which as many as three quarters of a million Greeks based in Anatolia, many in the Cappadocia region, were systematically killed. One of the abandoned villages formerly occupied by the Cappadocian Greeks can be visited in the Ihlara Gorge. The underground city of Derinkuyu was rediscovered in 1963, and by 1969 parts of the archaeological site were opened to tourists. Archaeologists continue to make new discoveries in the region.

Fig. 15.13 Üchisar Castle is associated with an outcrop of ignimbrite. The castle was probably first inhabited by the Hittites but subsequently used by recent civilizations

15.8 Current Hazards

Some of the underground shelters in the Cappadocia region are still in use, and the smaller hotels include rooms carved out of the volcanic rock. Permanent residents of the underground excavations experience a high incidence of mesothelioma, ascribed to inhalation of fibres of the mineral erionite (a type of asbestos) found in the volcanic ashes. The greatest threat to the region, including the city of Kayseri, is probably heating and melting of the ice fields on the upper slopes of Mount Erciyes if a major eruption were to occur.

Fig. 15.14 Numerous structures have been carved out of the cliff faces at Ihlara Gorge, including dovecotes

References

Aydar, E., Schmitt, A. K., Çubukçu, H. E., Akin, L., Ersoy, O., Şen, E., et al. (2012). Correlation of ignimbrites in the central Anatolian volcanic province using zircon and plagioclase ages and zircon compositions. *Journal of Volcanology and Geothermal Research, 213–214,* 83–97.

Bozkurt, E., Winchester, J. A., & Piper, J. D. A. (2000). Tectonics and Magmatism in Turkey and surrounding Areas. *Geological Society of London Special Publication, 173,* 521 p.

Göz, E., Kadir, S., Gürel, A., & Eren, M. (2014). Geology, mineralogy, geochemistry, and depositional environment of a Late Miocene/Pliocene fluviolacustrine succession, Cappadocian Volcanic Province, central Anatolia, Turkey. *Turkish Journal of Earth Sciences, 23,* 1–26.

Ocakoglu, F. (2004). Mio-Pliocene basin development in the eastern part of the Cappadocian Volcanic Province (Central Anatolia, Turkey) and its implications of regional tectonics. *International Journal of Earth Sciences, 93,* 314–328.

Şen, E., Kürkcüoğlu, B., Aydar, E., Gourgaud, A., & Vincent, P. (2003). Volcanological evolution of Mount Erciyes stratovolcano and origin of the Valibaba Tepe ignimbrite (Central Anatolia, Turkey). *Journal of Volcanology and Geothermal Research, 125,* 225–246.

Sorkhabi, R. (2011). Cappadocia, Turkey: Civilizations in a volcanic terrain. *Geoexpro, 8,* 64–67.

Temel, A., Gundogdu, M. N., Gourgaud, A., & Le Pennec, J. L. (1998). Ignimbrites of Cappadocia (Central Anatolia-Turkey), petrology and geochemistry. *Journal of Volcanology and Geothermal Research, 85,* 447–471.

Viereck-Goette, L., Lepetit, P., Gürel, A., Ganskow, G., Çopuroğlu, I., & Abratis, M. (2010). Revised volcano-stratigraphy of the Upper Miocene to Lower Pliocene Urgüp Formation, Central Anatolian volcanic province, Turkey. *Geological Society of America Special Paper, 464,* 85–112.

Zorlu, K., Tunusluoglu, M. C., Gorum, T., Nefeslioglu, H. A., Yalcin, A., Turer, D., et al. (2011). Landform effect on rockfall and hazard mapping in Cappadocia (Turkey). *Environmental Earth Sciences, 62,* 1685–1693.

Abstract

The Lake District in the northwest of England contains dramatic scenery associated with the deeply dissected Cumbrian Mountains. The Lake District National Park is the most widely visited rural area in the British Isles, with tourists attracted by outdoor activities including a network of mountain footpaths. Lakes including Coniston Water, Buttermere, Derwent Water, Ullswater, and Windermere are widely known. Scafell Pike is the highest summit in England. Human settlement in the Lake District is traced to Neolithic times. The mining heritage dates from the Romans who constructed transport routes for exporting lead and silver. Mining of copper and graphite was of particular importance during the Elizabethan times, and the Victorian period saw development of extensive copper and lead workings. The high quality, dark blue-grey slate from the Lake District roofs famous buildings in London. The combination of mountainous landscapes and tranquil valleys has inspired a long association with English literature and artists, and the region includes cultural centres associated with literary greats such as William Wordsworth and John Ruskin. The geomorphology of the Lake District is dominated by the ice-sculptured Cumbrian Mountains. Deeply-incised valleys radiate outward from the mountainous core. Valleys contain finger-lakes which developed as the Late Pleistocene ice sheets and glaciers retreated. Scree slopes and rocky summits (pikes) formed due to ice and frost erosion. Hanging valleys with small Alpine lakes and fast-flowing streams are an integral part of the topography. The geological setting is dominated by an inlier of Lower Palaeozoic strata, one of a number of disjointed terrains collectively known as the British Caledonides. Each of the Lower Palaeozoic inliers was subjected to the Caledonide Orogeny. The orogeny peaked in the Early Devonian. The oldest component of the Lake District inlier is the Skiddaw Group, a thick sequence of Lower Ordovician mudstones and shales. These rocks are associated with the rounded massifs in the northern part of the region. The mudstones and shales are unconformably overlain by the Borrowdale Volcanic Group, a succession of mostly sub-aerial calc-alkaline lavas, ashes, and volcaniclastics associated with a Mid-Ordovician volcanic island arc. The Borrowdale volcanics constitute the rugged mountains in the core of the inlier. Some of the volcanic ashes have a prominent cleavage and form high-quality slate deposits. The Windermere Supergroup, which is associated with the gently rolling hills that characterize the southern part of the Lake District, marks a marine transgression in the Upper Ordovician and Silurian. Thick sequences of limestone and shale were deposited in shallow seas rich in marine fossils. The regional unconformity separating the Borrowdale Volcanics from the Coniston Limestone is demarcated by a marked change in the topography. Two groups of intrusive igneous rocks occur in the Lake District inlier. Small bodies of Mid-Ordovician granite and gabbro are associated with many of the ore deposits. The buoyancy of a low-density, Devonian-age, granitic batholith located at depth is in part responsible for uplifting the mountainous core of the Lake District.

Keywords

Borrowdale volcanics • Copper • Glaciation • Graphite • Landscapes • Slate

Photographs not otherwise referenced are by the author.

The original version of this chapter was revised: Belated corrections have been incoporated. The corrections to this chapter are available at https://doi.org/10.1007/978-3-030-54693-9_18

16.1   Introduction

The Lake District in northwest England contains some of the most dramatic scenery in the British Isles. The landscape is dominated by the rugged, ice-sculptured Cumbrian Mountains. Glacial landforms include large finger-lakes in deeply eroded valleys. Windermere is the largest lake in England and Scafell Pike (978 m) is the highest peak. The region is well known for hanging valleys that contain Alpine lakes and waterfalls. The Lake District National Park (area of 2,362 km^2) is the most widely visited rural area in the British Isles, with over 15 million visitors annually. Many tourists visit for outdoor activities associated with the mountains and lakes, but others are attracted to the tranquil landscapes (Fig. 16.1). The national park recently achieved World Heritage status, despite human activities, including forestry and mining, adversely impacting the environment. Tourism is the mainstay of an otherwise rural economy. Organized tours to the region commenced more than 200 years ago with attractions primarily directed at the association with literary greats, notably William Wordsworth.

The Cumbrian Mountains cover an area of approximately 100 km north-south and 75 km east-west (Fig. 16.2). The finger-lakes are located in deep valleys that radiate from the mountainous core. The NNW-trending Eden Valley contains the eponymous River Eden, the largest river in the region. The Eden Valley separates the Cumbrian Mountains from the Pennine Hills. The largest towns in the national park are Ambleside, Coniston, Keswick, and Windermere. Towns on the perimeter of the region include Carlisle, Cockermouth, Kendal, Penrith, and Whitehaven. The region is served by both railways and roads, although infrastructure development within the national park has been controlled since the early part of the 20thC.

The geological setting of the Lake District is dominated by an inlier of Lower Palaeozoic strata. A significant component of the inlier is the Borrowdale Volcanic Group and these resistant rocks are associated with the rugged mountains in the core of the national park. The gentler landforms on the perimeter of the Lake District are underlain by more readily eroded metasedimentary rocks. The Lake District has an extensive mining heritage which

Fig. 16.1 Coniston Water viewed from Brantwood House, the home of Victorian writer and philosopher John Ruskin, is an example of the tranquil views which have attracted visitors to the Lake District views for many years. The gentle landforms near the lake are associated with metasedimentary rocks. The rugged hills in the background, including the broad massif of the Old Man of Coniston, consist of the resistant Borrowdale volcanics

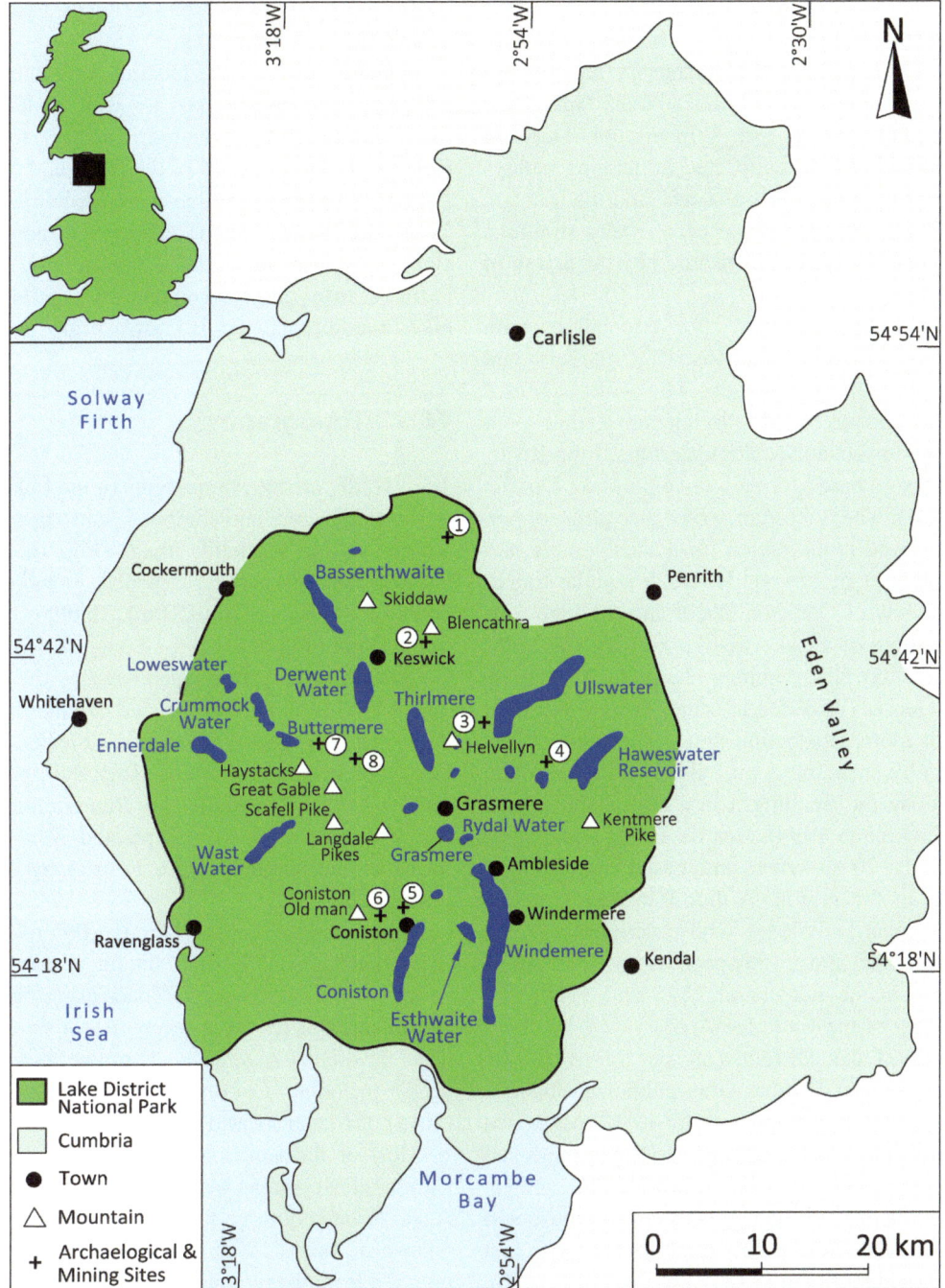

Fig. 16.2 Location map showing the Lake District National Park in Cumbria, northwest England. Also shown are the larger of the finger-lakes and some of the higher summits. Archaeological and mining sites are as follows: 1-Carrock Fell; 2-Castlerigg; 3-Greenside Lead mine; 4-High Street; 5-Tilberthwaite Slate quarries; 6-Coniston Copper mines; 7-Honister Slate mine; 8-Seathwaite Graphite mine

dates to the Roman times. Mining of lead, copper, and graphite was of particular importance. High quality slate continues to be produced from the region and dimension stone is obtained from quarrying of granite.

16.2 Human Settlement and Literary Connections

Cumbria has a long history of human settlement dating from the Neolithic times. The region was an important province of Roman Britain, was affected by Viking raiders and settlers, and has an equal claim to Cornwall as to the origin of the King Arthur legends. The long history of human settlement is reflected by the complexity of language for natural features. Examples include usage of the suffixes "*mere*" or "*water*" for the large lakes, e.g., Windermere, Derwent Water. The smaller Alpine lakes located in hanging valleys are known as "*tarns*". A stream is a "*beck*" and the hills are "*fells*". A "*pike*" describes the presence of rocky summits, often piles of boulders broken and fractured by the action of frost and ice.

The Lake District is well known for associations with English literature and artists. Famous 19[th]C writers and poets, including William Wordsworth, who wrote some of his most famous poems whilst living at Rydal, near Ambleside, as well as Samuel Taylor Coleridge, John Keats, and many others, were based for periods in the Lake District (e.g., Lindop 1994). The Victorian writer and philosopher, John Ruskin also lived in the region for a considerable part of his life, including at Brantwood House, a popular tourist attraction with a view of Coniston Water and the high fells (Fig. 16.1). The allure of a Lake District holiday was sufficient for Victorian novelist Thomas Hardy to decline an invitation from George V, observing that, despite the rain, "he probably got more satisfaction out of Coronation Day (22nd June 1911) by spending it on Lake Windermere than he would have done by spending it in a seat at the (Westminster) Abbey" (Lindop 1994). Beatrix Potter and Arthur Ransome are popular 20[th]C writers who based many of their children's books in the region. Arthur Ransome's stories included detailed local knowledge with descriptions of the copper and slate mines, and events such as the freezing of Lake Windermere in the winters of 1895 and 1929, i.e., followed the ending of the Little Ice Age.

Tourism to the Lake District can be traced to the mid-18[th]C, when the poet Thomas Gray publicized his scenic tour of 1769. By the early part of the 19[th]C, organized tours which included access by railways to Windermere had become established. The Lake District has been the subject of numerous guidebooks. William Wordsworth contributed one of the earliest and most widely-read guidebooks, first published in 1810 and reprinted in 1835 (Wordsworth 1835). Wordsworth was well acquainted with the region, having been born in Cockermouth and educated at Hawkshead (near Esthwaite Water). Wordsworth's guidebook included detailed notes on well known footpaths and historical sites, as well as descriptions of the principal physiographic features. Wordsworth also commented on the geology, observing that whilst the higher peaks consist mainly of schist (i.e., metamorphosed volcanics, specifically volcanic ashes) with subordinate granite, the lower-lying land in the south was associated with limestone and other sedimentary

formations. Brantwood has a small geological collection assembled by John Ruskin.

A feature of the Lake District National Park is that all of the peaks can be accessed by well trodden footpaths, and many are suitable (in good weather) for family outings (Fig. 16.3). Descriptions of the paths and climbing routes by Arthur Wainwright, which were written between 1955 and 1966 and include some geological notes, have become classics (Wainwright 1966). Wainwright's guides form the basis of many new formats with detailed descriptions of hikes encompassing well known peaks (e.g., Marsh 2011).

16.3   Topography

The two distinctive components of the Lake District, i.e., the mountainous core and smoother landscapes of the perimeter, are apparent in a satellite image (Fig. 16.4). The mountainous core reveals numerous peaks, including, in addition to Scafell Pike, Helvellyn (950 m), Skiddaw (931 m), Old Man of Coniston (803 m), and the Langdale Pikes (736 m). Great Gable (899 m) is often considered as the most iconic peak due to its free-standing form and extensive views. In 1923, a memorial to World War I was established on the summit of Great Gable by the Fell and Rock Climbing Club, where a service is held annually on Remembrance Sunday. The Cumbrian Mountains are separated from the gentler landforms of the southern Lake District by a regional unconformity (Fig. 16.5).

The excessive annual rain (>5,000 mm/annum in some areas) maintains the levels of the finger-lakes and Alpine lakes, which would otherwise be decreasing due to the limited catchment. This results in severe risk of storm flooding. Floods have devastated several small towns and villages in recent years, including Cockermouth (in 2009) and Glenridding (2015), the latter located near Ullswater.

Most of the well known landscapes associated with the Cumbrian Mountains were shaped by the Late Pleistocene Ice Ages. Relatively large finger-lakes, such as Windermere and Ullswater, occur in steep-sided glacial valleys that radiate outward from the mountainous core. Many hill slopes include steep scree slopes, which formed due to ice- and frost-dominated erosion during the Ice Ages. The screes located above Wast Water, at Wasdale, were introduced to an international audience by the cover photograph to a book on physical geography (Holmes 1965) (Fig. 16.6a). Other glacial features which can be readily identified are hanging valleys, many with small Alpine lakes and fast-flowing streams. The streams typically enter the primary valleys by a series of cascades and small waterfalls (Fig. 16.6b). Both the primary and secondary valleys contain extensive glacial moraines.

Fig. 16.3 The Lake District has been a popular destination for hiking, or fell walking, for many years. The background includes views of Derwent Water and the Skiddaw massif

16.4 Regional Geology

The geology of the Lake District was investigated by a number of famous geologists during the early years of the science (e.g., review of Oldroyd 2002). The dominant feature of the area is an inlier of Lower Palaeozoic strata (Fig. 16.7). The Lake District inlier is one of a number of disjointed Lower Palaeozoic terrains collectively known as the British Caledonides. Inliers are typically enclosed by Upper Palaeozoic strata. The strata in the Lake District inlier generally dip towards the southeast. The Lake District inlier is subdivided, from the base upwards, into the Skiddaw Group (Lower Ordovician), the Borrowdale Volcanic Group (Mid-Ordovician), and the Windermere Supergroup (Upper Ordovician and Silurian). The Skiddaw Group occurs in the northern part of the area, essentially to the north of Keswick. The Borrowdale Volcanic Group is constrained to the mountainous core. The Windermere Supergroup occurs in the southeastern part in the vicinity of Coniston Water and Windermere. The Borrowdale volcanics are separated from the Windermere Supergroup by a regional unconformity. The Eycott Volcanic Group is a subordinate feature in the northern part of the Lake District. Two groups of intrusive igneous rocks occur in the Lake District, plutons of Ordovician-age granite and gabbro and protuberances of a large Devonian-age granitic batholith.

16.4.1 British Caledonides

During the Neoproterozoic, the great southern mass of the supercontinent of Gondwana was separated from the two northern palaeo-continents of Baltica and Laurentia by the Iapetus Ocean (e.g., Windley 1977). The partial dismantling of Gondwana caused the separation of the microcontinent of Avalonia. Sediments and volcanic deposits accumulated in discrete Lower Palaeozoic geosynclines on the margins of

Fig. 16.4 The topography of the Lake District is dominated by a mountainous core with relatively gentle landforms on the perimeter. The dashed line shows the regional unconformity demarcating the contact of the SE-dipping Borrowdale Volcanics and the overlying Windermere Supergroup to the south. The Solway Firth is a regional suture which separates the Avalonian and Laurentian terrains of the British Caledonides. *Source* Landsat 7 ETM+ image for 2000 of Cumbria processed by Philip Eales of Planetary Visions/DLR

Fig. 16.5 The regional unconformity between the Borrowdale Volcanic Group (rugged hills in foreground) and the Windermere Supergroup (gentle landforms proximal to the lake) is associated with a marked change in the topography. View looking southeast from Coppermines Valley and overlooking Coniston Water

the different landmasses (Dewey 1969). Avalonia (a volcanic island arc) drifted northward and, as the Iapetus Ocean closed, collided with Baltica and Laurentia. The Caledonides developed as isolated terrains on the three palaeo-continental landmasses (Fig. 16.8). The Caledonides of the Lake District, North Wales, and the southern part of Ireland (together with similar strata in the extreme eastern parts of North America) were originally part of Avalonia. The Caledonides of Scotland and the northern part of Ireland (and large parts of eastern North America) were originally part of Laurentia. The Caledonide terrains of Scandinavia and parts of Central Europe were originally part of Baltica.

16.4.2 Caledonian Orogeny

The Caledonian Orogeny was a long drawn-out period of multiple continental collisions (McKerrow et al. 2000). The orogeny commenced in the Late Cambrian and persisted until the Early Devonian (490–390 Ma). The main phase of the orogeny in most terrains occurred in the Early Devonian. The collision of Baltica and Laurentia occurred at approximately 425–400 Ma. The collision of Avalonia and Laurentia (i.e., in the Lake District) occurred at 420–405 Ma. Collision caused the Avalonian oceanic crust to be subducted beneath the continental margins. The occurrence of Cambrian strata in most of the Lower Palaeozoic terrains is associated with the early onset of subduction. The absence of Cambrian strata, a diagnostic feature of the Lake District, is consistent with subduction of Avalonian oceanic crust during the Lower Ordovician. Collision of the ancient continental landmasses is preserved in the British Isles as major NE-SW trending sutures. The Solway Firth is associated with the suture where Avalonia (to the south) and Laurentia (to the north) were accreted. The disjointed nature of the Caledonide terrains prior to collision is supported by

a

b

Fig. 16.6   a The steep scree slopes above Wastwater (Wasdale) were formed by glacial erosion during the Late Pleistocene Ice Ages; **b** The town of Coniston is situated at the convergence of several hanging valleys which contain streams with small waterfalls

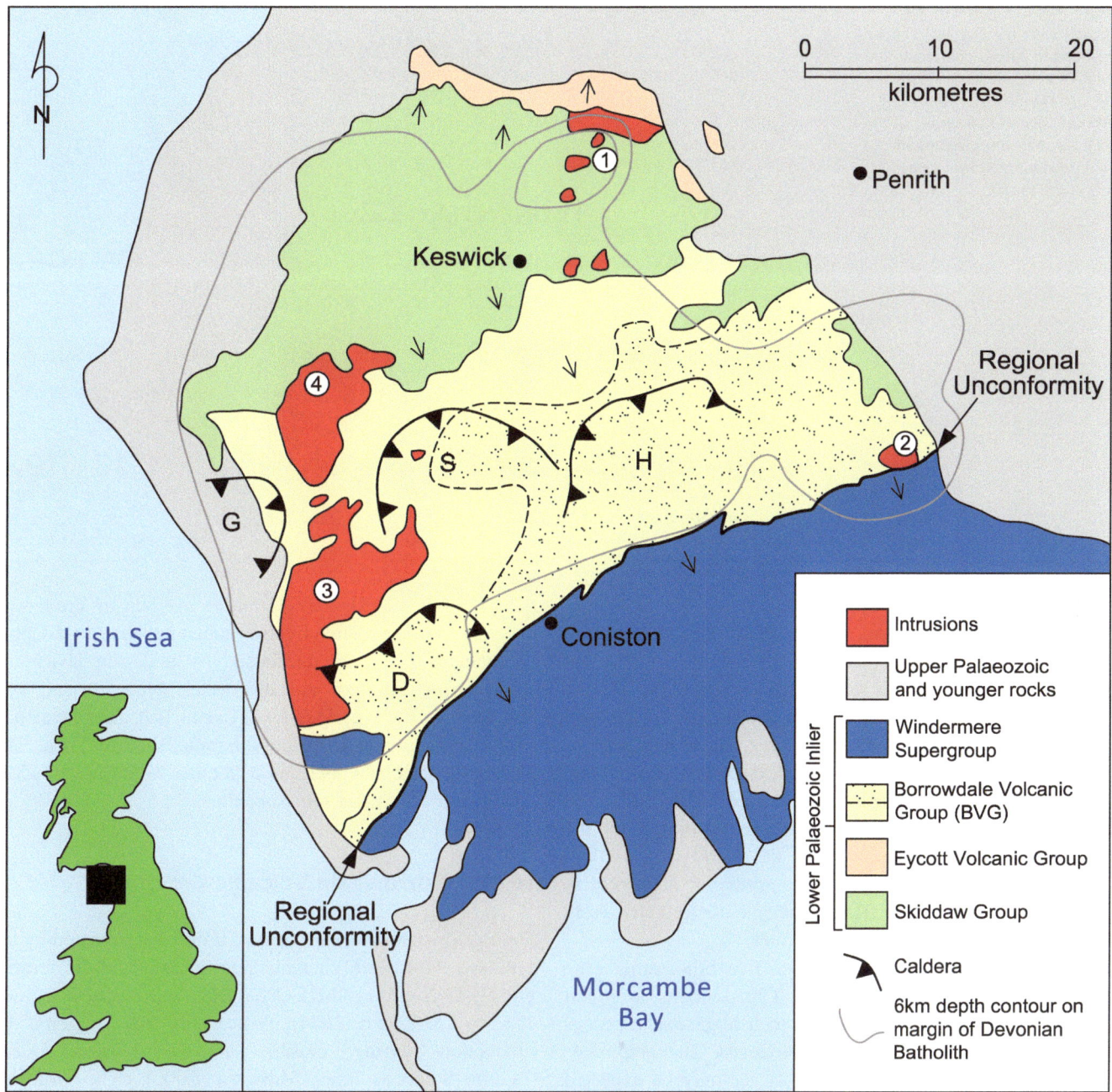

Fig. 16.7 Geological map of the Lake District inlier showing the dominance of Lower Palaeozoic strata. The Borrowdale Volcanic Group (BVG) is subdivided into lower (plain) and upper (stippled) components, the latter includes four calderas, Dutton (D), Gosforth (G), Helvellyn (H), and Scafell (S). Intrusive bodies include the Ordovician-age Carrock Fell gabbro (1), the Devonian-age Skiddaw granite (1); the Devonian-age Shap granite (2), the Ordovician-age Eskdale granophyre (3) and the Ordovician-age Ennerdale microgranite (4). *Source* Simplified from maps of the British Geological Survey

palaeontological studies. Of particular significance is the recognition of graptolite fossils linked to specific provinces (Zalasiewicz et al. 2009).

16.4.3   Skiddaw Group

The Skiddaw Group of Lower Ordovician turbidites is restricted to the northern part of the Lake District inlier

Fig. 16.8 Location of the Caledonide terrains (pale brown) relative to palaeo-continents at the end of the Caledonian Orogeny (Early Devonian). *Source* By Woudloper—Own work, CC BY-SA 1.0, https://commons.wikimedia.org/w/index.php?curid=5038110

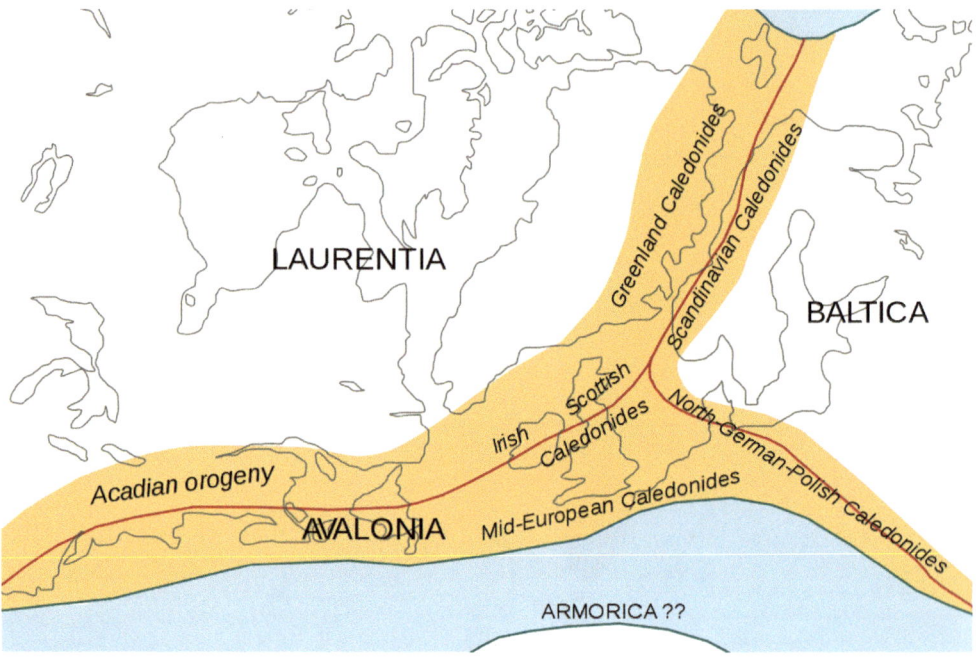

(Fig. 16.7). The turbidites are dominated by repetitive sequences of mudstone and siltstone and have a thickness of approximately 5,000 m (Millward and Stone 2012). The mudstone and siltstone is readily eroded and rather friable. These rocks have been extensively quarried in the past for production of rather poor quality slate. Most buildings in Keswick are constructed and roofed by Skiddaw Slate (high-quality slates from the region were "exported"). The Skiddaw Group is associated with large, rounded hills, the most well known examples being the Skiddaw and Blencathra massifs, to the north and northeast of Keswick, respectively (Fig. 16.9). These rocks outcrop extensively around Buttermere and Derwent Water.

The Skiddaw Slate encompasses five lithostratigraphic divisions of the Lower Ordovician. Type sections are found in North Wales. Subdivisions are to a large extent recognized on the basis of graptolite biozones. The graptolites reveal a high degree of provinciality, as may be expected from the Lower Ordovician sequences which developed on different sides of a palaeo-ocean (Zalasiewicz et al. 2009).

16.4.4 Eycott Volcanic Group

The Skiddaw Group was uplifted and eroded prior to onset of the Mid-Ordovician volcanism. In the northern part of the Lake District, the Skiddaw Group is overlain by the Eycott Volcanic Group. The Eycott volcanics are Mid-Upper Ordovician. They are dominated by andesitic lavas and sills interbedded with volcanic ashes, breccias, and volcaniclastics. The type locality is Eycott Hill, near Penrith, a scenic area protected by the Cumbria Wildlife Trust. The outcrop at Caldbeck Fells is a popular hiking destination, remote from the main tourist centres.

16.4.5 Borrowdale Volcanic Group

The Borrowdale Volcanic Group (BVG) is dominated by the explosive volcanism characteristic of an island arc terrain. The BVG is entirely Mid-Ordovician and is dated at 460–450 Ma (Millward 2004). Volcanism was triggered by subduction during closure of the Iapetus Ocean (Beddoe-Stephens and Millward 2000). The volcanic deposits are dominated by mostly sub-aerial calc-alkaline lavas, ashes, and volcaniclastics. The BVG is divided into two groups. The lower part is dominated by andesite lava, erupted from an intra-arc extensional rift, observed in the hills southwest of Keswick. The names of some peaks in this area reflect the geology, e.g., Green Gable (andesitic lavas) and Red Gable (pink microgranite).

a

b

Fig. 16.9 a The rounded massif of Skiddaw is typical of outcrops of the Skiddaw Group (viewed from the south); **b** This view of Derwent Water shows the town of Keswick located beneath the rounded slopes of Blencathra

The upper part of the BVG is dominated by volcanic ashes and volcaniclastics with abundant dacitic-rhyolitic ignimbrites. This section includes massive caldera sequences centred on the high peaks of Scafell and Helvellyn (Fig. 16.7). The calderas in the southwest part of the region, near Dutton and Gosforth, are younger. The latter area is largely obscured by younger rocks and the interpretation of a caldera sequence is based on drill cores. A young analogue to the upper part of the BVG may be the San Juan Volcanics in Colorado, USA (Beddoe-Stephens and Millward 2000). The San Juan Volcanics contain a sequence of nested calderas which developed during emplacement of a granitic batholith at depth.

The prominent striations or coarse banding in many of the volcanic ashes found in the lower part of the BVG is characteristic of subaerial volcanism (Fig. 16.10). Ignimbrite layers in the upper part of the BVG reveal columnar jointing (Fig. 16.11a, b). Finely laminated volcanic ashes are characteristic of subaqueous deposits in specific sections of the BVG (Fig. 14.11c).

The outcrops of the BVG on the Coniston Fells were mapped in detail by Mitchell (1940). This area was part of a major volcanic centre and the thick, intercalated sequences of lavas (andesite and rhyolite) and volcaniclastics can be traced for tens of kilometres along strike. The uppermost component of the BVG in this area is the Yewdale Breccia, a thick unit of agglomerate and ignimbrite which forms the prominent cliffs above Coniston (Fig. 16.5).

16.4.6   Windermere Supergroup

The Mid-Ordovician volcanism associated with the BVG was followed by a marine transgression and deposition of the Windermere Supergroup (Millward and Stone 2012). The earliest component is the Late Ordovician Dent Group (450 Ma). Two formations are recognized in the hill slopes north of Coniston, the Coniston Limestone (at the base) and the Ashgill Shale. The limestone is more resistant than the overlying shale, and is associated with a subsidiary topographic change on the slopes beneath the rugged topography associated with the BVG. The limestone and shale were deposited in shallow, subtropical seas. The Dent Group is particularly rich in shelly fossils. Stockdale Quarry, a historic site near Kendal, has yielded numerous trilobites, brachiopods, corals, and gastropods.

The principal component of the Windermere Supergroup is a thick succession of Silurian metasedimentary rocks. The shale, mudstone, chert, and greywacke contain abundant fossils. The paleoenvironment of the Silurian rocks is compared with a shallow tropical sea, such as in the Seychelles or Bahamas. The Silurian strata are relatively deeply eroded, and form the majority of the gently rolling hills that characterize the southern Lake District (Fig. 16.5).

The Late Ordovician and Silurian strata associated with the Windermere Supergroup are notably rich in graptolites, although the fossils do not exhibit the provinciality of the Lower Ordovician sequences (e.g., the Skiddaw Group). This is consistent with contraction of the Iapetus Ocean in the Upper Ordovician-Silurian (Zalasiewicz et al. 2009). The change from a localized palaeofaunal source (Lower Ordovician) to a uniform source may have occurred in the Mid-Ordovician, rather than the Upper Ordovician, but the Borrowdale volcanism has obscured the evidence.

16.4.7   Ordovician Intrusions

The magmatism of the Mid-Ordovician is linked with a suite of intrusive rocks in the Lake District that are broadly coeval with the BVG (Stone et al. 2010). They include small plutons of granite which have an approximate age of 450 Ma. Granitic plutons of this age are prominent in the northwest part of the region, where they may intrude both the Skiddaw Group and the BVG (Fig. 16.7). Examples include the Ennerdale pluton (pink-coloured microgranite) and the Eskdale pluton (granophyre). Microgranite of this age also occurs in the Threlkeld pluton, between Keswick and Penrith. The gabbroic intrusion at Carrock Fell, which has a slightly older age of 470 Ma, displays classic igneous layering (Wager and Brown 1968). Thin magnetite bands are intercalated with layers of magnetite-bearing gabbro which vary in colour index between leucogabbro and melagabbro.

16.4.8   Devonian Granite

The Lake District inlier is thought to be underlain by a large, Devonian-age granitic batholith. The main volume of this batholith is concealed, but small protuberances occur in the northern and eastern parts of the Lake District. These include the Skiddaw granite and the Shap granite (Fig. 16.7). The Shap granite is quarried as an attractive pink-coloured dimension stone. The buoyancy of the low-density granite is in part responsible for uplifting the mountainous core of the Lake District (e.g., Oldroyd 2002). The formation of the Devonian batholith is ascribed to a trans-tensional tectonic regime during the closing stages of the Caledonian Orogeny.

Fig. 16.10 **a** Striated or banded outcrops are typical of the terrestrial deposits of volcanic ash found in the lower part of the Borrowdale Volcanics (Seathwaite, Borrowdale); **b** A close up of the same outcrop shows the irregularity of the banding

Fig. 16.11 a Columnar jointing in an ignimbrite layer, Yewdale Breccia, in the upper part of the Borrowdale Volcanics (Coniston Fells); **b** Columnar joints in an ignimbrite layer, upper part of Borrowdale Volcanics (Tilberthwaite); **c** Finely laminated volcanic ash characteristic of subaqueous deposits of the Borrowdale Volcanics (Honister slate mine)

16.5   Late Pleistocene Glaciation

Many of the physiographic features of the Lake District landscapes were reshaped by the Ice Ages, cyclical events which occurred during the Late Pleistocene (Box 3.1). The Lake District was probably covered by ice sheets several kilometres thick during the Last Glacial Maximum (20,000 BP). Most of the classic glacial features, first recognized by Victorian geologists such as Adam Sedgwick and William Buckland, can be observed in the region (Oldroyd 2002). Some of the more photogenic outcrops are illustrated, together with concise explanations of how they formed, by Smith (2008). Glacial moraines include extensive deposits of boulders and gravels derived from retreating glaciers. Moraines have formed fertile deposits in some of the larger valleys, despite the occurrence of relatively small streams (Fig. 16.12a). Relatively large drumlins (smoothly rounded mounds of compacted glacial till) and eskers (elongate ridges of sand and gravel deposited at the base of glaciers) can be observed in the main valleys. Fields of small drumlins occur in some of the hanging valleys and high moorland (Fig. 16.12b). Glacial features include rocky hilltops, examples including the Langdale Pikes and Scafell Pike. Steep scree slopes are notably well exposed at Wasdale (Fig. 16.5a).

Amongst the most easily recognized glacial features are hanging valleys, i.e., second-order valleys located between the high peaks and the main valleys that contain the finger-lakes (Fig. 16.6b). Hanging valleys are associated with tributary glaciers which had lower erosive power than the primary glaciers. Hanging valleys are invariably capped by corries or cirques, i.e. deep, half-bowl-shaped recesses carved by the tributary glaciers into the mountainside. Many of the hanging valleys in the Lake District contain small Alpine lakes (Fig. 16.13a). The Alpine lakes feed small streams that generally tumble down the steep hill slopes to the main valleys by a series of cascades and waterfalls (Fig. 16.13b).

16.6   Finger Lakes

The finger-lakes which are so characteristic of the Lake District occur in the deeply glaciated valleys that radiate outwards from the mountainous core (Fig. 16.4). The most scenic of the lakes is possibly Buttermere (Fig. 16.14). The largest lake is Windermere with a surface area of 14.8 km^2

a

b

Fig. 16.12 a Steep-sided glaciated valleys in the Lake District contain relatively small streams, such as Styhead Gill, Borrowdale, and yet are partially filled by fertile deposits of alluvium; **b** Drumlin fields occur in many of the high moorlands of the Lake District (northwest of the Langdale Pikes)

a

b

Fig. 16.13 a The tarn below Goat Fell, part of the massif associated with the Old Man of Coniston, is an example of a hanging valley with an Alpine lake (Coniston Water visible in the background); **b** Streams typically exit hanging valleys via a series of cascades and small waterfalls, as seen below Levers Water on the Coniston Hills

Fig. 16.14 Buttermere is one of the most scenic lakes in the Lake District National Park

and the deepest lake is Wastwater (area of 2.9 km² and maximum depth of 76 m). Other relatively large lakes include Ullswater (area of 8.9 km²) and Coniston Water (area of 4.9 km²). Derwent Water (area of 5.4 km²) and Lake Bassenthwaite (area of 5.3 km²) are unusual in that they exhibit sub-rounded shapes as they were once joined. Alluvial deposits have built up where the River Derwent enters the lower of the two lakes.

Studies of lake sediments have yielded information on climatic changes as well as of the post-glacial history. Detailed investigation of Windermere, including use of digital terrain modelling, has enabled a reconstruction of ice-margin movements (Pinson et al. 2013). The results of this study suggest a complex and active valley glacier withdrawal from Windermere, rather than previously held views of rapid ice-stagnation.

16.7 Archaeology and Mining Heritage

Stone tools were produced in the Lake District from the volcanic rocks during the Neolithic period, including from the popular climbing area around the Langdale Pikes.

A number of ancient stone circles occur, of which the largest is Castlerigg, near Keswick (Fig. 16.15a). The standing stones at Castlerigg were sourced from volcanic rocks. They must have been specially selected, as they were transported several kilometres from the outcrop of the BVG (Fig. 16.15b). Non-ferrous mining within the Lake District can be traced back to the Romans (lead and silver). Mining of copper and graphite was of particular importance in the Lake District during the Elizabethan times (latter part of the 16thC). Copper and lead were mined extensively in the Victorian times (mid-late 19thC). Lead and tungsten were mined in the 20thC. High-quality slate has been exploited in the Lake District at least since the Elizabethan times. The industrial revolution spawned an extensive coal and iron ore industry from the younger strata enclosing the Lake District. Hikers that use popular trails observe the profusion of spoil heaps and often see mining in a negative light, without appreciating that most trails are old mining routes.

Most mines in the Lake District are now closed. There are a number of mining museums in the Lake District, including Threlkeld which is located between Keswick and Penrith, where mine plans and details of processing plants, as well as

a

b

Fig. 16.15   a The Neolithic stone circle at Castlerigg, near Keswick, is located on Skiddaw slates at the foot of Blencathra; **b** The standing stones consist of volcanic rocks derived from outcrops of the Borrowdale Volcanic Group in the rugged hills several kilometres to the south

many of the ore types can be seen. The history of the Lake District mines by Postlethwaite (1931) includes detailed underground plans and sections.

16.7.1   Lead Mines

During Roman times, Cumbria was situated on the north-west frontier of Roman Briton (the western extremity of Hadrian's Wall can be observed near Carlisle). The region was heavily militarized and several Roman forts are protected as heritage sites. Hardknott Fort is located in a remote, upland setting on a road linking the Roman settlement of Penrith and the harbour of Ravenglass. The road and harbour were primarily used for exporting minerals, including lead and silver. "High Street" is a section of the Roman road located on a relatively broad, yet high ridge (the summit of High Street at 828 m is the highest point in the eastern part of the Lake District) located between the Roman forts at Penrith and Ambleside (Fig. 16.16). The use of hill routes, rather than the valleys preferred by modern infrastructure may appear illogical, but the valleys were swampy and thickly wooded prior to being drained and settled in Elizabethan times. High Street was used for annual summer fairs, the most recent in 1835.

The most well known lead mine in the Lake District is the Greenside mine, located in the hills at an elevation of 600 m to the west of the village of Glenridding, at the southern extremity of Ullswater. The Greenside mine may have first opened in the 1700s. During large parts of the 1800s, as well as in the mid 20thC, Greenside was the largest and richest lead mine in England. The site includes refurbished buildings and extensive spoil heaps. The lead (together with by-product silver) was derived from the mineral galena contained within a large quartz vein (which strikes approximately N-S) and steeply-dipping pockets of breccia. The vein is emplaced in andesite of the BVG, but the mineralization is considerably younger, and may be related to hydrothermal fluids associated with intrusion of the younger granites.

In 1959, the Greenside mine was used by the Atomic Weapons Research Establishment for an experiment in detecting underground seismic explosions. The mine was finally closed in 1961, but sections have been reopened for underground exploration. The site which includes the spoil heaps (which may be unstable) has been designated as a scheduled ancient monument.

16.7.2   Coppermines Valley, Coniston

Mining of copper-rich quartz veins in the Coniston area was formerly a major industry. In Elizabethan times, German miners were brought over to modernize operations, and some of their legacies, together with later operations can be examined. The majority of the copper was used to clad the underwater sections of large sailing ships, including those of the British Navy under Sir Francis Drake. An undertaking to drain Levers Water, a small lake in a hanging valley perched several kilometres from the mines, bears testament to the skills of the miners during the Elizabethan times, despite not being completed. There was a rejuvenation of activity in Victorian times, when some of the mines at Coniston were enlarged into major operations. Most of the mines in this area closed in the 1890s. The historical trail in the Coppermines Valley includes the Bonsor mine, the largest of the Victorian mines in this region (Postlethwaite 1931; Holland 1981). The valley contains large spoil heaps, and numerous workings dot the hillsides (Fig. 16.17a). Mine workings include open cuts, on-vein drives, and adits, the latter used to access deeper veins and for drainage. (Some of the open cuts, such as "Simon's Nick", are potentially dangerous, especially when visibility is poor.) The majority of the copper mineralization occurs in quartz veins emplaced in the BVG (Fig. 16.17b). Samples of quartz veins with pyrite, chalcopyrite, chalcocite, barites, and bismuthinite occur on the mine dumps. Bismuthinite is a relatively rare mineral, observed as cm-sized clusters of silver-coloured, feathery grains. The ore minerals are not restricted to quartz veins, but also occur in volcanic agglomerates, possibly indicative of a petrogenetic relationship with the BVG (Scoon 1976).

16.7.3   Graphite

The graphite mined from workings near Keswick led to the development of a world-famous pencil industry based in the Lake District. Coloured pencils sold under names such as Lakeland and Derwent are well known to older generations of geologists. The Pencil Museum, Keswick, contains information on the industry, including the smuggling of graphite in Elizabethan times when the mineral was highly prized. (The origin of the expression "Black Market" may date from this activity.) The graphite occurs in quartz vein systems located in the BVG. The quartz-graphite veins are aligned on subvertical fault systems (Fig. 16.18a). The veins

a

b

Fig. 16.16 **a** High Street is a Roman road and wall (right) located on the crest of several intersecting cirques; **b** View from High Street looking west over Hayeswater, an Alpine lake located in a hanging valley at the base of Gray Crag (footpath on the ridge), with the massif of Helvellyn in the background (left)

Fig. 16.17 **a** The large spoil heaps in the Coppermines Valley attest to the extensiveness of the Victorian age copper mines near Coniston; **b** The copper mineralization is contained in quartz veins emplaced in the Borrowdale volcanics

were accessed from adits driven into the hillside (Fig. 16.18b). The principal source of graphite was from Seathwaite, one of only two known occurrences where economic quantities of graphite are found in volcanic rocks (Ortega et al. 2010). Formation of graphite in volcanic rocks is hindered by the low carbon content of magmas and degassing processes.

16.7.4 Tungsten

Prior to closure of the tungsten mine at Carrock Fell in 1981, the site was used by Cardiff University for training students in mining geology. The mineralization at Carrock Fell is associated with quartz vein systems within a section of the Skiddaw granite (Devonian) that has been subjected to intense greisenization (Shaw 2015). The granite is intruded into slates of the Skiddaw Group, which had been previously

contact metamorphosed to hornfels by the Carrock Fell gabbro (Ordovician). The main adit is currently blocked and the spoil heaps, which contain some interesting rocks specimens, are protected.

16.7.5 Slate

Deformation associated with the peak of the Caledonide Orogeny resulted in formation of a strong cleavage in many of the Lower Palaeozoic volcanic ashes, shales and mudstones. In some localities, the cleavage is sufficiently pronounced as to form slate. The conversion of a volcanic ash or shale to slate involves the alignment of the relatively small mineral grains in one direction. The best quality slate occurs when the cleavage planes are parallel to the axial planes of folds, which explains why miners refer to the occurrence of pockets, rather than laterally-extensive beds or strata. Some mining of slate still occurs in the Lake District, although the quantities being removed are tiny compared to Victorian times.

The Honister mine is located in a remote area of the northern Lake District, near the crest of the Honister Pass, west of Keswick. Large quantities of high-grade slate have been mined from the extensive underground workings (Fig. 16.19a). The oldest evidence of the slate mining here dates from 1643. The Honister mine has recently been reopened and includes a successful tourist component, which includes underground visits (Fig. 16.19b). The dark blue-grey "Honister slate" is associated with thick sequences of water-lain pyroclastics and volcanic ashes in an area with a relatively high level of deformation. The pyroclastics are very finely banded and can include deposits of volcanic agglomerate (Fig. 16.20). Honister slate is reputed to be the most durable in the world, and has been used to roof St Paul's Cathedral and other famous buildings in London. The slate has a lifespan of more than 400 years (far longer than slate from the southern part of the Lake District or North Wales); the longevity is crucial in areas exposed to acid rain. The slate occurs in pockets developed within specific beds of the pyroclastics and ashes, and there is considerable waste.

The quarries and underground caverns on the slopes of the Old Man of Coniston yield a good quality, light green slate. The slate occurs in a thick sequence of andesitic pyroclastics, within the upper part of the BVG. Small-scale operations are still working here. The slate quarries at Tilberthwaite, northeast of Coniston, yield a pale blue-green slate, marketed as "Coniston slate". The quarries at Tilberthwaite operated for approximately 150 years prior to closure in the 1990s (Fig. 16.21a). The slate at Tilberthwaite

a

Approximate trace
of graphite vein

b

Fig. 16.18 a The alignment of spoil heaps at Seathwaite, southwest of Keswick, traces the surface expression of the subvertical veins of quartz-graphite; **b** Entrance to one of the old graphite mines in the Borrowdale Volcanics, Seathwaite

Fig. 16.19 **a** Some of the adits (tiny dark features in the photograph) at the Honister slate mine are situated on almost inaccessible hill sides; **b** The main entrance to the Honister slate mine, used for both tourism and production, shows the steeply-dipping axial planar cleavage typical of the Borrowdale volcanics in this region

Fig. 16.20 **a** A slab of thinly bedded subaqueous volcanic ash shows the finely banded nature of the high quality Honister slate (left), although the presence of fractures and agglomerate (right) can affect the value; **b** Volcanic agglomerate from the Honister mine makes an attractive dimension stone

Fig. 16.21 **a** Quarries at Tilberthwaite, near Coniston, exploited pockets of slate from the upper part of the Borrowdale volcanics; **b** The slate at Tilberthwaite includes finely-bedded volcaniclastics with prominent sedimentary textures

is associated with terrestrial deposits of volcanic ashes and volcaniclastics (Fig. 16.21b). A relatively low quality slate was quarried from the metasedimentary rocks of the Windermere Supergroup to the south of Coniston.

References

Beddoe-Stephens, B., & Millward, D. (2000). Very densely welded, rheomorphic ignimbrites of homogenous intermediate calc-alkaline composition from the English Lake District. *Geological Magazine, 137,* 155–173.

Dewey, J. H. (1969). From micro to macro: plate tectonics in the Lake District - a tangle of theories (Chap. 10). In: D. Oldroyd (Ed.), Earth, Water, Ice and Fire: Two Hundred Years of Geological Research in the English Lake District (2002). *Memoir Geological Society London, 25,* 141–152.

Holland, E. G. (1981). Coniston copper mines: A field guide (p. 119). Milnthorpe, Cumbria: Cicerone Press.

Holmes, A. (1965). Principles of physical geography (2nd ed.). London: Thomas Nelson and Sons.

Lindop, G. (1994). A literary guide to the lake district (p. 424). London: Chatto and Windus.

Marsh, T. (2011). The high fells of lakeland: Mountain walks. Pathfinder Guides (p. 94). UK: Crimson Publishing, Richmond, Surrey.

McKerrow, W. S., MacNiocaill, C., & Dewey, J. F. (2000). The Caledonian Orogeny redefined. *Journal of the Geological Society, 157,* 1149–1154.

Millward, D. (2004). Stratigraphic framework for the upper Ordovician and lower Devonian volcanic and intrusive rocks in the English Lake District (4p.). British Geological Survey Research Report, RR/01/07.

Millward, D., & Stone, P. (2012). Stratigraphic framework for the Ordovician and Silurian sedimentary strata of northern England and the Isle of Man (122p.). British Geological Survey Research Report, RR/12/04.

Mitchell, G. H. (1940). The Borrowdale Volcanic Series of Coniston, Lancashire. *Quarterly Journal Geological Society London, 96,* 301–319.

Oldroyd, D. (2002). Earth, water, ice and fire: Two hundred years of geological research in the english lake district. *Memoir Geological Society London, 25,* 328p.

Ortega, L., Millward, D., Luque, F. J., Barrenechea, O., Huizenga, J.-M., Rodas, M., et al. (2010). The graphite deposit at Borrowdale

(UK): a catastrophic mineralizing event associated with the Ordovician orogeny. *Geochimica Cosmochimica Acta, 74,* 2429–2449.

Pinson, L. J. W., Vardy, M. E., Dix, J. K., Henstock, T. J., Bull, J. M., & Maclachan, S. E. (2013). Deglacial history of glacial lake Windermere, UK: Implications for the central British and Irish ice sheet. *Journal of Quaternary Science, 28,* 83–94.

Postlethwaite, J. J. (1931). Mines and mining in the Lake District (3rd ed.). Whitehaven.

Scoon, R. N. (1976). Volcanogenic mineralization at Coniston, Cumbria (172p.). Unpublished BSc (Hons.) Thesis, Cardiff University.

Shaw, R. P. (2015). The underground geology of part of the Carrock Tungsten mine (22p.). British Geological Survey Minerals and Waste programme Open Report OR/15/033.

Smith, A. (2008). The ice age in the Lake District. The landscapes of Cumbria No. 3 (60p.). Keswick: Rigg Side Publications. ISBN 0-9544679-2-2.

Stone, P. Millward, D., Young, B., Merritt, J.W., Clarke, S.M., McCormack, M., & Lawrence D.J.D. (2010). Lake District Batholith, Caradoc magmatism, Ordovician, Northern England. British Geological Survey 5th Edition, Nottingham.

Wager, L. R., & Brown, G. M. (1968). Layered Igneous Rocks (p. 588). Edinburgh: Oliver and Boyd.

Wainwright, A. (1966). The Wainwright guides: Pictorial guide to the lakeland fells (Vol. 7). 50th Anniversary Volumes published by WainwrightGuides.co.za.

Windley, B. F. (1977). The evolving continents (p. 399). Chichester: Wiley.

Wordsworth, W. (1835). Wordsworth's Guide to the Lakes (5th ed., 212 p.). Reprinted and edited by E. de Selincourt in 1906. Oxford University Press.

Zalasiewicz, J. A., Taylor, L., Rushton, A. W. A., & Loydell, D. K. (2009). Graptolites in British stratigraphy. *Geological Magazine, 146,* 785–850.

Famous Layered Igneous Intrusions

Abstract

Layered igneous intrusions are large plutons characterized by sub-horizontal rock layers. In a broad sense, the mineralogy and petrology of the igneous layers appears to change in a regular fashion with height, but in detail patterns can be extraordinarily complex. The two examples of layered igneous intrusion described here, the relatively small Skaergaard Intrusion in Greenland and the Eastern Limb of the giant Bushveld Igneous Complex in South Africa may represent end-members in terms of size, shape, and origin. There is no consensus on the formation of igneous layering, despite the cumulus theory based on gravitative crystal settling having been superseded to some extent by growth of crystals on the temporary floor of either a magma chamber or a crystal mush. The Skaergaard Intrusion occurs in the volcanic rifted margin of eastern Greenland. The intrusion is part of the North Atlantic Igneous Province (Eocene) and has an age of 55 Ma. The intrusion is funnel shaped with subvertical sidewalls and has a maximum width of 8 km and thickness of 3,200 m. Despite the northerly location, the Skaergaard Intrusion is one of the most widely studied geological sites on Earth. Field and geochemical evidence in support of an important principle of petrology, in situ magmatic differentiation of basaltic magmas, was first described here. The intrusion is dominated by the Layered Series, a sequence of gabbro, troctolite, ferrogabbro, and ferrodiorite. Three zones are identified, each of which reveals lithologies with differing compositions and mineralogy. The most spectacular example of igneous layering is the rhythmic alternation of thin, modally graded layers in the ferrogabbro of the Upper Zone. The Eastern Limb of the Bushveld Igneous Complex is associated with an erosional embayment into the interior plateau of southern Africa. The most important component is the Rustenburg Layered Suite, a thick sill-like unit of layered igneous rocks which has a thickness of approximately 10 km and extent of more than 180 km. The Rustenburg Layered Suite is part of a Palaeoproterozoic intrusive complex (age of 2.055 Ga) which includes an older component of extrusive rhyolite and a younger component of granite. The igneous layering of the Rustenburg Layered Suite has resulted in a remarkable landscape of alternating ridges and deeply-incised valleys. Five or more zones are recognized. Zones are characterized by specific lithologies with marker layers that can be traced for tens of kilometres along strike. The occurrence of intercalated layers of chromitite, pyroxenite, and anorthosite is a notable feature of the Critical Zone, the most widely studied part of the intrusion. The contrast between the irregular basal contact, exaggerated in some areas by syn-Bushveld domes, and the planar nature of the roof (which does not include a chill) is emphasized. The dominant process in formation of the igneous layering is addition of batches of magma from a range of parental or source magmas, rather than processes operating within a large reservoir of convecting magma. The Rustenburg Layered Suite contains important layered orebodies, including the famous platinum-rich Merensky Reef and economically viable layers of chromitite and vanadium-rich Ti-magnetite. An unusual feature is the occurrence of discordant bodies, including platinum-rich dunite pipes. The discovery of the platinum deposits in the Eastern Limb is an important historical event and, together with the spectacular geological exposures, may justify establishing a geopark.

Keywords

Chromitite • Dunite pipe • Igneous layering • Merensky Reef • Platinum • Ti-magnetite

Photographs not otherwise referenced are by the author.

The original version of this chapter was revised: Belated corrections have been incoporated. The corrections to this chapter are available at https://doi.org/10.1007/978-3-030-54693-9_18

17.1 Introduction

Layered igneous intrusions are large plutons characterized by sub-horizontal rock layers. In a broad sense, the mineralogy and petrology of the igneous layers appears to change in a regular fashion with height, but in detail patterns can be extraordinarily complex. In some localities, the igneous layering can be so pronounced that it can resemble sedimentary bedding. The occurrence of alternating rock layers with markedly different compositions is an intriguing feature (Fig. 17.1). Mineral layering is an important component of the layered igneous intrusions and can occur either as a discrete feature, or as part of the rock layering. The intrusive bodies can vary in size from areas of several hundreds of square kilometres to tens of thousands of square kilometres. The three-dimensional shape of the plutons is similarly highly variable. The two examples described here, the relatively small and funnel-shaped Skaergaard Intrusion in Eastern Greenland and the giant, sill-like Bushveld Igneous Complex in South Africa may represent end-members in terms of size, shape, and origin.

Many layered igneous intrusions occur in ancient cratonic settings where the intricately layered nature can be ascribed to the stable tectonic setting, but others occur in younger, rifted terrains. Layered igneous intrusions can be considered as part of "Large Igneous Provinces" and are similar to flood basalt provinces in terms of the huge volumes of magma involved. The composition of the layered igneous intrusion generally encompasses ultramafic and mafic components. Average compositions are difficult to ascertain, but the majority of layered igneous intrusions are associated with basaltic magmas.

Despite its location north of the Arctic Circle, the Skaergaard Intrusion (latitude 68 °N) is one of the most widely studied geological sites on Earth. The intrusion has been mapped in considerable detail. The mountainous outcrop has been swept clean of soils and vegetation by ice sheets and glaciers. In situ magmatic differentiation of basaltic magmas, an important principle of igneous petrology, is supported here by detailed field and geochemical evidence (Wager and Deer 1939). The differentiation of the basaltic magma which filled the funnel-shaped intrusion follows the Fenner trend of iron-enrichment (rather than the Bowen trend of increasing silica). The systematic trends observed in the layered rocks were for many years thought to have developed as a consequence of crystal settling (Wager and Brown 1968).

Fig. 17.1 The intercalation of thin layers or stringers of chromitite (black) in a thick layer of anorthosite (fawn) at the Dwarsriver National Monument, Eastern Bushveld (Geosite G10), is a classic example of igneous layering. The thin layers and stringers of chromitite bifurcate from the base of the metre-thick layer of UG1 chromitite (top left)

The principles of crystal settling (the "cumulus" theory) were thought to be applicable to all layered mafic intrusions, including the Bushveld Igneous Complex. As the role of crystal settling in tholeiitic magmas was questioned, the cumulus theory has been superseded to a large extent by the process whereby crystals grow at the temporary base of a convecting reservoir of magma (Bottinga and Weill 1970; McBirney and Noyes 1979; Naslund and McBirney 1996). Moreover, the Bushveld Igneous Complex has been found to be far more complex than was originally thought (e.g., Tankard et al. 1982; Von Gruenewaldt et al. 1985; Eales 2014).

17.2 Skaergaard Intrusion

The Skaergaard Intrusion is located on the shores of the Kangerdlugssuaq Fjord, a region of eastern Greenland that includes several tidewater glaciers and rocky islands. The glaciers may be >1 km in thickness. The Danish word "Skærgård" describes the setting of a rocky coastline protected by numerous small islands and rocks (Brooks 1997).

The Skaergaard Intrusion forms prominent *nunataks* (i.e., peaks that project through the ice sheet) with elevations of over 1,200 m. The most famous example is Wager Peak (Fig. 17.2). Icebergs and ice floes associated with the ice pack driven southward by the East Greenland current, the main outflow from the Arctic Ocean, are a feature of the fjord. An appreciation of sea conditions is crucial for short expeditions, as although the fjord provides a sheltered refuge, it can become blocked if the southerly movement of the ice pack is disrupted. Moreover, calving of the glaciers may exasperate ice conditions in the fjord. The region supports a varied sea life and is noted for the occurrence of narwhals. Polar bears are reportedly scarce, although expeditions require an armed guide.

The Kangerdlugssuaq region of Greenland was first visited by European explorers in 1900, but despite being uninhabited, both then and subsequently, there is evidence of older Inuit settlements in the 13–19[th]C. Archaeological evidence suggests both European Vikings and North American Inuit migrated to southern Greenland during the 10[th]C when the climate was far milder than currently experienced (part of the Medieval Warm Period). Average

Fig. 17.2 Wager Peak reveals the bare rock faces that characterize large parts of the Skaergaard Intrusion. The view exposes part of the Lower Zone, the entire Middle Zone, and part of the Upper Zone. The "Triple Group" is just visible below the snow line. The peak is named after geologist Lawrence Rickard Wager ("Bill"), a world-renowned mountaineer who ascended to a few hundered metres of the summit of Everest in 1933

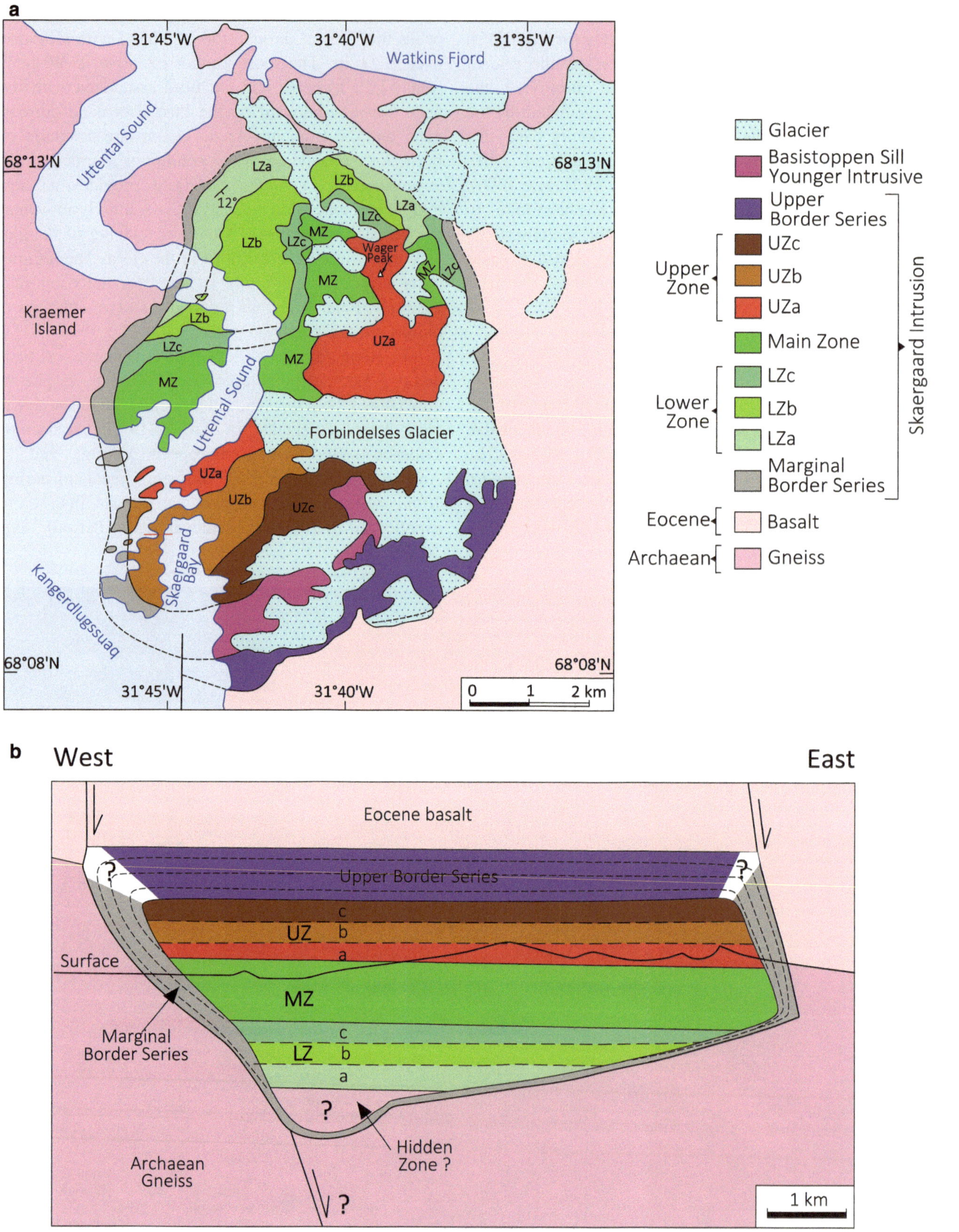

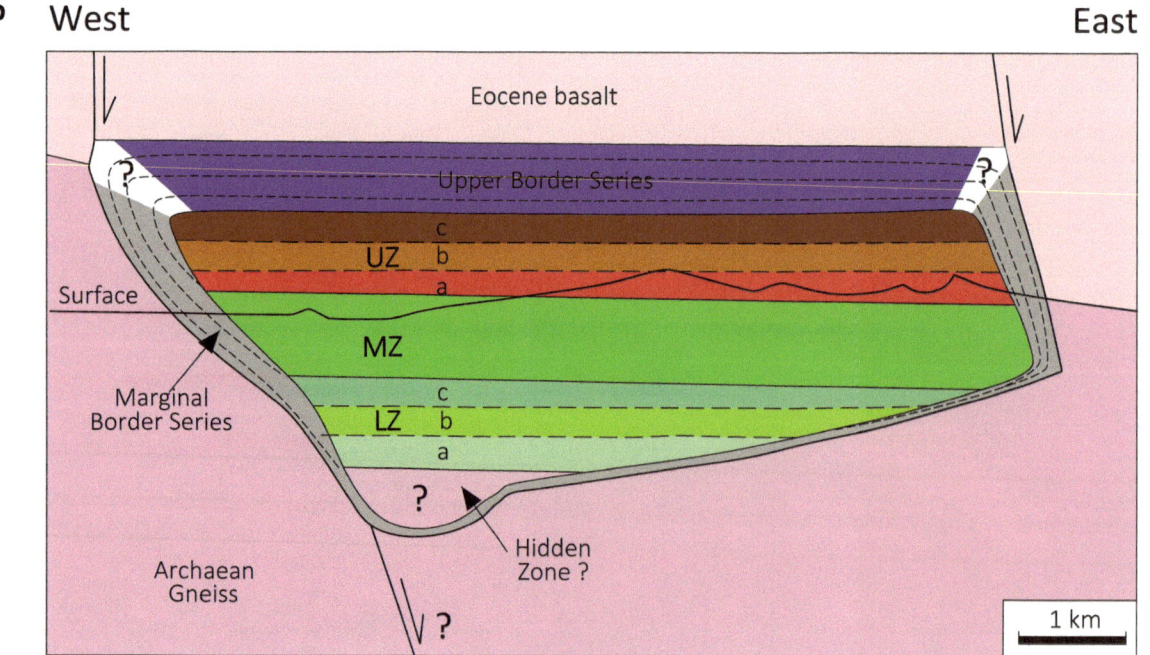

Fig. 17.3 **a** Generalized map of the Skaergaard Intrusion; **b** W-E section of the Skaergaard Intrusion. *Source* Irvine et al. (2001)

temperatures may have been several degrees warmer than currently recorded. The Inuit from southeast Greenland have recently had several attempts at re-colonizing the Kangerdlugssuaq area, which they consider has a spiritual significance, and there has been some opposition to potential mining in the region.

17.2.1 Geological Research

The extensive geological research undertaken on the Skaergaard Intrusion is reviewed by Brooks (1997). The intrusion was discovered in 1930 by Lawrence Rickard Wager, during the British Arctic Air Route Expedition. Wager undertook the first of many detailed field programmes in 1932. The British East Greenland Expedition, led by Wager, spent most of 1935 and 1936 at Skaergaard, despite winter temperatures that regularly drop as low as -30 °C. The expedition, which included two Inuit families, established temporary housing in the locality of Home Bay. Many of the samples from these earlier expeditions are catalogued at Oxford University. Expeditions to Skaergaard generally commence from Reykjavik, Iceland, and the crossing of the Denmark Strait with the pack-ice-dominated, and typically clear waters of the Arctic Ocean, which contrast with the grey North Atlantic, are an unforgettable sight. Even in summer, the Kangerdlugssuaq Fjord contains pack ice and ice bergs of variable sizes. Hikes to examine some localities include steep treks and traverse of ice sheets, such as the crevassed Forbindelses Glacier.

17.2.2 Geological Setting

The Skaergaard Intrusion is located in the volcanic rifted margin of eastern Greenland, part of the Eocene-age North Atlantic Igneous Province (Irvine et al. 2001). The age of the intrusion has been determined as 55 Ma. The intrusive body crystallized from basaltic magma emplaced at the contact between Archaean granite gneiss and a thick sequence of Eocene basalts (Fig. 17.3a). The intrusion is funnel shaped, with subvertical sidewalls (Fig. 17.3b). The maximum width of the intrusion is 8 km and approximate thickness is 3,200 m. The contact with the wall rocks is exposed in several localities, including on Kraemer Island (Fig. 17.4). The magmatism of the North Atlantic Igneous Province, which includes basaltic lavas and sills, as well as basaltic layered intrusions, was probably driven by a hot spot. The hot spot is currently located beneath Iceland (Andersen et al. 2002).

Many intrusions in the North Atlantic Igneous Province contain anomalous concentrations of platinum-group elements (PGE), although most appear to be sub-economic (they are both small and low grade). In the 1980s, the "Triple Group" of layers in the Skaergaard Intrusion was found to contain significant concentrations of gold and palladium (Fig. 17.4). The Triple Group is traceable over more than 60% of the intrusion. A drilling programme has been completed and a significant resource (over 5 million ounces of gold at a cut-off of 2 g/t) published. Mineralization is reported to be readily extractable by fine grinding and flotation. The economic viability of the project has recently been re-examined.

17.2.3 Igneous Layering

The principal component of the Skaergaard Intrusion is the Layered Series (Wager and Brown 1968; McBirney and Naslund 1979; Irvine et al. 2001). The regularity of the igneous layering has enabled the Layered Series to be subdivided into zones and subzones that can be traced over large parts (although not all) of the outcrop (Fig. 17.3a). The Lower Zone and Upper Zone are each subdivided into three subzones, but the Middle Zone is relatively uniform (Fig. 17.3b). Each zone is characterized by discrete rock types. The Lower Zone is dominated by gabbro and troctolite, the Middle Zone consists largely of ferrogabbro, and the principal lithologies in the Upper Zone are ferrogabbro and ferrodiorite. The Layered Series is enclosed on the basal contact and sidewalls by the Marginal Border Series and is overlain by the Upper Border Series. The average lithology of the marginal rocks is ferrogabbroic (McBirney and Naslund 1990). The iron-rich nature of the intrusion baffled geologists in earlier times and a "Hidden Zone", comprised of magnesian-rich ultramafic rocks, was postulated to occur at depth. The size of the ultramafic component has, however, decreased in the minds of recent researchers and may be entirely absent.

Spectacular examples of macrorhythmic modal layering occur in the ferrogabbro of the Middle Zone on Kraemer Island (Fig. 17.5a). The rhythmic alternation of thin, modally graded layers in ferrogabbro of the Upper Zone has been intensively investigated (Fig. 17.5b). The basal parts of the modally graded layers are rich in dark minerals (iron-rich olivine and Fe-Ti oxides) and the upper parts are dominated by light-coloured plagioclase. Trough layering is an unusual feature of the Upper Zone (Fig. 17.6). The repetition of the trough layering is significant and has a bearing on the origin, as discussed by Irvine et al. (2001).

Fig. 17.4 The contact between a remnant of Eocene basalt and Archaean granite gneiss is exposed on Kraemer Island. Gently dipping igneous layering of the Layered Series of the Skaergaard Intrusion is visible in the background. Ice floes and a small berg are visible in the fjord

Most layered igneous intrusions contain discordant bodies, i.e., features that transgress the primary igneous layering. Examples in the Skaergaard include small bodies of mafic pegmatite (Irvine et al. 2001). The occurrence of irregular bodies of anorthosite which probably have a similar composition to the anorthosite layers is also of interest (Fig. 17.7a). The occurrence of displaced blocks and breccias in the Marginal Border Series has been widely discussed (Irvine et al. 1998, 2001) (Fig. 17.7b). The Basistoppen Sill is a well known adjunct to the Skaergaard Intrusion, located high on the mountain slopes above the Kangerdlugssuaq Fjord. The Basistoppen Sill is cited as a locality in which the unusual process of disequilibrium partial melting may have occurred (Naslund 1986).

17.3 Eastern Bushveld Complex

The Bushveld Igneous Complex, located in the northeastern part of South Africa, is one of the world's largest igneous provinces (Fig. 17.8). Three principal components are recognized (South African Committee for Stratigraphy 1980). The different components of the intrusion have been radiometrically dated (e.g., Walraven et al. 1990; Scoates and Friedman 2008; Mungall et al. 2016). The extrusive rhyolite (known locally as felsite, together with subordinate granophyre) of the Rooiberg Group is the oldest component (2.061 Ga). The Rustenburg Layered Suite is an intermediate component (2.055 Ga). The Lebowa Granite Suite (together with subordinate granophyre) is the youngest component (2.054 Ga). The igneous rocks are emplaced within the interior of a large Palaeoproterozoic basin associated with the Transvaal Supergroup (approximate age of 2.4 Ga). The felsite and granite are primarily restricted to the centre of the basin, but the Rustenburg Layered Suite outcrops in three arcuate limbs or lobes, Eastern, Western, and Northern. The Rustenburg Layered Suite was intruded beneath the Rooiberg Group, into the uppermost part of the underlying Transvaal Supergroup. Parts of the Northern Limb transgress onto the cratonic basement. Intrusion of the Rustenburg Layered Suite triggered development of an extensive metamorphic aureole in the underlying Transvaal

Fig. 17.5 a Macrorhythmic
modal layering of ferrogabbro in
the Middle Zone, Kraemer Island;
b Rhythmic alternation of thin,
modally graded layers in uniform
ferrogabbro in the Upper Zone,
Home Bay. The basal parts of the
modally graded layers are rich in
dark minerals (iron-rich olivine
and Fe-Ti oxides) and the
light-coloured upper parts are
dominated by plagioclase

Supergroup. All three components of the Bushveld Igneous
Complex (i.e., felsite, layered igneous rocks, and granite)
may have formed in a relatively short interval, possibly due
to a mantle plume (Hatton and Schweitzer 1995).

The most spectacular outcrops of the Bushveld Igneous
Complex occur in the Eastern Limb, as originally described by
Wagner (1929) and Hall (1932). The Eastern Limb correlates
with a physiographic region known as the "Middleveld". The
Middleveld forms an erosional embayment into the interior
plateau (or "Highveld") of southern Africa and is separated
from the coastal plateau ("Lowveld") by the Eastern Escarpment (Fig. 17.9). The erosional embayment commenced
forming during break-up of the ancient supercontinent of
Gondwana. A major thermal event associated with rifting of

Fig. 17.6 a Trough layering
(Trough G) in the Upper Zone,
Home Bay. The dark layers are
rich in Fe-rich olivine and Fe-Ti
oxides; **b** Neil Irvine pointing out
trough layering in the Upper Zone
on the expedition to the
Skaergaard Intrusion in 2001

the Atlantic Ocean and Indian Ocean (at approximately 180–35 Ma) uplifted much of the African Plate. Uplift initiated intense cycles of erosion. Superimposed drainage of the Olifants and Steelpoort Rivers has produced a mountainous terrain in the Eastern Limb which includes localized escarpments and arcuate ridges (Scoon and Viljoen 2019).

The Rustenburg Layered Suite in the Eastern Limb forms a 10 km thick sill-like body with a strike length of

Fig. 17.7 a A small discordant
body of anorthosite (pale brown)
situated in a layered sequence of
troctolite (dark brown) and
anorthosite (pale brown) in the
Lower Zone, Uttental Plateau;
b Blocks or xenoliths of gabbroic
troctolite (pale grey) in
rhythmically layered magnetite
gabbro (dark grey) in the Middle
Zone, Kraemer Island

approximately 180 km (Fig. 17.8). The width of the sill
decreases in the extremities. Igneous layering generally dips
at relatively shallow angles, either west (southern sector) or
southwest (northern sector), i.e., into the centre of the
Transvaal Basin.

17.3.1 Human Habitation and Mining

The Middleveld has a long history of human habitation and
includes important cultural and archaeological sites. The
interior plateau and valleys are relatively densely populated,

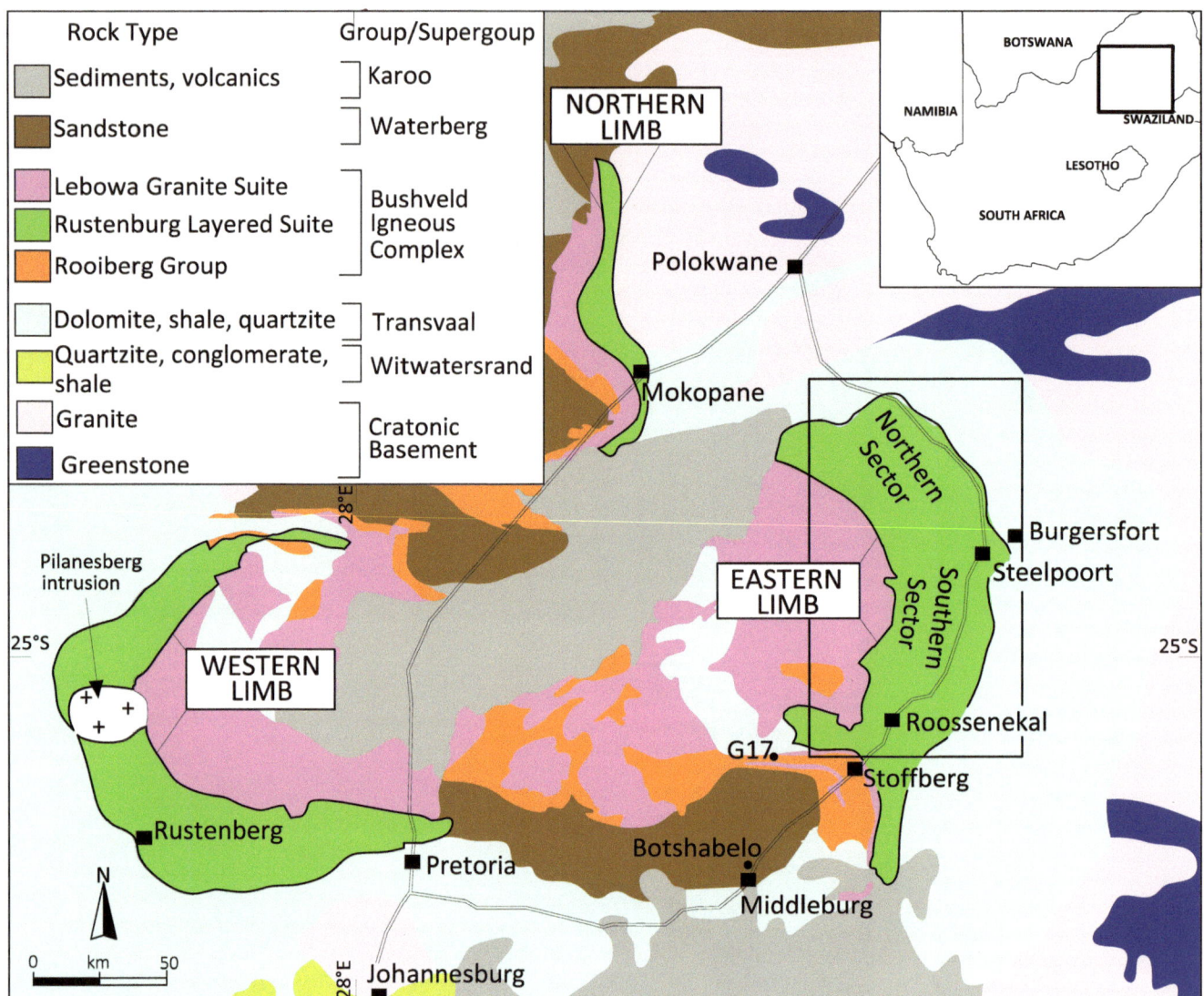

Fig. 17.8 Regional geological map of the Bushveld Igneous Complex showing the three main limbs of the Rustenburg Layered Suite (satellite bodies not shown). Each limb is an inclined sheet, or sill-like feature, which subcrop at depth beneath the felsites and granites in the interior of the Transvaal Basin. The boxed area of the Eastern Limb contains most of the well known geosites. *Source* Simplified from the 1:1,000,000 scale map published by the Council for Geoscience

but the mountain ranges (e.g., Leolo Mountains) are remote wildernesses. The regional towns of Burgersfort and Steelpoort are accessed by main roads from Johannesburg and Pretoria (Fig. 17.8). The Eastern Limb supports a world-famous mining industry and includes large deposits of chromium, PGE, vanadium, dimension stone, and industrial minerals such as andalusite and magnesite (Willemse 1969; Viljoen 2016). Most deposits occur in primary igneous layers (locally known as "reefs"), others constitute discordant bodies, and some industrial minerals occur in the metamorphic aureole.

The earliest humans in southern Africa, the Khoisan, may have lived and hunted in the Eastern Limb for tens of thousands of years. Evidence of habitation by Bantu-speaking peoples dates to the African Iron Age (1600 AD to 1000 AD). Archaeological evidence has been found at Tsate (or Dsjate), which includes a memorial centre, in a broad valley on the eastern slopes of the Leolo Mountains. Between 1830 and 1879, King Sekwati and his son, King Sekhukhune, united a number of rural communities into the formidable Pedi Empire. They established their capital at Thaba Mosega in the Tsate Valley. For many years, Sekhukhune warriors successfully defended this mountain fortress against attacks from tribal groups, as well as from European settlers (Fig. 17.10). However, in 1879 a combined assault from British troops and Swazi warriors forced their subjugation.

Fig. 17.9 A three dimensional satellite image (two times vertical exaggeration) of part of the Eastern Limb of the Bushveld Igneous Complex looking north showing the deeply-incised Groot and Klein Dwarsriver Valleys. Most of the famous geosites occur in the Steelpoort Valley and the Olifants Valley. The Bushveld Escarpment and Leolo Mountains are localized features of the Eastern Limb. The extreme irregularity of the floor contact between the towns of Steelpoort (S) and Burgersfort (B) is related to a trough-like body of Lower Zone. The mountainous terrain to the north and east of the Olifants Valley is associated with resistant rocks in the metamorphic aureole. *Source* Landsat 7 image for the year 2000 sourced from the University of Maryland Global Land Cover Facility and processed by Philip Eales of Planetary Visions/DLR

Fig. 17.10 Statue of King Sekhukhune situated on a koppie comprised of gabbronorite, at the Tsate Memorial in the Eastern Limb (Geosite G29)

Prior to the Sekhukhune uprising, in 1861, Alexander Merensky of the Berlin Missionary Society, together with his wife and young family, was granted permission by King Sekwati to build a mission station near Tsate. The uprising, however, caused the Merensky family to flee across the Steelpoort River, to seek safety at Roossenekal. The family finally settled at Botshabelo ("place of refuge"), near Middleburg, where they established a mission station that includes "Fort Merensky" (built in 1865).

Mapoch's Caves, near Roossenekal, is a provincial heritage site of archaeological significance located in resistant outcrops of gabbronorite. Mapoch was one of the earliest leaders of the Ndebele peoples. Extensive stone walling was constructed to join small cliffs and large boulders. In 1882, some of the Ndebele people fled to the caves after local skirmishes with both settlers and the Sekhukhune people.

Three geologists, Arthur Lewis Hall (1872–1955), Hans Merensky (1871–1952), and Percy Albert Wagner (1885–1929) greatly advanced our understanding of the geology of the Eastern Limb. Hall was born in England, partly educated in Germany, and spent most of his life in South Africa. Merensky was born at Botshabelo (the son of Alexander) and educated in Germany. Wagner was born in South Africa and educated in Cape Town. Field mapping undertaken over 29 years enabled Hall to produce the only comprehensive field guide to the Eastern Limb (Hall 1932). Recent guidebooks with selected geosites include those by Sharpe (1986) and Scoon (2016). The overview by Scoon and Viljoen (2019) includes descriptions of many of the well known geosites.

17.3.2   Discovery of Platinum in the Eastern Limb

The discovery of economic concentrations of platinum in the Eastern Limb, by Hans Merensky and his Lydenburg Platinum Syndicate, is widely known (e.g., Scoon and Viljoen 2019). Three days after panning concentrates in an ephemeral stream on the Maandagshoek property (on 15th August 1924), Merensky and his team located high-grade platinum mineralization in ultramafic rocks on the adjacent property of Mooihoek (Fig. 17.11). The Lydenburg Platinum Syndicate soon located two more pipes (Driekop and Twyfelaar). A fourth pipe (Onverwacht) was found by geologist F.W. Blaine. In September 1924, "the far more important finding" of a mineralized layer, soon to be named the Merensky Reef, was made at Maandagshoek (Merensky 1925). The Merensky Reef was delineated over >100 kilometres of strike in the Eastern Limb and historical workings and exploration trenches dating from the exploration period can still be examined. Most of the current platinum mines in the Eastern Limb exploit the UG2 Reef, (one of the upper group of chromitite layers) as the Merensky Reef is somewhat lower-grade. The fineness of the ore minerals in the UG2 is such that this reef could not be exploited until after the advent of processing technologies in the late 1970s. This has resulted in a rather unusual situation in that the UG2 Reef was discovered by Merensky in 1908 (i.e., before discovery of the pipes and reef), but this did not at the time constitute an economic orebody. Despite their small size, the dunite pipes provided the impetus to the mining of platinum in the Eastern Limb (an earlier venture on quartz veins in the Northern Limb proved to be an economic failure), as the coarse-grain size of the ore minerals was amenable to beneficiation by conventional means.

17.3.3   Dunite Pipes

The dunite pipes were emplaced perpendicular to the primary igneous layering (Wagner 1929; Scoon and Mitchell 2011). The principle component is a relatively large pipe-like body of dunite that outcrops as small hills or koppies (Fig. 17.12a). The platinum is restricted to small core-zones within the dunite, occurring in assemblages dominated by coarse-grained hortonolite dunite and hortonolite wehrlite (hortonolite is an iron-rich variety of olivine) (Fig. 17.12b). The core-zones are also described as "iron-rich ultramafic pegmatite", a lithology unique to the Rustenburg Layered Suite (Viljoen and Scoon 1985). Three of the dunite pipes discovered in the historical period of the "platinum rush" were mined during these early years. The high-grade Mooihoek and Onverwacht pipes proved extremely profitable, albeit the relatively low-grade Driekop pipe was less so. Onverwacht is the oldest underground platinum mine in the world (with the exception of small-scale workings in the Urals, Russia) (Fig. 17.13).

17.3.4   Zones and Igneous Layering

The Rustenburg Layered Suite is subdivided by an informal stratigraphy into five or more zones which in turn may be divided into subzones (Fig. 17.14). Each zone and subzone is characterized by specific lithologies. Laterally-persistent marker layers are a notable feature. The sill-like form of the Rustenburg Layered Suite is consistent with the stratigraphically highest zone, i.e., the Upper Zone, being the most laterally persistent. The contrast between the extreme irregularity of the basal contact and the planar nature of the roof is emphasized. Structurally complex areas of syn-Bushveld doming are a notable feature (Fig. 17.15). In a generalized way, the zones (with the exception of the Marginal and Lower Zones) become increasingly differentiated with height and the Rustenburg Layered Suite follows the same Fenner trend of iron-enrichment as does the Skaergaard Intrusion. There is, however, no Upper Border Group equivalent and in detail the petrography of the igneous layering is extraordinarily complex. Layered orebodies occur in specific zones. Thus the chromium and PGE deposits are restricted to the Critical Zone and the vanadium-rich Ti-magnetite deposits are mostly found in the Upper Zone. Some of the layered orebodies constitute well known marker layers, e.g., the Merensky Reef.

The **Marginal Zone** is sporadically developed in the Eastern Limb and varies from a thin rind to a well-defined unit several kilometres in thickness (Sharpe 1981). A chill may occur close to the base of the Rustenburg Layered Suite

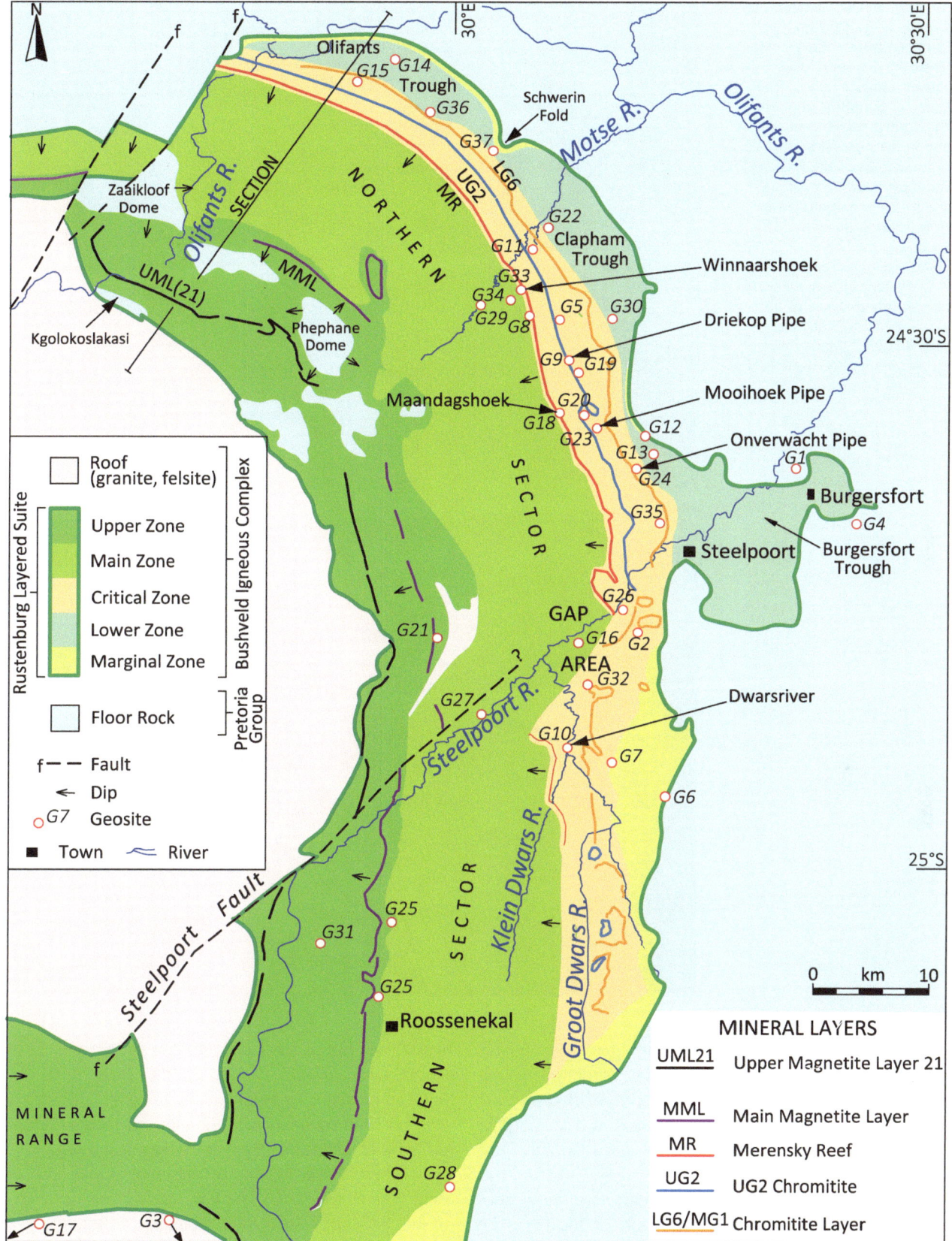

Fig. 17.11 Geological map of part of the Eastern Limb of the Bushveld Igneous Complex showing zones, reefs, and selected geosites described by Scoon and Viljoen (2019). *Source* Simplified from the 1:250,000 scale maps published by the Council for Geoscience

Fig. 17.12 a The platinum-rich core-zone at the Driekop pipe (was mined from a glory hole accessed by a vertical shaft (headgear remains intact). The majority of the hill is comprised of the normal (unmineralized) dunite in the pipe. The background shows the Bushveld Escarpment which consists of resistant gabbronorite (Geosite G9); **b** Relict of the 1926–1930 glory hole at the Onverwacht pipe includes reticulate veining of secondary magnesite in the barren dunite. A protuberance of the platinum-rich core-zone can be observed (slightly darker rock with less veining) (Geosite G24)

(although the contact with the metasediments is poorly exposed). The Marginal Zone contains some interesting lithologies, including migmatite (indicative of two parental magmas: Sharpe and Irvine 1983) and a rheomorphic breccia ascribed to partial melting of the quartzite floor (Fig. 17.16).

The **Lower Zone** is restricted to three troughs or sub-chambers in the Eastern Limb. The Burgersfort locality is the largest, outcrop in the Clapham Trough is rather poor, and the Olifants River Trough is the most well known. Subchambers are constrained by floor-rock domes which were initiated by heating of the metasediments during early intrusion of ultramafic magmas (Uken and Watkeys 1997). The Olifants River locality reveals spectacular igneous layering (Cameron 1978). The less resistant olivine-rich rocks of the harzburgite subzone alternate with hard layers of

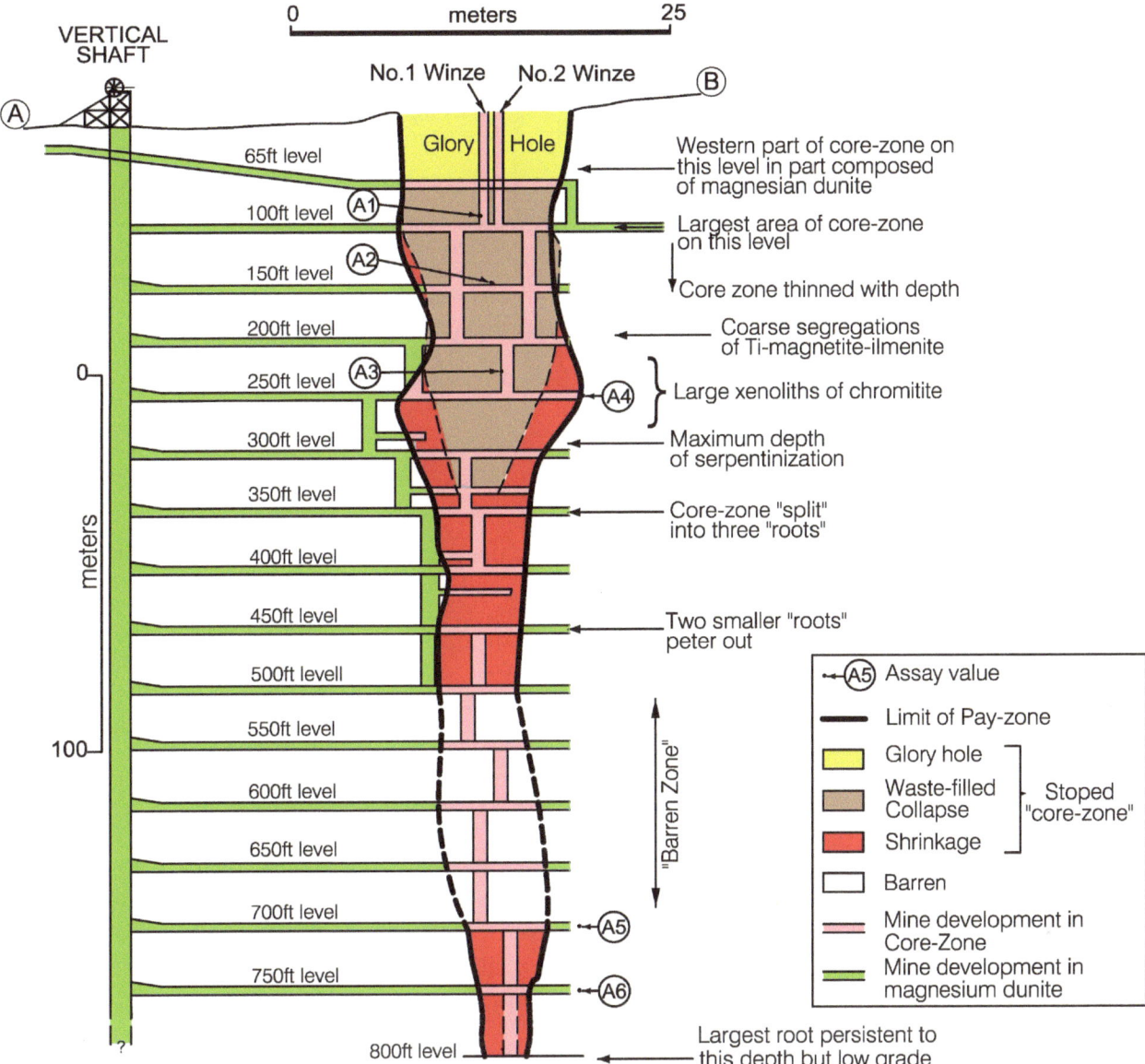

Fig. 17.13 Section of the Onverwacht platinum mine aligned approximately N-S and looking east (four times horizontal exaggeration). The high-grade ore included examples such as A1 (214 g/t Pt in core-zone rich in phlogopite); A2 (31 g/t average of central part of core-zone); A3 (1,186 g/t at contact between core-zone and lump of chromite); A4 (29 g/t average of 211 m² area of core-zone); A5 (17 g/t average of 39 m² area of core-zone); A6 (9.5 g/t average of 39 m² area of core-zone). The shaft was developed to a final depth of 320 m. *Source* Original mine plan of Onverwacht Platinum Limited with additional details after Wagner (1929)

orthopyroxenite of the lower bronzitite and upper bronzitite subzones (Fig. 17.17).

A significant feature of the **Critical Zone** is the occurrence of well-defined layers of chromitite. Layers of chromitite in the Lower Critical Zone are mostly intercalated with feldspathic orthopyroxenite, whereas those in the Upper Critical Zone are associated with feldspathic orthopyroxenite, norite, and anorthosite. Feldspathic harzburgite is a minor component of both subzones. The boundary between the subzones is demarcated by the Middle Group anorthosite

(Fig. 17.18). The chromitite layers are subdivided into three groups (lower, middle, and upper). Some of the lower and middle group chromitite layers (notably the LG6 and MG1) are exploited for their chromium content. The lowermost of the upper group of chromitite layers, the UG1, reveals spectacular igneous layering (Fig. 17.1). The UG2 chromitite is exploited as a primary PGE reef (Fig. 17.19). The origin of chromitite layers remains controversial. The problem of the "chromium budget" (i.e., the anomalous percentage of chromite in the Critical Zone, as compared

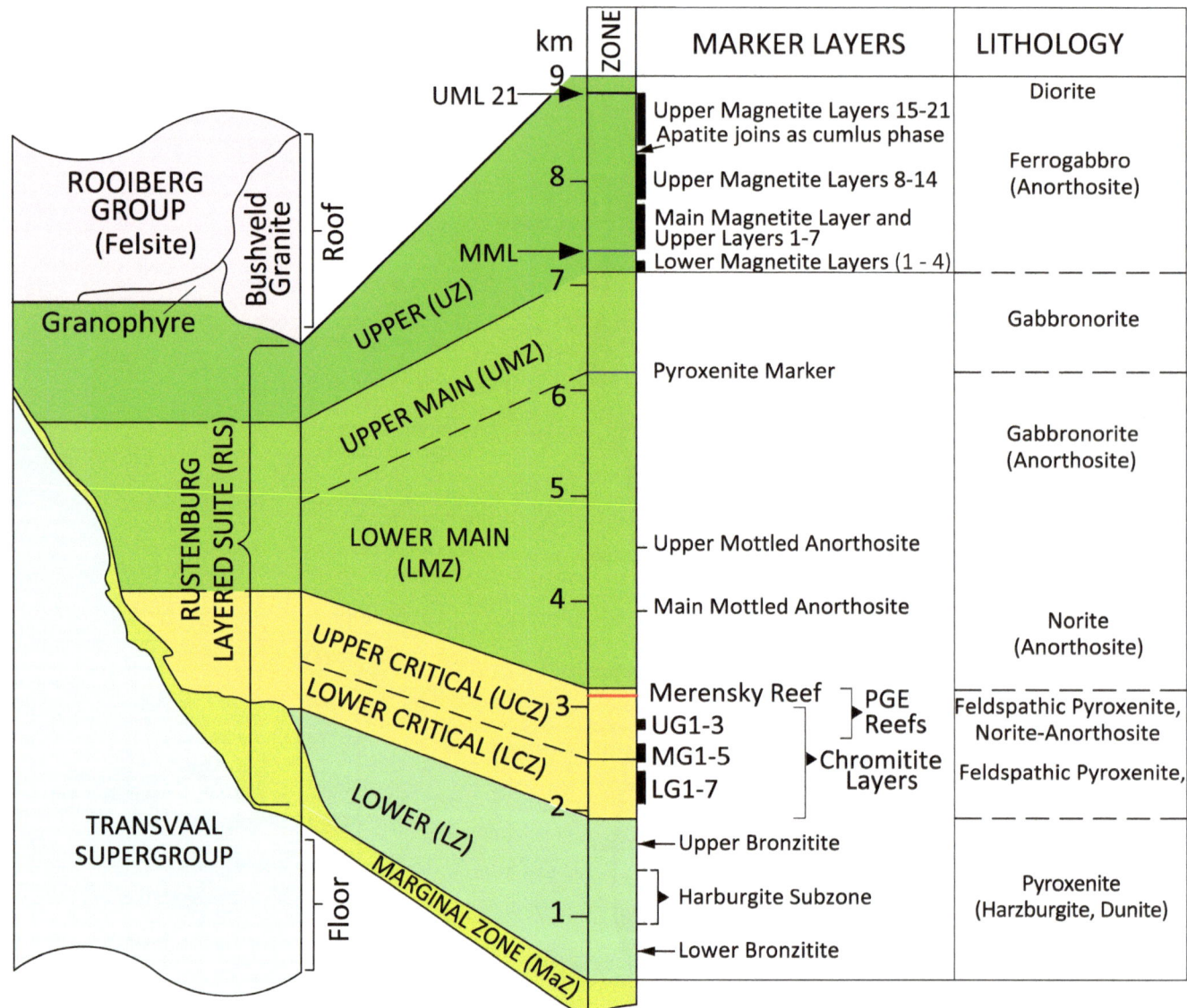

Fig. 17.14 Generalized stratigraphic column for the Rustenburg Layered Suite showing zones, lithologies, reefs and marker layers. The irregular basal contact is emphasized (the lowermost zones are not developed in many areas) and only the Upper Zone (noting the planar nature of the roof contact) is persistent over the entire strike length of the Eastern Limb

with possible parental magmas) is addressed by a hypothesis that invokes repeated injection of batches of ultramafic magma charged with chromite phenocrysts (Eales and Costin 2012). The chromite phenocrysts were sufficiently abundant as to form crystal slurries, thus providing a possible explanation of the enigmatic layering associated with the UG1.

The Merensky Reef outcrops in the northern part of Eastern Limb in a broad valley associated with the Motse River (colloquially known as the "Platinum Valley": Fig. 17.9). In this part of the intrusion, the Merensky Reef consists of multiple layers of feldspathic orthopyroxenite together with two thin chromite stringers. At the

Winnaarshoek locality, the irregular nature of the hanging-wall contact is in marked contrast to the regularity of the uppermost of the two chromite stringers (Fig. 17.20). A novel hypothesis that addresses both regional variations and the irregularity of the contact at Winnaarshoek suggest the Merensky Reef crystallized from ultramafic magma sheets injected as sills into the earlier-formed norite and anorthosite (Mitchell and Scoon 2007). Heat associated with the magma sheets partially melted the wall rocks. The thin chromite stringers (which may be laterally continuous over tens of kilometres) are interpreted as magmatic-reaction features, rather than primary layers (Scoon and Costin 2018).

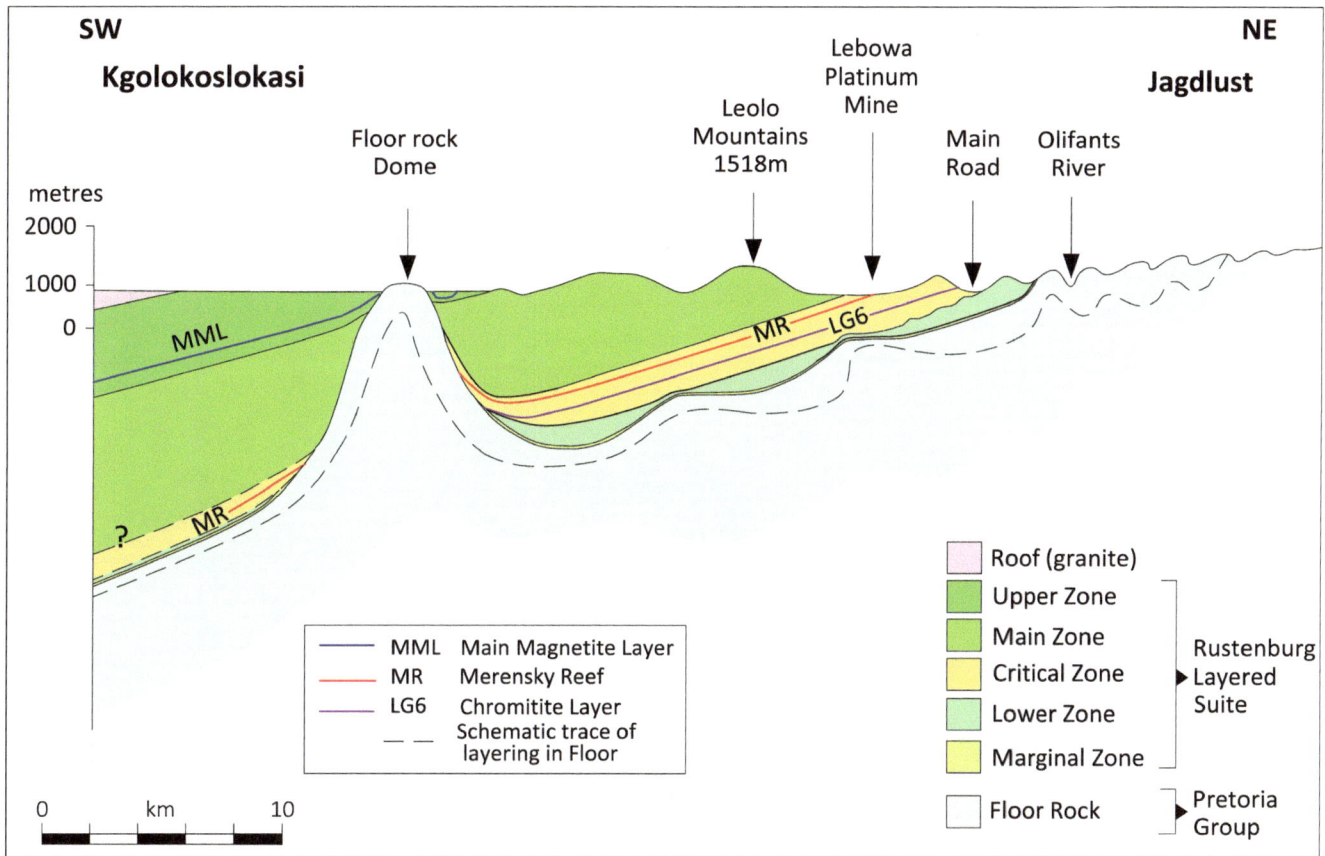

Fig. 17.15 Simplified cross section of the Rustenburg Layered Suite in the Eastern Limb showing the irregular basal contact associated with syn-Bushveld domes

The **Main Zone** is dominated by norite and anorthosite in the lowermost part (which is broadly similar to the uppermost part of the Critical Zone) with a monotonous sequence of gabbronorite, albeit punctuated by prominent layers of anorthosite, in the central and upper parts (Mitchell et al. 1998). The gabbronorite forms the most resistant lithology in the Eastern Limb and outcrops extensively in the Leolo Mountains. The boundary between the subzones of the Main Zone is positioned at the height of the Pyroxenite Marker.

The **Upper Zone** marks a significant mineralogical change in the layered sequence as the gabbronorite gives way to ferrogabbro. Disseminated Ti-magnetite is a significant component of the ferrogabbro. The ferrogabbro is intercalated with layers of monomineralic Ti-magnetite (Wager and Brown 1968). Four groups of Ti-magnetite layers are recognized. The lowermost group of layers are the richest in vanadium and the uppermost layers are the richest in titanium. The Main Magnetite Layer is exploited, primarily for vanadium, but also constitutes a source of iron ore with by-product titanium (Fig. 17.21). Ti-magnetite is also found in association with discordant bodies of iron-rich

ultramafic pegmatite, e.g., the Kennedys Vale body (Geosite G27). Many of the pipe-like bodies of Ti-magnetite cluster beneath the Ti-magnetite layers in the Upper Main Zone. The layers and discordant bodies of Ti-magnetite can be linked to immiscible Fe-Ti oxide melts (Bateman 1931). Suitable melts may have been segregated from the basaltic magma in deeper staging chambers (Scoon and Mitchell 2012). The dense Fe-Ti-oxide melts were injected into the Upper Zone where they spread laterally, possibly as sills, to form layers and/or drained downward to form pipes in areas of structural weakness.

17.4 Origin of Igneous Layering

For many years, the competing processes of crystal settling and fractional crystallization of crystals grown on the temporary floor of a magma chamber have dominated the scientific literature on layered igneous intrusions. The changes in the mineral layering of the Layered Series of the Skaergaard Intrusion have been supported by cryptic geochemical

Fig. 17.16 a The migmatite in the Marginal Zone consists of micropyroxenite (dark grey) and fine-grained norite (pale grey) (Geosite G6); **b** A rheomorphic breccia in the Marginal Zone contains blocks of micropyroxenite (dark grey) embedded in a pinkish matrix of quartz-rich granophyre (Geosite G12)

layering, with the observed patterns of iron-enrichment with height substantiated by experimental petrology. In other words, the patterns of mineral layering match the experimental results expected in an upward-cooling magma sheet. There is no evidence (and no need) to postulate magma replenishment. The debate may be summarized as between gravity settling (e.g., Wager and Brown 1968) and in situ nucleation (e.g., McBirney and Naslund 1990). The gravity settling hypothesis has been modified in recent years to include the possibility of crystal-rich density currents which periodically swept down the steep sidewalls and onto the floor (Irvine et al. 1998). The hypothesis of in situ nucleation implies that layering develops as a consequence of differing nucleation rates. Heat is drawn away from the floor of the chamber by convection.

Fig. 17.17 Aerial view of igneous layering in the Lower Zone (view looking north over the Olifants River Trough, Geosite G14). The deeply-eroded harzburgite subzone (pale) is sandwiched between resistant ridges associated with the Lower Bronzitite and Upper Bronzitite subzones. The pale colour of the olivine-rich layers is related to secondary magnesite. Photograph by M J Viljoen

Fig. 17.18 The MG2 and MG3 chromitite layers straddle an important boundary in the centre of the Critical Zone, Annex Grootboom (Geosite G2). The Lower Critical Zone consists entirely of ultramafic lithologies (and is dominated by feldspathic orthopyroxenite), but the Upper Critical Zone includes well-defined layers of norite and anorthosite. Photograph by M J Viljoen

Fig. 17.19 The PGE mineralization in the UG2 Reef at Hackney (Geosite G11) is largely constrained to the layer of chromitite. Features of note include the irregular contact with the underlying pegmatoidal feldspathic orthopyroxenite, the small xenoliths of anorthosite (white) in the chromitite (black), and the relatively planar upper contact of the chromitite layer. Photograph by M J Viljoen

The above hypotheses are unlikely to be compatible with the dominant processes involved in formation of the much larger, sill-shaped layered igneous intrusions. The intercalation of ultramafic layers with norite and anorthosite in the Critical Zone of the Rustenburg Layered Suite adds an additional complexity. The dominant process in formation of the Eastern Limb is repeated recharge of a giant, composite sill (Mitchell and Scoon 2007; Mitchell et al. 2019). Comparison with plumbing systems associated with giant stratovolcanoes and flood basalt provinces is important. The incremental emplacement of thousands of thin magma sheets, rather than the concept of a large, convecting magma chamber, is consistent with field relationships. The relatively small magma reservoir was episodically inflated by batches of magma fed from deeper staging chambers. As the sill inflated, magma was forced to spread laterally, a mechanism consistent with restriction of the Lower Zone to isolated troughs and with emplacement of the Upper Zone during a period of maximum inflation. Magma replenishment persisted to the height of Upper Magnetite Layer 21. The growth mechanism of the Rustenburg Layered Suite may have been similar to that envisaged in some granitic plutons (Miller et al. 2011). Multiple lineages of parental magma are identified to explain the isotopic inflexions in the Rustenburg Layered Suite. There are almost unlimited variations within these lineages due to the complexity of processes operating in the deeper staging chambers. The height-related trends in the Eastern Limb are, therefore, ascribed, not to fractional

Fig. 17.20 The planar nature of the upper chromite stringer in the Merensky Reef at Winnaarshoek (Geosite G33) contrasts with the irregularity of the contact between the feldspathic orthopyroxenite (brown) and the overlying layer of leuconorite (pale grey). The spotted appearance of the leuconorite is characteristic of this lithology (dark spots are orthopyroxene; white matrix is plagioclase). Photograph by Qin Wang

Fig. 17.21 The Main Magnetite Layer (black) at Magnetite Heights (Geosite G21) has a width of approximately 2 m and is underlain by a prominent layer of anorthosite (fawn-white)

crystallization as they are in the Skaergaard Intrusion, but primarily to changes in the composition and physical properties of intruding parental magmas.

References

Andersen, J. C. Ø., Power, M. R., & Momme, P. (2002). Platinum-group elements in the Palaeogene North Atlantic Igneous province: In: L. J. Cabri (Ed.) The geology, geochemistry, mineralogy, and mineral beneficiation of platinum-group elements. *Canadian Institute of Mining, Metallurgy, and Petroleum Special Volume, 54*, 637–667.

Bateman, A.M. (1951). The formation of late magmatic oxide ores. *Economic Geology, 46*, 404–426.

Bottinga, Y., & Weill, D. F. (1970). Densities of liquid silicate systems calculated from partial molar volumes of oxide components. *American Journal of Science, 372*, 438–475.

Brooks, K. (1997). The Skaergaard Intrusion: sixty years of petrological research (p. 170). Introduction and abstracts: Danish Lithosphere Centre and Institute of Petrology, University of Copenhagen.

Cameron, E. N. (1978). The Lower Zone of the eastern Bushveld Complex in the Olifants River trough. *Journal of Petrology, 19*, 437–462.

Eales, H. V. (2014). An introduction to the geology and setting of the Bushveld Complex (2nd Edition) (p. 201). Pretoria: Council for Geoscience.

Eales, H. V., & Costin, G. (2012). Crustally contaminated komatiite: primary source of the chromitites and marginal, lower, and critical zone magmas in a staging chamber beneath the Bushveld Complex. *Economic Geology, 107,* 645–665.

Hall, A.L. (1932). The Bushveld Igneous Complex of the central Transvaal. *Geological Survey of South Africa Memoir, 28,* 560 p.

Hatton, C. J., & Schweitzer, J. K. (1995). Evidence for synchronous extrusive and intrusive Bushveld magmatism. *Journal of African Earth Sciences, 21,* 579–594.

Irvine, T. N., Andersen, J. C. Ø., & Brooks, C. K. (1998). Included blocks (and blocks within blocks) in the Skaergaard intrusion: geologic relations and the origins of rhythmic modally graded layers. *Geological Society of America Bulletin, 110,* 1398–1447.

Irvine, T.N., Andersen, J.C.Ø. and Brooks, C.K. (2001). Excursion guide to Skaergaard Intrusion Kangerdlugssuaq area, East Greenland, 66 p.

McBirney, A. R., & Naslund, H. R. (1990). The differentiation of the Skaergaard intrusion. *Contributions to Mineralogy and Petrology, 104,* 235–247.

McBirney, A. R., & Noyes, R. M. (1979). Crystallization and layering of the Skaergaard Intrusion. *Journal of Petrology, 20,* 487–554.

Merensky, H. (1925). How we discovered platinum. *Mining and Industrial Magazine of Southern Africa, 1,* 265–266.

Miller, C. F., Furbish, D. J., Walker, B. A., Claiborne, L. L., Koteas, G. C., Bleick, H. A., et al. (2011). Growth of plutons by incremental emplacement of sheets in crystal-rich host. Evidence from Miocene intrusions of the Colorado River region, Nevada, USA. *Tectonophysics, 500,* 65–77.

Mitchell, A. A., Eales, H. V., & Kruger, F. J. (1998). Magma replenishment processes, and the significance of poikilitic textures, near the base of the Lower Main Zone of the western Bushveld Complex, South Africa. *Mineralogical Magazine, 62,* 435–450.

Mitchell, A. A., & Scoon, R. N. (2007). The Merensky reef at Winnaarshoek, Eastern Bushveld Complex: a primary magmatic hypothesis based on a wide reef facies. *Economic Geology, 102,* 971–1009.

Mitchell, A. A., Scoon, R. N., & Sharpe, M. R. (2019). The Upper Critical Zone in the Swartklip Sector, northwestern Bushveld Complex, on the farm Wilgerspruit 2JQ: II. Origin by intrusion of ultramafic sills with concomitant partial melting of host norite-anorthosite cumulates. *South African Journal of Geology, 122,* 143–162.

Mungall, J. E., Kamo, S. L., & McQuade, S. (2016). U-Pb geochronology documents out-of-sequence emplacement of ultramafic layers in the Bushveld Igneous Complex of South Africa. *Nature Communications, 7*(13385), 1–13.

Naslund, H. R. (1986). Disequilibrium partial melting and rheomorphic layer formation in the contact zone of the Basistoppen sill, East Greenland. *Contributions to Mineralogy and Petrology, 93,* 359–367.

Naslund, H. R., & McBirney, A. R. (1996). Mechanisms of formation of igneous layering. *Layered Intrusions* (pp. 1–44). Amsterdam: Elsevier.

Scoates, J. S., & Friedman, R. M. (2008). Precise Age of the Platiniferous Merensky Reef, Bushveld Complex, South Africa, by the U-Pb Zircon Chemical Abrasion ID-TIMS Technique. *Economic Geology, 103,* 465–471.

Scoon, R. N. (2016). Eastern Limb of the Bushveld Complex: 35th International Geological Congress, Cape Town, South Africa, Field Trip Guide Pre-7, 89 p.

Scoon, R. N., & Costin, G. (2018). Chemistry, morphology and origin of magmatic-reaction chromite stringers associated with anorthosite in the Upper Critical Zone at Winnaarshoek, Eastern Limb of the Bushveld Complex. *Journal of Petrology, 59,* 1551–1578.

Scoon, R. N., & Mitchell, A. A. (2011). The principal geological features of the Mooihoek platiniferous dunite pipe, Eastern Limb of the Bushveld Complex, South Africa, and similarities with Replaced Merensky Reef at the Amandelbult mine. *South African Journal of Geology, 114,* 15–40.

Scoon, R. N., & Mitchell, A. A. (2012). The upper zone of the bushveld complex at Roossenekal: geochemical stratigraphy and evidence of multiple episodes of magma replenishment. *South African Journal of Geology, 115,* 515–534.

Scoon, R. N., & Viljoen, M. J. (2019). Geoheritage of the Eastern Limb of the Bushveld Igneous Complex, South Africa - a uniquely exposed layered igneous intrusion. *Geoheritage, 11,* 1723–1748.

Sharpe, M. R. (1981). The chronology of magma influxes to the eastern compartment of the Bushveld Complex as exemplified by its marginal border groups. *Geological Society of London Journal, 138,* 307–326.

Sharpe M. R. (1986). Bushveld Complex: Excursion Guidebook, Geocongress '86 South Africa, 143 p.

Sharpe, M. R., & Irvine, T. N. (1983). Melting relations of two Bushveld chilled margin rocks and implications for the origin of chromitite. *Carnegie Institute Washington Yearbook, 82,* 295–300.

South African Committee for Stratigraphy. (1980). Stratigraphy of South Africa, Part I (Compiler L.E. Kent). *Geological Survey of South Africa Handbook, 8,* 690 p.

Tankard, A. J., Jackson, M. P. A., Eriksson, K. A., Hobday, D. K., Hunter, D. A., & Minter, W. E. L. (1982). Crustal evolution of Southern Africa, 3.8 billion years of earth history (523 p.). New York, Heidelberg, Berlin: Springer Verlag.

Uken, R., & Watkeys, M. K. (1997). Diapirism initiated by the Bushveld Complex. *Geology, 25,* 723–726.

Viljoen, M. J. (2016). The Bushveld Complex: Host to the worlds' largest platinum, chromium and vanadium resources. In: M. G. C. Wilson & R. P. Viljoen (Eds.), *The Great Mineral Fields of Africa. Episodes, 39,* 239–268.

Viljoen M. J. & Scoon R. N. (1985). The distribution and main geologic features of discordant bodies of iron-rich ultramafic pegmatite in the Bushveld Complex. *Economic Geology, 80,* 1109–1128

Von Gruenewaldt, G., Sharpe, M. R., & Hatton, C. J. (1985). The bushveld complex: Introduction and review. *Economic Geology, 80,* 803–812.

Wager, L. R., & Brown, G. M. (1968). Layered igneous rocks (p. 588). Oliver and Boyd: Edinburgh.

Wager, L. R., & Deer, W. A. (1939). Geological investigations in East Greenland. Part 3. The petrology of the Skaergaard intrusion, Kangerdlugssuaq, East Greenland. *Meddr. Gronland, 105,* 352 p.

Wagner, P. A. (1929). The platinum deposits and mines of South Africa (p. 326). Oliver and Boyd: Edinburgh.

Walraven, F. J., Armstrong, R. A., & Kruger, F. J. (1990). A chronostratigraphic framework for the north-central Kaapvaal craton, the Bushveld Complex and Vredefort structure. *Tectonophysics, 171,* 23–48.

Willemse, J. (1969). The geology of the Bushveld Igneous Complex, the largest repository of magmatic ore deposits in the World. *Economic Geology Monograph, 4,* 1–22.

Correction to: R. N. Scoon, _The Geotraveller_,
https://doi.org/10.1007/978-3-030-54693-9

In the original version of the book, belated corrections provided by the author have been incorporated througout the book. This book has been updated.

The updated version of these chapters can be found at
https://doi.org/10.1007/978-3-030-54693-9_5,
https://doi.org/10.1007/978-3-030-54693-9_6,
https://doi.org/10.1007/978-3-030-54693-9_10,
https://doi.org/10.1007/978-3-030-54693-9_12,
https://doi.org/10.1007/978-3-030-54693-9_13,
https://doi.org/10.1007/978-3-030-54693-9_14,
https://doi.org/10.1007/978-3-030-54693-9_15,
https://doi.org/10.1007/978-3-030-54693-9_16,
https://doi.org/10.1007/978-3-030-54693-9_17 and
https://doi.org/10.1007/978-3-030-54693-9.

© Springer Nature Switzerland AG 2021
R. N. Scoon, _The Geotraveller_,
https://doi.org/10.1007/978-3-030-54693-9_18

Glossary

Accreting plate boundary The name given in plate tectonics to the boundary between two plates that are moving apart with new oceanic crust froming between them. cf. **Converging plate boundary or collison zone**

Agglomerate Coarse-grained volcanic breccia i.e. a rock comprised of angular fragments of pyroclastic material typically found near a vent

Albertine Rift The Western Branch of the EARS named after Lake Albert in the northern part of the structure

Alkali or Alkaline Any strongly basic substance such as a hydroxide or carbonate of one of the alkali metals such as sodium (Na) or potassium (K)

Alkaline Basalt A group of volcanic rocks strongly undersaturated in silica and which contain more alkali metals (particularly sodium and potassium) than considered average for basalts

Alkaline Lake See Soda Lake

Amphibole A group of dark-coloured, rock-forming ferromagnesian minerals, including hornblende

Amphibolite A granular textured metamorphic rock dominated by amphibole and plagioclase (typically with no quartz)

Andalusite An aluminium silicate mineral that forms at moderate temperatures and pressures in metamorphosed terrains

Andesite A dark-coloured, or grey-green, fine-grained extrusive igneous rock that may contain prominent phenocrysts of plagioclase and/or a mafic phase such as hornblende or pyroxene. Named from the Andes Mountains

Anorthosite A plutonic rock with <5% quartz and <10% mafic minerals dominated by plagioclase feldspar. Anorthosites occur as large massifs in mobile belts and as stratiform layers in layered mafic igneous intrusions

Anticline An upfolded structure in which the oldest rocks occur in the core and the youngest rocks in the outer part. cf. **syncline**

Asthenosphere The layer or shell of the Earth underlying the lithosphere and which is relatively weak and easily deformed, and where many magmas are derived from

Badlands Severely eroded topography, typically with thin soils and poorly vegetated

Basalt A general term for dark-coloured (melanocratic)-mafic igneous rocks dominated by extrusive lavas and ashes. The dominant minerals are calcic plagioclase and clinopyroxene. Basalts are subdivided using chemical discriminations such as the TAS diagram

Basement A general term to describe the rock sequences-possibly undifferentiated and typically older than the near-surface rocks that underlie the region of interest

Basement Terrain The lowest (and typically the oldest) of the mappable rock units in the region of interest

Basin A low-lying or depressed area often downfolded with no surface outlet, such as a lake basin or a groundwater basin. Elongate basins are known as grabens

Batholith A large plutonic mass or accreted group of plutons generally related to bodies of granite

Block Faulting The dominant structure resulting from an area subjected to extensional tectonism i.e. including parallel faults which may develop to include a series of grabens and horsts

Breccia A coarse-grained clastic rock, composed of angular fragments of broken rock held togther by a cement, and often associated with faults or other tectonic processes

Brine Highly saline fluids typically warm or hot and rich in alkali and related elements such as calcium, sodium, potassium, and chlorine, that are generally only very minor constituents of water

© Springer Nature Switzerland AG 2021
R. N. Scoon, *The Geotraveller*,
https://doi.org/10.1007/978-3-030-54693-9

Calcareous Pertaining to material dominated by calcium carbonate

Calcrete Surficial deposits of sand and gravel cemented by calcium carbonate precipitated from infiltrating groundwater

Caldera A large basin-shaped volcanic depression more or less circular formed by a catastrophic process generally associated with collapse of the *entire* cone induced by rapid depletion of the subsurface magma chamber (from the Latin *calderia* or "boiling pot")

Carbonate A mineral that contains the compound CO_3 such as calcium carbonate ($CaCO_3$). Thus carbonate rocks (limestone) are primarily composed of carbonate minerals, skeletal material (corals flint), or organisms (shells)

Carbonatite A rare type of igneous rock dominated by carbonates rather than silicates rich in calcium and containing anomalously high amounts of potassium and/or sodium

Chert A hard, microcrystalline rock comprised almost entirely of silica (SiO_2). A class of sedimentary rocks ascribed to chemical precipitation

Chromite A dark brown or black member of the spinel group of oxide minerals in which chromium is the major constituent. Rocks conssiting of >50% chromite are known as **chromitites**

Cichlid An extensive group of fishes in which new species are continuously being discovered; one of the most extensive family of vertebrates. Includes the commercially-important variety tilapia and important in studies of speciation in evolution due to their great diversity with estimates of between 2,000 and 3,000 species. Typically found in freshwater but they can also thrive in hypersaline and alkaline lake

Cleavage In structural geology this applies to the property of a rock to split along secondary, planar fractures (other than bedding) resulting from deformation or metamorphism. Cleave also applies to minerals that break along preferred crystallographic planes

Compression The shortening of tectonic blocks or a specific geological terrain

Conglomerate A coares-grained clastic rock sedimentary rock, composed of rounded to subangular peebles in a fine-grained matrix or cement

Convergent plate boundary The boundary between two tectonic plates that are moving toward each other. In the case of a collision between continental and oceanic plates synomous with subduction zone

Clastic Pertaining to a rock or sediment composed of broken fragments (eroded) from a pre-exsiting rocks and that have been transported some distance from their origin. The most abundant clastic rocks are sandstone and shale

Craton The oldest most stable component of continental crust that has remained undeformed for several billions of years. Cratons are far more deep-rooted than the younger components of the crust such as mobile belts

Crust The outermost layer or shell of the Earth defined by factors such as seismic velocity, density, and composition and located above the Mohorovičić discontinuity **Continental crust** is discriminated from **oceanic crust** by being considerably thicker (35–60 km as compared to 5-10 km for oceanic crust), less dense and compositionally far more heterogeneous. Continental crust is dominated by granitic and sedimentary rocks; oceanic crust by basaltic rocks

Dacite A fine-grained extrusive igneous rock with a broadly similar composition to andesite but is more siliceous with less calcic plagioclase and abundant quartz

Debris Avalanche Deposit (DAD) Deposits of unsorted, mostly volcanic debris (loose material) that result from the partial or total collapse of volcanic cones; debris is transported by avalanches entirely independent of water

Deformation A general term for the processes of folding, faulting, shearing, compression, or extension, and and geological terrains of rocks as a consequence of forces associated with a dynamic Earth

Detrital Pertaining to minerals or rocks formed from **detritus** (loose material) typically transported on surface and deposited in sedimentary basins

Dimension Stone A general description of rocks used in cladding buildings and ornamental purposes. The most common dimension stones are marble, granite, and gabbronorite

Diorite A plutonic igneous rock with approximately equal proportions of plagioclase and pyroxene and in which the plagioclase is more sodic than calcic. Can be interpreted as a differentiated form of gabbro, intermediate between the mafic rocks and granite

Distal When applied to rocks pertains to distance or remoteness from the inferred source area. cf. **Proximal**

Dolerite An intrusive igneous rocks dominated by approximately equal proportions of plagioclase and ferromagneisn minerals and with an ophitic texture related to the shallow depth of intrusion i.e. hypabassal, found in dykes and sills, rather than plutons. The composition is broadly similar to a gabbro in deep-seated plutons. Equivalent to the American usage of **diabase**

Dolomite Applied to a sedimentary rock dominated by the mineral dolomite, a calcium magnesian carbonate, typically but not always interbedded with thick sequences of limestone. Named from the Dolomite Mountains, northern Italy

Dome An uplifted or anticlinal structure, circular or elliptical in outline, in which the rocks dip away from the centre in all directions (geology) or a generalized description of an elevated region of plateaus rather than fold mountains (geomorphology)

Dunite A plutonic igneous rock with >90% olivine

Dyke An igneous intrusion tabular in shape that cuts across the bedding of the country rocks, and typically relatively small and shallow. Most dykes are located vertical or near-vertical and vary in thickness from a few centimetres to hundreds of metres

Erosion Surface A land surface typically near-level shaped and rounded by the action of erosion and which correlates with a specific period, or age

Escarpment A linear more or less continuous steeply inclined slope, locally a cliff, facing in one direction that extends for several or many tens of kilometres and formed by erosion and/or faulting. In the context of the EARS, escarpments invariably represent the surface expressions of major boundary faults

Extrusive Igneous rocks that have been erupted onto the Earth's surface (synonym: volcanic)

Fault A fracture or a zone of fractures along which there has been significant displacement of the rock sequences on either side

Fault Breccia The filling of a fault zone by angular fragments derived from the crushing and shattering of the adjacent rock sequences

Feldspar The most abundant group of rock-forming silicate minerals comprised of two broad groups: plagioclase (calcium-iron-aluminium silicates) and alkali feldspar (sodium-potassium aluminium silicates). They are typically leucocratic and are the dominant component of granites; also found in many basalts

Feldspathoid A group of rare rock-forming minerals that are too poor in silica to form feldspar and including sodium and potassium members nepheline and leucite, respectively. They can never been found with quartz and are restricted to undersaturated alkaline igneous rocks

Felsic A generalised term for light-coloured minerals (feldspar, quartz) or light-coloured igneous rocks

Flood Basalt Eruptions of relatively fluid basaltic lavas typically from multiple fissures rather than discrete conduits on a regional scale that smooth out irregularities in older land surfaces

Flysch A marine sedimentary facies characterized by thinly bedded deposits of marls, shales, and muds. The sediments are generally poorly fossiliferous and have been deposited in discrete basins during the latter parts of geosynclines. In the Alpine tectonic zones, the flysch has accumulated from erosion of older scarps in front of the rising nappes, during the latter stages of the orogeny (cf. **Molasse**)

Foidolite A general term for igneous rocks dominated by feldspathoid minerals

Fold Mountain Mountains formed by large-scale folding and uplift, typically associated with major orogenies and collison of textonic plates

Formation Rock strata of intermediate rank in the hierarchy of stratigraphic subdivisions which form mappable units over large areas and formalized for international correlations. The strata is unified by consisting of a dominant lithology, often with distinct fossils. Formations may be combined into **groups** or subdivided into **members**

Fractionation The separation of chemical elements by natural processes such as preferential concentration into a magma (partial melting) specific minerals (fractional crystallization of a magma), or differential solubility (rock weathering)

Fumarole A small vent of volcanic origin from which gases or vapours are emitted typically associated with geothermal fields in volcanic systems that are either dormant or waning

Gabbro A leucocratic, plutonic igneous rock dominated by approximately equal proportions of plagioclase feldspar and clinopyroxene

Gabbronorite A description of a plutonic rock that merges between the two end members of gabbro and norite i.e., comprised of approximately equal proportions of plagioclase, clinopyroxene, and orthopyroxene

Geomorphology The general study of landforms and the surface of the Earth

Geopark A Geopark is a unified area that advances the protection and use of geological heritage in a sustainable way and promotes the economic well-being of the people who live there. There are Global Geoparks and National Geoparks (Wikipedia)

Geothermal The internal heat of the Earth related to the breakdown of radioactive elements an essential component of sustaining plate movements and possibly all forms of life

Geothermal Energy Heat that can be extracted from the Earth. Most areas of localized high heat flow in the crust are ascribed to volcanism and to a lesser extent radioactive decay of large plutons

Geothermal Features Related to heat associated with magma chambers and active or dormant volcanoes, e.g., fumaroles, hot springs

Geyser A type of hot spring that intermittently erupts jets of hot water and steam, derived from geothermal heating of groundwater which periodically converts to steam in a restricted space, prior to being forced to surface through a series of fractures. A group of geysers can be described as a **geyser basin** and the deposits or mounds at the bases of geysers consist of **geyserite**

Glacial Epoch A period in the Earth's history when extensive ice sheets and glaciers occur on large parts of the continental landmass

Glacial Period An interval during an **Ice Age** when temperatures are generally coldest and ice sheets and glaciers are at their maximum. The coldest peak time is described as a **Glacial Maximum**

Gneiss A foliated metamorphic rock consisting of bands of granular minerals (often leucocratic feldspars and quartz) that alternate with bands of flaky or platy minerals (often melanocratic mica and hornblende). The most common form of gneiss has a granitic composition and is referred to as **granite-gneiss**

Gondwana The ancient supercontinent of the Southern Hemisphere that assembled in the Late Palaeozoic and named after the Gondwana system of India (Carboniferous to Jurassic), which in turn is named after the Gonds, the oldest inhabitants of the Indian subcontinent

Graben An elongate, depressed unit of the crust bounded by faults on the long sides. **Half-grabens** are depressed units in which boundary faults are developed on one side only. The geomorphologic expression of a graben is a rift valley, of a half-graben is a step or inclined valley, and of a boundary fault is an escarpment

Granite A granular plutonic rock dominated by **leucocratic minerals** (feldspar and quartz), with subordinate amounts of mica and hornblende, which typically occurs in large bodies (plutons). The dominant constituent of continental crust

Granophyre A porpyritic textured igneous rock (either extrusive or intrusive), broadly similar in composition to granite or rhyolite, in which the phenocrysts and groundmass penetrate each other creating a graphic texture resembling hieroglyphs

Greenstones A field term applied to dark green, metamorphosed, typically basic igneous rocks characterized by the presence of chlorite or epidote

Greenstone Belt Elongate terrains dominated by greenstones, typically associated with complex volcano-sedimentary piles of Archaean or Palaeoproterozoic age

Gregory Rift The Eastern Branch of the EARS named after geologist and explorer J. W. Gregory

Hominid A broad group that includes all modern Great Apes and their extinct ancestors (e.g. *Proconsul africanus*) as well as humans and their ancestors

Hominin A more refined group of primates that includes modern and extinct humans (*Homo sapiens*), all of our immediate ancestors, the genus *Hominina* (with its three branches *Homo erectus or Homo ergaster*; *Homo heidelbergenesis*; *Homo habilis*) and the general *Australopithecine* (including *Australopithecus africanus* and *Australopithecus afarensis*) and *Panina* (chimpanzees). This definition is preferred by zoologists and geneticists as the above species probably share >99 per cent of their DNA. Anthropologists and others may restrict the definition to only *Homo sapiens* and our direct ancestors (*Hominina*)

Homo sapiens The scientific name for the only extant human species. **Homo** is the human genus which includes extinct species of hominin. **Sapiens** (or wise) is the only surviving species of this genus

Horst An elongate, uplifted unit of the crust bounded by faults on its long sides which is commonly observed as a geomorphological feature

Hydrothermal Pertaining to hot water and deposits formed from hot brines

Ice Age A loosely used synonym of glacial epoch, a period of extensive glacial activity

Ice Cores The material recovered from drilling of a glacier or ice sheet which may provide detailed information on palaeo-climates

Igneous A rock or mineral that crystallized from molten or partially molten material (magma). Igneous rocks are one of the three main categories of rocks (cf. metamorphic and sedimentary) and can be subdivided into plutonic, hypabyssal and volcanic

Ignimbrite A volcanic rock of lapilli-rich tuff resulting from violent eruptions i.e. pyroclastic flows often associated with formation of calderas

Ilmenite A titanium iron oxide mineral, either black or silvery. Often associated with magnetite as a common accessory mineral in mafic igneous rocks

Inlier An area or group of rocks surrounded by rocks of younger age cf. an **outlier**

Intracratonic A basin located entirely within a craton

Intrusive Igneous rocks that occur within pre-existing rock sequences within the Earth's interior i.e. derived from magma that solidifies prior ro reaching the Earth's surface

Isotope Various forms of an element which has the same number of protons in their nucleus but differing numbers of neutrons (and thus differing atomic weight). A well-known example is uranium with can occur with differing atomic masses including U^{235} and U^{238}

Lahar Superficial volcanic deposits mobilized by water, including mudflows and debris, with primary lava blocks and epiclastic material i.e. cemented fragments from earlier formed rock sequences

Leucocratic Applied to light-coloured igneous rocks typically with >65–70% felsic minerals

Limestone A type of sedimentary rock dominated by calcium carbonite that commonly forms as a chemical precipitate in warm, shallow water or from a concentration of shell and coral fragments

Lineation A generalized term for any linear structure in either a rock or a geological terrain

Lithosphere The solid portion of the Earth comprised of the Crust and part of the Upper Mantle, typically to a combined depth of approximately 100 km. The lithosphere is distinguished from the underlying **asthenosphere** by being relatively rigid and less readily deformed i.e. in plate tectonics theory, the rigid lithospheric plates ride upon the weaker, elastic asthenosphere

Mafic A generalised term for dark-coloured, ferromagnesian minerals (pyroxene, olivine, amphibole) or moderately dark-coloured igneous rock

Magma Naturally occurring molten rock material generated within the Earth, characterized by being highly mobile, and which may be transported upwards into the shallow crust to from either intrusive or extrusive igneous rocks

Magma Chamber A naturally occurring reservoir of magma, typically located in the crust at a depth of a few kilometres to tens of kilometres, formed from molten material which ascended from much deeper in the lithosphere. Many volcanoes are fed from magma chambers within which a complex array of magmas may be fractionated

Magnesite A white or pale grey magnesium carbonate mineral often found in association with altered dolomite (sedimentary) or peridotite (plutonic)

Magnetite A black, opaque iron oxide mineral. The magnetite found in mafic igneous rocks generally contains abundant titanium (Ti) and is known as either Ti-magnetite or titanomagnetite

Mantle The annular zone of the Earth located between the crust and the core, hot yet solid, and dominated by iron- and magnesian-silicates and oxides. Part of the Upper Mantle is incorporated with the Crust into the lithosphere as it includes a disconformity as significant as the Mohorovičić discontinuity

Mantle Plume An upwelling of abnormally hot rock derived in the asthenosphere and which may extend as buoyant columns creating huge volumes of magma by the process of decompression melting in the lithosphere. Mantle plumes may feed giant volcanic provinces such as that associated with the EARS and flood basalts

Marble A granular rock consisting primarily of recrystallized carbonates (calcite and/or dolomite) which has formed from metamorphism of limestone or dolomite

Melanistic Dark-coloured morphs of some species of animals e.g., leopards known colloquially as black panthers

Melanocratic Applied to dark-coloured igneous rocks typically with >65–70% mafic minerals

Member Subdivision of a formation in stratigraphic units.

Mesocratic Applied to igneous rocks typically with approximately equal proportions of felsic and mafic minerals

Metamorphic Pertaining to the process of **metamorphism** (changing or altering the primary minerals) and describing rocks that form due to combinations of pressure, heat, fluid flow, and shearing stress within the Earth's crust, such as gneiss, schist, and marble. Metamorphic rocks may be further classified as meta-igneous or meta-sediments, dependent on their primary origin

Mica A group of silicate minerals which have a sheet-like or platy form. Found as a primary component of granite and some volcanic rocks, as well as in metamorphic rocks such as schist and gneiss

Microcontinent A relatively small continental landmass typically located to a discrete **microplate** (i.e. a very small tectonic plate) that may assemble at a later stage to form a large continent

Milankovitch Cycle A description of cyclic variations of climate that respond to eccentricities of the Earth's orbital geometry including angle of tilt of the axis possibly in increments of 100,000, 43,000, 24,000, and 19,000 years

Mobile Belt A long relatively narrow crustal terrain of tectonic activity on a regional scale

Mohorovičić discontinuity The boundary surface or sharp sesmic-velocity layer that separates the crust from the mantle

Molasse A sedimentary facies of partly marine and partly terrestrial origin consisting of thick sequences of mostly sandstone and conglomerate which can be fossiliferous and often characterized by prominent cross-bedding. In the Alpine tectonic zones, the molasse is derived from erosion of older scarps in isolated basins after the peak of the orogeny. The molasse is younger than the associated deposits of flysch

Nappe A sheet like body of rock which has typically moved laterally due to tectonism associated with orogenies. Nappes can result in older rocks overlying younger strata

Natrocarbonatite A sodium-rich type of carbonatite i.e. a volcanic rock that consists mainly of carbonate minerals (rather than silicates) and has crystallised from a molten carbonate magma (rather than a silicate magma)

Natron A naturally occurring mineral salt, sodium carbonate decahydrate ($Na_2CO_3.10H_2O$), which is very soluble in water and which in nature is typically mixed with sodium bicarbonate ($NaHCO_3$)

Nephelinite A fine-grained volcanic or hypabyssal rock of silica undersaturated basaltic character primarily composed of **nepheline** (a sodic feldspathoid) and pyroxene and lacking olivine and feldspar

Norite A plutonic igneous rock dominated by approximately equal proportions of plagioclase feldspar and orthopyroxene. Distinctly leucocratic

Obduction The overriding or over thrusting of oceanic crust onto the leading edges of continental lithospheric plates

Obsidian A black or dark-coloured volcanic glass usually of rhyolite composition i.e. highly siliceous and characterized by conchoidal fractures. Used by ancient cultures for stone tools

Olivine A dark ferromagnesian silicate mineral found in mafic igneous rocks

Ophiolite A group of mafic and ultramafic igneous rocks, including gabbro and peridotite with broadly basaltic compositions, that originate in deep marine basins i.e. asscoiated with oceanic crust

Orogeny Episodes of mountain building often resulting from collision of continental plates which causes chains or belts of fold mountains to form

Palaeo- A prefix denoting great age or remoteness in terms of time, including prehistoric

Pangea A loosely assembled supercontinent that existed at approximately 250-200 Ma and including almost all of the Earth's continents and microcontinents

Partial Melting A process by which the relatively low-temperature components of the crust or upper mantle melt to generate a magma leaving behind a refractory component

Peridotite A plutonic rock with greater than 90 modal % mafic minerals i.e. ultramafic and cheifly comprised of olivine and pyroxene

Phenocryst A term applied to large, prominent crystals in igneous rocks. In lavas, phenocrysts formed earlier (in the magma chamber or conduit) and were transported upward in the magma

Phonolite A fine-grained extrusive basaltic rock primarily composed of alkali feldspar and with nepheline as the main feldspathoid

Phreatomagmatic An explosive-style of volcanic eruption that extrudes both magmatic gases and steam typically caused by the interaction of magma with groundwater or ice

Plagioclase A group of light coloured sodium, potassium, and calcium feldspars which constitute one of the most common rock-forming minerals. Plagioclase is a major component of gabbro and norite, as well as many granites

Plate Any of the mobile parts into which the lithosphere is fractured

Plateau A term used in geomorphology to describe a nearly flat land surface of great extent and at a relatively high altitude i.e. synonymous with an elevated plain

Plate Tectonics The study of the mechanisms by which plates can separate, collide, shear (slide sideways by means of transform faults), subduct (slide beneath each other) or obduct (thrust up over one another) through geological time

Plinian-style eruption An explosive style of volcanism named after Pliny the Younger who documented the catastrophic 79 AD eruption of Vesuvius and characterized by sustained eruptions that include high eruptive columns that generate copious amounts of pumice. They may represent the ejection of thousands of cubic kilometres of magma

Plutonic An igneous rock formed at great depth; thus a **pluton** is a large intrusive body of coarsely crystalline rocks which cooled slowly. Although plutons originally form at depth, they can crop out on surface due to erosion stripping away the surface cover

Primate A group of mammals that includes lemurs, monkeys, apes, and humans

Proto- A prefix denoting first or earliest form of

Proximal When applied to rocks pertains to being located close to the inferred source area. cf. **Distal**

Pumice A volcanic rock, typically glass foam with abundant vesicles (cavities), often sufficiently buoyant to float on water, and generally ejected from vents. The composition can be siliceous (rhyolite) or basaltic

Pyroclastic Volcanic material formed by the fragmentation of magma and/or pre-existing rock (including conduit walls) typically ejected from a vent by explosive activity and containing particles of chilled melt, (including pumice), earlier-formed crystals (phenocrysts), and lithic material.

Products include ash particles, **lapilli** (small fragments), volcanic bombs, breccia, and agglomerate. The amount of vesicular material in the **pyroclasts** determines the formation of products that include pumice, scoria or cinder, and spatter

Pyroxene A dark ferromagnesian silicate mineral found in mafic igneous rocks comprised of two subgroups identified by their monoclinic (clinopyroxene) or orthorhombic (orthopyroxene) form. The monoclinic pyroxenes are further distinguished by the presence of substantial calcium (a well known example is augite). **Pyroxenites** (subcategories include clinopyroxenite and orthopyroxenite) are plutonic igneous rocks dominated by pyroxene

Quartz A common mineral comprised of crystalline silica (SiO_2)

Quartzite A rock with a granular texture dominated by quartz and formed by recrystallization of sandstone by regional or thermal metamorphism

Radiometric The measurement of geologic time by the study of isotopic abundances of specific radioactive elements that disintegrate at a known rate

Rhyolite An extrusive igneous rock that commonly exhibits flow textures and is characterized by highly siliceous compositions. In comparison to basalts rhyolite magma is notably viscous and much richer in volatiles, often resulting in explosive eruptions

Rifting The separation of geological terranes including the drifting or pulling-apart of continents during formation of oceanic basins. In continental separation, rifting eventually results in formation of oceans and seafloor spreading centres

Rift Platform A geomorphological description of the uplifted blocks or horsts on either side of a rift valley

Rift Valley A geomorphological description of a linear, depressed block

Roll back The process by which the active component of a subduction zone (typically an oceanic trench) migrates away from the collison boundary due to the extensional tectonics of an accretionary margin

Salt A general description of a compound that is produced from a chemical reaction between an acid and a base and in which the acid's hydrogen atoms are replaced by metal atoms of the base. Most salts are crystalline ionic compounds such as sodium chloride (NaCl) or common table salt. Other naturally occurring salts include **soda, natron, and trona**

Sandstone A sedimentary rock primarily comprised of sand-sized grains of quartz and lesser amounts of feldspar; typically the product of erosion of detrital sediments and deposited by either water or wind

Saturated A chemical classification of igneous rocks that are dominated by silica and contain common rock-forming minerals such as quartz and feldspar. cf. **Undersaturated**

Savanna An area dominated by grassland with scattered trees (typically acacia)

Schist A strongly foliated crystalline rock formed by dynamic metamorphism that splits into thin flakes or slabs. A common variety is mica schist which has a markedly pronounced foliation.

Scoria Pyroclasts or small drops of vesicular magma ejected into the air and deposited as irregular fragments often forming poorly consolidated cones near a volcanic vent. Synonymous with cinder

Sedimentary One of the three primary groups of rocks describing compacted or lithified sediment i.e. solid fragmented material that originates from weathering of older rocks and is transported or deposited by air, water, or ice, or that accumulates by other natural agents such as chemical precipitation from solution or accretion by organisms. Sedimentary rocks form from layers of unconsolidated material on the Earth's surface at near-ambient temperatures

Seismic Energy released during an earthquake or similar vibration of the Earth. **Seismicity** is the phenomenon of Earth movements

Serpentine A group of common rock forming silicate minerals with a greasy or soapy feeling. Serpentinite rocks i.e. consisting almost entirely of serpentine, and commonly derived from alteration of ferromagnesian silicates, are an important component of ophiolite intrusions

Shale A fine-grained, detrital sedimentary rock formed by the consoldation of clay, silt, or mud. Shale has a finely laminated structure which is readily deformed into a schist or slate with a pronounced cleavage during metamorphism

Shear The deformation resulting from stresses that cause parts of a body, or rock terrains, to slide relative to the adjacent terrains in a direction parallel to their plane of contact

Shear Zone A tabular zone of rock that has been crushed and brecciated by many parallel fractures due to shear strain

Silicate An extensive group of minerals composed of varying proportions of silicon (Si) and oxygen (O) which are typically bonded with elements that include aluminium (Al), magnesian (Mg), iron (Fe), calcium (Ca), sodium (Na), and potassium (K). Silicates make up most of the crust and mantle

Sill An igneous intrusion with a tabular form, or an elongate layer emplaced parallel with the country rocks, typically relatively small and shallow

Soda A commercial term for sodium carbonate (Na_2CO_3) and/or sodium decahydrate ($Na_2CO_3.10H_2O$), naturally occurring salts also known as washing soda

Soda Ash A crystalline substance of sodium bicarbonate ($NaH.CO_3$) commonly known as baking soda, occurring in nature rarely as nahcolite. Produced from natural salt deposits, typically comprised of natron and trona

Soda Lake A loosely used term to describe an alkali lake whose waters contain a high content of various sodium salts

Spatter Blebs of lava ejected close to the vent that weld together on impact typically forming small cones

Species Any group of organisms that can interbreed and produce fertile offspring

Strain The change in the shape or volume of a body by application of stress. cf. **shear**

Strata Compacted layers or tabular bodies of sedimentary rocks further classified as a stratigraphic unit implying a chronological sequence

Stratigraphy The science of rock strata including an interpretation of the original succession and chronology-typically using fossil assemblages. In recent times geochronology or radiometric dating enables igneous rock units; i.e. lavas and ashes to be incorporated into a stratigraphic sequence

Strombolian A style of volcanism characterized by periodic, relatively small-scale eruptions of basaltic ash lapilli, and scoria, typically within or close to the crater

Structure The general disposition, attitude, arrangement, or relative position of rock masses or sequences of a region

Subduction The process of one lithospheric plate descending beneath another. This generally occurs along a long, narrow belt, or **subduction zone**

Supercontinent An amalgamation of large and small continental plates into a gigantic continental masssuch as Pangea

Suture Regional feature marking lineaments where discrete tectonic blocks or lithospheric plates and microplates join

Syncline A down-folded structure in which the youngest rocks occur in the core and the oldest rocks in the outer part. cf. **anticline**

TAS Diagram A chemical plot of total alkalis versus silica used to discriminate the alkali-rich basaltic group of volcanic rocks

Tectonics A branch of geology dealing with the broad architecture of the outer part of the Earth, specifically detailing the structural or deformational features and including the historical evolution of such features, including **tectonic plates**

Tephra A general term for volcanic material associated with pyroclastic deposits and volcanic ash

Terrain The description of a specific geologic region or area often bounded by faults or unconformities and with a different geologic history to adjacent regions

Trachyte A group of fine-grained generally porphyritic extrusive igneous rocks (volcanic) having alkali feldspar and minor mafic minerals as the main constituents

Triple Junction A point where three lithospheric plates join

Troctolite A plutonic rock of the gabbroic group in which the ferromagnesian silicate associated with plagioclase is olivine (rather than pyroxene)

Tuff Rock formed from fine-grained wind-borne or water-laid volcanic ash

Travertine A dense, finely crystalline form of limestone formed from rapid precipatation of calcium carbonate from solution, typically geothermal ground water, including at the mouth of a spring

Trona A naturally occurring mineral salt sodium sesquicarbonate dihydrate ($Na_3CO_3.HCO_3.2H_2O$) which is generally found mixed with natron

Ultramafic A generalized term for dark-coloured igneous rocks dominated by ferromagnesisn silicates (typically olivine and pyroxene). Most examples fall into the peridotite grouping of plutonic rocks

Unconformity A major break or time gap in the rock record and where rocks of a given time are absent between the underlying (typically older) and overlying (typically younger) sequences. Related to either non-formation or erosion having removed the missing strata

Undersaturated A chemical classification of igneous rocks that are relatively poor in silica and contain silica-poor minerals particularly feldspathoids (such as nepheline) and in which silica-rich minerals such as quartz and feldspar are absent

Uplift The process by which a structurally high area in the crust is produced by tectonic movements that raise the rock sequence (as in a dome or plateau)

Volcanic Rock types or activity associated with a volcano. Volcanic rocks are igneous rocks extruded onto the surface of the Earth e.g. as lavas or ash fall

Volcaniclastic A generalized description of all types of volcanic particles or rocks regardless of their origin i.e. not only those from a vent (pyroclastic) but also rock fragments released by erosion of earlier-formed volcanic material

Vulcanian Pertaining to volcanic activity similar to that of the active volcano on the island of Vulcano one of the Aeolian Islands located off the southwest coast of Italy, characterized by the explosive ejection of lithic fragments (pre-existing rock blocking the conduit as well as new lava and volcanic bombs) and irregular, cannon-like eruptions

Warp A slight flexure or bend (upward or downward) in the Earth's crust usually on a regional scale

Wehrlite A type of peridotite dominated by approximately equal amounts of olivine and clinopyroxene

Xenolith A foreign inclusion in an igneous rock typically derived from the earlier-formed country rock for example, the sidewalls of a conduit or pluton

Zeolite A generic term for a large group of white, yellow, or colourless aluminium, calcium, sodium, and potassium silicate minerals occurring in basalts (infilling vesicles) and more commonly in clays and silts (lake and deep sea sediments). The process of **zeolitization** describes the introduction or replacement of pre-existing minerals by members of the zeolite group, often in association with hydrothermal fluids

NOTE Explanations of geological eras, periods, and epochs are excluded as they are defined in Map B.

References

Bates, R. L., & Jackson, J. A. (1980). Glossary of Geology (p. 751). Virginia: American Geological Institute.

Redfern, R. (2002). Origins. The evolution of continents, oceans, and life. Wiedenfeld and Nicolson, London, 360 p.

Index

A

Absaroka Mountains, 55
Absaroka Volcanics, 55, 57
Academy of Assos, 285
Accretionary margin, 143, 248
Achaea, Achaia, 221
Achilleion, 283, 284
Achilles, 284
Acitrezza, 163, 165
Acrocorinth, 219, 220, 228, 229, 231
Acropolis, 198–204, 250, 285, 287, 288, 291
Acropolis of Athens, 193, 198–200
Active Faults, 227, 273, 275, 298, 303, 304, 307, 309, 319
Active margin, 249
Adamas, 253–257, 259
Adamas Graben, 254, 257
Adell, Paul, 133
Adriatic Microplate, 143, 145, 146, 169, 172
Adriatic Sea, 139, 140, 145, 166
Aegean Basin, back-arc basin, 207, 248–250
Aegean Sea, 139, 140, 145, 158, 198, 201, 206, 220, 223, 229, 236, 237, 245, 246, 248, 254, 273–280, 283, 285, 286, 293, 295
Aegina, 224, 247
Aeneid, 153, 172
Aeolian Volcanic Arc, 146, 148
Aeolus, Greek god of the wind, 140, 160
African elephant, 91, 93
African Erosion Surface, 95
African Great Lakes, 90, 92, 103
African Humid Period, 99
African Iron Age, 362
African Plate, 95, 139–143, 145, 146, 219, 276, 360
African Queen, 103
Afromontane forest, 89–93, 104, 109, 111, 122
Agamemnon, Greek history, 225
Agglomerate, 105, 107, 152, 155, 245, 250, 253, 260, 261, 266, 267, 269, 338, 345, 347, 350
Agios Nikolaos, 240, 250, 252, 253
Ahwahneechee, people, 38
Akrotiri, 245, 246, 266, 268
Akrotiri Peninsula, 264, 266, 267
Alaşehir Graben, 301, 303, 304
Alberta, 69, 70
Albertine Rift, 89–92, 94–101, 103–105, 107–109
Albert Nile, 91, 92, 98, 99
Alexander the Great, 170, 223, 284, 315
Alicudi, island, 149

Alluvium, 95, 158, 198, 199, 201, 210, 219, 221, 224, 225, 231, 232, 235, 249, 254, 281, 341
Alpheios, 232
Alpheios Graben, 219, 224, 231, 234
Alpheios River, 231, 233, 234
Alpine lake, 37, 38, 40, 42, 46, 69, 70, 72, 80, 82, 86, 213, 214, 327, 330, 340, 342, 346
Alpine massifs, zones, 195
Alpine moorland, 37, 39
Alpine Orogeny, 139, 141–143, 145, 193–195, 198, 199, 210, 223, 224, 241, 248, 276, 285, 303
Alunite, 256, 258
Amalfi, 170, 174
Amalfi Coast, 172
Ambleside, 328, 330, 345
Amphibolite, 95, 100, 278, 296
Amphitheatre, 157, 158
Amygdaloidal, 105, 107
Anatolia, 272, 274, 275, 295, 298, 315, 324
Anatolian Microplate, 195, 248, 273, 276–278, 303, 315
Ancient Corinth, 219, 220, 227–231
Ancient Greeks, 139, 140, 145, 158, 163, 166, 202, 247, 256, 274, 285, 288, 292
Ancient Thera, 269
Andalusite, 362
Andesite, 246, 250, 251, 253, 255, 258, 260, 261, 266, 269, 290, 317, 336, 338, 345
Angel Glacier, 75
Anorthosite, 353, 354, 358, 361, 367, 369, 372, 373
Anticline, 19, 25, 27, 240
Antiquities, 193–195, 199, 200, 202, 204–206, 211, 219–223, 225, 227, 230, 231, 233–235, 237, 240, 242, 273–276, 278–288, 292, 295, 296, 298, 299, 302, 303, 307–309
Apennine-Mahgreb Thrust, 143, 146
Apennines, mountains, 143, 144, 169, 172
Aphrodisias, 273, 274, 280, 295–299
Aphrodite, Greek god, 296, 299
Arabian Microplate, 142, 273, 276, 277, 315, 317
Arcadia, 219, 222, 233, 234
Arcadian Mountains, 223, 226, 233
Archaean, 95, 122, 357, 358
Archaic period, 227
Archangel Gabriel, 299
Archangel Michael, 276
Arches, 11, 19, 21–23, 25–29, 31, 46, 238, 240, 261
Arches National Park, 5, 19–31, 33, 34
Archimedes, Greek mathematician, 158
Areopagus, 199, 201

Argive Plain, 225
Argolid, 219, 221, 222, 224, 225
Argos Graben, 224, 225
Aristotle, 285
Arizona, 1–3, 8, 11, 15
Arso Lava, 176
Artemision, 273, 292, 295
Artifacts, 134, 181
Arusha, 118, 122
Arusha National Park, 118
Ash, 51, 56, 64, 120–123, 135, 150, 156, 157, 175, 180, 181, 185–190, 245, 255, 256, 258–262, 265, 267, 269, 270, 319, 320, 323, 327, 336, 347
Ashfall deposits, 186
Asia Minor, 219, 222, 275, 276, 284, 298
Asklepieion, 226, 227
Assos, 273, 274, 279, 284–287
Asthenosphere, 3, 95
Athabasca Falls, 77, 79
Athabasca Glacier, 69, 73, 77, 80, 81
Athabasca River, 73, 75, 77
Athenians, 140, 239
Athens, 193, 198–202, 219, 225, 226, 246
Athens Basin, 193, 198, 199
Athens Schist, 198, 199
Atlantic Ocean, 97, 142, 145, 360
Atlantis, 172, 245, 246, 262, 269
Atlas Mountains, 142
Atomic Weapons Research Establishment Attica, 345
Atticacycladic Zone, 248, 278
Australopithecine, 133, 135
Australopithecus afarensis, 117, 133, 135
Avalonia, 331, 333
Avellino Pumice, 180, 182

B
Back-arc basin, 139, 140, 143–145, 198, 223, 248, 273, 277, 278, 303
Badlands erosion, 9, 23, 32, 133, 134, 322, 323
Baker, Samuel and Florence, 103
Bakırçay Graben, 278, 285, 287, 289
Balanced Rocks, 19, 20, 23, 30, 33
Balkan Peninsula, 142, 145, 208, 222, 235, 277, 278
Baltica, 331, 333
Banded iron formation, 95
Banff, 69, 70, 73, 74, 82, 85
Banff National Park, 69, 70, 73, 77, 80, 82
Barchan, 128, 129
Basal Pumice, 180, 182
Basal surges, 169, 185, 186, 188, 267
Basalt, 51, 56, 57, 148, 149, 153, 158, 266, 314, 317, 357, 358
Basaltic, 52, 56, 60, 123, 127, 136, 148–150, 163, 169, 172, 177, 181, 249, 313, 314, 316, 318, 320, 353, 354, 357, 369
Basaltic trachyandesite, 123, 149
Basement, 37, 40, 46, 49, 57, 64, 89, 95, 105, 107, 119, 128, 130, 172, 182, 197, 224, 233, 264, 278, 281, 296, 303, 315, 358
Basement Complex, 33, 92, 95, 122, 315
Basistoppen Sill, 358
Bathymetric, bathymetry, bathymorphological, 141, 143, 145, 253, 269
Batieia, 284
Battle of Marathon, 201
Battle of Salamis, 202
Battle of Thermopylae, 193
Bauxite, 203
Bears, 37, 40, 51, 53, 345, 355
Beartooth Mountains, 53, 54, 56

Behram, 284
Beni Gap, 99
Benioff Zone, 146, 172
Besika Bay, Beşik Bay, 281, 283, 284
Biga Peninsula, 280, 284
Bimodal volcanism, 60, 245, 249
Bison, 51, 53, 54
Bitumen, bituminous deposits, 205, 242, 243
Black Sea, 278, 281
Blaine, F.W., 364
Block faulting, 53
Blue Mountains, 103
Boise, Charles, 132
Borehole, 158, 185, 207
Borrowdale Volcanic Group, 327, 328, 331, 333, 335, 336, 344
Borrowdale Volcanics, 327, 328, 331, 332, 339, 340, 347–350
Bosphorus, 274, 278
Botanical zones, 105
Bowen trend, 354
Bow Lake, 80
Bow River, 73, 82
Brantwood, 328, 330
Breccia, 177, 180, 200, 226, 294, 295, 336, 338, 340, 345, 358, 366, 370
Breccia dyke, 135, 136
Britain, British Isles, 327, 328, 333
British Caledonides, 327, 331, 332
British Navy, 178, 345
Bronze Age, 158, 160, 181, 211, 219, 221, 222, 225, 256, 258, 262, 267, 273, 280, 281, 283, 284
Brushy Basin Member, 34
Buckland, William, 340
Buganda-Toro System, 95, 100, 104, 105
Bujagali Falls, 100
Bunyaraguru Crater Field, 105
Bunyoro Escarpment, 100
Burgersfort, 362, 363, 366
Burgess Shales, 87, 88
Burton, Richard, 100
Burundi, 100, 103
Bushveld Escarpment, 363, 366
Bushveld Igneous Complex, 353–355, 358, 359, 362, 363, 365
Buttermere, 327, 336, 340, 343
Buttes, 1, 11, 13, 131
Büyük Menderes Graben, 301, 303, 304
Büyük Menderes River, 302
Bwindi Impenetrable Forest National Park, 89, 91, 107, 109
Byzantine, period, 288, 308, 322

C
Calabria, 140, 144–146, 163, 166
Calabrian earthquake, 145, 146
Calc-alkaline basalt, 148
Calcareous ashes, 117, 122, 124, 133
Calcite, 199, 225, 226
Calcium, 375, 380, 381, 383
Calcium carbonate, 28, 64, 301–304, 307, 308
Caldera, 51–53, 56–59, 64, 117, 119, 120, 122–124, 126–128, 135, 146, 149, 150, 152, 153, 158, 169–171, 173, 175, 176, 179, 180, 190, 245, 246, 262, 264, 267, 270, 271, 313, 316, 335, 338
Caledonian Orogeny, 141, 333, 336, 338
Caledonide Orogeny, 327, 347
Caledonides, 333, 336
California, 37, 39
Caligula, 170

Calypso Trench, 145
Cambrian, 69, 72, 73, 75, 77–79, 82, 86–88, 333
Cambrian Explosion of Life, 69, 87
Campanian Graben, 169, 172
Campanian Ignimbrite, 169, 175, 177, 185
Campanian Plain, 169–172
Campanian Volcanic Field, 146, 148, 170, 172
Campi Flegrei, volcano, 146, 169–173, 175, 177, 182
Canada, 19, 21, 69–72, 77, 80, 86
Canadian Pacific Rail, 70, 82, 85–87
Canadian Rocky Mountains., 72
Canadian Rocky Mountains, 69, 70, 73
Çanakkale, 274, 280, 284
Canyonlands National Park, 1, 2, 9, 11, 19, 22
Canyons, 1, 2, 9–11, 20, 28, 33, 37–39, 41–44, 46, 51, 52, 57, 64
Cape buffalo, 91, 93
Cape Pelekouda, 246, 261
Cape Vani, 257, 260
Cappadocia, 313–315, 321, 324, 325
Cappadocia Plateau, 313–317, 321
Cappadocia Volcanic Province, 313, 314, 316, 317
Capri, 170–172, 174
Carbon, dating, 124, 281
Carbonates, 28, 85–87, 141, 144, 203
Carbon dioxide, 62, 303, 304, 309
Carboniferous, 1, 6, 9, 10, 19, 22, 23, 73, 82
Carian marble, 273, 299
Carlisle, 328, 345
Carnegie Museum of Natural History, 32
Carrock Fell, 329, 335, 338, 347
Carthaginians, 158
Castalian Spring, 204
Castle Geyser, 64, 65
Castlerigg, 329, 343, 344
Catania, 139, 140, 145, 152–154, 157, 163
Catastrophic flood events, 319
Catastrophic volcanism, 51, 52, 56, 60, 117, 118, 140, 146, 169, 172
Cathedral formation, 77, 79, 85, 88
Cathedral Mountain, 85–87
Cave, 75, 77, 82, 158, 160, 161, 199, 204, 205, 210, 211, 213, 225, 226, 235, 238–240, 253, 261, 323, 363
Cavell Glacier, 75
Cave of Pólis, 160
Cave of the Nymphs, 160
Çavuşin, 318
Cedar Mesa Sandstone, 7, 10, 11
Cenozoic, 1, 6, 21, 41, 46, 69, 71, 72, 89, 90, 95, 98, 105, 117, 119, 193, 194, 197, 198, 203, 223, 224, 264, 276, 278, 303
Central African Craton, 95, 100, 117, 119, 120
Central African heritage, 105
Central Hellenic Basin, 193
Central Hellenic Shear Zone, 195, 223, 224, 226, 230
Chalcedony, 31, 34
Chamois, 214
Chancellor Formation, 85, 87
Charybdis, 140, 163, 166
Chert, 36, 95, 120, 197, 203, 211, 338
Chimpanzee, 89, 92, 93, 105, 111
Chinle Formation, 7, 8, 10, 15, 22
Christian, centres, 273–275
Chromitite, 353, 354, 367, 372
Cichlid fish, 92
Cilento and Vallo di Diano National Park, 158
Circe, 163
Cirque, 69, 72, 77, 78, 214, 340, 346
Clark, Galen, 39
Classical Greece period, 199, 211, 222, 227, 234, 285

Clay, 20, 62, 69, 80, 104, 124, 130, 133, 198, 199, 231, 235, 236, 258, 259
Climate, climatic cycles, 1, 2, 10, 14, 19, 22, 27, 32, 37, 43, 80, 89, 91, 99, 118, 120, 128, 134, 135, 172, 202, 203, 237, 253, 295, 302, 313, 314, 318, 355
Cockermouth, 328, 330
Codrington, Admiral, 239
Cog Group, 72, 82, 83, 85
Collisional tectonism, 303
Colluvial, 304
Colombia Icefield, 77
Colorado Lineament, 4, 9
Colorado Plateau, 1–6, 12, 19–21, 23, 27, 32, 319
Colorado River, 1, 2, 9–11, 22, 25, 26, 64
Colossae, 273–276, 280, 295, 298
Columbia River, 51, 56, 73
Columbia River basalts, 51, 56, 314
Columbo Seamount, 245, 254, 264, 270, 271
Columnar joints, 52, 60, 165, 261, 318, 338, 340
Conglomerate, 14, 34, 95, 130, 193, 198, 200, 207, 214, 216, 217, 225, 229, 232, 235, 240, 241, 249, 254, 278, 303, 315
Congo River, 99
Coniston, 327–329, 334, 338, 342, 345, 347, 350
Coniston Fells, 338, 340
Coniston Water, 327, 328, 330, 331, 333, 342, 343
Continental divide, 52, 53, 69–71, 77
Continental rifting, 95, 120
Cook's Meadow, 40
Copper, 8, 327, 329, 330, 343, 345, 347
Coppermines Valley, 333, 345, 347
Cordilleras, 3, 40, 53, 69, 71, 72
Cordilleria, microcontinent, 72
Corinth, 219, 220, 224–230
Corinth, Isthmus, 150, 219–221, 223, 227, 229, 230, 247
Corinth Basin, 203, 204, 219, 224, 226, 229, 230
Corinth Canal, 219–221, 229, 230, 232, 233, 236
Corinthians, 140, 158, 219, 223, 227
Corrie, 340
Cotton soil, 95
Coyote, 53
Crater, 89, 105, 107, 109, 111, 113, 114, 118, 121, 139, 140, 150, 152, 153, 155–157, 169, 176, 179, 190, 191, 225, 247, 252, 255, 260, 261, 269, 271
Crater field, 89, 94, 104, 105
Craton, 87, 95, 119
Cretaceous, 6, 22, 33, 40, 72, 73, 95, 96, 142, 193, 194, 198, 203, 207, 220, 223–226, 228–230, 233, 235, 238, 240, 243, 248
Crete, 245, 248, 258, 262, 267, 269, 272, 276
Critical Zone, 353, 364, 367, 369, 372
Crustal thinning, 5, 37
Crystal settling, 353–355, 369
Cumbria, 329, 332, 336, 345
Cumbrian Mountains, 327, 328, 330
Cutler Group, 6, 7, 14, 22
Cyanobacteria, 62, 120, 121
Cycladic archipelago, 246–248
Cyclopes, 163

D
Dacite, 148, 250, 253, 255, 266
Dardanelles, 160, 226, 273, 274, 277–282, 284
Dardanelles Campaign, 273, 284
Darwinian evolution, 89, 92
Dead Horse State Park, 1
Death Valley National Park, 40
Debris avalanche deposit, 136, 255, 261

Debris deposits, 135
De Chelly Sandstone, 1, 7, 12–15
Deformation, 1, 3, 20, 21, 33, 73, 88, 120, 198, 225, 273, 274, 276, 278, 304, 347
Delicate Arch, 19, 20, 23, 27, 29–31
Delos, 247, 248
Delphi, 193, 202–207, 227, 310
Delphic Fault, 205
Delphic landscape, 204, 206
Delphic Oracles, 204, 205
Deltas, 214, 273, 275, 280
Democratic Republic of Congo, 89–91, 97, 103, 104, 108, 110, 111
Denizli, 274, 278, 295, 296, 298, 299, 301–304
Denizli Earthquake, 301, 310
Denizli Fault, 304
Denmark Strait, 357
Dent Group, 338
Derinkuyu, underground city, 313, 314, 323, 324
Derwent Water, 327, 330, 331, 336, 337, 343
Desert landforms, 1, 2, 19
Desiccation events, 99
Devonian, 40, 73, 76, 327, 331, 333, 335, 336, 338, 347
Dewey Bridge Member, 19, 26, 28, 30, 31, 33
Diakofto, 226
Diatomite, 261, 262
Differential erosion, 1, 10, 19, 27, 30, 217, 313, 314, 319–321
Differentiation, 148, 353, 354
Dikili-Çandarli Volcanic Suite, 285
Dinosaur fossils, 19, 20, 22, 32–34, 36
Dinosaur Museum, 32, 35
Dinosaur National Monument, 19–22, 31–35
Diolkos, 219, 220, 229, 230, 233
Dodecanese Islands, 245–247
Dodona, 210, 211
Dolomite, 86, 144, 172, 223, 224, 233, 235–238
Domes, 10, 25, 37, 38, 40, 41, 43, 45–47, 57, 250, 255, 260, 266, 269, 353, 366, 369
Doric columns, style, 199, 202
Dragon Lakes, 213
Drake, Sir Francis, 345
Driekop, 364, 366
Dümrek River, 281
Dunite pipe, 353, 364
Dwarsriver, 354, 363

E
Ear of Dionysus, 160, 162
Earthquakes, 64, 66, 97, 139, 140, 145, 146, 153, 157, 158, 160, 163, 177, 182, 185, 193, 199, 205, 219, 222, 225, 228, 230, 231, 233, 235, 241, 254, 257, 268, 271, 273, 274, 277, 281, 288, 295, 296, 298, 299, 301, 304, 308, 310
East African Rift System, 89, 90, 94–97, 117–119
Eastern Limb, 353, 359, 360, 362–366, 368, 369, 372
Ecosystems, 37, 39, 89–91, 104, 109, 117, 122, 242
Eden River, 328
Eden Valley, 328
Egypt, 225, 239
Elanairobi Ridge, 126, 127
El Capitan monolith, 37
Eldon Formation, 82, 86
Elgin, Lord, 199
Elis, 219, 221, 230, 233
Elizabethan period, times, 327, 343, 345
Elk, 53
Emerald Lake, 85–87

Empakaai Crater, 123, 126, 127, 135
Empakaai Volcano, 126
Endulen, 133
Engare Sero Gorge, 135, 136
Engaruka-Natron Crater Field, 120
Entrada Formation, 8, 21, 27, 28
Eocene, 55, 69, 72, 73, 211, 223, 224, 230, 233, 235, 236, 240, 353, 357, 358
Ephesus, 273–276, 278, 280, 285, 288, 292–296
Epidaurus, 219, 220, 226, 227
Epirus, 194, 208
Eratosthenes, Greek polymath, 281
Erciyes Volcano, 313, 316
Ercolano, 182, 183
Erosion, 1, 2, 10, 14, 15, 19, 21, 26–30, 32, 37, 38, 41–44, 46, 60, 64, 66, 71, 72, 82, 89, 90, 95, 119–121, 172, 193, 199, 201, 203, 214, 226, 236, 239, 273, 280, 284, 304, 307, 313, 314, 316, 318, 327, 330, 334, 360
Ethiopia, 133
Ethiopian Dome, 95
Ethiopian Rift, 95–97
Ethylene, 205
Etna Volcano, 139, 140, 144, 146–149, 152–158, 163, 172
Euboean Graben, 193, 201
Euboean Gulf, 198, 205, 207
Eurasian Plate, 139–143, 145, 146, 172, 195, 198, 223, 226, 248, 273, 276, 277
Eurotas Graben, 224, 235–237
Evaporites, 6, 10, 11, 23, 25, 145, 249, 277
Evolution, evolutionary period, 92
Exfoliation weathering, 28, 37, 46
Explosion craters, 105, 120
Extensional faults, tectonism, 145, 169, 317
Eyasi Half-graben, 120, 123
Eycott Hill, 336
Eycott Volcanic Group, 331, 336

F
Fairy chimneys, 313, 314, 320
Fault, 22, 23, 26, 27, 61, 71, 97, 100, 104, 120, 144–146, 153, 172, 197, 198, 205, 224, 227, 228, 230, 231, 235, 238, 250, 251, 277, 278, 285, 296, 298, 299, 301–306, 309, 310, 315–317, 345
Fauna, 33, 51, 53, 88–91, 105, 107, 117, 135
Felsite, 358, 359, 362
Fenner trend, 354, 364
Filicudi, island, 149
Finger lake, 73, 90, 92, 94, 100, 327–330, 340
Fins, 28, 29
First Dark Age, 99
Fjord, 355, 358
Flamingo, 117, 120
Flood basalt, 51, 56, 148, 354, 372
Flora, 33, 89–91, 105, 107, 117, 135
Fluvial, 7, 8, 11, 82, 130, 172, 198, 304, 316
Flysch, 193, 197, 198, 210, 211, 214, 224, 226, 230
Foidite, 109, 148, 172
Fold Mountains, 3, 37, 40, 69, 71, 73, 139, 142, 145, 208, 276
Footprints, 117, 133, 135
Fore-arc basin, 144, 248
Forest, 8, 37, 39, 85, 89, 91, 93, 100, 102–105, 107, 109, 111, 122, 134, 214
Forester, C.S., 103
Fort Portal, 97, 105
Fort Portal-Kasekere Crater Field, 94, 105
Fossa cone, 139, 140, 149, 150, 152

Fossils, 19, 32, 33, 36, 69, 87, 88, 111, 117, 128, 130, 133, 135, 202, 281, 327, 335, 338
Fragmentation index, 146, 147
Franchti Cave, 225
Frescoes, 182, 268, 288, 295
Front Range, 70, 72, 73, 80, 82
Fumaroles, 51, 52, 61, 62, 126, 149, 150, 176, 179, 254, 260
Fylakopi, 257, 258, 261
Fyriplaka Volcano, 245, 246, 253–255, 258, 260, 261

G
Gabbro, 327, 331, 335, 338, 347, 353, 357, 361
Gabbronorite, 363, 366, 369
Galena, 345
Gallatin Range, 53
Gallipoli Peninsula Historical National Park, 273, 274, 284
Gases, hallucinogenic, poisonous, 182, 205, 207
Gavrovo Zone, 197, 223, 233, 235
Gazelle, 128
Gediz Graben, 304
Gelai Volcano, 120
Geomorphology, 46, 142, 143, 239, 280, 327
Geoparks, 193, 194, 235
Geothermal activity, 52, 57, 60, 66, 105, 254
Geothermal water, 51, 249, 250, 301–303
Geotourism, 245, 246, 249, 253
Geotrails, 245, 246, 255, 256, 260, 261, 269
Geyser, 51, 52, 57–59, 61, 62, 64, 65, 105
Geyser, Old Faithful, 51, 64, 65
Giant Sequoia, 37, 46, 50
Glacial deposit, 43, 56
Glacial moraine, 56, 72, 82, 330, 340
Glacial Trail, 75, 77
Glacier, 37, 38, 43, 46, 56, 57, 69, 70, 72, 73, 75, 77, 78, 80–83, 85, 99, 105, 214, 327, 340, 343, 354, 355, 357
Glaronisia Islet, 261
Glenridding, 330, 345
Gneiss, 53, 72, 100, 292
Gold, 219, 225, 233, 281, 357
Gol Mountains, 122, 123, 134, 135
Gombe Stream National Reserve, 105
Gondwana, 95, 119, 141, 331, 359
Goodall, Jane Goma, 105
Göreme, 313, 318–320, 323
Göreme Historical National Park, 313, 314, 316, 318, 320–322
Gorge, 75–77, 79, 86, 100, 101, 103, 128, 130, 135, 193, 204, 213, 214, 226, 229, 234, 235, 238, 318, 322
Graben, 96, 105, 120, 122, 169, 172, 193, 194, 198, 207, 210, 214, 219–221, 223–225, 230, 231, 234–236, 238, 254, 256, 260, 273–275, 277–281, 284, 285, 292, 293, 295, 296, 298, 301–305, 315, 316
Gran Cratere, 150, 152
Grand Canyon, 2, 9, 64
Grand Teton, 53
Grand Teton National Park, 53, 55
Granite, 37, 40, 41, 45–47, 49, 53, 72, 119, 278, 327, 329–331, 335, 338, 345, 347, 353, 358, 359, 362
Granite batholith, 37, 38, 40, 327, 331, 338
Granite domes, 46
Granite gneiss, granitic gneiss, 95, 100, 117, 119, 122, 128, 278, 285, 296, 357, 358
Granite pluton, 37, 38, 40, 46, 71, 338, 372
Graphite, 199, 327, 329, 343, 345, 347, 348
Graptolite, 335, 336, 338
Great Gable, 330
Greco-Persian Wars, 193

Greece, 139, 140, 158, 160, 193–198, 201–203, 205, 207–211, 219–223, 225–227, 232, 234, 238–240, 246, 248, 249, 256, 262, 269, 274, 276, 278, 279, 292, 310, 324
Greek Dark Ages, 222, 223, 227
Green Gable, 336
Green Lahar, 255, 261
Greenland, 252, 267, 353–355, 357
Green River, 1, 2, 9, 11
Greenside mine, 345
Greenstones, 95, 117, 119, 122
Green Tuff, 158
Gregory, John Walter, 117
Gregory Rift, 90, 94–97, 99, 105, 117, 118, 120, 130
Grevena, 208
Grotta di Terra, 177, 181
Groundwater, 51, 61, 63, 82, 301
Grumeti River, 100
Gulf of Argos, 225
Gulf of Corinth, 198, 202, 220, 223, 226–231
Gulf of Laconia, 236
Gulf of Malia, 205, 207, 208
Gulf of Naples, 169–173, 176, 178, 186, 188
Gulf of Patras, 220, 223, 226
Gythion, 236

H
Half Dome Monolith, 37, 38, 43, 46, 47
Half-graben, 94, 97, 120, 227
Hall, Arthur Lewis, 364
Hamilton, William, 172
Hanging valley, 37, 42, 46, 48, 69, 72, 73, 77, 82, 214, 327, 328, 330, 334, 340, 342, 345, 346
Hardy, Thomas, 330
Harzburgite, 366, 367, 371
Hawaii, 146, 147, 153
Hawaiian field, 146
Headward erosion, 100, 103
Hebgen Lake, 64, 66
Hellenic Microplate, 142, 145, 194, 195, 219, 223, 226, 248, 276, 277
Hellenic Trench, 139, 143, 145, 223, 241, 245, 248, 249, 273, 276, 277
Hellenic Volcanic Arc, 145, 146, 195, 224, 245–248
Hellenistic period, 205, 207, 219, 221, 223, 226–229, 285, 287, 295, 302
Helvellyn, 330, 335, 338, 346
Herculaneum, 169, 182–184, 186–188, 268
Herodotus, Greek historian, 193, 201, 206, 242, 243
Hexamillion, 229
Hierapolis, Hierapolis-Pamukkale, 301–303
Hierapolis Fault, 301, 304, 310
High Street, Roman road, 306, 345, 346
Hill of Kronos, 232, 234
Hill of the Nymphs, 199
Hittites, people, 315, 322, 325
Holes, 28–30, 126, 225, 258, 366
Holocene, 37, 38, 42, 43, 46, 57, 69, 73, 80, 97, 99, 105, 120, 121, 128, 158, 172, 207, 230, 239, 251, 265, 279, 281, 304
Homer, Greek poet, 139, 160, 222, 274
Homer's School, 160, 164
Homeric times, 281, 284
Hominins, 105, 111, 117, 128, 130–132, 134, 135
Homo erectus, 117, 130, 133, 134
Homo habilis, 117, 128, 133, 134
Honaker Trail Formation, 6, 22
Honeycomb weathering, 28
Honister mine, 347, 350
Honister Pass, 347

Horst, 89, 94, 97, 99, 105, 225, 236, 248, 254, 274, 278, 292, 293, 296, 297, 303, 304, 307
Hot spot, 51, 56, 59, 108, 109, 357
Hot springs, 51, 52, 62, 64, 66, 104, 105, 158, 193, 205–207, 228, 242, 250, 254, 301, 302, 304–308
Hudson Bay, 73
Human evolution, 117, 128
Hydrocarbons, 205
Hydrothermal fluids, 51, 57, 269, 345

I

Icefields Parkway., 74
Icefields Parkway, 69–71, 73–75, 80
Ice sheets, 37, 38, 43, 56, 64, 72, 73, 313, 327, 340, 354, 355, 357
Idaho, 51–53, 56
Igneous layering, 338, 353, 354, 357, 358, 361, 364, 366, 367, 369, 371
Ignimbrite, 57, 123, 128, 158, 267, 313, 316–319, 321, 322, 325, 338, 340
Ihlara Gorge, 313, 314, 318–320, 322, 324, 326
Iliad, 160, 219, 222, 225, 273, 274, 281, 283
Ilium, city, 160, 283
Illinois Glaciation, 42, 77
Indian Ocean, 95, 99, 360
Inlier, 108, 235, 278, 296, 303, 327, 328, 331, 335, 338
Inselberg, 122
Ioannina, 194, 208, 210
Ionian Basin, 145, 153
Ionian-Paxon Zone, 197
Ionian Sea, 139, 140, 145, 158, 198, 209, 220, 223, 224, 229–231, 236, 239, 249
Ionian Thrust, 241, 242
Iran, 142
Iron-rich ultramafic pegmatite, 364, 369
Irvine, Neil, 356–358, 360, 366, 370
Ischia, 170–172, 176, 177, 180, 181
Ischia Volcano, 169, 170, 176
Ishasha Plains, 104
Ishasha River, 104
Island arc (Volcanic island arc), 40, 139, 143, 144, 146, 149, 327, 333
Island Arc Volcanism, 336
Island in the Sky, 9–11
Isthmia, 220, 229
Italian Island, 139
Italian Islands, 139, 140, 144, 146, 148
Italy, 139, 140, 143–146, 160, 166, 169, 170, 204
Ithaca, 139, 140, 160, 163, 164, 222, 241
Izmir, 274, 276, 284
Izmir-Ankara Fault Zone, 273, 277, 278, 285

J

Jackson, 53, 55, 227
Jackson Hole, 56
Jackson Hole Block, 55
Jasper, 69, 70, 73, 77, 85
Jasper National Park, 69, 70, 73–75
Jenny Lake, 53, 55
Jinja, 91, 92, 99, 100, 103
Joints, jointing, 15, 19, 27, 28, 37, 38, 46, 47, 75, 77, 199, 214, 216
Jurassic, 1, 2, 6, 8, 10, 19–23, 26, 27, 33, 34, 69, 72, 73, 95, 119, 141, 203, 207, 210, 211, 223–227, 229, 231, 233, 235, 237, 278

K

Kafu River, 95, 97, 99
Kagera River, 99, 100

Kagera Valley, 99
Kaimeni Hora, 251, 254
Kalagala Island, 100
Kalamata, 219, 220, 235–238
Kalamata Graben, 220, 224, 235, 238
Kalambaka, 194, 214, 215
Kalarrytes, 197, 211, 212
Kameni Islands, 245, 246, 262, 263, 269, 270
Kanakas, 257, 259
Kangerdlugssuaq Fjord, 355, 357, 358
Kangulumira, 100
Kaolin, 245, 257–259
Karagwe-Ankolean System, 95, 108
Kara Menderes River, 281, 283
Karst, 204, 214
Karstic landscape, 158, 161, 204, 213, 219
Katakolon, 231
Katonga River, 95, 97, 99, 104, 107
Kattwinkel, Wilhelm, 130
Katwe-Kikorongo Crater Field, 105–107
Kavirondo Gulf, 99
Kavomalias Geopark, 235
Kayenta, 2, 8, 10–12, 21–23
Kayenta Formation, 2, 8, 10, 21–23
Kaymakli, underground city, 313, 314, 323
Kayseri, 313, 314, 322, 325
Kaystros River (or Cayster River), 288
Kazdağı National Park, 279, 284, 285
Kazinga Channel, 93, 104, 105
Kefalonia, 241
Kefalonia Fault, 223, 241
Kendal, 328, 338
Kennedys Vale, 369
Kenya, 90, 96, 99, 117, 120, 133, 134
Kenyan-Tanzanian Dome, 95
Keri, 242
Keri Lake, 242, 243
Kerimasi Volcano, 120, 130
Kerna Fault, 205
Kesik Cut, 283, 284
Keswick, 328, 331, 336–338, 343–345, 347, 348
Khoisan, 362
Kibale National Park, 89, 91, 93, 104, 105
Kibaran Belt, 95
Kibati gate, 109, 110
Kibiro Geothermal field, 104
Kibyra, 295, 298, 299
Kicking Horse Pass, 85–87
Kigezi Highlands, 107
Kijura Escarpment, 103, 104
Kikeronga Salt Pan, 105, 106
Kimeu, Kamoya, 132, 133
Kipina Monastery, 194, 211, 212
Kizilirmak River, 314, 316
Klepsydra Spring, 198, 199
Knickpoint, 73, 85, 87
Korongo, Frida Leakey, 132
Kraemer Island, 357–359, 361
Kranae, 236
Küçük Menderes Graben, 273, 274, 278, 292, 293, 295
Kyabirwa Falls, 100

L

Laconia, 219, 222, 235
Laetoli, 117, 118, 123, 125, 133–135
Laetoli Group, 133

Lagana Bay, 240, 242
Lake, alkaline, 99, 117, 120, 122, 124, 126
Lake, soda, 120
Lake Agnes, 82, 83
Lake Albert, 89–92, 99, 102, 103, 105
Lake Albert Basin, 99, 100, 104
Lake Averno, 176, 179
Lake District National Park, 327–330, 343
Lake Edward, 89–91, 99, 104, 105, 107
Lake Edward Basin, 99, 104
Lake Eyasi, 123
Lake George, 104, 105
Lake Kivu, 91, 109, 110, 112
Lake Kyoga, 92, 94, 99
Lake Lerna, 225
Lake Louise, 69, 82, 83
Lake Magadi, 120, 124
Lake Manyara, 117, 118, 122–124
Lake Masek, 128
Lake Mburo, 89, 91, 105, 108
Lake Mburo National Park, 89, 91, 105, 108
Lake Natron, 117, 118, 120–123, 126, 132, 134–136
Lake Ndutu, 125, 128, 130
Lake Obweruka, 99
Lake Tanganyika, 99
Lake Victoria, 89–92, 95, 98–100, 102, 103, 111, 122
Lake Victoria Basin, 89–91, 99
Lake Victoria Terrain, 119, 122, 128
Lamar River, 53
Lamia, 205
Landforms, 1, 19, 20, 51, 56, 69–72, 117, 170, 193, 194, 207, 224, 235, 246, 250, 255, 260, 269, 273, 275, 280, 295, 307, 313, 316, 319, 322, 328, 330, 332, 333
Landscape Arch, 19, 29–32
Landscapes, 12, 32, 38, 40, 46, 59, 66, 73, 85, 90, 105, 128, 150, 152, 193, 207, 253, 281, 313, 314, 327, 328, 330, 340, 353
Laodicea, 273, 274, 276, 280, 295, 298, 299
Laramide Orogeny, 5, 19, 21, 22, 33, 36, 53, 55, 69, 71, 72, 75, 88
La Sal Mountains, 6, 20, 22–24
Last Glacial Maximum, 42, 43, 73, 107, 340
Laterite, 95
Laurentia, 141, 331, 333
Lava domes, 57, 158, 245, 250, 252, 253, 255, 256, 285
Lava flows, 56, 57, 60, 105, 109, 111, 112, 121, 123, 135, 136, 140, 150, 152–155, 157, 245, 250, 252, 254, 255, 265–267, 269
Lava lake, 89, 109–111, 113, 114, 121, 176
Lava plugs, 150
Lava tubes, 153, 157
Layered igneous intrusion, 353, 354, 358, 369, 372
Leaky, Louis, 130
Leaky, Mary, 111, 130
Lechaion, 228, 230
Lemagrut Volcano, 123, 127, 133
Lentia caldera, 150
Leolo Mountains, 362, 363, 369
Lesbos, 158, 274, 285, 286
Libya, 157
Limestone, 6, 10, 64, 72, 73, 75–77, 79, 82, 144, 158, 160–162, 164, 172, 174, 193, 197–201, 203–207, 210–214, 219, 220, 223–236, 238–243, 248, 249, 254, 259, 264, 269, 278, 285, 301–304, 315, 322, 327, 330, 338
Lion Gate, 225
Lipari, 149, 150, 152
Lithosphere, 3, 56, 95, 153, 173, 248
Little egret, 92, 93
Little Ice Age, 37, 38, 43, 57, 69, 73, 75, 77, 80, 82, 99, 279, 330

Loolmalasin Cone, 126
Loolmalasin Volcano, 123
Loutraki, 228
Lower Geyser Basin, 57
Lower Zone, 355, 357, 361, 363, 366, 371, 372
Lycabettus Hill, 199, 200
Lycian Nappes, 277, 278
Lydenburg Platinum Syndicate, 364
Lydia, 274, 275, 285, 287
Lyell, Charles, 43, 202, 285

M
Maandagshoek, 364
Maasai, people, 118, 120–122, 126–128, 135
Macedonia, 194, 208, 225
Madison River, 53
Magma, 51, 52, 56, 57, 59, 60, 105, 110, 114, 120, 121, 123, 127, 128, 146–150, 153, 163, 169, 172, 173, 176, 177, 249, 250, 256, 258, 265–267, 271, 285, 317, 347, 353–355, 357, 366, 368, 369, 372, 373
Magma chamber, 51, 57, 59, 110, 123, 176, 177, 245, 265, 266, 372
Magna Graecia, 139, 140, 158
Magnesite, 362, 366, 371
Magnesium carbonate, 379
Magnetite layer, 369, 372, 373
Main Ice Age, 42, 43, 77, 82, 86
Main Range, 69, 72, 73, 75, 77, 82
Main Zone, 369
Makuzu, 109
Malae Peninsula, 235, 236
Maligne Canyon, gorge, 69, 77
Maligne Lake, 73
Malta, 145
Mammoth Springs, 51, 60, 64
Mantle plume, 51, 95, 359
Mantling Ash, 250, 252
Mapoch's Caves, 363
Maramagambo Forest, 104
Mara River, 100
Marble, 144, 172, 193, 197–199, 203, 204, 207, 211, 214, 219, 223–226, 233–236, 254, 258, 278, 290, 292–299, 301, 304, 307, 309, 310, 315
Margherita Peak, 105
Marginal Zone, 364, 370
Marine terrace, 219, 221, 224, 226–228, 231
Mariposa Battalion, 39
Mariposa Grove, 39, 46, 50
Marl, 193, 198, 199, 211, 214, 225, 229, 232, 235, 236, 240, 242
Mascali, 157
Matapan Peninsula, 235
Mavri Petra Volcano, 246, 251, 254
Maze, 11
Meander, 9, 288
Meandering rivers, 14, 34, 36
Medicine Lake, 73, 75
Medieval Warming, Warm Period, 279
Mediterranean Basins, 142, 145
Mediterranean Ridge, 143, 248, 249
Mediterranean Sea, 139, 140, 219, 220
Megalopolis, 234
Melian Earth, 257
Meltemi, 237, 262
Menalaion, 235
Menderes Massif, 195, 273, 277, 278, 280, 292, 296, 301–304, 315
Menderes River, 288, 295

Mercato Pumice, 180, 182
Merced River, 40, 46, 49
Merensky, Alexander, 363
Merensky, Hans, 364
Merensky Reef, 353, 364, 368, 373
Mesa, 1, 2, 9–11, 13–16, 269, 285, 287, 289, 313, 316, 318
Mesoproterozoic, 95
Mesothelioma, 313, 325
Mesozoic, 2–5, 33, 37, 38, 40, 53, 57, 144, 145, 158, 159, 164, 170, 172, 174, 197, 201, 203, 210, 211, 224, 248, 249, 264, 269, 278, 315
Messenians, 140
Messina, 139, 140, 146
Messinia, 219, 222
Messinian Earthquake, 145, 146
Messinian salinity crisis, 145, 249
Meteora, 193, 194, 214, 216, 217
Meteora Formation, 214, 215
Methana, 146, 249, 250
Methana Peninsula, 224, 245–254
Methana Volcanic Complex, 245, 246, 249, 250, 255
Methoni, 237, 238
Metsovo, 211, 213
Mexican Hat, 11, 12
Mgahinga Gorilla National Park, 89, 91, 108
Mica, 378, 379
Mica schist, 264, 278, 292, 296
Microcontinent, 72, 331
Middle East, 99, 140, 142, 270
Middleveld, 359, 361
Miette Group, 82
Migmatite, 366, 370
Migration, 56, 59, 117, 128, 130, 134, 140, 296, 304
Milazzo, 149
Milligan, Spike, 190
Milos, 146, 225, 245–247, 249, 253–261
Milos Mining Museum, 255
Miloterranean Geo Walks, 256–259, 261
Minoan, culture, period, 253, 262, 269
Minoan Event, 245, 246, 264–269
Miocene, 3, 5, 89, 95, 111, 145, 193, 194, 198, 199, 207, 214, 215, 226, 240, 249, 276, 285, 296, 304, 313, 315, 316
Miseno Peninsula, 170
Misenum, 171, 181
Mississippi-Missouri Rivers, 53
Mississippi Valley-type, 85
Mitten Butte, 1, 12, 15, 16
Moab, 2, 9, 11, 20–22, 24
Moab Fault, 19, 21–24, 26
Moab Tongue Member, 19, 27, 29, 31
Mobile belt, 95, 119
Moenkopi Sandstone, 7, 10, 12, 13, 15
Monastery, 194, 211
Monemvasia, 236, 238
Monkey, 105, 111
Monolith, 1, 2, 11–14, 19–21, 23, 31, 38, 42, 43, 46, 107, 108, 313, 314, 317, 318, 320, 321, 323, 324
Montagna Grande, 158
Montana, 52, 53
Monte Aria, 150
Monte Nuovo, eruption, 175, 176
Monument Uplift, 4–8
Monument Valley, 1–3, 5–8, 11–16
Mooihoek, 364
Moose, 53
Moraine (glacial), 43, 56, 72, 75, 82, 330, 340

Moraine Lake, 82, 84, 85
Morrison Formation, 19–21, 23, 33–35
Morsynus Graben, 296, 297
Mosaic, 123, 182, 186
Motse River, 368
Mountain Gorilla, 89, 91–93, 108, 111
Mount Arachneo, 224, 226, 227
Mount Bülbül (or Mount Preon), 292
Mount Burgess, 87
Mount Chelmos, 226, 229
Mount Chelmos National Park, 226
Mount Chelona, 250
Mount Columbia, 77
Mount Edith Cavell, 75, 78
Mount Elgon National Park, 90
Mount Epomeo, 176
Mount Epomeo Green Tuff, 176, 177, 180
Mount Erciyes, 313–317, 325
Mount Fairview, 82
Mount Geraldine, 77, 78
Mount Goodsir, 85
Mount Ida, 284
Mount Kallidromom, 205, 207
Mount Karisimbi, 108
Mount Kenya, 99, 105, 117
Mount Meru, 118
Mount Mgahinga, 108
Mount Mikeno, 108
Mount Muhabura, 108
Mount Parnassus, 202
Mount Parnes, 198
Mount Parnon, 234
Mount Penteli, 198, 199
Mount Pion, 274, 292, 293, 295
Mount Profitis, 264
Mount Robson, 73
Mount Sabinyo, 108
Mount Smolikas, 213
Mount Stanley, 105
Mount Stephen, 87
Mount Victoria, 82, 83
Mount Visoke, 108
Mount Washburn, 53
Mount Whyte, 82
Mozambique, 89
Mozambique Belt, 119, 120, 122, 128, 134
Mud pools, mudpots, 51, 62, 63, 150, 158
Mudstone, 1, 2, 6–8, 19, 33–35, 73, 87, 327, 336, 338, 347
Mud Volcano, 62
Muir, John, 39
Mule deer, 40
Munge River, 124, 126
Murchison, Roderick, 103
Murchison Falls, 91, 100, 101, 103
Murchison Falls National Park, 89–91, 93, 100, 102
Mussolini, 157
Mycenae, 219, 220, 222, 225, 281
Mycenaean civilization, 160, 219, 221, 222, 225
Mykonos, 248
Mystras, 220, 235, 237
Mythology, 139–141, 160, 219, 225, 236

N
Naabi Gate, 122
Nafplio, 220, 225

Naisuri River, 128
Naples, 139, 140, 146, 158, 169, 170, 172, 176, 182, 191
Napoleonic Wars, 176
Nappes, 197–199, 223
Nasera Rock, 134
National Garden Shaft, 200
National Museum of Athens, 225
Natrocarbonatite, 117, 120, 121
Natron, 120
Natron-Manyara Half-graben, 120
Natural Bridge, 27, 85, 87
Navajo, people, 11, 12
Navajo Sandstone, 10, 11, 19, 23, 26, 30, 33
Navajo Tribal Park, 1, 2, 5, 11–13
Navarinou Bay, 220, 238–240
Ndali-Kasenda Crater Field, 90, 105
Ndebele, 363
Neapolis Archaeological Park, 158, 161
Neapolitan volcanoes, 147, 169–172, 175
Neapolitan Yellow Tuff, 169, 175, 177
Necropolis, 306, 308, 309
Nelson, Admiral, 176
Neogene, 37, 38, 41, 43, 51, 149, 169, 170, 172, 198, 214, 219–221, 227, 231, 239, 276, 278, 301, 303, 304, 316
Neogene-Quaternary, 46, 122, 210, 226, 228, 230, 234, 278, 281, 284
Neolithic, 170, 221, 225, 268, 321, 327, 329, 343, 344
Neoproterozoic, 40, 69, 73, 80, 82, 119, 122, 248, 331
Neo-Tethys, 141, 142
Nepheline, nephelinite, 105, 109, 117, 120, 121, 127, 148, 172
Nestor, Greek history, 225
Nestor's Palace, 220, 239
Nevsehir, 318, 322
New Acropolis Museum, 199
Ngoitokitok hot springs, 124
Ngorongoro Caldera, 117, 118, 122–127, 135
Ngorongoro Conservation Area, 117, 118, 122, 123, 125
Ngorongoro Highlands, 117–119, 122, 123, 126, 128, 129, 131
Ngorongoro-Lengai Geopark, 118
Ngorongoro Volcano, 122, 128, 130, 135
Nicolosi, 152, 157
Nile River, delta, 102, 103
Nisyros, 247
Norite, 367, 369, 370, 372
Norris Geyser Basin, 60, 61
North American Craton, 69, 72
North American Plate, 37, 40, 51, 56
North Anatolian Fault Zone, 277
North Atlantic Igneous Province, 353, 357
Northern Tanzania divergence, 117, 120
North Pindus National Park, 214
North Saskatchewan River, 70, 80
Nutrient-rich grasses, 117, 128
Nyabaronga River, 100
Nyamulagira Volcano, 97, 108, 109
Nyanzian Greenstones, 95
Nychia, 253, 255, 256, 258
Nyinambuga Crater, 90, 105
Nyiragongo Volcano, 89, 91, 109, 110, 113, 114

O
Observatory, Etna, 157
Observatory, Vesuvius, 169
Obsidian, 31, 51, 57, 60, 225, 245, 246, 253, 255, 256, 258, 321
Obsidian Cliff, 57, 60
Oceanic crust, 139, 141, 142, 153, 173, 248, 277, 285, 333

Oceanic lithosphere, 248, 249
Odysseus, 139, 140, 160, 163, 222
Odysseus' Palace, 160
Odyssey, 140, 160, 163, 172, 222
Ogol Lavas, 127
Olbalbal Depression, 122
Oldeani Volcano, 127
Old Man of Coniston, 328, 330, 342, 347
Oldoinyo Lengai, 117, 118, 120–124, 126–128, 130, 133
Oldupai Gorge, 117, 118, 122, 125, 127, 128, 130–134
Olifants River, 366
Olifants River Trough, 366, 371
Oligocene, 95, 111, 198, 199, 240, 249, 278, 303
Ol Karien Gorge, 135
Olmoti Caldera, 123, 126
Olmoti Volcano, 127, 130
Olympia, 219, 220, 223, 230–235
Olympic events, 232
Omineca Belt, 71
Onverwacht, 364, 366, 367
Onyx, 321
Ophiolite, ophiolite complex, 141, 198, 207, 210, 211, 214, 223, 278, 285, 315
Oplontis Villa, 182, 184, 188, 189
Ordovician, 73, 77, 80, 327, 331, 333, 335, 336, 338, 347
Oregon, 51, 56, 314
Organ Rock Mudstone, 7, 10–14
Orogeny, 72, 142, 194, 198, 223, 248, 276, 327, 333
Ortigia, island, 158
Ottoman Empire, 210, 211, 239, 324
Owens Dam, 100

P
Pacific Ocean, 73, 146
Paestum, 139, 140, 158, 159
Palaeolithic, 175, 245, 246, 258
Palaeontologist, 32, 87, 88, 130
Palaeoproterozoic, 53, 95, 353, 358
Palaeozoic, 2, 3, 5, 33, 40, 53, 55, 70, 95, 141, 197, 224, 233, 273, 315, 327, 328, 331, 333, 335, 347
Paliorema mine, 256
Pamukkale, 301–308, 310
Pamukkale Fault, 296, 301, 303, 304
Panarea, island, 149
Pantelleria, island, 139, 140, 158
Pantelleria Volcano, 144, 146–149, 158
Pantellerite, 148
Papafrangas, 261
Papingo, 211, 213
Paraa, 103
Paradox Basin, 4
Paradox Formation, 6, 9–11, 19, 23
Paranthropus, 132
Parco dell 'Etna, 152
Parian Marble, 233, 299
Park Avenue Canyon, 30
Parnassus Mountains, 193, 202–206
Parnassus Zone, 197, 203, 223, 248, 249
Parthenon, 193, 199, 202
Paşabağ, 313, 314, 320
Pasha, Ali, 210
Pass of Thermopylae, 193, 205, 208
Patras, 219, 220, 224, 226, 228–230
Paul, the Apostle, 275, 276, 284
Pausanias Submarine Volcanic Field, 253

Pedi Empire, 362
Peisistratean Aqueduct, 200
Pelagonian Zone, 197, 198, 203, 207, 248
Pelikata Hill, 160
Peloponnese, 140, 158, 195, 201, 219–225, 227, 229, 230, 234, 235, 245, 246, 249, 281
Pencil Museum, 345
Peneois Graben, 214, 215, 217
Peneois River, 214
Peninj, 132, 134
Penrith, 328, 336, 338, 343, 345
Pentelic Marble, 193, 199, 202, 299
Perama Cave, 210
Pergamum, 273, 274, 276, 279, 284, 285, 287–291, 295
Pergamum Altar, 287, 289, 290
Periander, Athenian leader, 229
Peridotite, 214
Peristeria Volcano, 264, 266
Perlite, 245, 258, 260, 261
Permian, 1, 2, 6, 7, 9, 10, 12, 14, 22, 23, 40, 73
Persian, invasion, soldiers, 160, 199, 201, 202, 206, 315
Petrified Forest National Park, 8
Pheidippides, Greek runner, 201
Philopappos Hill, 199
Phlegraean district, 169, 170, 172
Phoenicians, 140, 223
Phonolite, 127, 148
Phreatomagmatic, 185, 186, 188
Pindus Mountains, 193, 194, 208–215
Pindus Zone, 197, 223, 233
Pinnacles, 1, 11, 13–16, 19, 23, 31, 41, 46, 82, 134, 194, 213–217, 313, 314, 316–318, 320, 321, 323
Piraeus, 199, 200, 202
Plains of Marathon, 193, 201
Plateau, 1, 11, 51, 52, 56, 72, 89, 107, 120, 133, 204, 239, 253, 256, 264, 265, 281, 285, 287, 288, 307, 310, 313–316, 353, 359, 361
Plateau Rhyolite, 57, 60
Platinum, 353, 357, 364, 366–368
Plato, Greek philosopher, 172, 245, 246, 269
Pleistocene, 37, 38, 46, 51, 52, 72, 73, 95, 99, 104, 105, 120, 121, 124, 134, 153, 158, 172, 202, 213, 224, 226, 227, 229–232, 234–236, 239–242, 249, 251, 255, 256, 258–262, 265, 278, 304, 327, 340
Pleistocene Ice Ages, 37, 42, 43, 46, 53, 54, 56, 69, 71, 75, 99, 135, 172, 194, 207, 214, 223, 227, 279, 330, 334
Pleistos Fault, 203
Pleistos River, 202, 205
Plinian Eruption, Plinian Event, 56, 140, 146, 147, 169, 173, 175, 180, 181, 186, 245, 265, 317
Pliny the Elder, Roman senator and natural scientist, 171, 181, 182, 188, 269
Pliny the Younger, Roman historian, 140, 169, 171, 172, 181, 186, 188
Pliny Trench, 249, 276
Pliocene, 95, 111, 117, 120, 122, 128, 130, 134, 139, 143, 145, 146, 158, 172, 198, 199, 210, 224, 226, 229, 231, 232, 234–236, 240–242, 245–247, 249, 250, 254, 255, 260, 276, 278, 285, 313, 316
Plumbing system, 51, 61–64, 66, 146, 153, 372
Pluton, 338
Plutonium, 301, 309, 310
Pompeii, 169, 182, 185, 188
Pompeii Pumice, 185, 187
Poros, 247
Port Athinias, 262, 264, 267
Port Frikes, 160
Port Pólis, 160
Poseidon, Greek mythology, 235, 236

Posidhonia, 229, 230, 233
Positano, 170
Potassium, 11, 105, 109, 148, 172, 256, 266
Potassium-rich alkaline lavas, 105, 148, 149, 173, 249
Potter, Beatrix, 330
Powell, J.W., 9
Pozzuoli, 169, 170, 175–179
Primates, 91, 104, 105, 111
Procida, 170, 171
Proconsul Africanus, 111
Pronghorn antelope, 53
Proterozoic, 3, 4, 33, 278, 292, 296, 303
Provatina Cave, 213
Pumice, 57, 60, 150, 158, 169, 175, 180, 182, 185, 188, 245, 259, 261, 262, 264, 267, 269, 313–316, 318–323
Punta Imperatore, 177, 180
Pylos, 220, 238, 239
Pyramid Thrust, 73
Pyrenees, mountains, 142
Pyrgos, 230, 231, 239
Pyroclastic, deposits, flows, rocks, 51, 52, 55–57, 105, 128, 146, 147, 150, 169, 177, 180–182, 185, 186, 188, 245, 246, 250, 252, 255, 258–262, 265, 267, 269, 270, 313, 316, 347
Pyroxenite, 353, 369
Pythia, House of Snakes, 204, 310
Pythian Games, 204

Q
Quarry Sandstone, 19, 34, 36
Quartz, veins, 41, 345, 347, 364
Quartzite, 40, 72, 73, 75, 77–79, 82, 83, 85, 95, 119, 128, 134, 135, 278, 292, 296, 366
Quaternary, 143, 144, 146, 158, 159, 169, 170, 172, 198, 199, 201, 210, 219, 221, 224, 225, 231, 234, 235, 241, 249, 277, 278, 303, 304, 313, 315, 316
Quaternary sands and gravels, 107, 214
Quaternary volcanoes, 118, 144, 247, 249, 264
Queen Elizabeth National Park, 89, 91, 93, 104–106

R
Rain God Mesa, 1, 13–15
RAMSAR locality, 104
Ransome, Arthur, 330
Reck, Hans, 130
Reck Skeleton, 130
Red Mountains, 53
Reggio Calabria, 146
Regional plateau, 13, 89–92, 94, 95, 99, 100, 107, 108, 117, 119
Rhodope-Serbomacedonian Massif, 278, 279, 281
Rhyolite, 51, 57, 60, 62–64, 123, 148, 149, 255, 256, 258, 266, 338, 353, 358
Ripon Falls, 100
River capture, 103
Rock flour, 69, 80
Rocky Mountains, 1–4, 22, 33, 40, 51–53, 56, 69–72
Rocky Mountain terrain, 5, 19, 21, 22, 32
Rogers, Albert Bowman, 70
Roman Empire, 158, 223, 228, 232, 274, 275, 299, 301, 302, 315
Roman Navy, 182
Romans, Roman times, 140, 145, 150, 158, 160, 185, 186, 256, 273–275, 281, 283, 295, 298, 299, 308, 310, 329, 345
Roman villa, 182, 186
Rooiberg Group, 358
Rosso Antico Quarries, 235

Royal Geographical Society, 103
Ruizi River, 107
Rusinga Island, 111
Ruskin, John, 327, 328, 330
Rustenburg Layered Suite, 353, 358, 360, 362, 364, 368, 369, 372
Rutherford, Adam, 135
Ruvyironza River, 100
Rwenzori Mountains, 89–91, 94, 99, 105, 107
Rwenzori Mountains National Park, 105
Rydal, 330

S
Sacred Harbour, 288, 293, 295, 296
Sadiman Volcano, 123, 127, 130, 133, 134
Sakarya Zone, 195, 277, 278, 285
Salei Plains, 122, 123
Salina, island, 149
Salinity, 145
Salt, 6, 9–11, 28, 105, 120, 135, 257
Saltation, 128
Salt dome, 10, 19, 20, 25
Salt pan, 105, 106
Salt Valley, 19, 23, 25–28
Salt Wash Member, 34
Sand dunes, 127
Sands, 6, 7, 11, 15, 19, 28–30, 36, 73, 80, 95, 103, 130, 230, 231, 241, 242, 266, 284, 340
Sandstone, 1, 2, 6–11, 14, 15, 19–24, 26–36, 72, 77, 82, 95, 130, 198, 214, 229, 230, 232, 235, 240–242, 278, 303, 315, 319
San Juan River, 2
San Juan Volcanics, 338
Santorini, 146, 245–249, 253, 254, 258, 262–266, 268, 270, 271
Santorini Caldera, 245, 246, 262, 263, 265, 266, 269, 271
Santorini Volcano, 172
Sarakiniko, 261, 262
Saronic Gulf, 220, 224, 229, 232, 246, 247, 249, 253
Savannah, 89–93, 100, 104, 105, 122, 134
Scafell Pike, 327, 328, 330, 340
Scamander River, 283, 284
Scamandrian Plain, 283
Scarrupo Di Panza Volcano, 177, 180
Schist, 40, 95, 119, 197, 198, 226, 254, 278, 315, 330
Schliemann, Heinrich, 225, 281
Schliemann Trench, 281, 282
School of Aphrodisias, 273, 299
Scoria, 123, 140, 150, 267
Scoria cone, 123, 181
Scree, 38, 82, 84, 210, 327, 330, 334, 340
Scylla, *Scylla*, 140, 163, 166
Sea level, stands, variations, 150, 172, 178, 185, 207, 208, 239, 278–281, 285, 303
Sea of Marmara, 274, 277–279, 281
Sedgwick, Adam, 340
Sedimentation, 6, 23, 158, 193, 207, 208, 214, 231, 239, 273, 275, 278, 280, 281, 288, 295
Seismicity, seismic activity, seismic event, 46, 55, 57, 61, 97, 145, 146, 198, 199, 205, 222–224, 226, 228, 230, 233, 241, 254, 269, 271, 274, 278, 296, 304, 310, 319
Sekhukhune, King, 362, 363
Sekhukhuneland, 363
Sekwati, King, 362, 363
Selçuk, 274, 288, 292
Sele Plain, 158, 159
Semliki National Park, 89, 91, 105
Semliki River, 99, 103, 104

Semliki Valley, 97, 105
Serengeti Plains, 100, 117, 118, 121–124, 127, 128, 130, 133–135
Serpentine, 214
Seven Churches, 275, 276, 284
Seven Wonders of the Ancient World, 233, 273, 292
Sfaktira Island, 220, 238, 239
Shafer Trail, 1, 11
Shale, 1, 2, 6–8, 10, 19, 35, 72, 73, 87, 88, 95, 211, 319, 327, 338, 347
Shap granite, 335, 338
Shield volcano, 123, 126
Shifting Sands, 127, 129
Shinarump Conglomerate, 1, 8, 12, 13, 15
Shipwreck Beach, 240
Shoebill stork, 103
Shoshonite, 149
Sicilian Earthquake, 145
Sicily, 139, 140, 143–146, 149, 153, 154, 158, 163, 166
Siculo-Calabrian Rift, 144, 145
Sierra Nevada, 37, 38, 40–43
Sierra Nevada Batholith, 40, 41, 46
Silt, siltstone, 14, 20, 33, 34, 336
Silurian, 327, 331, 338
Silver, 85, 210, 327, 343, 345
Simpson Pass Thrust, 73, 77
Skaergaard Intrusion, 353–358, 360, 364, 369, 373
Skiddaw, 330, 331, 335–338, 344, 347
Skiddaw Group, 327, 331, 335–338, 347
Slate, 85, 87, 211, 327, 329, 330, 336, 340, 343, 344, 347, 349, 350
Slickrock Member, 19, 26–33
Slovenia, 204
Smith, Admiral, 166
Smithsonian Global Volcanism Programme, 146
Snake River, 53, 288
Snake River Plateau, 53, 56
Sodium, 120, 121, 124, 148, 149, 153, 172
Sodium carbonate, 120, 135
Sodium-rich alkaline magmas, 105
Solfatara, 176, 179
Solway Firth, 332, 333
Somma-Vesuvius, massif, 169–171, 173, 175, 177, 185, 186, 188
Somma-Vesuvius district, 172
Somma Volcano, 172, 177
Sorrento Peninsula, 170, 172, 185, 186
Sousaki Volcano, 247
South Africa, 320, 353, 354, 358, 364
South Saskatchewan River, 73
Sparta, Ancient, 220, 235, 236
Spartans, 140, 202, 219, 223, 239
Sparti, 201, 219, 235–237
Speciation, 92
Species, 19, 32, 34, 40, 46, 88, 89, 91, 92, 104, 105, 108, 109, 111, 117, 120, 128, 134, 135, 249
Speke, John Hanning, 100
Spercheios Delta, 208
Spercheios River, 207, 208
Spiral tunnels, 85
Spoil heaps, 256, 258, 343, 345, 347, 348
Spring of Herodotus, 242
Stabaie, 171, 182, 188
Staging chamber, 146, 172, 176, 245, 369, 372
Stalactites, 210
Stalagmites, 210
Stavrós, 160, 163, 164
Steam, 22, 61–63, 65, 152, 155, 269
Steam vent, 51, 62
Steelpoort, 362, 363

Steelpoort River, 360, 363
Stephen Formation, 87
Sterea Ellas, 194, 202
Stockdale Quarry, 338
Stone tools, 31, 51, 60, 130, 134, 225, 245, 256, 343
Strabo, Greek historian, 245, 246, 252
Strabo Trench, 249, 276
Straits of Çanakkale, 281
Straits of Gibraltar, 145
Straits of Messina, 145, 146, 163, 166
Straits of Rion and Andírrion, 226
Straits of Sicily, 140, 158
Stratovolcano, 60, 109, 139, 144, 146, 150, 152, 313, 314, 316, 317, 372
Stromboli, 140, 146, 148–152, 163
Strombolian, volcanism, 139
Stromboli Island, 139, 146, 149–151
Stromboli Volcano, 139, 140, 163
Subducted slab, 146, 169, 172, 245, 248, 249, 266
Subduction, 139, 143, 146, 153, 169, 172, 195, 223, 248, 249, 273, 275–277, 333, 336
Subduction zone, 40, 73, 140, 144–146, 219, 248, 249, 272, 273, 276, 278
Sub-Plinian event, 169, 175, 180
Subspecies, 111
Sulfur Mountain, 82
Sulfur Mountain Thrust, 73, 82
Sulphur, 62, 150, 152, 156, 176, 179, 256, 259, 260
Sulphur Caldron, 62
Sunwapta Falls, 77
Supercontinent, 6, 95, 119, 141, 331, 359
Supervolcano, 56
Sutures, regional, 140, 273, 274, 276–278, 332
Syracuse, 139, 140, 145, 158, 161
Syracuse Limestone, 158, 161–163, 165
Syracuse Plateau, 158, 160

T
Tacitus, Roman historian, 172, 181
Tahoe Glaciation, 42, 43
Talus (see scree), 82
Tambora Event, 267
Tanzania, 90, 95, 96, 105, 126, 128
Tarangiri National Park, 128
TAS diagram, 147, 148
Taurus Mountains, 276
Taygetos Mountains, 235, 236
Tegea, 234
Temple of Aphrodite, 296
Temple of Apollo, 204–207, 227
Temple of Apollo and Plutonium, 301, 309, 310
Temple of Athena, 159, 284, 285, 287
Temple of Hera, 233
Temple of Rome and Augustus, 193, 198, 199, 203
Temple of Trajan, 288, 290
Temple of Zeus, 233
Tenaya Glaciation, 42
Tenaya Lake, 40, 42
Tethys Ocean, 139, 141, 142, 145, 198
Tetons Block, 55
Tetons Fault, 55, 56
Theatre, Great, 274, 292, 293, 295, 296
Theatre of Delphi, 204, 206
Theatre of Dionysus, 198, 200, 204
Thelma and Louise, 9

Theodosius, Christian Roman Emperors, 232
Thera, 262–269, 271
Therasia, 262–265
Thermoacidophiles, 62
Thermopylae, 205, 208, 209
Thesalian Plain, 194, 214
Theseus, 219, 225, 226
Thessalonica, 211
Thessaly, 194, 205, 207, 214
Tholeiite, 148
Tholeiitic basalt, 148, 149
Thrace, 274, 275, 284
Thrace Basin, 273, 277–279
Thrust, 71–73, 82, 169, 172, 193, 197, 210, 223, 241, 277, 278
Tidwell Member, 34
Tilberthwaite, 329, 340, 347, 350
Ti-magnetite, 353, 364, 369
Tioga Glaciation, 42, 43
Tioga Lake, 40
Tioga Pass, 40
Tools, stone, 31, 51, 60, 130, 134, 225, 245, 256, 343
Totem Pole, 1, 15, 16
Tourkovounia Limestone, 198–201, 204
Trachilas Volcano, 253, 255, 261
Trachyandesite, 149
Trachyte, 149
Trans-Canada Highway, 70
Transform fault, 139, 144, 146, 223
Transvaal Supergroup, 358
Travertine, 51, 66, 82, 205, 208, 301–310
Treks, trekking, 135, 357
Triassic, 1, 2, 6–8, 10, 12, 14, 15, 22, 23, 82, 203, 211, 212, 226, 236, 238, 241, 242
Triple Group, 355, 357
Tripoli, 221, 234
Troad, 280
Troctolite, 353, 357, 361
Troezen, 219, 220, 225, 226
Troia Bay, 281, 284
Trojan Wars, 160, 273, 274, 281
Trona, 120
Troy, 139, 140, 160, 163, 225, 236, 273, 274, 279–285
Tsate Valley, 362
Tsunami, 139, 140, 145, 146, 153, 219, 222, 223, 226, 228, 230, 253, 256, 258, 265, 267, 269, 271, 295
Tufa, 303
Tuff, 36, 57, 105, 120, 130–133, 226, 257, 258, 260, 261
Tungsten, 343, 347
Tunisia, 158
Tuolumne Glacier, 43
Turbidites, 335
Turkey, 139, 140, 142, 160, 195, 246, 248, 270, 275–278, 281, 284, 295, 302, 314, 315, 324
Turquoise Lakes, 80
Tyrrhenian Sea, 139, 140, 144–146, 149, 158, 160, 170, 172, 176

U
Ubendian Belt, 95
Üchisar, 318, 321
Üchisar Castle, 313, 321, 325
UG2 Reef, 364, 372
Uganda, 89–92, 94, 96, 100, 103–105, 108, 111
Uganda kob, 91, 93
Ugandan Gneiss Complex, 95, 100, 101, 104, 105, 107, 108
Uhuru Falls, 103

Uinta Mountain Group, 33
Ullswater, 327, 330, 343, 345
Ultra-Plinian event, 245, 265
Uncompahgre Uplift, 4, 7
Unconformity, 6, 7, 13, 15, 304, 327, 331–333
Underground cities, rock dwellings, 313–315, 320–322, 324
Upheaval Dome, 10
Uplift, 1, 3, 4, 7, 9, 21, 27, 32, 37, 38, 43, 46, 55, 89, 90, 95, 99, 107,
      119, 175, 178, 204, 226–228, 230, 278, 284, 303, 304, 360
Upper Geyser Basin, 60
Upper Zone, 353, 355, 357, 359, 360, 364, 368, 369, 372
Ürgüp, 318
Ürgüp Basin, 316
U-shaped valley, 53, 54, 56, 69, 70, 72, 73
Ustica, island, 149, 163
Utah, 1–3, 5–7, 9, 13, 19, 20, 27, 29, 31, 35

V
Vadar Zone, 195, 278
Vanadium, 8, 362, 369
Velia archaeological site, 160
Venetian, 223, 229, 235, 237, 238, 241
Verdoline Pumice, 180, 182
Verdolini Quarry, 175, 177
Vernal, 20, 32
Vernal Falls, 46, 49
Vesuvius, 146, 169, 170, 172, 173, 176, 180–183, 190, 191
Vesuvius 79 AD eruption, 140, 169, 170, 172, 181, 190
Vesuvius National Park, 169, 188
Victoria Nile, 89, 91, 92, 98–103
Victorian period, times, 327
Vikos-Aoös Geopark, 213
Vikos Gorge, 213
Virgil, Roman poet, 153, 172
Virunga Mountains, 89, 91, 93, 94, 97, 108, 110, 111
Virunga National Park, 90, 91, 108–110
Volcanic ashes, 52, 57, 64, 117, 120, 127–130, 133, 146, 152, 158,
      169, 180, 182, 183, 185, 226, 250, 259, 260, 264, 267, 268,
      313–316, 318–320, 322–325, 327, 330, 336, 338–340, 347, 350
Volcanic cone, 52, 94, 117, 118, 139, 146, 149, 150, 170, 247, 265,
      316
Volcanic deposits, 52, 56, 57, 172, 176, 181, 185, 248, 264–266, 313,
      318, 331, 336
Volcanic eruption, 109, 139, 140, 146, 185, 245, 246, 267, 268, 285,
      315
Volcanic glass (see obsidian), 31, 51, 57, 60, 245, 246, 258
Volcanic island arc, 40, 139, 143, 144, 146, 149, 327, 333
Volcanic rocks, 43, 51, 57, 105, 120, 147, 148, 150, 158, 172, 250,
      253, 256, 325, 343, 344, 347
Volcanism, 52, 55, 56, 66, 92, 95–97, 105, 108, 109, 111, 117, 118,
      120, 122, 126, 128, 135, 139, 146, 148–150, 153, 158, 172,
      245, 246, 249, 250, 253, 264–266, 269, 271, 277, 278, 285,
      313, 316, 336, 338
Volcanoes, active, 51, 52, 89, 97, 105, 108, 111, 118, 121, 140, 146,
      149, 150, 157, 158, 246, 247, 249, 254, 271
Volcanoes National Park, 90, 91, 108, 110
Vrachionas Mountains, 240
Vulcan, Roman god of fire, 139, 141, 172
Vulcanello Islet, 150
Vulcano, island, 139–141, 146, 149, 150, 152

W
Wager, Lawrence Rickard, 357
Wager Peak, 355
Wagner, Percy Albert, 364

Wainwright, Arthur, 330
Walcott, Charles, 87
Wall of Bones, 34
Wapta Mountain, 86
Wast Water, 330
Waterfalls, 22, 37, 38, 46, 51, 64, 66, 69, 70, 73, 77, 82, 89, 126, 135,
      285, 328, 330, 334, 340, 342
Waterspout, 166
Wawona, 38, 39
Wawona Tree, 46, 50
Wawona Tunnel, 43
West Anatolian Volcanic Province, 278
Western Canada Basin, 71
Whirlpool, 139, 140, 145, 166
Whitehaven, 328
White Nile, 91, 100
White Rim Sandstone, 1, 2, 7, 10, 11
Wildebeest, 128, 130
Wildwaters, 100
Windermere, 327, 328, 330, 331, 340, 343
Windermere Group, 72
Windermere Supergroup, 327, 331–333, 338, 350
Windows, 20, 23, 29, 30
Windy Hill Member, 34
Wingate Formation, 8
Winnaarshoek, 368, 373
Winnowing, 128
Wisconsin Glaciation, 42, 43
Wolf, 53
Wordsworth, William, 327, 328, 330
World Heritage status, 193, 301, 302, 305, 328
Wrangham, Richard, 106
Wyoming, 51–53, 56

X
Xenoliths, 41, 105, 361, 372
Xylokastro, 228, 230

Y
Yampa River, 32
Yellow-billed stork, 92, 93
Yellowstone Caldera, 56, 57
Yellowstone Canyon, 51, 57
Yellowstone Lake, 51, 62, 64
Yellowstone National Park, 51–54, 56, 58, 60, 63
Yellowstone Plateau, 52, 53, 56, 57
Yellowstone River, 51, 53, 64, 66
Yellowstone Tuff, 57
Yellowstone Volcano, 51, 52, 56, 59
Yoho National Park, 69, 70, 73, 74, 85–87
Yosemite Canyon, 37–40, 42–44, 46–49
Yosemite Falls, 37, 46, 49
Yosemite Glacier, 49
Yosemite National Park, 37–39, 41

Z
Zafferana, 157
Zagoria Villages, 211
Zakynthos, 219–221, 223, 224, 230, 240–243
Zakynthos Town, 220
Zanclean flood event, 145
Zebra, 128, 130
Zelve, 321
Zinjanthropus boisei, 117, 128, 132–134

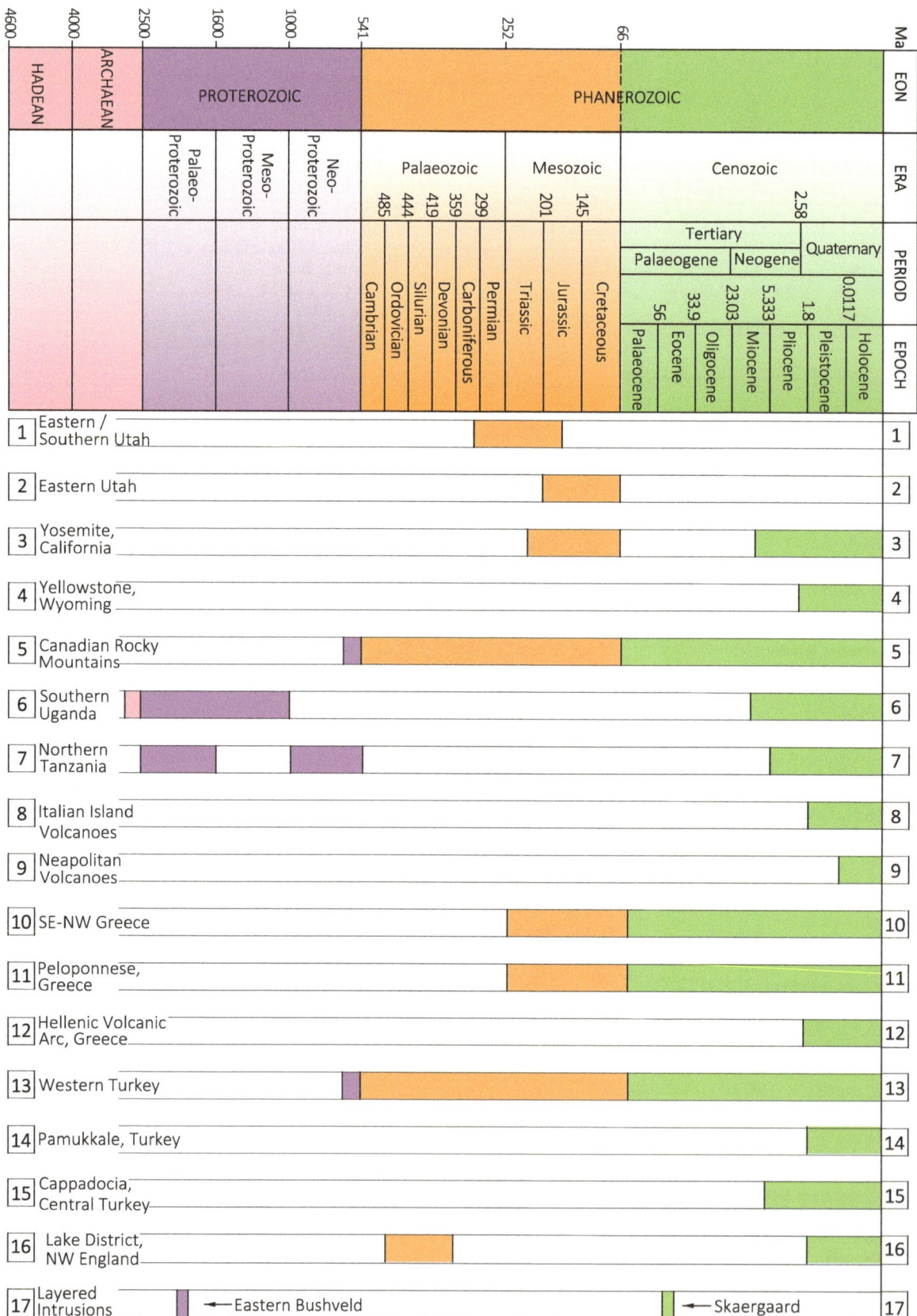

Map B Stratigraphic chart. The Lower Carboniferous in North America is referred to as the Mississippian and the Upper Carboniferous is the Pennsylvanian period. The base of the Pleistocene is located at 1.8 Ma (older usage), rather than 2.58 Ma as is currently recommended to indicate the start of the Quaternary